Plant-Microbe
Interactions

Plant-Microbe Interactions

Molecular and Genetic Perspectives
Volume 2

Tsune Kosuge
University of California, Davis

Eugene W. Nester
University of Washington, Seattle

Macmillan Publishing Company
NEW YORK

Collier Macmillan Publishers
LONDON

Macmillan Publishing Company
866 Third Avenue, New York, NY 10022

Collier Macmillan Canada, Inc.

Printed in the United States of America

Printing number: 1 2 3 4 5 6 7 8 9 10   Year: 7 8 9 0 1 2 3 4 5 6

Library of Congress Cataloging-in-Publication Data
Main entry under title:

Plant-microbe interactions.

 (Molecular and genetic perspectives; v. 2)
 Includes index.
 1. Plants—Microbiology. 2. Micro-organisms,
Phytopathogenic. 3. Plant viruses. I. Kosuge, Tsune.
II. Nester, Eugene W. III. Series.
QR351.P577 1987   581.2'3   85-21837
ISBN 0-02-947990-8

Contents

Preface

In the two years since the first volume of this series was published, significant advances have been made in the study of the molecular basis of a number of important metabolic processes in plants and microorganisms. Several microbial genes have been transferred and expressed in plants, and the isolation and sequencing of genes from microorganisms and plants continue at an ever-increasing pace. With this technology, biological processes can be explored to depths not previously possible. Meanwhile, promising new experimental systems for studying plant-microbe interactions are being developed. Thus, the opportunities for advancing our understanding of the molecular basis of many important processes in plant-microbe interactions continue to expand.

In the opening chapter, Dr. Shaw discusses specialized functions encoded by plasmid-borne genes, which help determine the relationships microorganisms establish with each other and with plants. Knowledge of these functions will help us understand the fundamental basis of the ecology of these organisms. Recent advances in the understanding of the pathogen-induced tumor formation in plants are discussed by Dr. Schröder. Seminal observations show that indoleacetic acid and cytokinin synthesis encoded by pathogen genes form the basis of tumorigenicity in pathogen-plant interactions. Another important area of research, fungal molecular biology, is addressed in the chapter by Dr. Lovett, who provides examples of molecular biological approaches being used in a fruitful way to study gene regulation and expression in fungi. The growing interest in the molecular biology of fungi reflects their importance as causal agents of plant disease. In the section on nitrogen fixation, Drs. Guerinot and Chelm discuss recent advances in the biochemistry and physiology of symbiotic nitrogen fixation and emphasize the profitable use of *Rhizobium* mutants to elucidate mechanisms that coordinate bacterial and legume metabolism in nodules. While the energy required to reduce elemental nitrogen to ammonia can be calculated on theoretical grounds, it is difficult to determine the energy costs of this process in nodules. Drs. Saari and Ludden address this complex problem from several different perspectives. From the practical standpoint of crop improvement, such calculations are important and provide insights on future experiments that will further define the energetics of this process in symbiotic interaction. In the final chapter in this section, Dr. Berry describes the cellular interfaces that delineate the interaction between the nitrogen-fixing actinomycete—Frankia—and the nonlegume—alder. Since biochemical and genetic approaches are being developed for the study of actinomycetes, the Frankia-Alnus symbiotic interaction has much promise for pioneering investigations.

Insights into the mechanisms of microbe-plant recognition are provided by Drs. Ralton, Smart, and Clarke, who discuss concepts common to several biological recognition phenomena including those in pathogen-plant interactions. Dr. Collmer's discussion on pectin-degrading enzymes evaluates their proposed role in pathogenesis, particularly in soft-rot diseases. The genes encoding these enzymes are being characterized, which should provide clues on the roles of the different isozymes of pectate lyase and polygalacturonase in pathogen-plant interactions. The final chapter on the establishment of microbes in plants addresses a different, yet important, problem of movement of viruses in plants. Drs. Meshi and Okada bring readers up-to-date in this important but complex area of study, which seems ripe for application of molecular genetic approaches.

Insights into the characteristics of a new group of viruses is provided by Dr. Van Etten, who describes phage-like viruses that attach and ultimately lyse eukaryotic algae. Research on this interesting group of pathogens will provide new information on the ecology of fresh water lakes and may provide new approaches for studying the biology of algae. In the second chapter in this section, Dr. Guilfoyle describes the molecular biology of retro-like viruses of plants. These interesting viruses, which utilize reverse transcriptase, are a unique group of plant pathogens much like their counterparts in animals. In a final chapter on viruses, Dr. Dickinson describes viruses that are pathogenic in fungi. These viruses are very diverse in their properties, and the understanding of their host-pathogen relationships will help with the understanding of the biology of fungi as well.

In the final section of this volume, the response of plants to stresses is discussed first by Dr. Boller, who describes the production of hydrolytic enzymes by plants in response to pathogen attack. Mounting evidence suggests that these and related enzymes produced by the plant may confer resistance to the inciting organisms. This discussion complements the chapter by Dr. Kuhn, who discusses plant responses to various stresses, with emphasis on the molecular biology of plant response to pathogens. Unifying concepts that relate plant responses to biological and environmental stresses are revealed in this chapter.

We are indebted to the members of the editorial board for their invaluable suggestions on the topics and authors for this volume. The important editorial contributions of Ms. Sharon Bradley, University of Washington, and Ms. Ann Martensen, University of California, Davis, are gratefully acknowledged. We are also indebted to Ann for her excellent graphics of the illustrations submitted by the authors. Finally, we express our sincere appreciation to the authors for their excellent and timely contributions. Their efforts have allowed us to proceed with volume 2 of this series.

Contributors

Alison M. Berry   Department of Environmental Horticulture, University of California, Davis, CA 95616

Thomas Boller   Botanisches Institut, Universität Basel, Schoënbeinstrasse 6, CH-4056 Basel, Switzerland

Barry K. Chelm   Department of Microbiology and Public Health, and DOE-Plant Research Laboratory, Michigan State University, East Lansing, MI 48824

Adrienne E. Clarke   Plant Cell Biology Research Centre, University of Melbourne, Parkville, Victoria 3052, Australia

Alan Collmer   Department of Botany, University of Maryland, College Park, MD 20742

Matthew J. Dickinson   Department of Chemical Pathology, Royal Postgraduate Medical School, Ducane Road, London, England

Mary Lou Guerinot   Department of Biological Sciences, Dartmouth
                      College, Hanover, NH 03755
Tom J. Guilfoyle   Department of Biochemistry, University of
                      Missouri, Columbia, MO 65203
David Kuhn   Department of Biochemistry, Purdue
                      University, West Lafayette, IN 47907
James S. Lovett   Department of Biological Sciences, Purdue
                      University, West Lafayette, IN 47907
Paul W. Ludden   Department of Biochemistry and Center for the
                      Study of Nitrogen Fixation, University of
                      Wisconsin, Madison, WI 53706
Russel H. Meints   School of Biological Sciences, University of
                      Nebraska, Lincoln, NE 68583-0722
Tetsuo Meshi   Department of Biophysics and Biochemistry,
                      Faculty of Science, University of Tokyo,
                      Hongo, Tokyo 113, Japan
Yoshimi Okada   Department of Biophysics and Biochemistry,
                      Faculty of Science, University of Tokyo,
                      Hongo, Tokyo 113, Japan
Julie E. Ralton   Plant Cell Biology Research Centre, University
                      of Melbourne, Parkville, Victoria 3052,
                      Australia
Leonard L. Saari   Department of Biochemistry, and Center for the
                      Study of Nitrogen Fixation, University of
                      Wisconsin, Madison, WI 53706
Joachim Schröder   Institut für Biologie II, Albert Ludwigs
                      Universität, D-7800 Freiburg, West Germany
Paul D. Shaw   Department of Plant Pathology, University of
                      Illinois, Urbana, IL 61801
Michael G. Smart   Plant Cell Biology Research Centre, University
                      of Melbourne, Parkville, Victoria 3052,
                      Australia
James L. Van Etten   Department of Plant Pathology, University of
                      Nebraska, Lincoln, NE 68583-0722
Yuannan Xia   Department of Plant Pathology, University of
                      Nebraska, Lincoln, NE 68583-0722

Plant-Microbe Interactions

SECTION I

Overview

Chapter 1

Plasmid Ecology

Paul D. Shaw

IMPLICIT IN THE TITLE of this chapter is the hypothesis that the presence of a resident plasmid in an organism gives the organism a selective advantage enabling it to occupy some ecological niche that would be less accessible or unavailable to plasmid-free strains. The objectives of this discussion are to examine the evidence for this hypothesis and to consider, primarily in relation to plant-microorganism interactions, the consequences for a microorganism of carrying genetic determinants on a plasmid. Research in microbial ecology has advanced considerably in the past few years and has been the subject of a recent symposium (Klug and Reddy, 1984). While considerable information is available on the roles of plasmids in legume root nodulation and plant tumor formation, plasmid involvement in other plant-microorganism ecosystems has not been well documented. Much of the research on the ecology of plasmid-containing organisms has been done with animals or in the laboratory using continuous-culture conditions, so many of the concepts regarding the association of microorganisms with higher plants are based these models.

Plasmids are extrachromosomal genetic elements found in many prokaryotes and more recently in eukaryotes. They are double-stranded, covalently closed, circular DNA (cccDNA) molecules usually found in the form of supercoils. While cccDNA has been detected in *Saccharomyces cerevisiae* (Guerineau, 1979) and in certain fungi associated with higher plants (Garber et al., 1984; Hashiba et al., 1984), the genetic determinants carried on these plasmids are largely undefined. This discussion will therefore be confined to bacterial plasmids and to bacteria-plant interactions. The role of the microbial chromosome cannot be ignored, as the expression of the plasmid genome is dependent on host functions, and both plasmid and chromosomal genes may be involved in the microorganism-plant association; however, this discussion will be concerned primarily with the plasmid genetic determinants that may be important in such associations.

PLASMID VERSUS CHROMOSOMAL GENETIC DETERMINANTS

A major question that arises in a consideration of the possible roles of plasmids in plant-microbe interaction is, What advantages are there for an organism to have genetic determinants on an extrachromosomal element rather than on the chromosome? The one most commonly mentioned is transmissibility; i.e., some plasmids are able to move from one cell to another by conjugation, and nontransmissible plasmids may be mobilized by self-transmissible plasmids. In addition, the transfer of chromosomal genes between bacterial cells may be mediated by conjugative plasmids. While chromosomal gene transfer can occur by one of three mechanisms, conjugation, transduction, or transformation, the evidence for conjugative transfer in natural environments (discussed in a subsequent section) is most convincing. The presence of a sex factor increases the rate of gene transfer by several orders of magnitude, so intra- or interspecific transfer of genetic information among members of a bacterial population is enhanced, thereby aiding the population in adapting to environmental variations. A theoretical treatment of potential plasmid and chromosomal gene transfer frequencies among *Rhizobium* species is given by Beringer and Hirsch (1984). Another factor in plasmid-mediated gene transfer involves transposable elements. These are genetic elements that are capable of transfer between two replicons by nonclassical (illegitimate) recombination mechanisms (Burkhari et al., 1977). If one of these transposons is present on one replicon (e.g., the chromosome) in a cell and if that cell also harbors a conjugative or transmissible plasmid, the potential exists for transposition of the element to the plasmid and subsequent dissemination of the transposon genes throughout the population. Because of their ease of detection, the first bacterial transposons to be described were those containing genes for drug or heavy-metal resistance, and the best documented cases of the transposition phenomenon are those associated with the spread of drug resistance plasmids (R factors) (Heffron et al., 1975). Many other transposable elements are likely to exist in bacteria, but if so they have remained largely undetected because they confer unrecognized phenotypes. One, detected because of the mutations it induced as a result of transposition, has been found in a plant-pathogenic pseudomonad (Comai and Kosuge, 1983).[b]

A second factor which might relate to the question of plasmid versus chromosomal genetic determinants involves control of replication and gene expression. While plasmid replication and transcription are dependent on host enzymes, plasmids have genes whose products are capable of modulating these processes. These functions are discussed in more detail below, but one, plasmid copy number, is pertinent to this issue. Chromosomal genes are usually present in a single copy in prokaryotes, but a bacterial cell may contain many (>100) copies of a plasmid. Thus plasmid-borne genes, and potentially the gene products, can be present in many copies in a single cell. Gene amplification is known to occur in both the bacterial chromosome and in extrachromosomal elements, but the response to environmental changes that initiate such amplifications is slow when compared to the ability of a cell to transcribe multiple copies of a gene already present. (For reviews of gene amplification in bacteria see Anderson and Roth, 1977; Clarke, 1984a). Another phenomenon related to replication and segregation is plasmid maintenance. Evidence discussed in a later section indicates that plasmids may be lost from most cells of a bacterial population unless environmental conditions are such that the plasmid confers a selective advantage on the bacterium. Although it is difficult to determine all the conditions under which a plasmid exerts a positive effect, the implication is that, under most conditions, the plasmid is not essential for survival of the organism and that, when plasmid-borne genes become dispensable, they can be deleted from the population. The cell is thus freed from the burden of maintaining superfluous DNA by a mechanism not so readily available if that DNA were part of the chromosome. As in the acquisition of new genetic information, the rate of deletion of unwanted genes may be enhanced if the information resides on a plasmid.

PLASMID-BORNE GENETIC DETERMINANTS

This section will deal with genetic determinants that are often plasmid-borne and how the resulting phenotypic characters affect or might affect the relationship between the host bacteria and plants. Some of these will be rather obvious, while others will be speculative. Beringer and Hirsch

(1984) have compiled a list of five known plasmid functions with examples of each plus a miscellaneous category. The five are resistances, catabolism, antagonism, gene transfer, and interactions between organisms. In the miscellaneous category are DNA restriction and modification enzymes, DNA polymerase, and pigments.

Replication, Segregation, and Transfer

All plasmids thus far investigated have at least one unique site, termed the origin of replication, that binds the host polymerases required for plasmid replication. The mechanisms involved in DNA replication have been reviewed (Tomizawa and Selzer, 1979) and will not be discussed here. One aspect of plasmid replication is the control of plasmid copy number. This has important implications for the bacteria because it is related to the level of the corresponding plasmid gene products. Although dependent on host polymerases for replication, the genes for control of copy number are on the plasmid, as evidenced by the isolation of plasmid mutants with an altered copy number (e.g., Womble et al., 1985). It is frequently the case that the number of plasmid copies per chromosome is inversely related to the size of the plasmid. For example, F [94.5 kilobases (kb)] is usually present in one or two copies, whereas pBR322 (4.4 kb) can be found in 40–50 copies per chromosome. Another factor associated with copy number control is segregation or partition, i.e., the number of copies of the plasmid each daughter cell receives when the cell divides. Genes that control partition and mechanisms that regulate the process have been investigated by Nördstrom (1984) and Aagaard-Hansen. In the case of plasmids present in one or two copies per chromosome, it is important that plasmid replication be tightly coupled to chromosomal replication and cell division so that each daughter cell receives at least one copy (Meynell, 1972). Replication of high-copy-number plasmids need not be under such stringent control because of the greater possibility that each daughter cell will receive at least one copy. Replication and segregation are thus two of the factors involved in plasmid stability. This has important implications for plasmids described in subsequent sections of this chapter, because many are large and usually present in a low copy number.

In addition to controlling replication and segregation, these genes also appear to be involved in the phenomenon of incompatibility (Inc). Plasmids can be classified according to the incompatibility group to which they belong (e.g., IncF, IncI, IncP); this simply means that under nonselective conditions two different plasmids within the same incompatibility group cannot be maintained in the same cell. Incompatibility among *Rhizobium* plasmids (Djordjevic et al., 1982) and Ti plasmids (Koekman et al., 1982) has been investigated, but it is not known if they are related to plasmids from other gram-negative bacteria. pEX8080 from *Pseudomonas syringae* pv. *phaseolicola* appears to belong to the IncP group (or IncP1 as defined for pseudomonads) (Poplawsky and Mills, 1983). Thiry (1984) and Thiry and Faelen (1984) have described a 28-kb IncN plasmid and an 88-kb IncM plasmid from *Erwinia uredovora*, an organism found commonly growing saprophytically in association with plants. In addition, a 280-kb plasmid, when cured from the bacteria, gave rise to strains that had lost the ability to produce a yellow pigment and thiamine prototrophy. Incompatibility properties of plasmids from other plant-associated bacteria have not been studied. Other plasmid-encoded functions that have a role in determining which plasmids can be stably maintained in a cell are surface exclusion and restriction and modification enzymes. Surface exclusion genes prevent the uptake of a plasmid by a recipient from a donor containing the same plasmid. Modification enzymes alter the structure of a resident plasmid, usually by methylation, so that the plasmid-encoded restriction endonuclease is unable to degrade the resident plasmid but is able to cleave the DNA of a foreign plasmid or bacteriophage.

As mentioned previously, the ability of plasmids to transfer genes from one cell to another is an important mechanism in the transfer of information throughout a population. Conjugative plasmids encode functions that enable them to transfer themselves from one cell to another, and some are also able to mobilize nontransmissible plasmids and chromosomal genes (Holloway, 1979) as well. Genes that are responsible for conjugal transfer are called *tra* genes. Some of the products encoded by these genes are cell surface proteins (for a review see Manning and Achtman, 1979). The organization of transfer genes has been studied in only a few plasmids. Approximately 20 genes (ca. 40% of the plasmid genome) of the IncF plasmid, F, are involved in transfer, and 12 of them are required for sex pilus formation (Willetts and Wilkins, 1984). Organization of the *tra* genes in the IncN plasmid pKM101 is different from that in F, and there appear to be about 11 genes involved, 4 of which are involved in pilus formation (Langer et al., 1981; Winans and Walker, 1985). Conjugal transfer appears to involve about 19 genes of RP4 (IncP) (Barth et al., 1978). Among bacteria found in association with higher plants, several are known to contain conjugative plasmids. These include Ti plasmids from *Agrobacterium tumefaciens* (when mating occurs in the presence of the appropriate opine) (Genetello et al., 1977; Hooykaas et al., 1979), Ri plasmids from *Agrobacterium rhizogenes* (White and Nester, 1980), some of the symbiotic plasmids of *Rhizobium* (Long, 1984), pBPW1 (Staskawicz et al., 1981; Obukowicz and Shaw, 1983) and pJP1 (Obukowicz and Shaw, 1985), from *P. syringae* pv. *tabaci*, pJP30 from *P. syringae* pv. *angulata* (Obukowicz and Shaw 1985), two plasmids from *P. syringae* pv. *syringae* (Currier and Morgan, 1981), pEX8080 from *P. syringae* pv. *phaseolicola* (Poplawsky and Mills, 1983), and plasmids from *Erwinia* spp. (Chatterjee and Starr, 1973; Coplin et al., 1981a,b McCammon et al., 1981; Thirty, 1984). Interspecific transfer has been demonstrated for pBPW1 (Obukowicz and Shaw, 1983), Ti plasmids (Hooykaas et al., 1977), and *Rhizobium* plasmids (Hooykaas et al., 1982). pBPW1 (Quinnet and Shaw, unpublished results) and certain *Erwinia* plasmids (Chatterjee and Starr, 1973) are able to mobilize chromosomal genes, and a plasmid from *P. syringae* pv. *phaseolicola* is able to integrate into the bacterial chromosome and to form primes (Szabo et al., 1981); i.e., it can excise with fragments of chromosomal DNA. With the exception of the Ti, Ri, and *Rhizobium* plasmids, little is known about the organization of *tra* genes in these other species.

The presence of conjugative plasmids in plant-associated microoranisms is suggestive of but not proof that gene exchange occurs in the organism's natural habitat. Plasmid transfer *in planta* has been demonstrated in *Pseudomonas* spp. (Lacy and Leary, 1975) and between pathogenic and nonpathogenic strains of *A. tumefaciens* (van Larebeke et al., 1975), but these transfers were accomplished by inoculating plants with mixed populations of bacteria. Definitive examples of plasmid transfer in a natural ecosystem (e.g., in soil, in aquatic systems, or on plant surfaces) are rare; nevertheless, there is considerable circumstantial evidence for such transfers, particularly the transfer of drug resistance (R factor) plasmids. A recent review by Freter (1984) addresses the medical aspects of plasmid transfer, and Slater (1984) and Clarke (1984a) discuss the transfer of metabolic plasmids as well as the potential for such transfers to generate, by genetic recombination, organisms with the ability to metabolize compounds that neither parent can use. The environment and the physiological state of the host and recipient bacteria are important factors in the conjugative transfer of genetic material (Freter, 1984). For example, it is known that in the laboratory DNA transfer occurs at a higher frequency when matings take place on a solid surface rather than in liquid culture, and evidence indicates that this may also be true in soil (Weinberg and Stotzky, 1972) and in the animal gut (Freter et al., 1983).

In addition to ecological and epidemiological evidence for plasmid transfer in nature, attempts have been made to simulate natural conditions in the laboratory and to use these studies to examine factors that affect the rate of plasmid transfer. Slater (1984) has described experiments showing the transfer of metabolic plasmids among *Pseudomonas* species growing under continuous-flow culture conditions. His as well as other results discussed in his review led him to propose that, under strongly selective conditions, gene transfer leading to new metabolic pathways can occur within the population of a microbial community.

As well as the more obvious role in producing genetic diversity, conjugative plasmid transfer is a third factor contributing to the stability of a plasmid in a microbial population; i.e., the rate of loss of a plasmid under nonselective conditions is counterbalanced by the rate of plasmid transfer. This aspect of plasmid ecology has been reviewed by Freter (1984). With the use of continuous-flow culture conditions, mathematical models have been developed that relate the transfer of conjugative plasmids and the mobilization of nonconjugative plasmids to plasmid loss under nonselective conditions (Stewart and Levin, 1980, Levin et al., 1979; Levin and Rice, 1980; and Levin and Stewart, 1980). Interpretations of these models led the authors to conclude that there were conditions under which plasmids could survive in a population even when there was no selective pressure to maintain them. This requires that environmental and nutritional conditions be such that the rate of plasmid transfer, either by conjugation or mobilization, is high enough to counteract plasmid loss due to the decreased fitness of the plasmid-containing cells. It was suggested, however (Stewart and Levin, 1980), that these conditions were sufficiently stringent that they were unlikely to occur in many natural situations and, therefore, that plasmids would be maintained in a population only under conditions that conferred a selective advantage on the host bacteria. Slater (1984), on the other hand, found that a plasmid-containing *Pseudomonas* sp. HB2002 strain was able to outcompete the parent plasmid-free strain under laboratory conditions considered to be nonselective. It was concluded that the plasmid must confer some selective but unknown advantage on the bacteria. Freter et al. (1983) describe *in vitro* and *in vivo* experiments and propose models indicating that plasmids in *Escherichia coli* could be maintained in the animal gut under nonselective conditions. In studies with a group of nonconjugative plasmids in *E. coli,* Helling et al. (1981) found that, in continuous culture in nonselective media, the plasmid-containing strain always lost in competition with the plasmid-free strain. The decrease in the plasmid-containing population was not continuous, however. There was a cyclic increase and decrease that was dependent on nutritional conditions. The authors attributed these multiphasic growth patterns to mutations that provided a transient increase in the fitness of the plasmid-containing cells so that their survival was prolonged. They felt that this was not due to a mutation during culture but was a result of the presence of a small proportion of mutant cells in the initial plasmid-containing cell population. This phenomenon might be related to recent proposals by Hall et al. (1983) and Li (1984) who suggest that bacteria may contain cryptic genes, i.e., genes that are not expressed, but that under specific conditions, mutational or recombinational events activate the genes and allow their expression. Proposed models indicate that the genes could be retained over a prolonged time period if occasional reactivation occurred. Godwin and Slater (1979) described continuous-culture experiments indicating that plasmid-free populations were always more competitive than isogenic plasmid-containing populations. They also found that an *E. coli* strain harboring a conjugative plasmid with four drug resistance markers could lose one or more of the resistance genes, *tra* genes, or the whole plasmid during prolonged culture under conditions of limiting carbon or phosphorus. Even after prolonged growth, however, a small fraction of the population still retained intact plasmids. The mechanisms that allowed plasmid retention under these conditions are unknown, but the authors point out the potential advantage to the population of maintaining the plasmid in a small number of cells. In more recent studies, Sterkenberg et al. (1984) demonstrated (by measuring the decrease in β-lactamase activity and loss of tetracycline resistance) that, under continuous-culture conditions in nonselective media, *Klebsiella aerogenes* did not maintain plasmid pBR322.

The question of plasmid retention in nature in the absence of selective pressure obviously has not been completely resolved. A major reason is the lack of information on the precise nutritional and environmental conditions that prevail in the natural habitat of the organisms. Nevertheless, it is often tacitly assumed that any plasmid that is conserved under conditions not known to be selective must carry genes that confer some unknown advantage on the host cells. The fact providing some validity to this assumption is that complete genetic maps are not known for many plasmids, particularly those from plant-associated bacteria. The uncharacterized regions of DNA could contain genes responsible for characters that confer a selective advantage. Even the interpretation of the role of known plasmid genetic determinants in defining parameters for establishing a bacterium as part of a functional ecosystem may be difficult because some plasmid-encoded gene products are known to be multifunctional [e.g., the determinant for tetracycline resistance encoded on pBR322 is also involved in potassium uptake (Dosch et al., 1984)]. Thus the phenotypic character under study may be a fortuitous property of a protein whose primary function is unknown, so any proposed rationale as to a function the protein might have in aiding in the survival of the organism may have little validity in the organism's natural environment.

Cell Surface Properties

The bacterial cell envelope represents the boundary between the cell and its environment, and the components of the envelope have profound effects on the response of the cell to environmental variations. The composition of the envelope varies considerably among bacterial species, but in general consists of a membrane and a peptidoglycan layer. Gram-negative bacteria have an inner membrane enclosing the cytoplasm and an outer membrane surrounding the peptidoglycan. Embedded in this outer membrane is a lipopolysaccharide layer. Each of these components has multiple functions; the membranes, for example, may contain proteins that serve as receptors for bacteriocins and bacteriophages, proteins involved in the transport of molecules into or out of the cell, and other proteins involved in conjugation. (For a review of bacterial membranes see Inouye, 1979.) The peptidoglycan has a structural function, and it determines cell shape. The lipopolysaccharide carries antigenic determinants and has been associated with other recognition phenomena (Lippincott and Lippincott, 1984) (see Chapter 8) and the adsorption of bacteria to cell surfaces (Pueppke, 1984). In addition, many bacteria have other proteinaceous components such as flagella and pili as environmentally exposed parts of their structure. It is primarily the protein components of the bacterial cell envelope for which there is evidence for plasmid-encoded genes, and it is this aspect that will be emphasized in this section. There is, however, no a priori reason why plasmids cannot encode determinants for other factors, such as glycosidases of glycosyl transferases that could affect envelope polysaccharides important in the interaction of bacteria with plant cell surfaces, but direct evidence for such functions is lacking.

Environmental effects on membrane proteins have been considered by Freter (1984). Outer membranes of *E. coli* grown in a nutrient-poor medium, for example, lack several proteins that are present in membranes of cultures grown in a rich broth, and the bacteria have reduced sensitivity to colicins and bacteriophages (Chai, 1983). It is not known if genetic determinants for these proteins are on plasmids, but plasmid-related phage receptor proteins have been described (see below). Examples of other outer membrane functions are iron uptake by *Vibrio anguillarum* mediated by a protein encoded on plasmid pJM1 (Tolmasky and Crosa, 1984), the aerobactin iron uptake system encoded on plasmid ColV-K30 and other plasmids in several gram-negative bacteria (Perez-Casal and Crosa, 1984), and potassium uptake mediated by the tetracycline resistance determinant of pBR322 (Dosch et al., 1984). Although of unknown function, the outer membrane of strains of *P. syringae* pv. *syringae* that contain pCG131 have one additional protein but lack two others that are present in plasmid-free strains (Hurlbert and Gross, 1983). Similarly, strains of *P. syringae* pathovars that contain pBPW1 have a protein in their outer membranes that is absent in plasmid-free strains (Beck-von Bodman and Shaw, 1985a). It is possible that these proteins also play a role in the transport of nutrients across the membrane of these bacteria.

Sex pili appear to be essential for the establishment of mating pairs of bacteria so that conjugal DNA transfer can take place. As discussed in an earlier section, genes for the pilus proteins are located on conjugative plasmids, e.g., F and pKM101. Sex pili also serve a secondary function as receptors for certain bacteriophages such as the F-specific RNA phage R-17 (Crawford and Gesteland, 1964). Sex pili or possibly other membrane proteins are also receptors for a group of phages (PRD1, PR3, and PR4) that are specific for cells containing plasmids from incompatibility groups P, N, and W (Bradley and Rutherford, 1975). Plasmid pBPW1 from *P. syringae* pv. *tabaci* BR2 apparently also encodes receptors for this group of phages (Obukowicz and Shaw, unpublished results), but it is not known if sex pili are involved, nor has the plasmid been classified as to its incompatibility group. A 28-kb IncN plasmid from *E. uredovora* confers sensitivity to phage GU5 (Thiry, 1984). The presence of plasmid pCG131 in *P. syringae* pv. *syringae*, on the other hand, confers resistance to phages Psp1 and Psy 4A as well as bacteriocin PSC-1B (Gonzalez and Vidaver, 1979). It is not known if these effects are related to the plasmid-induced changes in the outer membrane protein composition mentioned previously. Ti plasmids from *A. tumefaciens* carry genes that confer resistance to phage AP_1 (van Larebeke et al., 1977) and, because they are conjugative, presumably genes for sex pili. The presence of Ti plasmids appears to affect the cell surface composition of *A. tumefaciens* (Smith and Hindley, 1978), and it has been proposed (Watley et al., 1978) that both plasmid and chromosomal genes may have an effect on the structure of the lipopolysaccharide present in the outer membrane of the bacteria.

It is obvious that the bacterial cell envelope is structurally and functionally very complex. Presumably, it contains recognition structures that allow bacteria to interact with plants, as well as components that determine how the bacteria respond to environmental factors. The presence of genetic determinants on plasmids and the ability to transfer these plasmids to other members of a population could provide the bacteria with the versatility to interact with many plant species and to respond to environmental changes rapidly. It is apparent from several of the chapters in the volumes in this series that much effort has gone into study of the structure and function of envelope components from plant-associated microorganisms, but there have been few attempts to relate the results of this research to the genetics of the microorganisms. The importance of collaborative research by molecular geneticists and microbial ecologists was stressed by Klug and Reddy (1984) in the preface of their book. In no area could such cooperation be more fruitful than in research on the bacterial cell envelope.

Utilization of Nutrients

One of the most remarkable things about microorganisms that inhabit soil or aquatic environments is their ability to utilize diverse organic compounds as nutrients. It has been suggested by many that no organic compound, either natural or synthetic, exists that cannot be used by some microorganism. Whether this is an accurate statement or not, organisms in nature and, more recently, laboratory constructs have amply proved their metabolic versatility. Degradation of organic compounds by microorganisms has received much attention from ecologists, biochemists, and geneticists, so pertinent reviews of the topic will be mentioned. A recent book edited by Gibson (1984) is largely devoted to the metabolic pathways involved in the catabolism of organic compounds having a variety of different structures. One chapter in this book by Clarke (1984b) and another review by the same author (Clarke, 1984a) address various aspects of the evolution of metabolic pathways. Another chapter by Slater and Lovatt (1984) and a review by Slater (1984) discuss the degradation of organic compounds by microbial communities and stress the importance of collaboration among the different species or strains within the community. Other topics reviewed include carbon and energy flow through soil microflora and microfauna (Elliott et al., 1984), the genetics of xenobiotic degradation (Eaton and Timmis, 1984), theoretical and experimental studies on biodegradation in natural ecosystems (Larson, 1984), and experiments for inducing evolutionary changes in azo-dye-degrading bacteria in the laboratory (Kulla et al., 1984).

Although several genera of bacteria capable of degrading organic compounds are known, evidence for the role of plasmids in this process comes largely from work on soil pseudomonads, particularly strains of *Pseudomonas putida*. Several of these metabolic plasmids, including the TOL (toluene and xylene) and NAH (napthalene) plasmids are described by Bayly and Barbour (1984). These authors speculate that plasmids are likely to be involved in the breakdown of other aromatic hydrocarbons. A major question concerning the breakdown of organic compounds is whether such breakdowns are in some way related to the interactions of plants and bacteria or whether the bacteria are simply garbage disposal units whose only role is in the recycling of carbon and nitrogen in plant ecosystems. One complicating factor is the propensity of genes for these metabolic enzymes to undergo rearrangement or modification when the bacteria are challenged by novel organic compounds (Clarke, 1984a,b; Slater and Lovatt, 1984). This makes it difficult to determine if a particular metabolic pathway has, or at one time had, the ability to utilize as substrates undefined products produced by living plants. Plasmid-encoded enzymes have been described, however, that are able to metabolize known plant products such as camphor (Fig. 1) (Rheinwald et al., 1973), salicylic acid (Chakrabarty, 1972), and nicotine (Thacker et al., 1978). Salicylic acid, as well as the plant products catechol and gentisic acid, are proposed intermediates in the degradation of naphthalene (Gibson and Subramanian, 1984). It is also not difficult to imagine that enzymes encoded on the OCT plasmid, which is involved in the degradation of aliphatic hydrocarbons, might also use plant waxes or cutins as substrates.

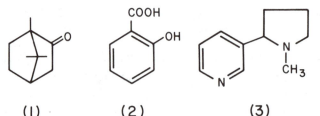

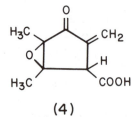

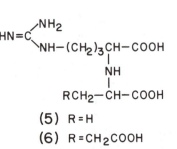

(1)   (2)   (3)

(4)

(5) R = H

(6) R = CH₂COOH

Figure 1. Plant and microbial products synthesized or degraded by plasmid-borne genetic determinants. Plant products camphor (1), salicylic acid (2), and nicotine (3) are metabolized by microbial enzymes encoded by plasmid-borne genes. Synthesis of methylenomycin (4) occurs by enzymes encoded by genes borne on a plasmid in *Streptomyces coelicolor*. The opines, octopine (5), and nopaline (6) are synthesized from arginine by enzymes encoded by genes on the Ti plasmid of *Agrobacterium tumefaciens*.

    In all the above examples, the role of the plasmid-containing bacteria is probably in the breakdown of dead plant tissue. Other plasmid-encoded enzymes are important in the elimination of synthetic materials, such as halogenated aromatic compounds, from the environment (Gibson, 1984, particularly the chapter by Slater and Lovatt). There is meager evidence for a similar function for plasmids in the interaction of bacteria with living plants. The suggestion has been made that the compounds present in plant exudates serve as attractants to soil bacteria such as *Pseudomonas lachrymans* (Chet et al., 1973) and *P. syringae* pv. *phaseolicola* (Mulrean and Schroth, 1979), and presumably such compounds are metabolizable by the bacteria. Structures for the compounds in the exudates remain hypothetical, and plasmid involvement in these processes has not been demonstrated. Among the plant-pathogenic bacteria, genes that encode enzymes for the degradation of opines have been found in Ti and Ri plasmids from *Agrobacterium* and for the breakdown of tryptophan to indole-3-acetic acid (IAA) in Ti and Ri plasmids and plasmids from oleander strains of *P. syringae* pv. *savastanoi* (see section on plant tumorigenesis for speculation on the possible origins of these enzymes). Little is known about the genetics of plasmids from other plant pathogens. It seems likely, however, that many of these cryptic plasmids carry genes for the degradation of compounds found on above- or below-ground plant surfaces. The various compounds produced by different plant species could be factors in determining which resident plasmids would be predominant in the bacterial population associated with a given plant species.

Production of Compounds

Antibiotics and Bacteriocins

Antibiotics are compounds produced by microorganisms that, at low concentrations, inhibit the growth or reproduction of other microorganisms. In spite of the common perception, there is little evidence that antibiotics are produced by organisms in their natural environment or that antibiotics play a role in the ecology of the soil microflora. The effect on a bacterial population of antibiotic production by one member of the population has been studied in the laboratory, but such studies, particularly on microorganisms associated with higher plants, are difficult. A review by Rothrock (1980) points out that there is little relationship between antibiotic production in culture and antibiosis in soil. The lack of evidence for antibiotic production in soil, however, does not prove that it does not occur, and because it is such a widely held concept as a teleogical rationale for antibiotic production by microorganisms, the subject will be treated briefly here.

Attempts to detect antibiotic production by the addition of large inocula of known antibiotic producers have generally been unsuccessful (Gottlieb, 1976), but when the soil had been previously sterilized Gottlieb and Siminoff, 1952), and when material such as soybean meal was added to unsterilized soil before addition of the antibiotic-producing organism (Gottlieb et al., 1952), antibiotics could be detected. Organic nutrients produced by thermal decomposition and organic nutrients added directly to soil presumably simulate the conditions of cultures grown in the laboratory. In culture, the production of antibiotics usually takes place in the late logarithmic or stationary phase of growth, the so-called idiophase (Bu'Lock, 1965). It is possible that under most conditions antibiotic-producing organisms are in a dormant form (e.g., spores of actinomycetes, fungi, or *Bacillus* species) and that vegetative growth and antibiotic production occur for only relatively short periods when environmental and nutritional conditions are suitable. These conditions include an adequate supply of water, appropriate temperatures, and metabolizable material such as fallen leaves or newly cut grass. Such conditions presumably enhance the growth and reproductive rate of all soil microorganisms, but might also be suitable for the production of antibiotics and thus increase the competitiveness of the producing organisms. In this context, it might be significant that the characteristic odor produced by soil after a warm spring rain is thought to be due to volatile compounds produced by actinomycetes. These are difficult problems to study, so convincing proof of the speculations may be a long time in coming.

The possible role of plasmids in the production of several antibiotics has been investigated, but convincing evidence for plasmid involvement is available in only two cases. Genes for methylenomycin (Fig. 1) production and resistance are present on SCP-1 from *Streptomyces coelicolor* A3(2) (Kirby and Hopwood, 1977) and pSV1 from *S. violaceus-ruber* (Aguilar and Hopwood, 1984). On the other hand, genetic evidence indicates that genes involved in the biosynthesis of chloramphenicol by *Streptomyces venezuelae* (Akagawa et al., 1975), streptomycin by *S. griseus* and *S. bikiniensis* (Hanssen and Kirby, 1983), and hydroxystreptomycin by *S. glaucescens* (Ono et al., 1983) are on chromosomal loci. Antibiotic-producing streptomycetes appear to have genes that are at least in part responsible for resistance to their own antibiotics. The facts that these resistance genes are often closely associated with antibiotic production genes (Shaw and Piwowarski, 1977) and that they are usually located on streptomycete chromosomes (Ono et al., 1983) argue against the hypothesis that antibiotic producers are the source of antibiotic resistance genes found in other bacteria (Benveniste and Davis, 1973). Streptomycetes, however, appear to differ from the well-studied gram-negative bacteria in that certain DNA sequences can be excised from the chromosome, be conjugatively transferred to another species, and there replicate as independent elements (Hopwood et al., 1984). Thus biosynthetic and resistance genes might be more mobile in a bacterial population than it would first appear. Similarities in the base sequence of the aminoglycoside phosphotransferases from *Streptomyces fradiae* and the comparable enzymes found in two different bacterial transposons tend to support the hypothesis of a common origin for the genes (Thompson and Gray, 1983).

Bacteriocins may be considered a particular class of antibiotics that differ from most others because of their high degree of specificity. They are usually proteins, produced by one strain of an organism, that are inhibitory to other strains of the same species (Jacob et al., 1953) or, in some instances, to related species. They are produced by many gram-positive and gram-negative bacteria and may function to reduce competition from nonproducing organisms. As with other antibiotics, evidence is rather sparse for such a role in nature. An example that supports this hypothesis, however, is agrocin-84. This is an atypical bacteriocin because it is an adenine derivative, not a polypeptide (Tate et al., 1979); but it has the same high degree of specificity that polypeptide bacteriocins have. It is produced by the avirulent *Agrobacterium* strain K84, and it inhibits many virulent strains of this organism. It is not known if the genes for its synthesis are on plasmids. Sensitivity of agrocin-84 is thus a Ti plasmid-encoded function that is potentially deleterious to sensitive strains of *Agrobacterium*. Conservation of genes responsible for this sensitivity suggests that, as in the case of bacteriocin receptors in other bacteria, the gene product might have as yet undisclosed functions in the virulent strains. There is evidence that production on plant surfaces protects the plants from tumorigenic *Agrobacterium* species (Kerr, 1980). Unfortunately, not all virulent strains of *Agrobacterium* (discussed in a later section) are sensitive to agrocin-84, and virulent strains may become resistant (Schroth et al., 1984), so it has not proved to be a very effective method for disease control.

In contrast to other antibiotics, evidence indicates that genetic determinants for many bacteriocins are located on plasmids. Those from gram-negative bacteria, e.g., ColEI, are among the most extensively studied plasmids. Little is known about bacteriocinogenic plasmids in plant-associated microorganisms. *Rhizobium leguminosarum* was reported to produce a plasmid-encoded bacteriocin (Hirsch, 1979), and a strain of *Erwinia herbicola*, a bacterium that grows saprophytically on many plant species, produces a bacteriocin (Gantotti et al., 1981) thought to be plasmid-encoded, but there is no evidence that these bacteriocins are functional in plant ecosystems. A report (Gonzales and Vidaver, 1979) that the loss of a 53-kb plasmid from *P. syringae* pv. *syringae* resulted in the loss of resistance to bacteriocin PSC-1B has not been confirmed. As bacteriocin resistance and biosynthetic genes are usually encoded on the same plasmid, however, this could represent another example of a bacteriocinogenic plasmid from a plant-pathogenic bacterium.

Phytotoxins and Pathogenicity

The detection in enteric bacteria of plasmids that contain genes for toxin production and the role of plasmids in crown gall (discussed below) suggested to several researchers that genes for phytotoxins or other characters associated with pathogenicity might be similarly located in plant-pathogenic bacteria. Research dealing with the isolation and characterization of plasmids from plant-pathogenic bacteria and a consideration of the role of plasmids in plant disease has been reviewed (Lacy and Leary, 1979; Shaw, 1985). The involvement of plasmids in plant tumorigenesis and the role of phytohormones in plant disease are considered in a later section of this chapter. This section will deal only with instances in which plasmids have been implicated in some aspect of plant disease other than plant tumor formation.

Initial work with *P. syringae* pv. *syringae* (Gonzales and Vidaver, 1979) and *P. syringae* pv. *phaseolicola* (Gantotti et al., 1979) supported the hypothesis that syringomycin and phaseolotoxin genes were on plasmids. Subsequent work (Jamieson et al., 1981; Gonzalez and Olsen,

1981; Currier and Morgan, 1983; Gonzalez et al., 1984) failed to confirm the earlier results. Likewise, plasmids do not appear to be involved directly either in toxin production or in other aspects of pathogenicity in *P. syringae* pv. *tabaci* or *P. syringae* pv. *angulata* (Obokuwicz and Shaw, 1985). Plasmid involvement in coronatine production by *P. syringae* pv. *glycinea* (Willis and Leary, 1984) and *P. syringae* pv. *atropurpurea* (Sato et al., 1983) has been proposed, but this has not yet been confirmed. A plasmid in *Pseudomonas solanacearum* has been implicated in diseases caused by this organism (Ofuya and Wood 1981). It was proposed that the plasmid encoded cell wall-degrading enzymes, but the proposal was based on rather tenuous evidence and has not yet been confirmed. With the exception of bacteria that induce plant tumors (see below), there is no convincing evidence for plasmid involvement in plant disease. It is probably, however, that future research will prove this statement to be in error.

Plant Tumorigenesis

The mechanism and genetics of plant tumorigenesis have been the subject of several recent reviews (Nester and Kosuge, 1981; Gelvin, 1984; Nester et al,. 1984), so they will not be treated in depth here. However, the role of plasmids in providing tumor-inducing bacteria with the facility for establishing an ecological niche is of such importance in any consideration of bacterial ecology that the topic must receive some attention. The best known tumor-inducing bacteria are the *Agrobacterium* species (or strains). *Agrobacterium tumefaciens, A. rhizogenes,* and *A. radiobacter* are common soil microorganisms that are, like many other members of the soil microflora, not particularly fastidious about their growth requirements, but the presence of the Ti plasmid in *A. tumefaciens* and the Ri plasmid in *A. rhizogenes* allows these bacteria to become established in an environment essentially to the exclusion of all competitors. *Agrobacterium tumefaciens* infects a wide variety of dicotyledonous plants and causes the disease called crown gall; *A. rhizogenes* causes hairy root disease; the plasmid-free *A. radiobacter* is avirulent. As these bacteria are for the most part identical except for their plasmid content, the basis of their taxonomic identity resides in the Ti and Ri plasmids. Plasmids that confer host specificity have performed a similar taxonomic function in the classification of legume-nodulating rhizobia, a group of bacteria (discussed in the next section) closely related to agrobacteria. These plasmids thus have become classical examples in the study of plasmid ecology.

Most *Agrobacterium* strains contain large plasmids (50–275 kb), but some also contain small plasmids (~15 kb) (Merlo and Nester, 1977). It is not uncommon for virulent strains to contain multiple plasmids, only one of which is involved in tumorigenesis (Currier and Nester, 1976).

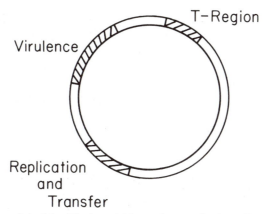

Figure 2. A model of the Ti plasmid from *A. tumefaciens.* Both the virulence (vir) and the T regions are needed for full expression of tumorigenicity in plants. However, only the T- region is stably integrated and maintained in the plant genome after transformation of the plant cell by the bacterium.

Agrobacterium radiobacter also contains large cryptic plasmids (Merlo and Nester, 1977). Many of the Ti and Ri plasmid genes essential for disease establishment and tumor induction and maintenance have been identified and mapped. They include the virulence genes, the T region, and genes involved in conjugal transfer and plasmid replication. The approximate loci of these genes on most Ti plasmids are shown in Fig. 2. Other genes, such as those involved in host specificity, bacteriocin sensitivity, incompatibility, phage exclusion, opine catabolism, and stability, have also been identified on Ti and Ri plasmids. These functions are discussed in detail in the above-mentioned reviews. In all, however, only about 50–60% of the DNA of Ti and Ri plasmids and essentially none of the DNA of cryptic plasmids has been characterized genetically. It has been suggested (Currier and Nester, 1976) that most of the genetic information on the virulence plasmids is not directly related to pathogenicity. There is considerable sequence homology among the DNA of plasmids responsible for virulence and between that of the virulence and cryptic plasmids found in *Agrobacterium* strains (Merlo and Nester, 1977; Drummond and Chilton, 1978). There is also sequence homology between the DNA of Ti plasmids and *Rhizobium* DNA (Prakash and Schilperoort, 1982). The results of such structural comparisons suggest that many of the functions involved in tumorigenesis have been conserved in these plasmids and that *Agrobacterium* and *Rhizobium* might have functions in common, but that other parts of the plasmids contain regions of considerable genetic diversity. These uncharacterized portions of the Ti and Ri plasmids may play as yet unidentified roles in the ecology of the organisms, possible in their activity as soil saprophytes.

The genes discussed above are all expressed in the bacteria. The aforementioned T region (a continuous region of about 12–23 kb in Ti plasmids and 20–50 kb in Ri plasmids) is expressed only poorly in the bacterium but is expressed after it becomes integrated into the plant chromosome. Genes of the T-region encode enzymes for the synthesis of the opines (e.g., octopine and nopaline, Fig. 1) utilized by the strain of bacteria that induced the tumor, two enzymes involved in the synthesis of IAA (Inze et al., 1984; Schröder et al., 1984; Thomashow et al., 1984), and an enzyme involved in cytokinin synthesis (Akiyoshi et al., 1984). The genes of the T region, particularly those for plant hormone synthesis, are not required for tumor induction, but they are necessary for maintenance of the tumorigenic state.

It might be inferred from the above discussion that only plasmid-encoded functions are required for the interaction of *Agrobacterium* species and plants. This is probably not the case, however, as there is evidence that chromosomal genes are also required. For example, Whatley et al. (1978) have proposed that either chromosomal or plasmid genes are required for adhesion of the bacteria to plant cell walls. Nester et al. (1984) point out that there is conflicting evidence for plasmid involvement in the binding of *Agrobacterium* to plant cells, but that no nonbinding avirulent strains with mutations in the Ti plasmid have been described.

Agrobacterium species thus appear to have the ability to form some type of association with higher plants. Acquisition of one of a particular class of plasmids, which may reduce their competitive fitness in soil, allows the bacteria to invade certain plant species and to utilize compounds (opines) that are not metabolizable by most bacteria. The genes on T-DNA, after integration into the plant chromosome, are turned on and begin to produce the specific opine utilized by the bacteria, as well as the plant hormones that induce plant cell proliferation and provide an extended habitat for the bacteria. Furthermore, the presence of the opine, because it enhances the frequency of conjugative plasmid transfer (Hooykaas et al., 1979), ensures that the plasmid will not be lost to future generations.

One of the more intriguing aspects of this research is the question of the source of the genes found on tumor-inducing plasmids. There appears to be no homology between the DNA of Ti plasmids and the DNA from uninfected plants; however, there is some homology between Ri DNA and DNA from uninfected *Nicotiana glauca* plants (White et al., 1983). Whether the source of this DNA is plant or bacterial is not known, but Nester et al. (1984) tend to favor the theory that the the homologous regions represent DNA sequences that have been conserved in both plants and bacteria. The problem of transcribing prokaryote DNA in a eukaryote system has been solved completely, but typical eukaryote transcription signals are present in T-DNA. Whether the source of these sequences is eukaryote DNA or whether they are the result of fortuitous mutations in bacterial DNA is not known. A knowledge of the products of genes that have been identified would help in our understanding, but only a few tumor plasmid gene products are known. The enzymes involved in IAA biosynthesis, tryptophan monooxygenase and indoleacetamide hydrolase, may be responsible for IAA synthesis in several bacteria (Comai and Kosuge, 1980), whereas in at least one plant (tomato) the hormone is synthesized via indolepyruvic acid and indoleacetaldehyde (Wightman, 1973). IAA biosynthesis genes are thus probably of bacterial origin. Octopine, one of several opines utilized by specific strains of *Agrobacterium*, was originally isolated by Morizawa in 1929 (quoted in Greenstein and Winitz, 1961) from octopus muscle where it may serve as a phosphagen in the manner of creatine in vertebrate muscle. Baldwin (1952), on the other hand, states that octopine is not present in normal octopus tissue but arises post mortem. He goes on to suggest that it might result from the condensation of pyruvate with arginine to form a Schiff base that can serve as a substrate for lactate dehydrogenase. The metabolism of a group of recently characterized mannityl opines (mannopine, mannopinic acid, agropine, and agraopinic acid) has been investigated (Chilton and Chilton, 1984). Based on the relative ability of *Agrobacterium* strains to metabolize these compounds and the structural resemblance among opines, it has been suggested that there may have been a common ancestral pathway from which all opine metabolic pathways are derived. Based on Baldwin's proposal, it is also possible that opine biosynthetic and catabolic enzymes might have been derived from the mutation of enzymes (e.g., lactate dehydrogenase or glutamate dehydrogenase) to proteins with altered substrate specificities.

Another class of plant tumors is induced by *P. syringae* pv. *savastanoi*. Strains of these bacteria cause gall formation on olive, oleander, and privet, and the diseases are called olive and oleander knot. The physiology and genetic bases of these diseases have been reviewed (Nester and Kosuge, 1981; Gelvin, 1984), so these topics will not be dealt with in depth here. Briefly, it has been demonstrated that the ability to induce gall formation is associated with the ability of the bacteria to produce IAA. The bacteria synthesize IAA from tryptophan utilizing the two enzymes tryptophan monooxygenase and indoleacetamide hydrolase whose activities also may be responsible for IAA synthesis in crown gall tumors (Fig. 3). The genes for these enzymes have geen located on plasmids pIAA1 (51 kb) (Fig. 4), pIAA2 (72 kb), and pIAA3 (90 kb) from three different oleander strains of *P. syringae* pv. *savastanoi* (Comai and Kosuge, 1982; Comai et al., 1982). The plasmids show homology in a region of about 6 kb that contains the IAA genes (Comai and Kosuge, 1983a). Strains lacking a pIAA plasmid do not produce IAA, and they are weak pathogens. The gene for tryptophan monooxygenase from pIAA1 has been cloned

1. Tryptophan Monooxygenase

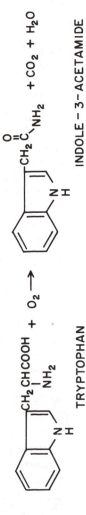 $CH_2CHCOOH$ $+ O_2 \longrightarrow$

$\underset{|}{\overset{}{N}H_2}$

TRYPTOPHAN

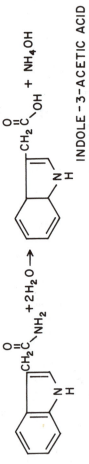 $CH_2\overset{\overset{\displaystyle O}{\|}}{C}_{NH_2}$ $+ CO_2 + H_2O$

INDOLE – 3– ACETAMIDE

2. Indoleacetamide Hydrolase

 $CH_2\overset{\overset{\displaystyle O}{\|}}{C}_{NH_2}$ $+ 2H_2O \longrightarrow$

 $CH_2\overset{\overset{\displaystyle O}{\|}}{C}_{OH}$ $+ NH_4OH$

INDOLE – 3–ACETIC ACID

Figure 3. The pathway for indoleacetic acid synthesis from tryptophan in *Pseudomonas savastanoi*. The enzymes and their genetic determinants are (1) tryptophan 2-monooxygenase, *iaaM*; (2) indoleacetamide hydrolase, *iaaH*. Both genes occur on a plasmid, pIAA (Fig. 4) in oleander strains of the bacterium.

22

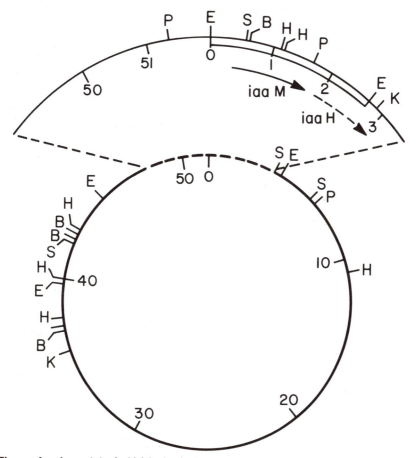

Figure 4. A model of pIAA1, the indoleacetic acid plasmid found in California oleander strains of *P. savastanoi.* The enzymes encoded by the genetic determinants are described in Fig. 3. The letters indicate the sites cleaved by various restriction endonucleases: P, Pst; E, EcoR1, S, Sal 1; B, BamH1; H, HindIII; K, Kpn. The numbers refer to base pair numbers (50 = 50kb) The start point, 0, is arbitrarily assigned to the leftward end of fragment M, a 2.75kb portion of pIAA1, which is delineated by double lines bounded by EcoRi (E) sites.

(Comai and Kosuge, 1982, 1983a). In contrast to the oleander strains, the olive strains of *P. syringae* pv. *savastanoi* carry the genes for IAA biosynthesis on the chromosome rather than on a plasmid (Comai et al., 1982; Comai and Kosuge, 1983). Hybridization studies indicated that the chromosomal locus of the IAA genes was not the result of integration of pIAA1.

IAA Biosynthetic Pathway

Comai and Kosuge (1983a) have speculated on the origin of the IAA
genes. They have proposed that the two enzymes responsible for the conversion of tryprophan to IAA may have been part of a pathway for tryptophan degradation used by the organism under saprophytic growth conditions. Mutations in a gene involved in regulating the pathway by induction in the presence of substrate and in a gene required for further metabolism of IAA would result in a strain that produced IAA constitutively.
Such a strain, after invasion of plant tissue, presumably would cause cell
proliferation and the potential for large increases in the bacterial population. This is a reasonable hypothesis, as IAA− strains are able to invade
the plant, but they do not cause tumors nor do they reach high population
levels. It might be interesting to examine *P. syringae* pathovars closely
related to *P. syringae* pv. *savastanoi* for their ability to catabolize IAA
and to look for comparable mutant genes in *P. syringae* pv. *savastanoi*.
The work on the interaction of *P. syringae* pv. *savastanoi* with plants has
led to a number of intriguing questions: Why do other bacterial
saprophytes that produce IAA not establish a niche for themselves in
higher plants? Are other plasmid or chromosomal genes also required for
the establishment or development of the disease? Is it a coincidence that
P. syringae pv. *savastanoi* strains that carry IAA genes on a plasmid are
able to parasitize both oleander and olive while those with IAA genes on
the chromosome infect only olive, or does the locus somehow affect host
specificity? Do the plasmid-borne and chromosomal IAA genes have a
common origin? These and undoubtedly other aspects of this problem will
be the subject of future research.

 Corynebacterium fascians causes a fasciation disease, i.e., a proliferation of shoots, in several plant species, and the role of plasmids in this
disease has been studied by Murai et al. (1980). The severity of symptoms
was correlated with the level of cytokinin production in bacterial cultures,
and the predominant cytokinin was identified as N-isopentenyladenine,
the same compound produced by an enzyme encoded on T-DNA
(Akiyoshi et al., 1984). All of the virulent strains examined contained
plasmids, but the highly virulent strain had a large ($>$100 mdal) plasmid
that was absent in the moderately virulent strain. No plasmids were detected in the avirulent strain. It was suggested that the disease symptoms
were a result of the cytokinin produced by the bacteria and that the large
plasmids were involved in cytokinin production. This hypothesis has not
yet been confirmed.

 Convincing evidence for the involvement of plasmids in plant disease
has thus far been confined to diseases characterized by abnormal growth
patterns. In every case the abnormal growth is a result, at least in part, of
altered levels of plant hormones. The genes responsible for the hormone

biosynthetic enzymes are encoded on bacterial plasmids in crown gall, hairy root, and oleander knot diseases, and on the bacterial chromosome in olive knot. The role of plant hormones in plant disease has been reviewed by Gelvin (1984) and by Schröder (Chapter 2, this volume). One aspect that has received little attention is the role of the plant in interactions with tumor-inducing bacteria. Plants probably differ in the levels and the ratio of the hormones required to maintain normal growth and development, so it might be expected that different plant species of the same species at different stages of development respond differently to invasion by hormone-producing bacteria. It is conceivable that some species might be able to deal with abnormal hormone levels (e.g., by altering the rate of endogenous production or degradation) so that typical symptoms are not observed. There is evidence that IAA/cytokinin ratios affect the morphology of *Agrobacterium*-induced tumors (Akiyoshi et al., 1983), and these ratios can also affect host specificity (Hoekma et al., 1984). Instances of plant invasion by tumor-inducing bacteria may go undetected because of the lack of obvious symptoms.

Legume Nodulation

While the Ti plasmids of *Agrobacterium* species have been eminently successful in providing an almost unassailable niche for their hosts, the Sym (symbiotic) plasmids have provided a similar service for *Rhizobium* species but in a more elegant fashion. In this case the interaction of the bacteria with a plant usually results in a symbiotic relationship rather than a plant disease. The similarities between *Rhizobium* and *Agrobacterium* and homologous regions shared by the plasmids from these two genera of bacteria have led to the speculation that both classes of plasmids were derived from a common ancestor (Prakash and Schilperoort, 1982). The *Rhizobium* species, however, have had to pay a price for this association with plants and have lost their identity as free-living bacteria. The various aspects of legume nodulation by *Rhizobium* have been reviewed (Long, 1984; Vance, 1983) (see Chapters 3 and 4). The ecological consequences for bacteria in having genes enabling them to establish a symbiotic relationship with a plant are self-evident, and the basis of proposing an advantage to the bacteria in carrying these genes on a self-transmissable or mobilizable plasmid makes use of the same arguments used to support the rationale for other plasmid-borne determinants. This section therefore will be limited to a brief summary of Sym plasmid-encoded functions and comments on what is not known about these plasmids.

Legume nodulation and nitrogen fixation by *Rhizobium* have been subjects of study by ecologists and biochemists for many years, but it was not until about 7 years ago (Johnston et al., 1978) that modern techniques of molecular biology were applied to the problem. Since that time information on these complex processes has increased by several orders of magnitude, and the results of the research illustrate the importance of cooperation among several disciplines in solving complex problems. The breakthrough was the discovery that, in fast-growing *Rhizobium* species, large plasmids were essential for nodulation. The use of techniques such as transposon mutagenesis and cloning of Sym plasmid DNA sequences has allowed the identification of genes required for host specificity, nodulation, and nitrogen fixation, and of equal importance, the defining of some of the physiological and biochemical processes involved in nodule formation and function (Long, 1984) (see Chapter 3). These plasmid-borne genes are essential but not sufficient, as evidenced by the lack of nitrogen fixation by *Agrobacterium* into which an intact Sym plasmid has been transferred (Hooykaas et al., 1982; Truchet et al., 1984; Hirsch et al., 1985). Genes on the bacterial chromosome, as well as plant genes, are also required for establishment of a fully functional nodule (Brewin et al., 1983). A tentative scheme for describing nodule development and a proposed system for denoting phenotypic characters associated with the developmental stages are discussed by Long (1984).

The role of plasmids in nodule formation by fast-growing *Rhizobium* species is well documented, but such is not the case with slow-growing *Rhizobium* (e.g., most isolates of *Rhizobium japonicum*) or with other nitrogen-fixing bacteria (e.g., *Frankia* species, discussed in Chapter 5). It is not known whether fast-growing rhizobia are unique in having genes for nodulation and nitrogen fixation on plasmids or whether the other species contain replicons equivalent to Sym plasmids that cannot be isolated by the usual plasmid isolation techniques. Genes that encode proteins of the nitrogen-fixing complex from many genera of nitrogen-fixing bacteria have extensive homology (Ruvkin and Ausubel, 1980). These DNA sequences and possibly their organization (Corbin et al., 1982) have been conserved in many species, but the organization of all the genes required for the formation of a functioning nodule might vary among species.

Even though we know a great deal about plasmid-borne genes involved in the interaction of legumes and rhizobia, there is still much to be learned. Considering the *Rhizobium* plasmids as a group, the lack of information about the molecular genetics of nodulation is trivial when compared to our ignorance of other plasmid-determined characters. At least one of the plasmids carries a gene for bacteriocin production (Hirsch, 1979), others have genes for conjugative transfer [e.g., *R. leguminosarum* (Hooykaas et al., 1982)], and all presumably have genes essential for maintenance and segregation. *Rhizobium* plasmids are usually very large (approximately 75–600 kb) and remain largely uncharacterized. Furthermore, many strains contain more than one plasmid, of which only the Sym plasmid has been studied. The Sym plasmid from *R. leguminosarum* strain P8, for example, is 600 kb in size, and this strain has three additional plasmids ranging in size from 200 to 500 kb (Krol et al., 1982). Thus the plasmid DNA in this strain is about one-third the size of the bacterial chromosome. The energy burden of maintaining this much DNA is very high, so if proposals discussed previously are correct, there must be selective pressure to maintain the cryptic plasmids as well as the large, uncharacterized regions of the Sym plasmids. The structural relationships among Sym plasmids and between Sym plasmids and the other *Rhizobium* plasmids have been investigated (Long, 1984), but functionally they remain very large black boxes. Ecological and genetic studies of the plasmid DNA *not* associated directly with nodulation should provide interesting research for years to come.

STRUCTURAL RELATIONSHIPS AMONG PLASMIDS

In the early studies, bacterial plasmids were detected primarily because they carried genetic determinants for easily recognizable phenotypic characters such as fertility and drug resistance. It soon became apparent, however, that plasmids could be isolated from many different bacterial species even though in many instances they conferred no apparent phenotype; i.e., they were cryptic. Many plasmids have been characterized physically and chemically, and such studies have led to speculation as to possible taxonomic relationships, evolutionary processes, and ecological implications for the host bacteria. The confidence that can be placed in such speculation is dependent on the techniques used to compare plasmid structures. Size comparisons and the patterns produced by electrophoresis of restriction endonuclease digests are not reliable predictors of plasmid relatedness. For example, plasmids with as much as 95% homology can give different restriction patterns (Thompson et al., 1974). Hybridization of restriction endonuclease fragments with labeled probes under known conditions of stringency can provide an estimate of the extent of DNA sequence homology, but the actual base sequences of two DNA molecules must be compared if it is important to determine their absolute relationship.

Given the fact that the plasmids from two different organisms are to some degree structurally related, the question remains as to what information can be gained from such a relationship. Presumably, identical DNA sequences contain the same information, and if they are expressed, they encode the same products. From an ecological standpoint then, the products are likely to perform the same or similar functions for the bacterium. DNA sequences that are conserved among isolates of the same or different bacterial species (assuming that plasmid sequences are retained only under selective conditions) presumably have information that provides a selective advantage to the bacteria when they are in a particular environment or part of a particular ecosystem. In this regard, it is interesting that a comparison of plasmid restriction patterns was recently used to establish the identity of two isolates of *Salmonella newport*, and these data were used as evidence to support the transfer of drug-resistant bacteria from antibiotic-fed animals to human beings (Holmberg et al., 1984).

In the case of plant-associated bacteria, the structural relationships of plasmids from strains of tumor-inducing bacteria and from *Rhizobium* species have already been discussed. Most homologous sequences were detected because they contained genes that conferred a common phenotype (e.g., nitrogen fixation, opine or hormone synthesis, transfer, virulence). There are other soil-borne organisms, however, in which structural relationships between resident plasmids have been found but for which no functions have been identified. Much of this work has been done with plant-pathogenic bacteria (Shaw, 1985). For example, different isolates within several pathovars of the plant pathogen *P. syringae* contain plasmids with varying degrees of sequence homology. These include *P. syringae* pv. *phaseolicola* (Gantotti et al., 1979; Jamieson et al., 1981; and Quant and Mills, 1981, 1982, 1984) and pv. *tabaci*, pv. *angulata,* and pv. *coronafaciens* (Piwowarski and Shaw, 1982; Beck-von Bodman and Shaw, 1985b). Plasmids from *P. syringae* pv. *phaseolicola* contain sequences that also hybridize with plasmids from pv. *glycinea*, and plasmids from pv. *tabaci*, pv. *angulata,* and pv. *coronafaciens* also contain homologous sequences. In fact, most *P. syringae* pv. *tabaci* and pv. *angulata* (both are pathogens on tobacco) isolates contain plasmids that were indistinguishable by restriction endonuclease digestion and hybridization (Beck-von Bodman and Shaw, 1985b). It appears that plasmid sequences have been extensively conserved among these pseudomonads but, as already discussed, the basis of the conservation does not appear to be related directly to pathogenesis. It is possible that the plasmids play a role in the saprophytic growth of the bacteria in some plant-associated ecosystem, but whatever their function they remain interesting objects for study by bacterial geneticists and ecologists.

The use of plasmids to establish taxonomic relationships among bacteria requires a good bit of caution because of the inherent instability of these replicons. The ability of plasmids to be transferred among members of a population, while usually of indeterminant frequency in nature, is sufficiently probable that plasmid-determined phenotypic characters are unreliable taxonomic characters [e.g., rhizobia and agrobacteria (Beringer and Hirsch, 1984)]. Their ability to undergo genetic recombination and the possibility of deletion of DNA segments or insertion of other segments by transposition further limit their usefulness. Such activities have been observed in many bacteria including plant-pathogenic pseudo-monads (Comai and Kosuge, 1983a; Szabo and Mills, 1984), and there are indications that they also occur in the soil (Shigyo et al., 1984).

SUMMARY

Extrachromosomal genetic elements are potentially capable of carrying any piece of genetic information found on a bacterial chromosome and some, such as those involved in replication and maintenance, that are not usually present on the chromosome. With the exception of functions such as those essential for survival of the plasmid, most plasmid-encoded genes were discovered because they conferred some recognizable phenotype; i.e., recognition of the phenotype led to the discovery of the gene, hence the plasmid. Unfortunatey, most of the ecological, physiological, and biochemical parameters that affect bacterial persistence in a specific plant-associated ecosystem have not been defined. As a result, most of the functions present on the extrachromosomal DNA found in these or-ganisms has remained cryptic. Correlations between a plasmid structure and a particular environment in which bacteria harboring the plasmid can be found are not sufficient to prove that the plasmid has a function for the bacteria in that environment; however, evidence supporting the hypoth-esis that the conservation of plasmid genes establishes the essentiality of these genes, while not completly convincing, is at least adequate to provide an impetus for the study of such plasmids. Such research, howev-er, taxes the imagination of scientists in their attempts to deduce possible functions for plasmid-encoded genes.

Those involved in the study of the roles of plasmids in plant-microbe interactions should not be discouraged. Techniques using molecular gene-tics have been available for only a short time, but they already prove to be powerful tools. Elucidation of the functions of Ti and Ri plasmids has provided answers to questions that have baffled plant pathologists and microbiologists for 60 years. In fact, the only bacterial diseases of plants for which there is some understanding at a genetic, physiological, and biochemical level are the diseases caused by *Agrobacterium* and by *P. syringae* pv. *savastanoi*. The other area in which molecular genetics has had an impact is root nodulation and nitrogen fixation by *Rhizobium*. The discovery of Sym plasmids has allowed the study of these processes at levels of physiological and biochemical sophistication not possible before. A collaborative effort among scientists from several disciplines will be required if these techniques are to be applied successfully in identifying the functions of the genes present in the vast amount of uncharacterized plasmid DNA present in bacteria associated with plants.

REFERENCES

AGUILAR, A., AND HOPWOOD, D. A. 1984. Determination of methylenomycin A synthesis by the pSV1 plasmid from *Streptomyces violaceus-ruber* SANK 95570, *J. Gen. Microbiol.* **128**:1893–1904.

AKAGAWA, H. OKANISHI, M., AND UMEZAWA, H. 1975. A plasmid in chloramphenicol production in *Streptomyces venezuelae*: Evidence from genetic mapping, *J. Gen. Microbiol.* **90**:336–346.

AKIYOSHI, D. E., MORRIS, R. O., HINZ, R., MISCHKE, B. S., KOSUGE, T., GARFINKEL, D. J., GORDON, M. P., AND NESTER, E. W. 1983. Cytokinin/auxin balance in crown gall tumors is regulated by specific loci in the T-DNA, *Proc. Natl. Acad. Sci. USA* **80**:407–411.

AKIYOSHI, D. E., KLEE, H., AMASINO, R. M., NESTER, E. W. AND GORDON, M. P. 1984. T-DNA of *Agrobacterium tumefaciens* encodes an enzyme of cytokinin biosynthesis, *Proc. Natl. Acad. Sci. USA* **81**:5994–5998.

ANDERSON, R. P., AND ROTH, J. R. 1977. Tandem genetic duplications in phage and bacteria, *Annu. Rev. Microbiol.* **31**:473–505.

BALDWIN, E. 1952. *Dynamic Aspects of Biochemistry*, 2nd ed. Cambridge, University Press.

BARTH, P. T., GRINTER, N. J., AND BRADLEY, D. E. 1978. Conjugal transfer system of plasmids RP4: Analysis by transposon 7 insertion, *J. Bacteriol.* **133**:43–52.

BAYLY, R. C., AND BARBOUR, M. G. 1984. The degradation of aromatic compounds by the meta and gentisate pathways: Biochemistry and regulation, in: *Microbial Degradation of Organic Compounds* (D. T. Gibson, ed.), pp. 253–294, Marcel Dekker, New York.

BECK-VON BODMAN, S., AND SHAW, P. D. 1985a. Plasmid pBW1, indigenous to *Pseudomonas syringae,* pv. *tabaci* strain BR2, specifies an outer membrane protein, 85th Annual Meeting of the American Society for Microbiology, Las Vegas, Nevada. Abstract H 161.

BECK-VON BODMAN, S., AND SHAW, P. D. 1985b. Plasmid DNA relationships in plant-pathogenic *Pseudomonas syringae* pv. *tabaci, angulata, coronafaciens,* and *striafaciens,* 85th Meeting of the American Society for Microbiology, Las Vegas, Nevada, Abstract H 160.

BENVENISTE, R., AND DAVIES, J. 1973. Aminoglycoside antibiotic-inactivating enzymes in actinomycetes similar to those present in clinical isolates of antibiotic-resistant bacteria, *Proc. Natl. Acad. Sci. USA* **70**:2276–2280.

BERINGER, J. E., AND HIRSCH, P. R. 1984. The role of plasmids in microbial ecology, in: *Current Perspectives in Microbial Ecology* (M. J. Klug and C. A. Reddy, eds.), pp. 63–70, American Society for Microbiology, Washington, D.C.

BRADLEY, D. E., AND RUTHERFORD, E. L. 1975. Basic characterization of a lipid-containing bacteriophage specific for plasmids of the P, N, and W compatibility groups, *Can. J. Microbiol.* **21**:152–163.

BREWIN, N. J., WOOD, E. A., AND YOUNG, J. P. W. 1983. Contribution of the symbiotic plasmid to the competitiveness of *Rhizobium leguminosarum*, *J. Gen. Microbiol.* **129**:2973–2977.

BU'LOCK, J. D. 1965. Aspects of secondary metabolism in fungi, in: *Biogenesis of Antibiotic Substances* (Z. Vanek and Z. Hostalek, eds.), pp. 61–71, Czechoslovak Academy of Sciences, Prague.

BUKHARI, A. I., SHAPIRO, J. A., AND ADHYA, S. L. (eds.) 1977. *DNA Insertion Elements, Plasmids, and Episomes*, Cold Spring Harbor Laboratory, Cold Spring Harbor, New York.

CHAI, T. J. 1983. Characteristics of *Escherichia coli* grown in bay water as compared with rich medium, *Appl. Environ. Microbiol.* **45**:1316–1323.

CHAKRABARTY, A. M. 1972. Genetic basis of the biodegradation of salicylate in *Pseudomonas*, *J. Bacteriol.* **112**:815–823.

CHATTERJEE, A. K., AND STARR, M. P. 1973. Transmission of lac by the sex factor E in *Erwinia* strains from human clinical sources, *Infect. Immun.* **8**:563–572.

CHET, I., ZILBERSTEIN, Y., AND HENIS, Y. 1973. Chemotaxis of *Pseudomonas lacrymans* to plant extracts and to water droplets from leaf surfaces of resistant and susceptible plants, *Physiol. Plant Pathol.* **3**:473–479.

CHILTON, W. S., AND CHILTON, M. D. 1984. Mannityl opine analogs allow isolation of catabolic pathway regulatory mutants, *J. Bacteriol.* **158**:650–658.

CLARKE, P. H. 1984a. Evolution of new phenotypes, in: *Current Perspectives in Microbial Ecology* (M. J. Klug and C. A. Reddy, eds.), pp. 71–78, American Society for Microbiology, Washington, D.C.

CLARKE, P. H. 1984b. The evolution of degradative pathways in: *Microbial Degradation of Organic Compounds* (D. T. Gibson, ed.), pp. 11–27, Marcel Dekker, New York.

COMAI, L., AND KOSUGE, T. 1980. Involvement of plasmid deoxyribonucleic acid in idoleacetic acid synthesis in *Pseudomonas savastanoi*, *J. Bacteriol.* **143**:950–957.

COMAI, L., AND KOSUGE, T. 1982. Cloning and characterization of iaaM, a virulence determinant of *Pseudomonas savastanoi*, *J. Bacteriol.* **149**:40–46.

COMAI, L., AND KOSUGE, T. 1983a. The genetics of indoleacetic acid production and virulence in *Pseudomonas savastanoi*, in: *Molecular Genetics of the Bacteria-Plant Interaction* (A. Puhler, ed.), pp. 362–366, Springer-Verlag, Berlin.

COMAI, L., AND KOSUGE, T. 1983b. Transposable element that causes mutations in a plant pathogenic *Pseudomonas* ps., *J. Bacteriol.* **154**:1162–1167.

COMAI, L., SURICO, G., AND KOSUGE, T. 1982. Relation of plasmid DNA to indoleacetic acid production in different strains of *Pseudomonas syringae* pv. *savastanoi*, *J. Gen. Microbiol.* **128**:2157–2163.

COPLIN, D. L., ROWAN, R. G., CHISHOLM, D. A., AND WHITMOYER, R. E. 1981a. Characterization of plasmids in *Erwinia sterartii*, *Appl Environ. Microbiol.* **42**:599–604.

COPLIN, D. L., FREDERICK, R. D., TINDAL, M. H., AND McCAMMON, S. L. 1981b. Plasmids in virulent and avirulent strains of *Erwinia stewartii*, in *Pro-*

ceedings of the Fifth International Conference on Plant Pathogenic Bacteria, Cali, Colombia, pp. 379–388.

CORBIN, D., DITTA, G., AND HELINSKI, D. R. 1982. Clustering of nitrogen fixation (fix) genes in *Rhizobium meliloti, J. Bacteriol.* **149:**221–228.

CRAWFORD, E. M., AND GESTELAND, R. F. 1964. The adsorption of bacteriophage R-17, *Virology* **81:**163–181.

CURRIER, T. C., AND MORGAN, M. K. 1981. Analysis of plasmids in *Pseudomonas syringae, Phytopathology* **71:**869. Abstract.

CURRIER, T. C., AND MORGAN, M. K. 1983. Plasmids of *Pseudomonas syringae*: No evidence of a role in toxin production or pathogenicity, *Can. J. Microbiol.* **29:**84–89.

CURRIER, T. C., AND NESTER, E. W. 1976. Evidence for diverse types of large plasmids in tumor-inducing strains of *Agrobacterium, J. Bacteriol.* **126:**157–165.

DJORDJEVIC, M. A., ZURKOWSKI, W., AND ROLFE, B. G. 1982. Plasmids and stability of symbiotic properties of *Rhizobium trifolii, J. Bacteriol.* **151:**560–568.

DOSCH, D. C., SALVACION, F. F., AND EPSTEIN, W. 1984. Tetracycline resistance element of pBR322 mediates potassium transport, *J. Bacteriol.* **160:**1188–1190.

DRUMMOND, M. H., AND CHILTON, M. D. 1978. Tumor-inducing (Ti) plasmids of *Agrobacterium* share extensive regions of DNA homology, *J. Bacteriol.* **136:**1178–1183.

EATON, R. W., AND TIMMIS, K. N. 1984. Genetics of xenobiotic degradation, in: *Current Perspectives in Microbial Ecology* (M. J. Klug and C. A. Reddy, eds.), pp. 694–703, American Society for Microbiology, Washington, D.C.

ELLIOTT, E. T., COLEMAN, D. C., INGHAM, R. E., AND TROFYMOW, J. A. 1984. Carbon and energy flow through microflora and microfauna in the soil subsystem of terrestrial ecosystems, in: *Current Perspectives in Microbial Ecology* (M. J. Klug and C. A. Reddy, eds.), pp. 424–433, American Society for Microbiology, Washington, D.C.

FRETER, R. 1984. Factors affecting conjugal plasmid transfer in natural bacterial communities, in: *Current Perspectives in Microbial ecology* (M. J. Klug and C. A. Reddy, eds.), pp. 105–114, American Society for Microbiology, Washington, D.C.

FRETER, R., FRETER, R. R., AND BRICKNER, H. 1983. Experimental and mathematical models of *Escherichia coli* plasmid transfer *in vitro* and *in vivo, Infect. Immun.* **39:**60–84.

GANTOTTI, B. V., PATIL, S. S., AND MANDEL, M. 1979. Apparent involvement of a plasmid in phaseolotoxin production by *Pseudomonas phaseolicola, Appl. Environ. Microbiol.* **37:**511–516.

GANTOTTI, B. V., KINDLE, K. L., AND BEER, S. V. 1981. Transfer of the drug-resistance transposon Tn5 to *Erwinia herbicola* and the induction of insertion mutations, *Curr. Microbiol.* **37:**511–516.

GARBER, R. C., TURGEON, B. G., AND YODER, O. C. 1984. A mitochondrial plasmid from the plant pathogenic fungus *Cochliobolus heterostrophus, Mol. Gen. Genet.* **196**:301–310.

GELVIN, S. B. 1984. Plant tumorigenesis, in: *Plant-Microbe Interactions, Molecular and Genetic Perspectives,* Vol 1 (T. Kosuge and E. W. Nester, eds.) pp. 343–377, Macmillan, New York.

GENTELLO, C., VAN LAREBEKE, N., HOLSTERS, M., DE PICKER, A., VAN MONTAGU, M., AND SCHELL, J. 1977. Ti plasmids of *Agrobacterium* as conjugative plasmids, *Nature* **265**:561–563.

GIBSON, D. T. (ed) 1984. *Microbiol Degradation of Organic Compounds,* Marcel Dekker, New York.

GIBSON, D. T., AND SUBRAMANIAN, V. 1984. Microbial degradation of aromatic hydrocarbons, in: *Microbial Degradation of Organic Compounds* (D. T. Gibson, ed.), pp. 181–252, Marcel Dekker, New York.

GODWIN, D., AND SLATER, J. H. 1979. The influence of growth environment on the stability of a drug resistance plasmid in *Escherichia coli* K12. *J. Gen Microbiol.* **111**:201–210.

GONZALEZ, C. F., AND OLSEN, R. H. 1981. Conjugal transfer of an indigenous plasmid of *Pseudomonas syringae, Phytopathology* **71**:220. Abstract.

GONZALEZ, C. F., AND VIDAVER, A. K. 1979. Syringomycin production and holcus spot disease of maize: Plasmid-associated properties, *Curr. Microbiol.* **2**:75–80.

GONZALEZ, C. F., LAYHER, S. K., VIDAVER, A. K., AND OLSEN, R. H. 1984. Transfer, mapping, and cloning of *Pseudomonas syringae* pv. *syringae* plasmid pCG131 and assessment of its role in virulence, *Phytopathology* **74**:1245–1250.

GOTTLIEB, D. 1976. The production and the role of antibiotics in the soil, *J. Antibiot.* **29**:978–1000.

GOTTLIEB, D., AND SIMINOFF, P. 1952. The production and role of antibiotics in soil. II. Chloromycetin, *Phytopathology* **42**:91–97.

GOTTLIEB, D. SIMINOFF, P., AND MARTIN, M. M. 1952. The production and role of antibiotics in soil. IV. Actidione and clavicin, *Phytopathology* **42**:493–496.

GREENSTEIN, J. P., AND WINITZ, M. 1961. *Chemistry of the Amino Acids,* Vol. 3, pp. 2529–2558. Wiley, New York.

GUERINEAU, M. 1979. Plasmid DNA in yeast, in: *Viruses and Plasmids in Fungi* (P. A. Lemke, ed.), pp. 539–593. Marcel Dekker, New York.

HALL, B. G., YOKOYAMA, S., AND CALHOUN, D. 1983. Role of cryptic genes in microbial evolution, *Mol. Biol. Evol.* **1**:109–124.

HANSSEN, R., AND KIRBY, R. 1983. The induction by *N*-methyl-*N'*-nitro-*N*-nitrosoguanidine of multiple closely linked mutations in *Streptomyces bikiniensis* ISP5235 affecting streptomycin resistance and streptomycin biosynthesis, *FEMS Microbiol. Lett.* **17**:317–320.

HASHIBA, T., HOMMA, Y., HYAKUMACHI, M., AND MATSUDA, I. 1984. Isolation of a DNA plasmid in the fungus *Rhizoctonia solani, J. Gen. Microbiol.* **130**:2067–2070.

HEFFRON, F., SUBLETT, R., HEDGES, R. W., JACOB, A., AND FALKOW, S. 1975.

Origin of the TEM beta-lactamase gene found on plasmids, *J. Bacteriol.* **122**:250–256.

HELLING, R. B., KINNEY, T., AND ADAMS, J. 1981. The maintenance of plasmid-containing organisms in populations of *Escherichia coli, J. Gen. Microbiol.* **123**:129–141.

HIRSCH, A. M., DRAKE, D., JACOBS, T. W., AND LONG, S. R. 1985. Nodules are induced on alfalfa roots by *Agrobacterium tumefaciens* and *Rhizobium trifolii* containing small segments of the *Rhizobium meliloti* nodulation region, *J. Bacteriol.* **161**:223–230.

HIRSCH, P. R. 1979. Plasmid-determined bacteriocin production by *Rhizobium leguminosarum, J. Gen. Microbiol.* **113**:219–228.

HOEKMA, A., DE PATER, B. S., FELLINGER, A. J., HOOYKAAS, P. J. J., AND SCHILPEROORT, R. A. 1984. The limited host range of an *Agrobacterium tumefaciens* strain extended by a cytokinin gene from a wide host range T-region, *EMBO J.* **3**:3043–3048.

HOLLOWAY, B. W. 1979. Plasmids that mobilize bacterial chromosome, *Plasmid* **2**:1–19.

HOLMBERG, S. D., OSTERHOLM, M. T., SENGER, K. A., AND COHEN, M. L. 1984. Drug resistant *Salmonella* from animals fed antimicrobials, *N. Engl. J. Med.* **311**:617–622.

HOOYKAAS, P. J. J., KLAPWIJK, P. M., NUTI, M. P., SCHILLPEROORT, R. A., AND RORSCH, A. 1977. Transfer of the *Agrobacterium tumefaciens* Ti plasmid to avirulent agrobacteria and to rhizobium *ex planta, J. Gen. Microbiol.* **98**:477–484.

HOOYKAAS, P. J. J., ROOBOL, C., AND SCHILPEROORT, R. A. 1979. Regulation of the transfer of Ti plasmids of *Agrobacterium tumefaciens, J. Gen. Microbiol.* **110**:99–109.

HOOYKAAS, P. J. J., SNIJDEWINT, F. G. M., AND SCHILPEROORT, R. A. 1982. Identification of the Sym plasmid of *Rhizobium leguminosarum* strain 1001 and its transfer to and expression in other rhizobia and *Agrobacterium tumefaciens, Plasmid* **8**:73–82.

HOPWOOD, D. A., HINTERMANN, G., KIESER, T., AND WRIGHT, H. 1984. Integrated DNA sequences in three streptomycetes form related autonomous plasmids after transfer to *Streptomyces lividans, Plasmid* **11**:1–16.

HURLBERT, R. H., AND GROSS, D. C. 1983. Isolation and partial characterization of the cell wall of *Pseudomonas syringae* pv. *syringae* HS191: Comparison of outer membrane proteins of HS191 with those of two plasmidless derivatives. *J. Gen. Microbiol.* **129**:2241–2250.

INOUYE, M. (ed.) 1979. *Bacterial Outer Membranes.* John Wiley, New York.

INZÉ, D., FOLIN, A., VANLIJSBETTENS, M., SIMOENS, C., GENETELLO, C., VANMONTAGU, M., AND SCHELL, J. 1984. Genetic analysis of the individual T-DNA genes of *Agrobacterium tumefaciens*: Further evidence that 2 genes are involved in indole-3-acetic acid synthesis, *Mol. Gen. Genet.* **194**:265–274.

JACOB, F., LWOFF, A., SIMINOVITCH, L., AND WOLLMAN, E. 1953. Definition de quelques termes relatifs a la lysogenie, *Ann. Inst. Pasteur* **84**:222–224.

JAMIESON, A. F., BIELESKI, R. L., AND MITCHELL, R. E. 1981. Plasmids and phaseolotoxin production in *Pseudomonas syringae* pv. *phaseolicola, J. Gen. Microbiol.* **122**:161–165.

JOHNSTON, A. W. B., BEYNON, J. L., BUCHANAN-WOLLASTON, A. V., SETCHELL, S. M., HIRSCH, P. R., AND BERINGER, J. E. 1978. High frequency transfer of nodulating ability between strains and species of *Rhizobium, Nature* **276**:634–636.

KERR, A. 1980. Biological control of crown gall through production of agrocin 84, *Plant Dis.* **64**:24–30.

KIRBY, R., AND HOPWOOD, D. A. 1977. Genetic determination of methylenomycin synthesis by the SCP1 plasmid of *Streptomyces coelicolor* A3(2), *J. Gen Microbiol.* **98**:239–252.

KLUG, M. J., AND REDDY, C. A. (eds.) 1984. *Current Perspectives in Microbial Ecology*, American Society for Microbiology, Washington, D.C.

KOEKMAN, B. P., HOOYKAAS, P. J. J., AND SCHILPEROORT, R. A. 1982. A functional map of the replicator region of the octopine Ti plasmid, *Plasmid* **7**:119–132.

KROL, A. J. M., HONTELEZ, J. G. J., AND VAN KAMMEN, A. 1982. Only one of the large plasmids in *Rhizobium leguminosarum* strain PRE is strongly expressed in the endosymbiotic state, *J. Gen. Microbiol.* **128**:1839–1847.

KULLA, H. G., KRIEG, R., ZIMMERMANN, T., AND LEISINGER, T. 1984. Experimental evolution of azo dye-degrading bacteria, in: *Current Perspectives in Microbial Ecology* (M. J. Klug and C. A. Reddy, eds.), pp. 663–667, American Society for Microbiology, Washington, D.C.

LACY, G. H., AND LEARY, J. V. 1975. Transfer of antibiotic resistance plasmid RP1 into *Pseudomonas glycinea* and *Pseudomonas phaseolicola in vitro* and *in planta, J. Gen. Microbiol.* **88**:49–57.

LACY, G. H., AND LEARY, J. V. 1979. Genetic systems in phytopathogenic bacteria, *Annu. Rev. Phytopathol.* **17**:181–202.

LANGER, P. J., SHANABRUCH, W. G., AND WALKER, G. C. 1981. Functional organization of plasmid pKM101, *J. Bacteriol.* **145**:1310–1316.

LARSON, R. J. 1984. Kinetic and ecological approaches for predicting biodegradation rates of xenobiotic organic chemicals in natural ecosystems, in: *Current Perspectives in Microbial Ecology* (M. J. Klug and C. A. Reddy, eds.), pp. 677–686, American Society for Microbiology, Washington, D.C.

LEVIN, B. R., AND RICE, V. A. 1980. The kinetics of transfer of nonconjugative plasmids by mobilizing conjugative factors, *Genet. Res.* **35**:241–259.

LEVIN, B. R., AND STEWART, F. M. 1980. The population biology of bacterial plasmids: A priori conditions for the existence of mobilizable nonconjugative factors, *Genetics* **94**:425–443.

LEVIN, B. R., STEWART, F. M., AND RICE, V. A. 1979. The kinetics of conjugative plasmid transmission: Fit of a simple mass action model, *Plasmid* **2**:247–260.

LI, W. H. 1984. Retention of cryptic genes in microbial populations, *Mol. Biol. Evol.* **1**:213–219.

LIPPINCOTT, J. A., AND LIPPINCOTT, B. B. 1984. Concepts and experimental approaches in host-microbe recognition, in: *Plant-Microbe Interactions: Molecular and Genetic Perspectives*, Vol. 1 (T. Kosuge and E. W. Nester, eds.), pp. 195–214, Macmillan, New York.

LONG, S. R. 1984. Genetics of *Rhizobium* nodulation, in: *Plant-Microbe Interactions, Molecular and Genetic Perspectives*, Vol 1. (T. Kosuge and E. W. Nester, eds.), pp. 265–306, Macmillan, New York.

MANNING, P. A., AND ACHTMAN, M. 1979. Cell-to-cell interactions in conjugating *Escherichia coli*: The involvement of the cell envelope, in: *Bacterial Outer Membranes* (M. Inouye, ed.), pp. 409–447, John Wiley, New York.

MCCAMMON, S. L., ROWAN, R. G., AND COPLIN, D. L. 1981. Properties of *Erwinia stewartii* plasmid pDG250 containing bacteriophage Mu cts pf 7701 and Tn*10* insertions, *Phytopathology* **71**:240. Abstract.

MERLO, D. J., AND NESTER, E. W. 1977. Plasmids in avirulent strains of *Agrobacterium*, *J. Bacteriol.* **129**:76–80.

MEYNELL, G. G. 1972. *Bacterial Plasmids*, MIT Press, Cambridge, Massachusetts.

MULREAN, E. N., AND SCHROTH, M. N. 1979. *In vitro* and *in vivo* chemotaxis by *Pseudomonas phaseolicola*, *Phytopathology* **69**:1039. Abstract.

MURAI, N., SKOOG, F., DOYLE, M. E., AND HANSON, R. S. 1980. Relationships between cytokinin production, presence of plasmids, and fasciation caused by strains of *Corynebacterium fascians*, *Proc. Natl. Acad. Sci. USA* **77**:619–623.

NESTER, E. W., AND KOSUGE, T. 1981. Plasmids specifying plant hyperplasias, *Annu. Rev. Microbiol.* **35**:531–565.

NESTER, E. W., GORDON, M. P., AMASINO, R. M., AND YANOFSKY, M. F. 1984. Crown gall: A molecular and physiological analysis, *Annu. Rev. Plant Physiol.* **35**:387–413.

NORDSTROM, K., AND AAGAARD-HANSEN, H. 1984. Maintenance of bacterial plasmids: Comparison of theoretical calculations and experiments with plasmid Rl in *Escherichia coli*, *Mol. Gen. Genet.* **197**:1–7.

OBUKOWICZ, M., AND SHAW, P. D. 1983. Tn*3* labeling of a cryptic plasmid found in the plant pathogenic bacterium *Pseudomonas tabaci* and mobilization of RSF1010 by donation, *J. Bacteriol.* **155**:438–442.

OBUKOWICZ, M., AND SHAW, P. D. 1985. Construction of Tn*3*-containing plasmids from plant-pathogenic pseudomonads and an examination of their biological properties, *Appl. Environ. Microbiol.* **49**:468–473.

OFUYA, C. O., AND WOOD, R. K. S. 1981. Cell wall degrading polysaccharidases and extrachromosomal DNA in wilts caused by *Pseudomonas solanacearum*, in: *Proceedings of the Fifth International Conference on Plant Pathogenic Bacteria, Cali, Colombia*, pp. 271–279.

ONO, H., CRAMERI, R., HINTERMANN, G., AND HUTTER, R. 1983. Hydroxystreptomycin production and resistance in *Streptomyces glaucescens*, *J. Gen. Microbiol.* **129**: 529–537.

PEREZ-CASAL, J. F., AND CROSA, J. H. 1984. Aerobactin iron uptake sequences in plasmid ColV-K30 are flanked by inverted Is1-like elements and replication regions, *J. Bacteriol.* **160**:256–265.

PIWOWARSKI, J. M., AND SHAW, P. D. 1982. Characterization of plasmids from plant pathogenic pseudomonads, *Plasmid* **7**:85–94.

POPLAWSKY, A. R., AND MILLS, D. 1983. Conjugal transfer and incompatibility properties of a plasmid from a plant pathogenic pseudomonad, *Phytopathology* **73**:826. Abstract.

PRAKASH, R. K., AND SCHILPEROORT, R. A. 1982. Relationship between Nif plasmids of fast-growing *Rhizobium* species and Ti plasmids of *Agrobacterium tumefaciens*, *J. Bacteriol.* **149**:1129–1134.

PUEPPKE, S. G. 1984. Adsorption of bacteria to plant surfaces, in: *Plant-Microbe Interactions: Molecular and Genetic Perspectives*, Vol. 1 (T. Kosuge and E. W. Nester, eds.), pp. 215–261, Macmillan, New York.

QUANT, R., AND MILLS, D. 1981. DNA homologies among plasmids of *Pseudomonas syringae* pv. *phaseolicola*, in: *Proceedings of the International Conference on Plant Pathogenic Bacteria, Cali, Colombia*, pp. 413–419.

QUANT, R., AND MILLS, D. 1982. *Pseudomonas* plasmid homology: A possible clue to plasmid evolution, *Phytopathology* **72**:1000. Abstract.

QUANT, R., AND MILLS, D. 1984. An integrative plasmid and multiple-sized plasmids of *Pseudomonas syringae* pv. *phaseolicola* have extensive homology, *Mol. Gen. Genet.* **193**:459–466.

RHEINWALD, J. G., CHAKRABARTY, A. M., AND GUNSALUS, I. C. 1973. A transmissable plasmid controlling camphor oxidation in *Pseudomonas putida*, *Proc. Natl. Acad. Sci. USA* **70**:885–889.

ROTHROCK, C. S. 1980. The importance of antibiosis in control of two soil-borne plant pathogens by selected *Streptomyces* species, M. S. Thesis, University of Illinois, Urbana.

RUVKIN, G. B., AND AUSUBEL, F. N. 1980. Interspecies homology of nitrogenase genes, *Proc. Natl. Acad. Sci. USA* **77**:191–195.

SATO, M., NISHIYAMA, K., AND SHIRATA, A. 1983. Involvement of plasmid DNA in the productivity of coronatine by *Pseudomonas syringae* pv. *atropurpurea*, *Ann. Phytopath. Soc. Japan* **49**:522–528.

SCHRÖDER, G., WAFFENSCHMIDT, S., WEILER, E. W., AND SCHRÖDER, J. 1984. The T-region of Ti plasmids codes for an enzyme synthesizing indole-3-acetic acid, *Eur. J. Biochem.* **138**:387–391.

SCHROTH, M. N., LOPER, J. E., AND HILDEBRAND, D. C. 1984. Bacteria as biocontrol agents of plant disease, in: *Current Perspectives in Microbial Ecology* (M. J. Klug and C. A. Reddy, eds.), pp. 362–396, American Society for Microbiology, Washington, D.C.

SHAW, P. D. 1985. Plasmids in phytopathogenic bacteria, in: *Recent Topics in Experimental and Conceptual Plant Pathology* (A. S. Singh, ed.), Oxford and IBH, New Delhi, in press.

SHAW, P. D., AND PIWOWARSKI, J. 1977. Effects of ethidium bromide and acriflavine on streptomycin production by *Streptomyces bikiniensis*, *J. Antibiot.* **30**:404–408.

SHIGYO, T., HOTTA, K., OKAMI, Y., AND UMEZAWA, H. 1984. Plasmid variability in the istamycin producing strains of *Streptomyces tenjimariensis*, *J. Antibiot.* **37**:635–640.

SLATER, J. H. 1984. Genetic interactions in microbial communities, in: *Current Perspectives in Microbial Ecology* (M. J. Klug and C. A. Reddy, eds.), pp. 87–93, American Society for Microbiology, Washington, D.C.

SLATER, J. H., AND LOVATT, D. 1984. Biodegradation and the significance of microbial communities, in: *Microbial Degradation of Organic Compounds* (D. T. Gibson, ed.) pp. 439–485, Marcel Dekker, New York.

SMITH, V. A., AND HINDLEY, J. 1978. Effect of agrocin 84 on attachment of *Agrobacterium tumefaciens* to cultured tobacco cells, *Nature* **276**:498-500.

STASKAWICZ, B. J., SATO, M., AND PANOPOULOS, N. J. 1981. Genetic and molecular characterization of an indigenous conjugative plasmid in the bean wildfire strain of *Pseudomonas syringae*, *Phytopathology* **71**:257. Abstract.

STERKENBERG, A., PROZEE, G. A. P., LEEGWATER, P. A. J., AND WOUTERS, J. T. M. 1984. Expression and loss of the pBR322 plasmid in *Klebsiella aerogenes* NCTC418, grown in chemostat culture, *Anton. Leeuwenhoek J. Microbiol.* **50**:397–404.

STEWART, F. M., AND LEVIN, B. R. 1980. The population biology of bacterial plasmids: A priori conditions for the existence of conjugationally transmitted factors, *Genetics* **87**:209–228.

SZABO, L. J., AND MILLS, D. 1984. Characterization of 8 excision plasmids of *Pseudomonas syringae* pv. *phaseolicola, Mol. Gen. Genet.* **195**:90–95.

SZABO, L. J., VOLPE, J., AND MILLS, D. 1981. Identification of F'-like plasmids of *Pseudomonas syringae* pv. *phaseolicola*, in: *Proceedings of the fifth International Conference on Plant Pathogenic Bacteria, Cali, Colombia,* pp. 403–411.

TATE, M. E., MURPHY, P. J., ROBERTS, W. P., AND KERR, A. 1979. Adenine N^6-substituent of agrocin 84 determined its bacteriocin-like specificity, *Nature* **280**:697–699.

THACKER, R. O., RORVIG, O., KAHLON, P., AND GUNSALUS, I. C. 1978. NIC, a conjugative nicotine-nicotinate degradative plasmid in *Pseudomonas convexa, J. Bacteriol* **135**:289–290.

THIRY, G. 1984. Plasmids of the epiphytic bacterium *Erwinia urwinia uredovara, J. Gen. Microbiol.* **130**:1623–1631.

THIRY, G., AND FAELEN, M. 1984. *Erwinia Uredovora* 20D3 contains an IncM plasmid, *J. Gen. Microbiol.* **130**:1633–1640.

THOMASHOW, L. S., REEVES, S., AND TOMASHOW, M. F. 1984. Crown gall oncogenesis: Evidence that a T-DNA gene from the *Agrobacterium* Ti plasmid pTiA6 encodes an enzyme that catalyzes synthesis of indoleacetic acid, *Proc. Natl. Acad. Sci. USA* **81**:5071–5075.

THOMPSON, C. J., AND GRAY, G. S. 1983. Nucleotide sequence of a streptomycete aminoglycoside phosphotransferase gene and its relationship to phosphotransferases encoded by resistance plasmids, *Proc. Natl. Acad. Sci. USA* **80**:5190–5194.

THOMPSON, R., HUGHS, S., AND BRODA, P. 1974. Plasmid identification using specific endonucleases, *Mol. Gen. Genet.* **133**:141–149.

TOLMASKY, M. E., AND CROSA, J. H. 1984. Molecular cloning and expression of genetic determinants for the iron uptake system mediated by the *Vibrio anguillarum* plasmid pJM1, *J. Bacteriol.* **160**:860–866.

TOMIZAWA, J., AND SELZER, G. 1979. Initiation of DNA synthesis in *Escherichia coli*, *Annu. Rev. Biochem.* **48**:999–1034.

TRUCHET, G., ROSENBERG, C., VASSE, J., JULLIOT, J. -S., CAMUT, S., AND DENARIE, J. 1984. Transfer of *Rhizobium meliloti* pSym genes into *Agrobacterium tumefaciens*: Host-specific nodulation by atypical infection, *J. Bacteriol.* **157**:134–142.

VAN LAREBEKE, N., GENETELLO, C., SCHELL, J., SCHILPEROORT, R. A., HERMANS, A. K., HERNALSTEENS, J. P., AND VAN MONTAGU, M. 1975. Acquisition of tumor-inducing ability by non-oncogenic agrobacteria as a result of plasmid transfer, *Nature* **255**:742–743.

VAN LAREBEKE, N., GENETELLO, C., HERNALSTEENS, J. P., PICKER, A. D., ZAENEN, I., MESSENS, B., VAN MONTAGU, M., AND SCHELL, J. 1977. Transfer of Ti plasmids between *Agrobacterium* strains by mobilization with the conjugative plasmid RP4, *Mol. Gen. Genet.* **152**:119–124.

VANCE, C. P. 1983. *Rhizobium* infection and nodulation: A beneficial plant disease, *Annu. Rev. Microbiol.* **37**:399–424.

WEINBERG, S. R., AND STOTZKY, G. 1972. Conjugative genetic recombination of *Escherichia coli* in soil, *Biol. Biochem.* **4**:171–180.

WHATLEY, M. H., MARGOT, J. B., SCHELL, J., LIPPINCOTT, B. B., AND LIPPINCOTT, J. A. 1978. Plasmid and chromosomal determination of *Agrobacterium* adherence specificity, *J. Gen. Microbiol.* **107**:395–398.

WHITE, F. F., AND NESTER, E. W. 1980. Hairy root: Plasmid encodes virulence traits in *Agrobacterium rhizogenes*, *J. Bacteriol.* **141**:1134–1141.

WHITE, F. F., GARFINKEL, D. J., HUFFMAN, G. A., GORDON, M. P., AND NESTER, E. W. 1983. Sequences homologous to *Agrobacterium rhizogenes* T-DNA in the genomes of uninfected plants, *Nature* **301**:348–350.

WIGHTMAN, F. 1973. Biosynthesis of auxins in tomato shoots, *Biochem Soc. Symp.* **38**:247–275.

WILLETTS, N. S., AND WILKINS, B. 1984. Processing of plasmid DNA during bacterial conjugation, *Microbiol. Rev.* **48**:24–41.

WILLIS, J. W., AND LEARY, J. V. 1984. Cloning of an entire small molecular weight plasmid of *Pseudomonas syringae* pv. *glycinea* in pBR329 and pRK404, *Phytopathology* **74**:838. Abstract.

WINANS, S. C., AND WALKER, G. C. 1985. Conjugal transfer system of the IncN plasmid pKM101. *J. Bacteriol.* **161**:402–410.

WOMBLE, D. D., DONG, X., LUCKOW, V. A., WU, R. P., AND ROWND, R. H. 1985. Analysis of the individual regulatory components of the IncFII plasmid replication control system. *J. Bacteriol.* **161**:534–543.

Chapter 2

Plant Hormones
in Plant-Microbe Interactions

Joachim Schröder

AUXINS AND CYTOKININS have profound effects on growth and differentiation in plants, and much work has focused on these topics. Nevertheless, their precise actions at the molecular level are poorly understood, and the same may be said about their biosynthesis and its regulation. These plant hormones are also synthesized by microorganisms interacting with plants, and since this interaction often leads to morphological changes in the plants, it is generally assumed that they play an important role. This has been demonstrated convincingly in some cases, but there are other instances where plant responses via increased hormone formation may have a decisive role. Hormone analysis of the separate partners is not likely to provide answers in these cases, since the specific interaction may be necessary to activate hormone synthesis. For a detailed understanding it would be ideal to identify the genes responsible for hormone biosynthesis, analyze their regulation, and determine the precise enzymic function of the proteins. There are some systems where a combination of genetic and biochemical techniques has begun to provide the background for such studies. So far, all these systems involve genes of bacterial origin, but there are some interesting implications for hormone biosynthesis in higher plants. This chapter focuses on recent developments in the most advanced systems, with emphasis on genes and enzymes specific for auxin and cytokinin biosynthesis. Other systems will be discussed within this framework, but it is not our intention to compile a complete catalog of plant-microbe interactions in which auxins or cytokinins are involved.

AGROBACTERIA AND PLANTS

General Description of the System

The genus *Agrobacterium* contains several members which induce formation of tumorous overgrowths on a large variety of plants, mostly dicotyledonous plants and gymnosperms (De Cleene and De Ley, 1976). *Agrobacterium tumefaciens* and *A. rhizogenes* are the members of the group which have been characterized in the most detail. In both cases the capacity to induce tumors resides in large plasmids, called Ti and Ri plasmids, and tumor induction is accompanied by the transfer of specific parts of the plasmids from bacteria into the plants cells where they are stably incorporated into nuclear DNA. This part of the plasmid is called the T region in the bacteria, and T-DNA after incorporation into plant DNA. The transferred DNA is responsible for new properties of the transformed cells. They synthesize a number of new substances which are selectively utilized by the inducing agrobacteria, and genes for catabolism are localized on the plasmids. The substances are called *opines,* and different plasmids are often classified according to the type of opine they induce in plant cells (Guyon et al., 1980; Tempé and Goldmann, 1982; Tempé and Petit, 1982). T-DNAs also cause a proliferation of transformed cells which cannot be controlled by the plant, and the visible results are the tumors induced by *A. tumefaciens* (Ti plasmids) and the root proliferation caused by *A. rhizogenes* (Ri plasmids). Recent reviews provide excellent, detailed descriptions of the systems (Gelvin, 1984; Hille et al., 1984; Nester et al., 1984; Schell et al., 1984a, b).

The interaction of agrobacteria with plants represents a natural case of genetic manipulation of plants by bacteria. The prospect of modifying plasmids in such a way that any desired gene can be transferred to plants without causing the disease has attracted many researchers, and this is one of the reasons why analysis of the system has progressed so far. Equally important is the fact that the system can be dissected: T-DNA-containing cells can easily be grown without contamination by agrobacteria, and therefore the action of T-DNA genes can be observed in a relatively simple system. Although such an analysis certainly does not cover all aspects of the interaction, this simplification has turned out to be important in the detection of genes which control growth and differentiation in plants by coding for enzymes of hormone biosynthesis.

T-DNA Genes and Plant Hormones

With the development of tissue culture techniques it became possible to cultivate tissues or cells of many different plants under sterile conditions, and it was discovered that most plant cells needed supplementation with phytohormones such as auxins and cytokinins for continued growth and division. When these techniques were applied to tumor tissues induced by agrobacteria (crown galls), it soon became clear that they did not require phytohormones (Braun, 1956, 1958), and this hormone independence was for some time the most characteristic property of the tumor cells. Detailed experiments on growth characteristics led to the postulate that a tumor-inducing principle (TIP) of unknown nature is transferred from the bacteria to the plant cells, that TIP is responsible for hormone independence, and that the new property is stable and is inherited in plant cells (Braun and Wood, 1962; Braun, 1978). The history of these experiments and early attempts to identify TIP have been described in a review (Braun, 1982).

TIP has now been identified as T-DNA, some T-DNAs of octopine and nopaline Ti plasmids have been well characterized both physically and transcriptionally (see reviewes cited above), and DNA sequences have been published (octopine T-DNA: Barker et al., 1983; De Greve et al., 1983; Gielen et al., 1984; Heidekamp et al., 1983; Klee et al., 1984; Lichtenstein et al., 1984; Sciaky and Thomashow, 1984; nopaline T-DNA: Depicker et al., 1982; Goldberg et al., 1984). All of the analyses indicate that the T-DNAs of octopine and nopaline plasmids share extensive DNA homology (a common core region) and that this conserved part contains six protein-coding regions (Fig. 1) which are transcribed in plant cells.

Since the six genes are conserved in different Ti plasmids, it is reasonable to assume that they have important functions. Genetic experiments with Ti plasmids mutagenized in the T region show that the functions of three of the genes (genes 5, 6a, and 6b) are difficult to define, since their inactivation does not always lead to obvious changes in the phenotypes of tumors. This is not true of the genes in the center of the common core region (genes 2, 1, and 4). Their inactivation causes well-defined changes in the morphological appearance of the transformed tissues (Garfinkel et al., 1981; Hoekema et al., 1984; Inzé et al., 1984; Joos et al., 1983; Leemans et al., 1982; Ream et al., 1983; Van Slogteren et al., 1983,

1984). Mutations in genes 1 and/or 2 induce attenuated tumors with an abundance of shoots, indicating that these genes encode functions suppressing shoot formation in wild-type tumors. Ti plasmids with mutations in gene 4 induce attenuated tumors which typically show root proliferation, indicating that the gene suppresses root formation in tissues containing complete T-DNA. Ooms et al. (1981) were the first to notice that the morphology of tumors with T-DNA mutants resembled the phenotypes obtained with normal tobacco cells supplied with different ratios of auxins and cytokinins (Figs. 1, 2). Shoot formation was observed with high cytokinin/auxin ratios, root formation was favored by high auxin/cytokinin ratios, and unorganized growth was obtained with a balanced addition of both hormones (Skoog and Miller, 1957). Without implying a mechanism, the correlation suggests that the combined action of genes 1 and 2 is "auxin-like," that the action of gene 4 is "cytokinin-

Figure 1. Top: T_L region of pTiAch5 and part of the T region of pTiC58. Lines with arrows represent the protein-coding regions of the six genes in the conserved part of the T regions. Arrows indicate the direction of transcription; the numbering of the genes is according to the size of the transcripts in plant cells (Schell et al., 1984). Bottom: Genetic definition of functions. Name: abbreviations used by the different groups; shi, roi, see Schell et al. (1984); tms, tmr, tml, see Nester et al. (1984).

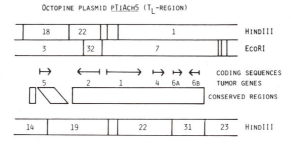

OCTOPINE PLASMID pTiAch5 (T_L-REGION)

NOPALINE PLASMID pTiC58 (PART OF T-REGION)

CONSERVED GENES	GENETIC DEFINITION OF FUNCTION	NAME
5	NO DISTINCT PHENOTYPE OF TISSUES	
2* 1*	COOPERATION IN SUPPRESSION OF SHOOTS, STIMULATION OF ROOTS; AUXIN INDEPENDENCE	SHI, TMS
4*	STIMULATION OF SHOOTS, SUPPRESSION OF ROOTS; CYTOKININ INDEPENDENCE	ROI, TMR
6A	NO DISTINCT PHENOTYPE OF TISSUES	
6B	LARGER TUMORS ON SOME PLANTS, WHEN GENE IS INACTIVE	TML

* THESE GENES ARE ALSO EXPRESSED INTO PROTEIN IN BACTERIA

EFFECTS OF T-DNA GENES IN CROWN GALL CELLS AND HORMONE
ADDITION TO NORMAL TOBACCO CELLS

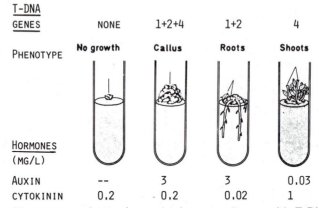

Figure 2. Phenotypes of transformed tobacco cultures with T-DNA genes
and of normal cultures supplied with different ratios of auxins and cytokinins.

like," and that all three genes together have an effect like a balanced addition of both hormones. A number of experiments with T-DNAs containing single genes support the "auxin-like" and "cytokinin-like" activity of these genes. They also show that the action of the genes can be mimicked by supplementing the tissues with the hormones. One of the important points emerging from these studies is that the three T-DNA genes must have a dual function—not only suppression of shoot and root formation but also a stimulating effect on growth and cell division.

For these effects to occur it is not necessary that the genes be in the same cells. Simultaneous infection of plants with different Ti plasmids affected in different genes shows that genes 1 and 2 result in auxin independence of the culture when present in different cells. Similarly, cells containing genes 1 and 2 and cells possessing gene 4 become hormone independent when cultured together. The results indicate that the gene products act not only inside the cells but also across cell borders. This is consistent with the finding that uncloned tumor tissues contain in large part T-DNA-free normal cells which are stimulated to undergo continued cell growth and division by cross-feeding from transformed cells.

An important implication of these data is that crown gall cells are likely to contain and excrete higher amounts of auxins and cytokinins than normal cells, and this has been confirmed in many, but not all, cases investigated (auxins: Amasino and Miller, 1982, Mousdale, 1981, 1982; Nakajima et al., 1981; Pengelly and Meins, 1982; Weiler and Spanier,

1981; cytokinins: Amasino and Miller, 1982; Einset, 1980; Nakajima et al., 1981; Weiler and Spanier, 1981). The involvement of specific T-DNA genes is suggested by the finding that the presence or absence of the genes correlates with high or low concentrations of the hormones (Morris et al., 1982; Akiyoshi et al., 1983). Taken together, most of the results suggest that T-DNA genes 1, 2, and 4 cause hormone independence by influencing directly or indirectly the concentrations and/or balances of auxins and cytokinins in transformed plant cells.

T-DNA-Encoded Hormone Biosynthesis

The next step in the analysis is identification of the protein gene products and elucidation of their precise functions. This is difficult for both practical and conceptual reasons. One practical problem is that there is no functional assay for the proteins, and this renders the task of purification for further analysis difficult. Another problem is that transcription of the genes is low [usually less than 0.01% of poly(A)-containing RNA], suggesting that the proteins are not abundant. The conceptual problems are mainly due to our lack of understanding of the molecular events controlling hormone concentration and action in growth and differentiation. How does one analyze the function of a protein interfering in a process which itself is not understood? The genetic and hormone data could be explained by several different models which are more or less difficult to test our present state of knowledge. It is most likely because of this that all the studies reported so far investigated only one of the possible models, namely, one in which T-DNA genes 1, 2, and 4 operate by coding for enzymes which are directly involved in auxin and cytokinin biosynthesis. All these investigations also made two additional assumptions: (1) the proteins are functional when expressed in bacterial systems, and (2) the enzymes catalyze steps in hormone biosynthesis which are already known in plants or bacteria.

Auxins

In crown gall cells the cooperation of genes 1 and 2 is responsible for auxin effects, including auxin independence. Both genes are expressed weakly in *Escherichia coli* protein, apparently with promoter activities from the T region, and the sizes of the proteins correspond to predictions based on the DNA sequence within the limits of accuracy using one-dimensional sodium dodecyl sulfate (SDS) gel electrophoresis (Schröder et al., 1983). If the proteins are enzymes involved in the formation of indole-3-acetic acid, one would expect at least one of them to utilize a substrate which is present in all plants successfully infected by agrobacteria.

One would also expect this reaction to initiate a pathway to the plant hormone and that the end product is indole-3-acetic acid. The second protein may be involved in this pathway, but it cannot be excluded that it operates independently; the genetic data do not distinguish between sequential and parallel action of the genes.

Figure 3 summarizes the most often discussed pathways to indole-3-acetic acid in plants and bacteria, and most of the enzyme reactions postulated have been detected in extracts from higher plants (for a review see Sembdner et al., 1980). However, all the reactions may also be assigned to other pathways in either primary or secondary metabolism, and it has not been possible to demonstrate unambiguously that a particular reaction sequence is specifically involved in auxin biosynthesis *in vivo* in higher plants. Indole-3-acetamide is of particular interest, since it is definitely involved in auxin biosynthesis in *Pseudomonas savastanoi* (Kosuge et al., 1966). The same substance was implicated as an intermediate in higher plants (Riddle and Mazelis, 1964, 1965), but this report was later generally ignored because the formation of indole-3-acetamide from tryptophan was considered a biologically insignificant reaction catalyzed by peroxidase *in vitro* under special conditions. Also, reports of the occurrence of this substance in higher plants have been challenged because of the presence of isolation artifacts (Zenk, 1961).

Functional assays for the enzyme activities postulated in Fig. 3 were first performed with genes 1 and 2 cloned in *E. coli* plasmids and expressed in *E. coli* protein. The results showed that the protein encoded in gene 2 (protein size, MW 49,000) catalyzed the hydrolytic conversion

Figure 3. Pathways from tryptophan to the auxin indole-3-acetic acid and conjugates of the hormone.

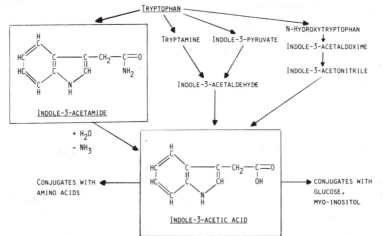

of indole-3-acetamide to indole-3-acetic acid. This was reported by two independent groups (Schröder et al., 1984; Thomashow et al., 1984), and the product of the reaction was identified by sequential high-performance liquid chromatography (HPLC) and mass spectrometry. The same enzyme activity was then demonstrated in tobacco crown gall cells, and it was shown that a habituated tobacco cell line was devoid of the activity (Schröder et al., 1984). The data indicate that one of the two "auxin genes" of the T-DNA codes for an enzyme directly involved in auxin biosynthesis. The use of indole-3-acetamide as a substrate, however, raises new questions. In view of the fact that its presence in normal plants has been neither proven nor disproven, it seems reasonable to assume that T-DNA genes 1 and 2 code for a two-step pathway from tryptophan through indole-3-acetamide to indole-3-acetic acid, as described for *P. savastanoi*. This assumption assigns a tryptophan monooxygenase function to the protein encoded in gene 1, and the following circumstantial evidence supports this hypothesis. The protein contains an adenine-binding domain, probably for FAD (Klee et al., 1984), like the *Pseudomonas* enzyme (Hutcheson and Kosuge, 1985); gene 1 and its protein share homology with the corresponding gene and protein from *Pseudomonas* (Yamada et al., 1985); and crown gall tissues containing gene 1 but not gene 2 accumulate indole-3-acetamide (Von Onckelen et al., 1985). Direct demonstration of monooxygenase activity is difficult, however, and has been successful only recently in bacterial cells expressing the gene (L. Thomashow, personal communication) and in crown gall cells (H. Van Onckelen, personal communication). Taken together, the results suggest that the two T-DNA genes express in transformed plant cells two enzymes catalyzing a pathway which has been demonstrated in bacteria but not in higher plants. It is therefore tempting to speculate that they are bacterial genes adapted to function in eukaryotic cells that they introduce a "foreign" pathway which cannot be controlled by plant regulatory mechanisms. This hypothesis is interesting, but by no means proven, and it will be necessary to reinvestigate a possible role of indole-3-acetamide in higher plants before a definite conclusion can be reached.

Since the function of gene 1 has not yet been fully explained at the enzyme level, it is of interest whether the hydrolase encoded in gene 2 is highly specific for indole-3-acetamide or whether it also accepts other substrates. This was investigated by Kemper et al. (1985), and the results (compare Fig. 3) indicate that the enzyme also hydrolyzes phenylacetamide (to phenylacetic acid), indole-3-acetonitrile (to indole-3-acetic acid), and esters of the auxin with glucose and myoinositol (to indole-3-acetic acid). All these reactions are of possible physiological significance: Phenylacetic acid has been proposed to be an auxin in plants (Wightman, 1973; Fregeau and Wightman, 1983); indole-3-acetonitrile and enzymes required for its biosynthesis have been described in several plants

(Bearder, 1980; Sembdner et al., 1980; Mahadevan, 1973); esters of in-
dole-3-acetic acid are widespread in plants and have been discussed as
possible storage forms of auxin (Sembdner et al., 1980; Cohen and Bandurski, 1982). The interesting point concerning these results is that in one
enzymatic step, all these reactions liberate the active auxin from less active precursors or storage forms, suggesting that the enzyme may have a
generalized function in providing auxins. The significance of these findings with respect to the gene-1 protein function is not clear at present,
since it is not known which of the several possible substrates for the
gene-2 protein is actually used *in vivo* or whether all of them can contribute to the auxin function of the gene.

Cytokinins

Genetic analysis of T-DNA functions indicates that gene 4 is sufficient to induce cytokinin effects and cytokinin independence in transformed plant cells. The gene is expressed in protein in bacterial systems,
and the size of the protein (MW 27,000) is as predicted by the DNA
sequence (Schröder et al., 1983). If the protein functions in hormone
biosynthesis, one would expect it to use substrates present in all plant
cells and that it either directly produces a cytokinin or initiates a pathway
to cytokinins which also involves plant enzymes.

Figure 4 summarizes in a simplified scheme the reactions discussed
for biosynthesis of isopentenyladenine- and zeatin-type cytokinins; we
have not attempted to include the large number of derivatives, intercon-
versions, and studies on cytokinin degradation since this would go beyond
the scope of this chapter (for a review see Letham and Palni, 1983). In the
formation of cytokinins two pathways have received the most attention:
the release of hormone-active molecules from tRNA by hydrolysis, and
the synthesis of N^6-(Δ^2-isopentenyl)-5' AMP from Δ^2-isopen-
tenylpyrophosphate and 5'AMP. The tRNA pathway has been discussed
for bacteria, including agrobacteria (Hahn et al., 1976), and for higher
plants (Holtz and Klämbt, 1978; Maass and Klämbt, 1981). In a strict
sense it does not represent de novo synthesis, since isopentenylation
occurs at the tRNA level where this modification has its specific function. It is also not plant-specific, since cytokinin-active substances are
present in RNAs of many nonplant organisms; nevertheless hydrolysis
of tRNA may well contribute to the concentration of free cytokinins in
higher plants (Letham and Palni, 1983). Enzymic isopentenylation of
free 5' AMP has been described in *Dictyostelium* (Taya et al., 1978;
Ihara et al., 1984), in *Corynebacterium fascians* (Murai, 1981), and in
cytokinin-autotrophic plant cells (Chen and Melitz, 1979; Chen, 1982),
including crown gall cells (Morris et al., 1982).

There is no indication that the gene-4-encoded protein operates by the

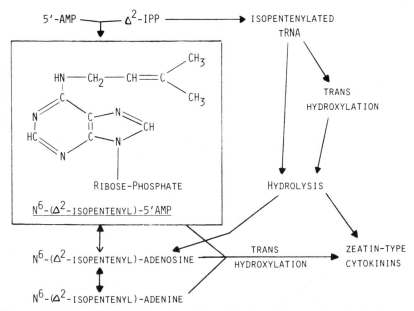

Figure 4. Simplified scheme for biosynthesis of isopentenyladenine and trans-zeatin type cytokinins. Δ^2 -IPP: Δ^2 -isopentenylpyrophosphate.

releasing cytokinins from tRNA. Instead, several independent investigations have shown that the protein catalyzes the formation of N^6-(Δ^2-isopentenyl),-5' AMP from free 5' AMP (Fig. 4). This was demonstrated with the gene product from octopine plasmid (Akiyoshi et al., 1984), with a nopaline plasmid (Barry et al., 1984), and with genes from both sources in the same report (Buchmann et al., 1985); in the last-mentioned case the product was identified by mass spectrometry. As in the first studies on the gene-2 protein, most experiments were performed with the gene expressed in *E. coli* protein, but in one case antibodies against a decapeptide derived from the DNA sequence were used to demonstrate that crown gall cells contained an immunoreactive protein of the expected size (MW27,000) and that isopentenyltransferase activity in extracts from such cells was inhibited by a reaction with the specific antibodies (Buchmann et al., 1985). The results correspond to the prediction: Both 5' AMP and Δ^2-isopentenylpyrophosphate are present in normal cells, and the product of the enzyme reaction is a cytokinin after the simple removal of phosphate by plant enzymes.

Precursor studies suggest that normal plants utilize the same reaction for cytokinin biosynthesis, and enzymic isopentenylation of free 5'AMP in extracts has been reported. However, the most convincing enzyme results were obtained with cytokinin-independent cell cultures (Chen and

Melitz, 1979; Chen, 1982). In this property they are identical to crown gall cells, and thus it is necessary to establish without doubt that gene 4 is not present before a definite conclusion can be reached. In a study comparing habituated and crown gall cells (Buchmann et al., 1985), isopentenyltransferase activity or a protein reacting with antibodies against the gene-4 protein was not detected in hormone-autotrophic (habituated) cells. These results, however, are not conclusive, since the protein isolation and fractionation procedure was specifically designed to identify the T-DNA-encoded protein. It is quite possible that other proteins with isopentenyltransferase activity were lost in other fractions of cell extracts.

Hormone Synthesis in Agrobacteria

Cultures of Agrobacteria produce auxins and cytokinins under suitable conditions, but the significance of this hormone production in the microbe-plant interaction is not quite clear. Bacteria free of Ti plasmids or containing Ti plasmids with the T-region deleted may cause swelling of the invaded tissues, but they do not induce tumors. Tumor formation requires the transfer of T-DNA and the expression of T-DNA genes in plant cells, and this appears to be the major determinant in the disease. However, a supportive role for agrobacterial hormone production could be envisioned in the following model: Efficient tumor induction requires a wound in plant tissue, which induces cell division, and transformation is believed to occur during this period (for a review see Braun, 1978). In noninfected tissue cell division ceases after a few rounds, but continues in infected tissue. It is possible that hormones produced by the bacteria participate in maintaining wound-induced cell division until the T-region is transferred and becomes active in the plant cells.

Indole-3-Acetic Acid

Agrobacteria can produce auxin in copious amounts, and most of the hormone is found in the culture medium (Beltra et al., 1978; Liu and Kado, 1979; Weiler and Spanier, 1981; Lobanok et al., 1982). Production is largely dependent on the amount of tryptophan supplied in the nutrient broth; high concentrations of this amino acid lead to high auxin production, especially after the cultures reach the stationary phase of growth. Under these conditions the differences between strains containing Ti plasmids and avirulent strains are small, suggesting that Ti plasmid genes are not major determinants. The pathway involved appears to utilize indole-3-pyruvate as an intermediate, since this substance can be readily isolated in high concentrations from cells and culture fluids (Kaper and

Veldstra, 1958; Liu et al., 1982), and a transaminase catalyzing the reaction has been identified (Sukanya and Vaidyanathan, 1964; Atsumi, 1980). This situation is reminiscent of that in *Klebsiella aerogenes* where tryptophan is utilized as a nitrogen source and indole-3-pyruvate is excreted into the medium (Parish and Magasanik, 1981). It is possible that tryptophan-induced formation of indole-3-acetic acid in agrobacteria primarily reflects utilization of the amino acid, and that further metabolism of indole-3-pyruvate to auxin represents additional reactions to supply a product which is useful in the interaction with plants. This may also explain why *agrobacteria* cultures produce the highest concentrations of auxin in the stationary phase of growth, although early log-phase cells are much more virulent (Weiler and Spanier, 1981).

There is evidence, however, that Ti plasmid genes can play a significant role in auxin formation. When bacteria are grown in a minimal medium, or at least without tryptophan supplementation, hormone formation is much reduced, and Ti plasmid–containing cells synthesize significantly more indole-3-acetic acid than strains without Ti plasmids (Liu and Kado, 1979). Transposon mutagenesis shows that a specific part of the Vir region in nopaline plasmid pTiC58 participates in elevated auxin production, and hormone formation has been positively correlated with the virulence of the bacteria (Liu et al., 1982). It is not known whether one or more genes are involved and whether they act directly in auxin formation. Nevertheless, the results suggest that Ti plasmid genes in some way contribute to auxin production, and this could be significant in the interaction of agrobacteria with plants.

Under these conditions, it is of interest whether the auxin genes of the T region (which are primarily defined by their action in transformed plant cells) are also active in agrobacteria. Transcripts from the T regions of genes 1 and 2 have been described (Janssens et al., 1984), and experiments with a cell-free system for coupled transcription or translation suggest that the proteins can be expressed in agrobacteria (Schröder et al., 1983). Assays for the indole-3-acetamide hydrolase encoded in gene 2 showed that the activity was present in free-living agrobacteria and that it was correlated with the presence of the gene (Schröder et al., 1984; J. Schröder unpublished observations). The results suggest that T-DNA genes may participate in auxin formation in the bacteria, but their actual contribution *in vivo* is not known.

Cytokinins

Agrobacteria also produce cytokinins, of both the isopentenyladenine and the *trans*-zeatin types, and they have been identified by various analytical techniques (Hahn et al., 1976; Kaiss-Chapman and Morris, 1977; Messens and Claeys, 1978; McCloskey et al., 1980; Weiler and Spanier,

1981; Morris et al., 1981). Although it is difficult to present a generalized picture, some interesting points emerge from more recent studies. It seems that isopentenyladenine-type cytokinins are synthesized by agrobacteria, regardless of whether Ti plasmids are present or not. Production changes with the growth of the culture, and quantitative results suggest that the presence of Ti plasmids leads to somewhat higher concentrations of cytokinins in the medium. In contrast to this, *trans*-zeatin production seems to be correlated with the presence of nopaline Ti plasmids (Regier and Morris, 1982; see, however, the exception reported by Weiler and Spanier, 1981). Akiyoshi and co-workers recently showed that a gene responsible for *trans*-zeatin production was localized close to, or within the Vir region of nopaline plasmid pTiT37 and that this gene was also functional after recloning in *E. coli* (Akiyoshi et al., 1985). A DNA sequence analysis by the same authors revealed that the gene shared homology with gene 4 of the T region and that the protein encoded in the Vir region also possessed isopentenyltransferase activity *in vitro*. Results from our laboratory (Schröder et al., in preparation) on the corresponding Vir region from nopaline plasmid pTiC58 support these findings, and a detailed kinetic comparison of the two isopentenyltransferases suggests that the enzymes have similar properties under most assay conditions *in vitro* (Schröder et al., in preparation). In view of these data it is puzzling that the Vir region gene leads to *trans*-zeatin production *in vivo*, and the mechanism of the hydroxylation step remains to be clarified.

AUXINS AND CYTOKININS IN OTHER PLANT-MICROBE INTERACTIONS

Rhizobium

Nodule-inducing rhizobia and tumor-inducing agrobacteria are closely related, and it is the presence of specific plasmids which mainly determines whether a member of the family Rhizobiaceae is classified as *Rhizobium* or *Agrobacterium* (see, for example, reviews by Nuti et al., 1982; Long, 1984). The development and maintenance of nodules obviously requires a highly regulated interaction of bacteria and plants, which may well be influenced or controlled by auxins and cytokinins. It is generally believed that hormone production by rhizobia plays a role (Dart, 1977; Newcomb, 1980), but actually there is little evidence supporting this assumption.

Several reports suggest that nodules have larger auxin concentrations than the parent tissue (Pate, 1958; Dullaart, 1970; Kretovich et al., 1972), but it is not known whether bacteroids contribute or not. Nodules appear to be sinks for auxin transported from the shoots to the roots, and much of the hormone is found as a conjugate with aspartic acid (Badenoch-Jones et al., 1983), suggesting that the metabolism of auxin may be important for the persistence of functional nodules. Indole-3-acetic acid production by wild-type *Rhizobium trifolii* and by mutants with defects in adhesion, root hair curling, and infection thread growth has been investigated by Badenoch-Jones et al. (1982), but these authors detected no obvious correlation between auxin production and any of the phenotypes. Interestingly, they reported that auxin formation in rhizobia was largely dependent on the concentration of tryptophan in the medium. This is reminiscent of the situation discussed above for agrobacteria; again it may be possible that tryptophan-induced auxin formation masks the action of other genes which are active only in the interaction with plants. The work of Syono et al. (1976) and Newcomb et al. (1977) supplies circumstantial evidence that in developing nodules both plant and bacterium contribute to cytokinin concentrations, but the details are not understood.

It seems that analysis of the roles of hormones in the interaction of rhizobia and agrobacteria with plants is hindered by the same problems. Auxin formation in the bacteria is mainly a function of the tryptophan concentration in the medium, not of the plasmids. Similarly, formation of cytokinin active substances occurs regardless of whether the plasmids specifying the interaction are present or not. For *Agrobacterium* systems there are indications that plasmid-coded genes participate in hormone biosynthesis or its control, but in most cases the precise roles are not clear. The situation is even less clear with *Rhizobium* systems. For a better understanding it may be useful to identify the genes essential for the interaction and then to investigate, using isolated genes, whether the gene products are involved in hormone biosynthesis or metabolism. If such experiments are successful, the correlation would support a hypothesis that hormone production by rhizobia plays a role in the interaction with plants.

Pseudomonas savastanoi

This gram-negative bacterium causes tumorlike overgrowths on olives, oleander, and privet. The bacteria produce large amounts of indole-3-acetic acid in culture, especially when provided with tryptophan. The capacity to induce tumors is correlated with the capacity to produce auxin (Wilson and Maggie, 1963; Magie, et al., 1963; Smidt and Kosuge, 1978; Surico et al., 1984), indicating that auxin is a major determinant in virulence. The bacteria also produce cytokinins, and subcloning of the genes and analysis of enzyme functions are in progress (T. Kosuge and R. Morris, personal communication).

The pathway to indole-3-acetic acid is clearly defined by two enzyme reactions which move from tryptophan through indole-3-acetamide, as an

intermediate, to auxin (Kosuge et al., 1966). Indole-3-acetic acid does not
seem to be the end product; at least some of it is further conjugated with
lysine and excreted in this form, and this reaction is strain-dependent
(Kuo and Kosuge, 1969; Hutzinger and Kosuge, 1968). The genes coding
for the enzymes are in some cases on defined plasmids (Comai and
Kosuge, 1980; Comai et al., 1982). The tryptophan monooxygenase gene
has been cloned and expressed in *E. coli* (Comai and Kosuge, 1982), and
the enzyme has been purified and characterized as FAD-containing protein (Hutcheson and Kosuge, 1985). In contrast to the *Agrobacterium*
system, there is no evidence that gene transfer is involved in tumor induction by *P. savastanoi*; it appears that hormones produced by the bacteria
are the main determinants in the disease. An interesting idea explaining
the presence of auxin genes was expressed by Comai and Kosuge (1983).
It is possible that they originate from genes for complete utilization of
tryptophan under saprophytic conditions, and that indole-3-acetic acid is
one of the intermediates. Under these conditions a simple loss of auxin
degradation under the selective pressure of a parasitic situation would
lead to accumulation of the plant hormone. This theory is intriguing and is
compatible with the limited information available, but it has not been
proven. As discussed before, it is tempting to speculate that tryptophan-
induced auxin production in agrobacteria and rhizobia could have similar
origins.

Corynebacterium fascians

This bacterium causes witches' broom in dicotyledonous plants. The
disease typically is characterized by the release of apical dominance and
outgrowth of lateral buds. The symptoms are duplicated by cytokinin
treatment, indicating that cytokinin production either by the bacteria or
by the plants plays an important role.

   All strains of *C. fascians* release cytokinins of the isopentenyladenine
and *trans*-zeatin type into the medium, and the patterns are similar
(Murai, 1981). The available data suggest that at least some of the hormones are produced by a tRNA pathway, but there is also clear evidence
that isopentenylation of free 5'AMP plays a major role. An isopentenyltransferase catalyzing this reaction has been isolated and purified from a
highly virulent strain (Murai, 1981); its properties are similar to those of
enzymes isolated from higher plant cells (Chen and Melitz, 1979; Chen,
1982). The presence of this enzyme, high cytokinin production, and virulence are correlated, suggesting that the hormones synthesized in the bacteria via isopentenylation of free 5'AMP are major determinants in virulence. Plasmid involvement in virulence has been claimed (Murai et al.,

1980; Murai, 1981), but there is no evidence for the transfer of genes to plants.

CONCLUSIONS AND PERSPECTIVE

For *P. savastanoi* and *C. fascians* the evidence is convincing that hormone biosynthesis in the bacteria is a decisive factor in the interaction with plants. In *Agrobacterium* systems the role of bacterial hormone production is less well defined, but some of the T-DNA genes transferred to and expressed in plant cells are responsible for hormone effects through coding for enzymes of auxin and cytokinin biosynthesis. Progress in recent years has been mostly due to approaches which combine molecular genetics and enzymology. This combination has been applied successfully to some bacterial genes, including T-region genes, and it seems likely that similar strategies will be used in the further analysis of plant-microbe interactions. This should be particularly useful in *Rhizobium* systems where identification of the genes responsible for the interaction with plants is in progress (see, for example, Long, 1984). The role of plant genes in hormone biosynthesis is unclear; however, there is no obvious reason why the techniques found to be useful with bacterial systems should not be successful with plant genes responsible for auxin and cytokinin formation.

ACKNOWLEDGMENTS

I thank D. Akiyoshi, S. Hutcheson, D. Inzé, T. Kosuge, and J. Schell for discussions and for sharing results prior to publication. Supported by Deutsche Forschungsgemeinschaft (SFB 206).

REFERENCES

AKIYOSHI, D. E., MORRIS, R. O., HINZ, R., MISCHKE, B. S., KOSUGE, T., GARFINKEL, D. J., GORDON, M. P., AND NESTER, E. W. 1983. Cytokinin/auxin balance in crown gall tumors is regulated by specific loci in the T-DNA, *Proc. Natl. Acad. Sci. USA* **80**:407–411.

AKIYOSHI, D. E., KLEE, H., AMASINO, R. M., NESTER, E. W., AND GORDON, M. P. 1984. T-DNA of *Agrobacterium tumefaciens* encodes an enzyme of cytokinin biosynthesis, *Proc. Natl. Acad. Sci. USA* **81**:5994–5998.

AKIYOSHI, D. E., REGIER, D. A., AND GORDON, M. P. 1985. Cloning and nucleotide sequence of the tzs gene from *Agrobacterium tumefaciens, Nucl. Acids Res.* **13**:2773–2791.

AMASINO, R. M., AND MILLER, C. O. 1982. Hormonal control of tobacco crown gall tumor morphology, *Plant Physiol.* **69**:389–392.

ATSUMI, S. 1980. Relation between auxin autotrophy and tryptophan content in sunflower crown gall cells in culture, *Plant Cell Physiol.* **21**:1031–1039.

BADENOCH-JONES, J., SUMMONS, R. E., DJORDJEVIC, M. A., SHINE, J., LETHAM, D. S., AND ROLFE, B. G. 1982. Mass spectrometric quantification of indole-3-acetic acid in *Rhizobium* culture supernatants: Relation to root hair curling and nodule initiation, *Appl. Environ. Microbiol.* **44**:275–280.

BADENOCH-JONES, J., ROLFE, B. G., AND LETHAM, D. S. 1983. Phytohormones, *Rhizobium* mutants, and nodulation in legumes, *Plant Physiol.* **73**:347–352.

BARKER, R. F., IDLER, K. B., THOMPSON, D. V., AND KEMP, J. D. 1983. Nucleotide sequence of the T-DNA region from the *Agrobacterium tumefaciens* octopine Ti plasmid pTi15955, *Plant Mol. Biol.* **2**:335–350.

BARRY, G. F., ROGERS, S. G., FRALEY, R. T., AND BRAND, L. 1984. Identification of a cloned cytokinin biosynthetic gene, *Proc. Natl. Acad. Sci. USA* **81**:4776–4780.

BEARDER, J. R. 1980. Plant hormones and other growth substances—their background, structures and occurrence, in: *Encyclopedia of Plant Physiology*, New Series, Vol. 9 (J. MacMillan, ed.), pp. 9–112, Springer-Verlag, New York.

BELTRA, R., SANCHEZ-SERRANO, J. J., AND SERRADA, J. 1978. Relationship between plasmids and plant tumorigenesis, *Proceedings of the 4th International Conference on Plant Pathogenic Bacteria, Angers* (Station de Pathologie Végétale et Phytobactériologie, ed.) pp. 199–205, INRA, Beaulouze, Angers, France.

BRAUN, A. C. 1956. The activation of two growth-substance systems accompanying the conversion of normal to tumor cells in crown gall, *Cancer Res.* **16**:53–56.

BRAUN, A. C. 1958. A physiological basis for autonomous growth of the crown gall tumor cell, *Proc. Natl. Acad. Sci. USA* **44**:344–349.

BRAUN, A. C. 1978. Plant tumors, *Biochim. Biophys. Acta* **516**:167–191.

BRAUN, A. C. 1982. A history of the crown gall problem, in: *Molecular Biology of Plant Tumors* (G. Kahl and J. Schell, eds.), pp. 155–210, Academic Press, New York.

BRAUN, A. C., AND WOOD, H. N. 1962. On the activation of certain essential biosynthetic systems in cells of *Vinca rosea*, *Proc. Natl. Acad. Sci. USA* **48**:1776–1782.

BUCHMANN, I., MARNER, F. -J., SCHRÖDER, G. WAFFENSCHMIDT, S., AND SCHRÖDER, J. 1985. Tumor genes in plants: T-DNA encoded cytokinin biosynthesis, *EMBO J.*, **4**:853–859.

CHEN, C. -M. 1982. Cytokinin biosynthesis in cell-free systems, in: *Plant Growth Substances 1982* (P. F. Wareing, ed.), pp. 155–163, Academic Press, New York.

CHEN, C. -M., AND MELITZ, D. K. 1979. Cytokinin biosynthesis in a cell-free system from cytokinin-autotrophic tobacco tissue cultures, *FEBS Lett.* **107**:15–20.

COHEN, J. D., AND BANDURSKI, R. S. 1982. Chemistry and physiology of bound auxins, *Annu. Rev. Plant Physiol.* **33**:403–430

COMAI, L., AND KOSUGE, T. 1980. Involvement of Plasmid deoxyribonucleic acid in indoleacetic acid synthesis in *Pseudomonas savastanoi, J. Bacteriol.* **143**:950–957.

COMAI, L., AND KOSUGE, T. 1982. Cloning and characterization of iaaM, a virulence determinant of *Pseudomonas savastanoi, J. Bacteriol.* **149**:40–46.

COMAI, L., AND KOSUGE, T. 1983. The genetics of indoleacetic acid production and virulence in *Pseudomonas savastanoi*, in: *Molecular Genetics of the Bacteria-Plant Interaction*, (A. Pühler, ed.), pp. 362–366, Springer-Verlag, Berlin.

COMAI, L., SURICO, G., AND KOSUGE, T. 1982. Relation of plasmid DNA to indoleacetic acid production in different strains of *Pseudomonas syringae* pv. *savastanoi, J. Gen. Microbiol.* **128**:2157–2163.

DART, P. 1977. Infection and development of leguminous nodules, in: *A Treatise on Nitrogen Fixation* (R. W. F. Hardy and W. S. Silver, eds.), pp. 367–472, John Wiley, New York.

DE CLEENE, M., AND DE LEY, J. 1976. The host range of crown gall, *Bot. Rev.* **42**:389–466.

DE GREVE, H., DHAESE, P., SEURINCK, J., LEMMERS, M., VAN MONTAGU, M., AND SCHELL, J. 1983. Nucleotide sequence and transcript map of the *Agrobacterium tumefaciens* Ti plasmid-encoded octopine synthase gene, *J. Mol. Appl. Genet.* **1**:499–511.

DEPICKER, A., STACHEL, S., DHAESE, P., ZAMBRYSKI, P., AND GOODMAN, H. M. 1982. Nopaline synthase: Transcript mapping and DNA sequence, *J. Mol. Appl. Genet.* **1**:561–573.

DULLAART, J. 1970. The bioproduction of indole-3-acetic acid and related compounds in root nodules and roots of *Lupinus luteus* L. and by its rhizobial symbiont, *Acta Bot. Neerl.* **19**:573–615.

EINSET, J. W. 1980. Cytokinins in tobacco crown gall tumors, *Biochem. Biophys. Res. Commun.* **93**:510–515.

FREGEAU, J. A., AND WIGHTMAN, F. 1983. Natural occurrence and biosynthesis of auxins in chloroplast and mitochondrial fractions from sunflower leaves, *Plant Sci. Lett.* **32**:23–34.

GARFINKEL, D. J., SIMPSON, R. B., REAM, L. W., WHITE, F. F., GORDON, M. P., AND NESTER, E. W. 1981. Genetic analysis of crown gall: Fine structure map of the T-DNA by site-directed mutagenesis, *Cell* **27**:143–153.

GELVIN, S. B. 1984. Plant tumorigenesis, in: *Plant-Microbe Interactions: Molecular and Genetic Perspectives* (T. Kosuge and E. W. Nester, eds.), pp. 343–377, MacMillan, New York.

GIELEN, J., DE BEUCKELEER, M., SEURINCK, J., DE BOECK, F., DE GREVE, H., LEMMERS, M., VAN MONTAGU, M., AND SCHELL, J. 1984. The complete nucleotide sequence of the TL-DNA of the *Agrobacterium tumefaciens* plasmid pTiAch5, *EMBO J.* **3**:835–846.

GOLDBERG, S. B., FLICK, J. S., AND ROGERS, S. G. 1984. Nucleotide sequence of the tmr locus of *Agrobacterium tumefaciens* pTiT37 T-DNA, *Nucleic Acids Res.* **12**:4665–4677.

GUYON, P., CHILTON, M.-D., PETIT, A., AND TEMPÉ, J. 1980. Agropine in "null-type" crown gall tumors: Evidence for the generality of the opine concept, *Proc. Natl. Acad. Sci. USA* **77**:2693–2697.

HAHN, H., HEITMANN, I., AND BLUMBACH, M. 1976. Cytokinins: Production and biogenesis of N^6-(Δ^2-isopentenyl) adenine in cultures of *Agrobacterium tumefaciens* strain B6, *Z. Pflanzenphysiol.* **79**:143–153.

HEIDEKAMP, F., DIRKSE, W. G., HILLE, J., AND VAN ORMONDT, H. 1983. Nucleotide sequence of the *Agrobacterium tumefaciens* octopine Ti plasmid-encoded tmr gene, *Nucleic Acids Res.* **11**:6211–6223.

HILLE, J., HOEKEMA, A., HOOYKAAS, P., AND SCHILPEROORT, R. A. 1984. Gene organization of the Ti plasmid, in: *Plant Gene Research: Genes Involved in Microbe-Plant Interactions* (D. P. S. Verma and T. Hohn, eds.), pp. 287–309, Springer-Verlag, Berlin.

HOEKEMA, A., PATER, B. S., FELLINGER, A. J., HOOYKAAS, P., AND SCHIL-PEROORT, R. A. 1984. The limited host range of an *Agrobacterium tumefaciens* strain extended by a cytokinin gene from a wide host range T-region, *EMBO J.* **3**:3043–3047.

HOLTZ, J., AND KLÄMBT, D. 1978. tRNA isopentenyltransferase from *Zea mays* L., *Hoppe-Seyler's Z. Physiol. Chem.* **359**:89–101.

HUTCHESON, S. W., AND KOSUGE, T. 1985. Regulation of 3-indoleacetic acid production in *Pseudomonas syringae* pv. *savastanoi:* Purification and properties of tryptophan 2-monooxygenase, *J. Biol. Chem.* **260**:6281–6287.

HUTZINGER, O., AND KOSUGE, T. 1968. Microbial synthesis and degradation of indole-3-acetic acid. III. The isolation and characterization of indole-3-acetyl-L-lysine, *Biochemistry* **7**:601–605.

IHARA, M., TAYA, Y., NISHIMURA, S., AND TANAKA, Y. 1984. Purification and some properties of Δ^2-isopentenylpyrophosphate: 5' AMP Δ^2-isopentenyltransferase from the cellular slime mold *Dictyostelium discoideum, Arch. Biochem. Biophys.* **230**:652–660

INZÉ, D. FOLLIN, A., VAN LIJSEBETTENS, M., SIMOENS, C., GENETELLO, C., VAN MONTAGU, M., AND SCHELL, J. 1984. Genetic analysis of the individual T-DNA genes of *Agrobacterium tumefaciens*: Further evidence that two genes are involved in indole-3-acetic acid synthesis, *Mol. Gen. Genet.* **194**:265–274.

JANSSENS, A., ENGLER, G., ZAMBRYSKI, P., AND VAN MONTAGU, M. 1984. The nopaline C58 T-DNA region is transcribed in *Agrobacterium tumefaciens, Mol. Gen. Genet.* **195**:341–350.

JOOS, H., INZÉ, D., CAPLAN, A., SORMANN, M., VAN MONTAGU, M., AND SCHELL, J. 1983. Genetic analysis of T-DNA transcripts in nopaline crown galls, *Cell* **32**:1057–1067.

KAISS-CHAPMAN, R. W., AND MORRIS, R. O. 1977. *Trans*-zeatin in culture filtrates of *Agrobacterium tumefaciens, Biochem. Biophys. Res. Commun.* **76**:453–459.

KAPER, J. M., AND VELDSTRA, H. 1958. On the metabolism of tryptophan by *Agrobacterium tumefaciens, Biochim. Biophys. Acta* **30**:401–420.

KEMPER, E., WAFFENSCHMIDT, S., WEILER, E. W., RAUSCH, T., AND SCHRÖDER, J. 1985. T-DNA encoded auxin formation in crown gall cells, *Planta,* **163**:257–262.

KLEE, H., MONTOYA, A., HORODYSKI, F., LICHTENSTEIN, C., GARFINKEL, D., FULLER, S., FLORES, C., PESCHON, J., NESTER, E. W., AND GORDON, M. P. 1984. Nucleotide sequence of the tms genes of the pTiA6NC octopine Ti plasmid: Two gene products involved in plant tumorigenesis, *Proc. Natl. Acad. Sci. USA* **81**:1728–1732.

KOSUGE, T., HESKETT, M. G., AND WILSON, E. E. 1966. Microbial synthesis and degradation of indole-3-acetic acid: The conversion of L-tryptophan to indole-3-acetamide by an enzyme system from *Pseudomonas savastanoi, J. Biol. Chem.* **241**:3738–3744.

KRETOVICH, V. L. ALEKSEEVA, I. I., TSIVINA, N. Z. 1972. Content of β-indolylacetic acid in root nodules and roots of lupine, *Sov. Plant Physiol.* **19**:421–424.

KUO, T. T., AND KOSUGE, T. 1969. Factors influencing the production and further metabolism of indole-3-acetic acid by *Pseudomonas savastanoi, J. Gen. Appl. Microbiol.* **15**:51–63.

LEEMANS, J., DEBLAERE, R., WILLMITZER, L., DE GREVE, H., HERNALSTEENS, J. -P., VAN MONTAGU, M., AND SCHELL, J. 1982. Genetic identification of functions of TL-DNA transcripts in octopine crown galls, *EMBO J.* **1**:147–152.

LETHAM, D. S., AND PALNI, L. M. S. 1983. The biosynthesis and metabolism of cytokinins, *Annu. Rev. Plant Physiol.* **34**:163–197.

LICHTENSTEIN, C., KLEE, H., MONTOYA, A., GARFINKEL, D., FULLER, S., FLORES, C., NESTER, E. W., AND GORDON, M. P. 1984. Nucleotide sequence and transcript mapping of the tmr gene of the pTiA6NC octopine Ti plasmid: A bacterial gene involved in plant tumorigenesis, *J. Mol. Appl. Genet.* **2**:354–362.

LIU, S. -T., AND KADO, C. I. 1979. Indoleacetic acid production: A plasmid function of *Agrobacterium tumefaciens* C58, *Biochem. Biophys. Res. Commun.* **90**:171–178.

LIU, S. -T., PERRY, K. L., SCHARDL, C. L., AND KADO, C. I. 1982. *Agrobacterium* Ti plasmid indoleacetic acid gene is required for crown gall oncogenesis, *Proc. Natl. Acad. Sci. USA* **79**:2812–2816.

LOBANOK, E. V., FOMICHEA, V. V., AND KARTEL, N. A. 1982. The synthesis of indolylacetic acid by oncogenic and nononcogenic strains of *Agrobacterium tumefaciens, Dokl. Akad. Nauk BSSR* **26**:565–566.

LONG, S. R. 1984. Genetics of *Rhizobium* nodulation, in: *Plant-Microbe Interactions, Molecular and Genetic Perspectives,* Vol. 1 (T. Kosuge and E. W. Nester, eds.), pp. 265–306, Macmillan, New York.

MAASS, H., AND KLÄMBT, D. 1981. On the biogenesis of cytokinins in roots of *Phaseolus vulgaris, Planta* **151**:353–358.

MAGIE, A. R., WILSON, E. E., AND KOSUGE, T. 1963. Indoleacetamide as intermediate in the synthesis of indoleacetic acid in *Pseudomonas savastanoi*, *Science* **141**:1281–1282.

MAHADEVAN, S. 1973. Role of oximes in nitrogen metabolism in plants, *Annu. Rev. Plant Physiol.* **24**:69–88.

McCLOSKEY, J. A., HASHIZUME, T., BASILE, B., OHNO, Y., AND SONOKI, S. 1980. Occurrence and levels of *cis*- and *trans*-zeatin ribosides in the culture medium of a virulent strain of *Agrobacterium tumefaciens*, *FEBS Lett.* **111**:181–183.

MESSENS, E., AND CLAEYS, M. 1978. Ti plasmid encoded production of cytokinins in culture media of *Agrobacterium tumefaciens*, in: *Proceedings of the 4th International Conference on Plant Pathogenic Bacteria, Angers* (Station de Pathologie Végétale et Phytobactériologie, ed.), pp. 169–175, INRA, Beaucouze, Angers, France.

MORRIS, R. O., REGIER, D. A., AND MacDONALD, E. M. S. 1981. Analytical procedures for cytokinins: Application to *Agrobacterium tumefaciens*, in: *Metabolism and Molecular Activities of Cytokinins* (J. Guern and C. Péaud-Lenoel, eds.), pp. 3–16, Springer-Verlag, Berlin.

MORRIS, R. O., AKIYOSHI, D. E., MacDONALD, E. M. S., MORRIS, J. W., REGIER, D. A., AND ZAERR, J. B. 1982. Cytokinin metabolism in relation to tumor induction by *Agrobacterium tumefaciens*, in: *Plant Growth Substances 1982* (P. F. Wareing, ed.), pp. 175–183, Academic Press, New York.

MOUSDALE, D. M. A. 1981. Endogenous indolyl-3-acetic acid and pathogen induced plant growth disorders: Distinction between hyperplasia and neoplastic development, *Experientia* **37**:972–973.

MOUSDALE, D. M. A. 1982. Endogenous IAA and the growth of auxin dependent and auxin autotrophic crown gall plant tissue cultures, *Biochem. Physiol. Pflanzen* **177**:9–17.

MURAI, N. 1981. Cytokinin biosynthesis and its relationship to the presence of plasmids in strains of *Corynebacterium fascians*, in: *Metabolism and Molecular Activities of Cytokinins* (J. Guern and C. Péaud-Lenoel, eds.), pp. 17–26, Springer-Verlag, Berlin.

MURAI, N., SKOOG, F., DOYLE, M. E., AND HANSON, R. S. 1980. Relationship between cytokinin production, presence of plasmids and fasciation caused by strains of *Corynebacterium fascians*, *Proc. Natl. Acad. Sci. USA* **77**:619–623.

NAKAJIMA, H., YOKOTA, T., TAKAHASHI, N., MATSUMOTO, T., AND NOGUCHI, M. 1981. Changes in endogenous ribosyl-*trans*-zeatin and IAA levels in relation to the proliferation of tobacco *Nicotiana tabacum* cultivar Hicks-2 crown gall cells, *Plant Cell Physiol.* **22**:1405–1410.

NESTER, E. W., GORDON, M. P., AMASINO, R. M., AND YANOFSKY, M. F. 1984. Crown gall: A molecular and physiological analysis, *Annu. Rev. Plant Physiol.* **35**:387–413.

NEWCOMB, W. 1980. Control of morphogenesis and differentiation of pea root nodules, in: *Nitrogen Fixation*, Vol. 2 (W. E. Newton and W. H. Orme-Johnson, eds.), pp. 87–102, University Park Press, Baltimore, Maryland.

NEWCOMB, W., SYONO, K., AND TORREY, J. G. 1977. Development of an ineffective pea root nodule: Morphogenesis, fine structure, and cytokinin biosynthesis, *Can. J. Bot.* **55**:1891–1907.

NUTI, M. P., LEPIDI, A. A., PRAKASH, R. K., HOOYKAAS, P. J. J., AND SCHILPEROORT, R. A. 1982. The plasmids of *Rhizobium* and symbiotic nitrogen fixation, in: *Molecular Biology of Plant Tumors* (G. Kahl and J. S. Schell, eds.), pp. 561–588, Academic Press, New York.

OOMS, G., HOOYKAAS, P. J. J., MOOLENAAR, G., AND SCHILPEROORT, R. A. 1981. Crown gall plant tumors of abnormal morphology induced by *Agrobacterium tumefaciens* carrying mutated octopine Ti plasmids: Analysis of T-DNA functions, *Gene* **14**:33–50.

PARISH, C. G., AND MAGASANIK, B. 1981. Tryptophan metabolism in *Klebsiella aerogenes:* Regulation of the utilization of aromatic amino acids as sources of nitrogen, *J. Bacteriol.* **145**:257–265.

PATE, J. S. 1958. Studies of the growth substances of legume nodules using paper chromatography, *Aust. J. Biol. Sci.* **11**:516–528.

PENGELLY, W. L., AND MEINS, F., JR. 1982. The relationship of IAA content and growth of crown gall tumor tissues of tobacco *Nicotiana tabacum* cultivar Turkish in culture, *Differentiation* **21**:27–31.

REAM, L. W., GORDON, M. P., AND NESTER, E. W. 1983. Multiple mutations in the T-region of the *Agrobacterium tumefaciens* tumor-inducing plasmid, *Proc. Natl. Acad. Sci. USA* **80**:1660–1664.

REGIER, D. A., AND MORRIS, R. O. 1982. Secretion of *trans*-zeatin by *Agrobacterium tumefaciens:* A function determined by the nopaline Ti plasmid, *Biochem. Biophys, Res. Commun.* **104**:1560–1566.

RIDDLE, V. M., AND MAZELIS, M. 1964. A role for peroxidase in biosynthesis of auxin, *Nature* **202**:391–392.

RIDDLE, V. M., AND MAZELIS, M. 1965. Conversion of tryptophan to indoleacetamide and further conversion to indoleacetic acid by plant preparations, *Plant Physiol.* **40**:481–484.

SCHELL, J., VAN MONTAGU, M., HOLSTERS, M., ZAMBRYSKI, P., JOOS, H., HERRERA-ESTRELLA, L., DEPICKER, A., HERNALSTEENS, J. P., DE GREVE, H., WILLMITZER, L., SCHRÖDER, J. 1984a. Ti plasmids as gene vectors for plants, in: *Eucaryotic Gene Expression* (A. Kumar, ed.), pp. 141–160, Plenum Press, New York.

SCHELL, J., HERRERA-ESTRELLA, L., ZAMBRYSKI, P., DE BLOCK, M., JOOS, H., WILLMITZER, L., ECKES, P., ROSAHL, S., AND VAN MONTAGU, M. 1984b. Genetic engineering of plants, in: *The Impact of Gene Transfer Techniques in Eucaryotic Cell Biology* (J. S. Schell and P. Starlinger, eds.), pp. 73–90, Springer-Verlag, Heidelberg.

SCHRÖDER, G., KLIPP, W., HILLEBRAND, A., EHRING, R., KONCZ, C., AND SCHRÖDER, J. 1983. The conserved part of the T-region in the Ti-plasmids expresses four proteins in bacteria, *EMBO J.* **2**:403–409.

SCHRÖDER, G., WAFFENSCHMIDT, S., WEILER, E. W., AND SCHRÖDER, J. 1984. The T-region of Ti plasmids codes for an enzyme synthesizing indole-3-acetic acids, *Eur. J. Biochem.* **138**:387–391.

SCIAKY, D., AND THOMASHOW, M. F. 1984. The sequence of the tms 2 transcript locus of the *A. tumefaciens* plasmid pTiA6 and characterization of the mutation in pTiA66 that is responsible for auxin attenuation, *Nucleic Acids Res.* **12:**1447–1461.

SEMBDNER, G., GROSS, D., LIEBISCH, H. -W., AND SCHNEIDER, G. 1980. Biosynthesis and metabolism of plant hormones, in: *Encyclopedia of Plant Physiology*, New Series, Vol. 9, *Hormonal Regulation of Development I* (J. MacMillan, ed.), pp. 281–373, Springer-Verlag, New York.

SKOOG, F., AND MILLER, C. O. 1957. Chemical regulation of growth and organ formation in plant tissues cultured *in vitro, Symp. Soc. Exp. Biol.* **11:**118–131.

SMIDT, M., AND KOSUGE, T. 1978. The role of indole-3-acetic acid accumulation by alpha methyl tryptophan-resistant mutants of *Pseudomonas savastanoi* in gall formation on oleanders, *Physiol. Plant. Pathol.* **13:**203–214.

SUKANYA, N. K., AND VAIDYANATHAN, C. S. 1964. Aminotransferases of *Agrobacterium tumefaciens:* Transamination between tryptophan and phenyl-pyruvate, *Biochem. J.* **92:**594–598.

SURICO, G., COMAI, L., AND KOSUGE, T. 1984. Pathogenicity of strains of *Pseudomonas syringae* pv. *savastanoi* and their indoleacetic acid-deficient mutants on olive and oleander, *Phytopathology* **74:**490–493.

SYONO, K., NEWCOMB, W., AND TORREY, J. G. 1976. Cytokinin production in relation to the development of pea root nodules, *Can. J. Bot.* **54:**2155–2162.

TAYA, Y., TANAKA, Y., AND NISHIMURA, S. 1978. 5′-AMP is a direct precursor of cytokinin in *Dictyostelium discoideum, Nature* **271:**545–547.

TEMPÉ, J., AND GOLDMANN, A. 1982. Occurrence and biosynthesis of opines, in: *Molecular Biology of Plant Tumors* (G. Kahl and J. S. Schell, eds.), pp. 427–449, Academic Press, New York.

TEMPÉ, J., AND PETIT, A. 1982. Opine utilization by *Agrobacterium*, in: *Molecular Biology of Plant Tumors* (G. Kahl and J. S. Schell, eds.), pp. 451–459, Academic Press, New York.

THOMASHOW, L. S., REEVES, S., AND THOMASHOW, M. F. 1984. Crown gall oncogenesis: Evidence that a T-DNA gene from the *Agrobacterium* Ti plasmid pTiA6 encodes an enzyme that catalyzes synthesis of indoleacetic acid, *Proc. Natl. Acad. Sci. USA* **81:**5071–5075.

VAN ONCKELEN, H., RÜDELSHEIM, P., INZÉ, D., FOLLIN, A., MESSENS, E., HOREMANS, S., AND SCHELL, J. 1985. Tobacco plants transformed with the *Agrobacterium* T-DNA gene 1 contain high amounts of indole-3-acetamide, *FEBS Lett.* **181:**373–376.

VAN SLOGTEREN, G. M. S., HOGE, J. H. C., HOOYKAAS, P. J. J., AND SCHIL-PEROORT, R. A. 1983. Clonal analysis of heterogeneous crown gall tumor tissues induced by wild-type and shooter mutant strains of *Agrobacterium tumefaciens*—Expression of T-DNA genes, *Plant Mol. Biol.* **2:**321–333.

VAN SLOGTEREN, G. M. S., HOOYKAAS, P. J. J., AND SCHILPEROORT, R. A. 1984. Tumor formation on plants by mixtures of attenuated *Agrobacterium tumefaciens* T-DNA mutants, *Plant Mol. Biol.* **3:**337–344.

WEILER, E. W., AND SPANIER, K. 1981. Phytohormones in the formation of crown gall tumors, *Planta* **153:**326–337.

WIGHTMAN, F. 1973. Biosynthesis of auxins in tomato shoots, *Biochem. Soc. Symp.* **38**:247–275.

WILSON, E. E., AND MAGIE, A. R. 1963. Physiological, serological, and pathological evidence that *Pseudomonas tonelliana* is identical with *Pseudomonas savastanoi, Phytopathology* **53**:653–659.

YAMADA, T., PALM, C. J., AND KOSUGE, T. 1985. Fine structure of tryptophan monooxygenase gene in *Pseudomonas syringae* pv. savastanoi, in: *Advances in Molecular Genetics of the Bacteria-Plant Interaction* (A. A. Szalay and R. P. Legocki, eds.), pp. 180–181, Cornell University Publishers, Ithaca, New York.

ZENK, M. H. 1961. 1-(Indole-3-acetyl)-β-D-glucose, a new compound in the metabolism of indole-3-acetic acid in plants, *Nature* **191**:493–494.

Chapter 3

The Molecular Biology of Fungal Development

James S. Lovett

FUNGI AS A WHOLE present an enormous array of different types, degrees, and complexities of development. In this chapter attention will be focused on a few species that have been extensively studied. The emphasis will be placed on an examination of the evidence from this group for the role of differential gene expression in development, mainly asexual sporulation. Wherever possible, the reader will be directed to reviews that cover broader subjects or give greater detail than can be provided here. Fungi are considered good eukaryotic organisms for developmental study because of their relatively simple morphology, ease and rapidity of growth, and, in a few cases, well-established haploid genetic systems. However, despite research at one level or another over at least a century, our understanding of fungal developmental regulation is still surprisingly limited. There are a number of reasons for this, some historical and some methodological. Historically, mycologists as a group have been much more preoccupied with the impressive morphological diversity of fungi than they have been with how this comes about. Nevertheless, a great deal of research has been done on the conditions that lead to the development of asexual and sexual reproductive structures in fungi. This kind of work has shown that various fungi respond to many environmental stimuli (e.g., carbon and nitrogen limitation, varied carbon/nitrogen ratios, light of various wavelengths, and aeration) with the production of reproductive structures (Smith and Berry, 1974; Dahlberg and Van Etten, 1982). A number of sexually reproducing species also secrete and respond to hormonal substances that coordinate sexual interactions and the development of reproductive structures. The molecular mechanisms of hormone action in fungi are still largely unknown, although several hormones have been chemically characterized and significant progress is being made with the peptide hormones of the yeast *Saccharomyces cerevisiae* (Gooday, 1983; van den Ende, 1983). Unfortunately, the wealth of factual information on the nutritional and other environmental influences that lead to development has so far resulted in few useful generalizations about regulation. Even such an apparently "simple" developmental change as the dimorphic switch from a hyphal form of growth to a yeastlike budding form, or vice versa, has so far defied causal analysis despite considerable attention (Sypherd et al., 1978; Stewart and Rogers, 1983; San-Blas and San-Blas, 1984). Historically, this lack of progress has arisen from a combination of our sometimes simplistic assumptions about developmental regulation and insufficient recognition of the fact that fungi, though developmentally simple, are far from lacking in complexity.

COMPLEXITIES OF DEVELOPMENTAL REGULATION

It has frequently been assumed that one could understand development by examining the activities of various metabolic enzymes, and, of course, many changes in enzymes and pathways have been described during fungal development (Smith and Berry, 1974). However, in most cases it has not been at all clear whether the changes described were a cause of development, a consequence of it, or simply a response to the nutritional milieu. It is difficult to make such distinctions, because far too often a condition that induces a developmental response, such as asexual sporulation induced by a medium step-down or other change, also requires the fungus to adapt metabolically (e.g., as a result of release from carbon catabolite repression, adjustment to nitrogen starvation). A good illustration comes from the study of induced yeast sporulation which requires a shift of diploid cells to an unbalanced medium with an acetate carbon source in the absence of available nitrogen. This problem has complicated many studies on fungi. These organisms typically grow in competitive nonhomeostatic environments and have evolved sensitive, efficient regulatory

mechanisms for adapting to the available nutrient sources. An understanding of the complexity and sensitivity of these systems has only
gradually emerged as a result of much genetic and biochemical work that
has identified interrelated regulatory systems and genes that may have
regulatory functions over several different pathways (Metzenberg, 1979;
Marzluf, 1981). Metabolic shifts in response to shifts in the organism's
environment may thus provide a significant background of change, some
of which may be required for the completion of development and some of
which may not. But it is against this complex background that the experi-
menter must attempt to detect specific developmental activities that are
unique to the process.

Protein turnover is another factor that has rarely been examined in detail in sporulating cultures, even though it might be expected to be extensive in step-down-induced sporulation systems with fungi (Timberlake et
al., 1973; Lovett, 1983) as it is in *Dictyostelium* development (Hames and
Ashworth, 1974). Extensive turnover provides an additional and significant level of background synthetic activity, only part of which might be
essential to concomitant developmental activities. Another limitation of
the analysis of fungal development arises from our incomplete knowledge
of cell wall architecture and the synthesis and assembly of cell wall components. Although much progress has been made in characterizing the
cell walls in a few fungi (Wessels and Sietsma, 1981; Cabib, 1981), a
complete molecular characterization indicating how the major polymers
are organized within hyphal walls is still lacking. Much also remains to be
learned about the biosynthesis and assembly of the major and minor wall
polymers. For example, mechanisms for the secretion of chitin synthetase
and its function in producing chitin microfibrils within the wall are just
becoming clear (Bartnicki-Garcia et al., 1979; Cabib et al., 1983), but it is
still unknown how this biosynthesis is integrated with the synthesis and
deposition of glucans, proteins, glycoproteins, and so on, to make the
organized wall. Much less is known about how the plasticity of the growing tip is maintained or how the direction of growth is controlled. These
are hard problems to attack, because hyphal growth is a vectorial process
in which the critical activities are largely confined to the hyphal apices
which represent a miniscule part of the total hyphal biomass. The
complexities due to the cell walls are not raised here to suggest that the
cell wall is the be all and end all of fungal development (though some
might agree with such a claim). Rather, it is to point out that fungal development, which is the formation of multicellular fungal structures, involves
the synthesis and assembly of cell walls as a major and crucial component
and that limitations in our understanding of the basic facts make it more
difficult to establish how regulation might be achieved.

Because the main structural parts of reproductive elements are usually hyphal filaments, modified hyphae, or other walled cells, one developmental question concerns whether such functionally specialized cells have a cell wall composition and/or architecture that is different from that of the vegetative hyphae and therefore require new synthetic activities. In some cases the answer is yes. Only in relatively few instances has it been established that the differences in composition actually involved new products and/or de novo synthesis of new biosynthetic enzymes, but obviously the presence of unique products implies stage-specific expression of the necessary genes.

Since the major structural and synthetic components, metabolic enzymes, and so on, of all the cells would be expected to be rather similar, much of visible development could largely involve the redirection of macromolecular synthesis and structural assembly in space and time (e.g., the first step in producing a conidiophore does not seem to differ much in its requirements from normal hyphal branching, except that it is vertical). If one accepts this interpretation, then other important objectives of a developmental analysis are to discover the regulatory factors that influence these redirections, to establish the extent to which truly new gene products are involved in achieving this regulation, and to determine how they are related to preexisting systems.

The evidence, from a few fungi that have been studied using deliberate production of and screening for developmental mutants, supports the interpretation that relatively few unique gene products are involved in sporulation. Many sporulation-defective mutants have been found to be pleiotropic, affecting not only sporulation but some additional aspects of growth or sexual reproduction as well. When genes with pleiotropic effects were removed from mutant collections, the number of sporulation-specific mutants remaining was usually quite small. Despite an extensive search, only two morphological phenotypes (bristle and abacus) were identified for *Aspergillus nidulans* conidiation. From such results Clutterbuck (1977) estimated that only 45–150 loci were specifically involved in this developmental process—less than 1% of the genome. A similarly low estimate for the number of genes indispensable for *S. cerevisiae* meiosis and sporulation, 48 ± 27 genes representing only 0.5% of the genome, has been provided by Esposito et al. (1972). Similarly, low estimate appears appropriate for *Neurospora crassa* (Matsuyama et al. 1974; Mishra, 1977; Perkins et al., 1982).

In contrast to the conservative estimates based on transmission genetics, hybridization experiments have indicated that a variable, but much larger, fraction of the fungal genome may be specifically transcribed and expressed during development (6.5–20%; see below). The estimates based on mutant studies are consistent with the production of relatively few unique gene products during sporulation. The estimates from

hybridization clearly imply that many more unique gene products are
produced, though such experiments cannot show whether they are essential. It is at present difficult to rationalize these contrasting predictions.

Phase-Specific Gene Products

Based on our present understanding, three basic kinds of gene products
may be involved in development: (1) a large number of enzymes and
structural proteins that are common to both growth and development, (2)
genes whose products may be phase-critical (Schmit and Brody, 1976;
Brody, 1981) in the sense that a higher level of a protein, intermediate, or
biosynthetic product may be required to complete an essential developmental event compared to the need during vegetative growth, and (3)
genes whose products are phase-specific and are produced only during
some stage of development. The distinction between types 1 and 2 is
rather arbitrary but useful conceptually. It seems evident, a priori, that
most major biosynthetic pathways and energy-generating systems would
be needed for the synthesis of new, specialized cells, including the basic
components of cell walls. In starvation- or step-down-induced sporulation this may involve both new synthesis and turnover of preexisting
proteins as well as other cell components. What could also change significantly during development is the regulation of existing pathways and
the flux of metabolites through them (Wright and Emyanitoff, 1983).

Phase-critical gene products may be normal end products or intermediates that are simply required at higher concentrations to complete some
essential event. A number of *Neurospora* mutants with abnormal growth
morphologies arising from defects in genes of basic carbon metabolism
provide a good analogy for phase-critical gene products (Mishra, 1977;
Clutterbuck, 1978). Examples of apparent phase-critical molecules are
provided by two *A. nidulans* mutants. Serlupi-Crecenzi et al. (1983)
isolated a pleiotropic aconidial mutant that was found to be auxotrophic
for arginine because of a defect in the ornithine transcarbamylase gene
(*ArgB*). Adequate arginine supplementation "cured" the conidiation
defect but did not correct the production of sterile cleistothecia. Yelton et
al. (1984b) found that *trpC* mutants of *A. nidulans* required a much higher
level of typtophan supplementation for conidiation than was needed for
mycelial growth. The results from studies of such mutants suggest that
rather significant changes might well occur in an unknown number of
preexisting biosynthetic pathways and that this could lead to differential
accumulation or depletion of specific products, e.g., storage of glutamic
acid in the amino acid pool of *N. crassa* conidia (Brody, 1981) and
glycogen accumulation and subsequent utilization during sporulation in *S.
cerevisiae* (Esposito and Klapholz, 1981).

Relatively few phase-specific gene products have been identified in fungi, although there certainly are some obvious candidates such as conidial rodlet wall layer proteins (Cole and Pope, 1981; Beever et al., 1979), sporopollenins (Gooday, 1981), melanins (Rast, et al., 1981), germination inhibitors (Macko, 1981), and ascospore surface antigens (Dawes et al., 1983) that are structural parts of differentiated spores. However, in only a few cases have the required enzymes or other direct gene products been studied and/or shown to be unique to a particular stage of development. The laccase I (ap-diphenol oxidase) that converts a yellow precursor to a green conidial pigment in *A. nidulans* (Clutterbuck, 1977) is synthesized *de novo* during conidiation (Law and Timberlake, 1980; Kurtz and Champe, 1982). The appearance of a second, distinct phenol oxidase, laccase II, appears to be similarly correlated with cleistothecial development in the same fungus (Hermann et al., 1983). These enzymes, and probably a cresolase (Clutterbuck, 1977), are products of secondary developmental genes in the sense that they function to produce pigments associated with normal development but are not essential for its otherwise normal progression. The ascospore surface antigen of yeast (Dawes et al., 1983) is also such a secondary product. An α-1, 4-glucosidase (Colonna and Magee, 1978) and a β-glucanase (del Ray et al., 1980) have also been identified as sporulation-specific products in *S. cerevisiae*. The activity of the α-glucosidase is clearly associated with the depletion of glycogen during yeast sporulation, but the role of the glucanase is less obvious.

Of even greater interest are mutants that may have specific, critical regulatory functions in sporulation. Mutants that may be of this type have been isolated from *Saccharomyces, Neurospora,* and *Aspergillus.* They include sporulation (*spo*) mutants in yeast that cause blocks at different stages (Esposito and Klapholz, 1981; Dawes, 1983), some aconidial *spo* mutants of *N. crassa* (Matsuyama et al., 1974), as well as bristle (*brl*) and abacus (*aba*) mutants (Clutterbuck, 1977; Martinelli, 1979), and early acting "competence" mutants (for sporulation induction) of *Aspergillus* (Yager et al., 1982; Butnick et al., 1984b). Although their roles are not known, some of these genes may regulate the switch from hyphal growth to a sporulation pathway or to a particular stage or process of development [e.g., the auxiliary and strategic loci, respectively, suggested by Clutterbuck (1978)]. In the case of *brl* and *aba*, development occurs but is incomplete and abnormal, suggesting abnormal regulation of the process once it has been initiated. Aconidial *spo* mutants that appear to function before an induction stimulus occurs (preinduction competence mutants) appear particularly interesting as potential switching or strategic loci that regulate the initial response of the fungus. Three such conditional mutants were blocked for both conidiation and sexual reproduction, which may suggest a single control for the initiation of both processes

(Butnick et al., 1984a, b). However, a few second-site revertants corrected the conidiation defect but not the sexual cleistothecial block, which suggests that the two developmental triggers are dissociable. The production of phenolic metabolites by the preinduction mutants, particularly diorcinol, was inversely correlated with sexual development in the revertants (aconidial suppressors). This raised the intriguing possibility that a single pathway, in which these compounds are intermediates, regulates both developmental systems at different points (Butnick et al., 1984b).

SIZE AND ORGANIZATION OF FUNGAL GENOMES

In a discussion of the role of differential gene expression in fungal development it is worthwhile first to consider the size and organization of the genomes in fungi. Data are available for representatives from five major classes of true fungi and for both plasmodial and cellular slime molds. Fortunately, these reports include most of the species that have been popular for developmental study (Table 1). From the nucleotide sequence complexity values and DNA contents per haploid genome, it is evident that all these organisms have genomes that are quite small compared to those of higher plants (Grierson, 1977) and animals (Davidson and Britten, 1983). The genomes of true fungi are all 3–16 times the size of the *Escherichia coli* chromosome, with a sequence complexity of 4.2×10^6. Fungal nuclei are proportionately small and, where the DNA per nucleus has been measured chemically, the sizes calculated from the complexity data (0.015–0.072 pg, for true fungi) are in reasonable agreement with the chemical estimates of DNA per haploid genome. The ploidy of the *Blastocladiella emersonii* zoospores that have the highest DNA content of the nonplasmodial organisms listed in Table 1 is not known. Based on microspectrophotometric measurements (Horgen et al., 1985), they are perhaps diploid or even polyploid.

The genome of the cellular slime mold *Dictyostelium discoideum* is similar in size to that of true fungi, but the plasmodial slime mold *Physarum polycephalum* has a 10-fold larger genome based on the reported sequence complexity. The organization of the genomic material in slime molds is, on the other hand, clearly different from that in true fungi. Both *D. discoideum* and *P. polycephalum* have a significant fraction of repetitive DNA sequences interspersed throughout their genomes. This type of organization, with short repetitive segments interspersed between unique coding sequences, or short-period interspersion, is common in higher plants and animals which typically contain large fractions of repetitive DNA (Grierson, 1977; Davidson and Britten, 1983). In contrast, true fungi appear to contain modest amounts of repetitive DNA, mostly in the middle repetitive class, and do not have a significant amount of

Table 1 Size and Complexity of Fungal Genomes.

Organism	DNA Complexity (NTP)[a]	Picograms DNA per Haploid Genome[b]	DNA per Nucleus[c]	Reference
True fungi (Oomycetes)				
Achlya ambisexualis (Chytridiomycetes)	4.19×10^7	0.046		Hudspeth et al. 1971
Allomyces arbuscula	2.58×10^7	0.028		Ohja et al. 1977
Blastocladiella emersonii (Zygomycetes)	1.86×10^7	0.02	0.074[d]	Johnson and Lovett, 1984a
Phycomyces blakesleeanus (Ascomycetes)	6.6×10^7	0.072		Harshey et al. 1979
Aspergillus nidulans	2.6×10^7	0.028	0.022	Timberlake, 1978
Neurospora crassa	2.7×10^7	0.03		Krumlauf and Marzluf, 1979
Saccharomyces cerevisiae (Basidiomycetes)	1.4×10^7	0.015		Hereford and Rosbash, 1977
Schizophyllum	1.33×10^7	0.015		Lauer et al., 1977
commune	3.4×10^7	0.038		Dons et al., 1979
	3.6×10^7	0.039		Ullrich et al., 1980
Slime molds(Myxomycetes)				
Physarum polycephalum (Acrasiales)	7.2×10^8	0.79	1.0	Fouquet et al., 1974
Dictyostelium discoideum	4.6×10^7	0.05		Firtel et al., 1976

[a]Base sequence values based on DNA reassociation kinetics.
[b]DNA per haploid genome calculated from the sequence complexity.
[c]Values for DNA per nucleus obtained by chemical measurement.
[d]A value for total zoospore DNA that includes mitochondrial DNA.

short-period interspersion. In most cases the genome contains predominantly single-copy (unique) sequences (82–98%) and a number of middle repetitive sequences roughly equivalent to what would be expected for multiple rRNa and tRNA genes (e.g., 60–200 copies per genome). There is some evidence for long-period interspersion in *Achlya* (Hudspeth et al., 1977) and *Phycomyces* (Harshey et al., 1979), analogous to the pattern found in insects.

The genomic organization with short-period interspersion characteristic of *Physarum* and *Dictyostelium*, as well as their satellite rDNAs, probably betray their closer affinity with animals than with true fungi. These organisms will not be considered here, although they have historically been included with fungi in most mycology tests and are often studied by mycologists.

Because fungi contain small amounts of repetitive DNA, the potential coding capacity per unit of nuclear DNA is large relative to that of higher eukaryotes; however, estimates of the total potential coding capacity of the single-copy DNA of ca. 9000–24,000 average-sized mRNAs are modest compared to the potential coding potential of higher plants and animals ($\geq 10^6$ mRNAs). On the other hand, fungi actually transcribe a considerably larger fraction of their single-copy DNA into mRNA (31–80% at any given stage) compared to higher organisms (ca. $\leq 3\%$) (Grierson, 1977; Davidson and Britten, 1983). Thus the disparity may be less extreme than might be suggested by the relative nuclear DNA contents.

It seems unlikely that the regulatory model of Davidson and Britten (1983) based on large numbers of highly repetitive interdispersed DNA sequences of animal cells can be applied directly to true fungi which lack significant repetitive DNA and short-period interspersion. This smaller genome and apparently simpler genomic organization, along with their simpler growth and developmental forms, make fungi attractive for the study of developmental regulation. The rapid progress being made in the cellular and molecular biology of yeasts and a few filamentous fungi certainly tends to reinforce such an optimistic view. In the following sections the extent to which this promise has been fulfilled in regard to the complex problem of fungal development will be examined.

TRANSCRIPTION AND PROCESSING OF RNA

Evidence for Heterogeneous Nuclear RNA in Fungi

There have been relatively few studies that have directly attempted to detect and measure a heterogeneous nuclear RNA (hnRNA) fraction in fungi. It is evident that there is no fungal equivalent to the large fraction of mammalian hnRNA that turns over rapidly in the nucleus. In higher-meta-zoan cells only a small fraction of this rapidly labeled RNA is both polyadenylated and transported to the cytoplasm for translation into cellular proteins. In *Dictyostelium* a hnRNA fraction has been identified that is about 25% larger than the cytoplasmic poly(A) RNA fraction (Firtel et al., 1976). The 250- to 450-nucleotide (NT) repetitive sequences removed by processing occur at the 5' end of each ~1500-NT transcript. Although this processing is analogous to the hnRNA of mammalian cells, it represents the turnover of a much smaller fraction of the initial nuclear transcripts. In the same species, a 3' poly(A) tail of approximately 125 NT is added via nuclear polyadenylation (Firtel et al., 1976).

In true fungi a distinct hnRNA fraction has not been found (Brambl et al., 1978; Van Etten et al., 1981). The pulse-labeled total RNA of fungi has generally been found to lack nonribosomal high-molecular-weight precursors when examined by polyacrylamide gel electrophoresis. However, this relatively insensitive method probably would not have detected a fraction that was only a few percent larger than the cytoplasmic mRNA fraction. Nevertheless, a careful comparison of the nuclear poly(A) RNA and cytoplasmic poly(A) RNA in *Achlya ambisexualis* by Timberlake (1977) revealed no difference in either the average size of the two populations or their kinetic complexity. Similar conclusions have been drawn for *Schizophyllum commune* by Wessels' group (Zantinge et al., 1981; Hoge et al., 1982a). Although these studies could not entirely eliminate the existence of a small fraction of hnRNA that turned over rapidly, they did show that such a fraction would have to be small and probably not much larger than the final transcripts. Hereford and Rosbash (1977) also concluded that, if an mRNA precursor fraction existed in *S. cerevisiae*, it would have to be present at a very low concentration and consist of molecules only marginally larger than the processed mRNA.

Introns and mRNA Processing in Fungi

Too few fungal genes have been isolated and sequenced to permit confident generalization concerning the frequency of introns in the single-copy coding sequences of fungal genomes or their shared characteristics. It appears that, unlike the situation in higher animals, introns are the exception rather than the rule in fungal genes. Excluding mitochondrial genomes, introns have so far been found in the *S. cerevisiae* actin gene (Ng and Abelson, 1980), 13 of the genes for yeast ribosomal proteins (Woolford et al., 1979; Rosbasch et al., 1981; Bollen et al., 1982), yeast *MAT A* genes (see Langford et al., 1984), a *Schizosaccharomyces pombe* α-tubulin gene (see Langford et al., 1984), 3 *N. crassa* histone genes (Woudt et al., 1983) and the NADP glutamate dehydrogenase gene (Kinnaird and Fincham, 1983), a cellobiohydrolase gene of *Trichoderma* (Shoemaker et al., 1983), and a developmentally expressed gene of unknown function in *S. commune* (Dons et al., 1984a). In yeast a single, relatively small intron (300–500 NT) is usually found near the 5' terminus of the coding sequence; in the case of the actin gene this intron occurs between the third and fourth codons. The actin intron, like those of other organisms, has a GT dinucleotide at the 5' end and an AG dinucleotide at the 3' end; it also contains an internal 5'-TACTAACA-3' sequence that appears to be required for it to be properly removed by splicing (Langford and Gallwitz, 1983). It is interesting that the 16 known *S. cerevisiae* introns show a remarkable similarity in their sequences at the 5' and 3' splice sites, as well as the internal consensus sequence (5'-GTATGT . . . TACTAACA . . . $^{C}_{T}$AG-3') of the intron. The few introns reported for filamentous fungi are all quite small [50–100 base pairs (bp)] and in 3 of 4 more than one is present within the coding sequence [e.g., 1 of 67 nucleotide pairs (NTP) is histone 3 and 2 of 68 and 69 NTP in the histone 4 gene of *Neurospora* (Woudt et al., 1983); 2 of 66 and 61 NTP in the NADP glutamate dehydrogenase gene of *Neurospora* (Kinnaird and Fincham, 1983); 3 of 53, 49, and 49 NTP in the 1G2 gene sequence of *Schizophyllum* (Dons et al., 1984b); and 2 of 63 and 69 NTP in the cellobiohydrolase of *Trichoderma* (Shoemaker et al., 1983)]. These introns all have typical GT and AG dinucleotides at the 5' and 3' splice sites, respectively, but do not have internal sites that correspond closely with the internal intron consensus sequences of yeast except for some similarity in the *Neurospora* histone introns.

Saccharomyces cerevisiae contains small nuclear RNAs (snRNAs) whose sizes and base sequences (Wise et al. 1983) are nearly identical to those of the snRNAs of other eukaryotes (Busch et al., 1982). Unlike other organisms, however, the genes for these snRNAs in yeast are present as single copies instead of multiple copies (Wise et al. 1983). Despite the accumulating evidence that such small RNAs may have a role in mRNA transcript processing in other eukaryotic cell types, the one actually tested for this function in yeast did not cause lethality when the single gene was disrupted by a large insertion, transformed into a diploid, and examined for its effect in haploid ascospores (Tollervey et al., 1983). It is too soon to tell whether this is consistent with the suggestion (Pikielny et al., 1983) that the internal intron consensus sequences of yeast serves the putative function of the snRNAs in mRNA splicing.

While relatively little is known about the details of transcription and processing of fungal mRNAs (Hopper, 1984), it seems safe to conclude that introns are small and, with the exception of the ribosomal protein genes, relatively rare. Further, despite the lack of a significant fraction of high-molecular-weight nuclear RNA precursors and extensive intranuclear turnover, processing must take place to splice out the introns that do occur, including some small transcribed regions upstream from the coding sequences of the structural genes. The failure to detect larger mRNA precursors by *in vivo* labeling experiments with filamentous fungi is probably due to their very rapid synthesis and processing, as well as to the fact that the fractions processed out are small. Such small differences would be hard to detect in the presence of the large amounts of rRNA precursors that are rapidly synthesized and accumulate at the same time.

So few developmentally expressed fungal genes have been identified, cloned, and sequenced that much of the discussion of their transcriptional regulation is preliminary. Some analogies could perhaps be drawn from the system for cross-pathway control of amino acid biosynthetic enzymes in yeast (Hinnebusch and Fink, 1983). In this system, genes for a number of enzymes that are synthesized at increased levels as a result of depletion of an amino acid product of another pathway have been found to have the following upstream consensus sequence:

$$A_{\frac{A}{T}}GTGACTC$$

The sequence may occur in several copies and appears to be responsible for the increased transcription rate of the responsive genes resulting from the cross-pathway "signal." It is certainly tempting to speculate that analogous upstream sequences are involved in the regulation of developmentally specific genes [which would be a modification of the Davidson and Britten model (1983) with the difference that it does not invoke significant stretches of reiterated sequences upstream of the regulated structural gene]. Other even more hypothetical comparisons could be drawn from the well-studied nitrogen catabolite repression systems of *Neurospora* and *Aspergillus* (Marzluf, 1981). Sporulation is often blocked in fungi as long as a preferred, readily utilized nitrogen source such as NH_4^+ is available, and this may be due in part to nitrogen catabolite repression of developmental genes. If so, the transcription of such developmental genes might be controlled by upstream sequences that recognize signals from both the catabolite repression control system and other sequences that recognize signals related to developmental triggers [analogous to the induction of a nitrogen catabolic pathway or an enzyme that requires both the depletion of nitrogen, signaled by the product of a major regulatory gene, and an inducer for the specific catabolic pathway (Marzluf, 1981)].

The clustering of several genes transcribed during *A. nidulans* conidiation (Gwynne et al., 1984) and sporulation in *S. cerevisiae* (Clancy et al., 1983; Percival-Smith and Segal, 1984), discussed below, may indicate a mechanism for regulating blocks of contiguous developmentally expressed genes. However, in the one well-characterized developmental gene cluster from *A. nidulans,* genes are present whose transcripts are not limited to the conidiation phase (Gwynne et al., 1984).

STUDIES ON DEVELOPMENTALLY SPECIFIC GENE EXPRESSION

Approaches

Two different approaches have been used to examine the extent of differential gene expression in fungi: the spectrum of the proteins produced and nucleic acid hybridization. The number of different proteins being made can be estimated either by labeling *in vivo* or by *in vitro* protein synthesis (translation) using total cellular RNA or poly(A) RNA. Wheat germ and rabbit reticulocyte preparations freed of endogenous mRNA can translate heterologous mRNAs reasonably efficiently. These systems have therefore been used to assess the qualitative and quantitative spectra of mRNAs present in fungal cultures and during various stages of development. This method, like *in vivo* pulse-labeling, lacks the sensitivity to detect the products of mRNAs that are present at very low concentrations in cells. This is because labeling periods sufficiently long to detect rare mRNA products result in so much incorporation into proteins translated from abundant mRNAs that the latter swamp out the former when the two-dimensional acrylamide gels are radioautographed.

Hybridization experiments using either single-copy DNA or cDNA transcribed from isolated poly(A) RNA have also been used to estimate the fraction of the genome expressed during fungal development. When large excesses of RNA with tracer quantities of radiolabeled single copy DNA or cDNA are used, this technique is more sensitive for estimating the numbers of different unique mRNAs being translated (expressed via transcription and presumably translation during the stage at which the RNA was isolated), because it can in theory detect nearly all of the rare mRNAs if sufficiently high concentrations and long incubations are used (i.e., high R_0t's). However, in practice, even this technique can detect only fairly significant changes in mRNA populations [e.g., ca. 10% of the mRNA mass or 17% of the sequence complexity (Rozek and Timberlake, 1980)]. Neither two-dimensional gel systems for proteins nor hybridization results identify the gene products that might be involved. Instead, they provide an estimate of the magnitude of the differential gene expression that may be involved in a developmental process. Further, it is important to keep in mind that this work represents quite different aspects of development in various fungi of four different classes and that the entire developmental potential has been examined for no single fungus. Thus a certain caution is needed in trying to derive useful generalizations.

Developmentally Specific mRNAs and Proteins

Achlya

Two developmental processes have been examined in the aquatic oomycete *Achlya*: the production of "male" antheridial branches from hyphae induced by the steroid hormone antheridiole, and asexual sporulation induced by starvation of hyphae through a shift-down from a complete medium to a dilute $CaCl_2$ solution. Sexual reproduction in *Achlya* spp., including antheridial induction, and its analysis by various techniques have recently been considered in detail by Timberlake and Orr (1984), and their review should be consulted for details and an extended discussion of the system.

Antheridial hyphae are induced by, and grow trophically toward, the source of antheridiole. The slender antheridial branches begin to appear on the growth hyphae ca. 2–4 hr after hormone addition in the experimental systems usually used. As might be expected, because numerous slender hyphal branches are produced in response to the hormone, the process requires continued RNA and protein synthesis. The rates of both processes also increase as a result of hormone addition (Silver and Horgen, 1974, Horgen et al., 1975; Timberlake, 1976). Attempts to detect the synthesis of new proteins during the first several hours by one- and two-dimensional gel analysis have yielded conflicting results. Michalski (1978), using *in vivo* labeling, reported a number of quantitative changes and the appearance of two new proteins 40–60 min after hormone stimulation. In contrast, Gwynne and Brandhorst (1980) found no obvious differences between proteins synthesized *in vitro* or *in vivo* when they compared the protein patterns of vegetative and induced cultures.

A study of the mRNA patterns present during hormonal induction of antheridial hyphae by hybridization of excess poly(A) RNA with tracer cDNA by Rozek and Timberlake (1980) resulted in similar conclusions. Uninduced hyphae contained two abundance classes of poly(A) RNA: One, comprising 52% of the poly(A) RNA mass, consisted of ca. 440 diverse sequences represented about 50 times per genome, and a second, representing 48% of the poly(A) RNA mass, contained 7000–8000 different sequences with 2–3 copies per genome. The values obtained for poly(A) RNA from hormonally induced cultures were not significantly different. The observation that no measurable accumulation or loss of a significant complex class of messages could be detected was consistent with two-dimensional gel studies on proteins. These workers concluded that stage-specific mRNA transcription and translation, if it occurs, must involve relatively small numbers of low-abundance transcripts.

Zoosporangial development of *Achlya* also requires continued RNA and protein synthesis, and considerable turnover of both RNA and protein accompanies the process (Griffin and Breuker, 1969; Timberlake et al., 1973; Horgen and O'Day, 1975; Sutherland et al., 1976; Gwynne and Brandhorst, 1982). Gwynne and Brandhorst (1982) examined both the spectrum of proteins synthesized and the number of mRNAs expressed. As in the case of antheridial hyphae, two-dimensional gel patterns of *in vivo* pulse-labeled or *in vitro* translated proteins showed some changes in the intensity of different polypeptides but few if any new proteins. On the other hand, translation of encysted spore poly(A) RNA *in vitro* indicated the presence of 16 mRNAs for unique proteins not found in translations from differentiating sporangia. Some vegetative transcripts were also clearly reduced or absent. Since it was not possible to isolate good RNA from late stages of sporulation (presumably because of excessive nuclease activity), some of the new mRNAs may have been transcribed at that time. The numbers of abundant (300) and unique (8000) mRNA sequences they estimated for vegetative cells were in good agreement with those reported by Rozek and Timberlake (1980). Three-quarters of the way through sporulation (6 hr after induction or T_6) the poly(A) RNA sequence complexity was only marginally higher (9.0×10^6 versus 8.8×10^6 NT), but the poly(A) RNA complexity of encysted zoospores (1.7×10^7 NT) was almost twice as high, representing about 15,000 different average-sized sequences (40.6% of the genome) versus 8000 sequences (21% of the genome) in vegetative hyphae.

Although many details remain unclear, it seems evident that considerable differential gene expression may be involved in the later stages of *Achlya* sporulation, a phase that involves active zoospore differentiation within the sporangium. The relative lack of specific gene expression in the antheridial induction system probably reflects the rather subtle differences between large growing hyphae and small chemotropically growing branches. Much larger differences in gene expression might be found for the considerably more complex differentiation of the developing oogonia antheridium formation, fertilization, and maturation of oospores, all of which occur subsequently.

Blastocladiella

Sporulation in a second aquatic fungus, the nonfilamentous chytridiomycete *B. emersonii,* has also been examined (for reviews see Lovett, 1975, 1983). Unlike the situation in *Achlya* spp., where starvation induces the formation of elongate zoosporangia at the hyphal tips, starvation of *Blastocladiella* leads to conversion of the multinucleate semispherical growth-phase cell with a tuft of basal rhizoids into a sporangium within which the zoospores are subsequently differentiated. This developmental process also requires continued RNA and protein synthesis, while significant turnover of both RNA and protein occur. Net rRNA and tRNA syntheses both stop early (Lovett, 1983); poly(A) RNA synthesis and turnover occur throughout all but the last ca. 40 min of the 3.5-hr process (Jaworski and Thompson, 1980) and preformed proteins turn over at a rate of 12% per hour (Lodi and Sonneborn, 1974). The number of proteins synthesized during *B. emersonii* sporulation has not been examined by two-dimensional gel electrophoresis, but there is considerable evidence for the formation of at least five new proteins: an acid protease, an alkaline protease, two gamma particle proteins, and a basic protein associated with the zoospore nuclear cap (a membrane-bound ribosomal aggregate associated with the zoospore nucleus). Both proteases appear after induction, and the alkaline protease is released into the medium at the end of sporangium differentiation (Lodi and Sonneborn, 1974; Correa et al., 1978). The evidence for unique synthesis of the proteases during sporulation is based on *in vitro* enzyme assays in homogenates and is thus indirect, but protease function appears to be essential during the first 60–90 min (Correa et al., 1978).

Gamma particles are 0.5-μm-diameter, membrane-enclosed particles with an electron-dense core found in the cytoplasm of the motile zoospores. It was proposed that they sequester inactive chitin synthetase during this motile, wall-less stage (Myers and Cantino, 1974; Cantino and Mills, 1979; Cantino and Mills, 1983). Precursors of the electron-dense gamma particle cores first appear in the cytoplasm of sporulating cells before zoospore cleavage (Barstow and Lovett, 1975) and aggregate to form the mature particles during zoospore differentiation. They disappear again rapidly during early stages of zoospore germination by vesiculation in a process interpreted as secretion to the cell surface where the chitin synthetase is released and activated for cell wall chitin synthesis (Myers and Cantino, 1974; Cantino and Mills, 1979). We isolated and purified the two major proteins from the gamma particles. Using antisera prepared against the pure proteins, we found (by Western blotting of proteins after separation on polyacrylamide gels) that the 41-kilodalton and 43-kilodalton (kd) gamma particle proteins appeared in the cytoplasm at the same time that the gamma particle precursors were detected by electron microscopy (Hohn et al., 1984; Bartstow and Lovett, 1975). They also disappeared from the cytoplasm again during zoospore germination, as did the gamma particles. The latter would not be expected for an enzyme essential in cell wall synthesis, such as chitin synthetase, that is normally associated with the plasma membrane when active. These and other experiments (Dalley and Sonneborn, 1982; Hohn et al., 1984) have indicated that gamma particles are not the site of chitin synthetase storage in zoospores, as proposed by Cantino and coworkers, but have not uncovered the function of the gamma particle proteins. Nevertheless, the proteins of the particles are clearly indicated as products of differential gene expression. The 57-kd basic protein associated with the zoospore nuclear cap organelle also appears to be restricted to the spore stage. It has not been studied extensively but cannot be detected in the total extracted proteins of growth-phase cells (G. Brown and J. S. Lovett, unpublished observations).

Significant levels of differential gene expression occur in Blastocladiella during both spore germination and sporulation (Johnson and Lovett, 1984a, b). The nonsynthetic zoospores contain cryptic mRNA sequences equivalent to 43% of the total genomic unique base sequence complexity (6400 average-sized mRNAs) in the form of both poly(A) RNA (45%) and nonpoly(A) RNA (36%) with only a partial overlap (15%). When zoospores germinate, 2000–4000 additional mRNA sequences can be detected, depending on the richness of the growth medium; they decrease to about 47% of the genomic complexity as the cells enter the growth phase. An even larger fraction of the genomic information (80% of the sequence complexity of ca. 12,000 different mRNAs) appears to be expressed during at least two-thirds of the way through sporulation; nearly half of this mRNA complexity is lost in late sporulation, since the zoospores contain only 6400 different sequences. This sporulation phase with high levels of gene transcription was associated with rapid turnover of newly synthesized poly(A) RNA (Jaworski and Thompson 1980). Most of the newly made poly(A) RNA conserved in the differentiated zoospores was synthesized during the last 20–40 min of sporulation, a period which would roughly coincide with the loss of half of the mRNA previously associated with the sporulation-phase polysomes (Johnson and Lovett, 1984b). Although many new mRNAs are apparently transcribed during zoospore germination, the process can proceed through the early stages without new RNA synthesis. Two-dimensional gels of proteins pulse-labeled during germination revealed only two proteins that failed to appear when new RNA synthesis was blocked with actinomycin D (Lovett, 1983). Presumably, only the products of the most abundant mRNAs were detected by the 15-min pulses with radioactive amino acids.

Botryodiplodia and Sclerotinia Storage Protein

Two fungi, the imperfect fungus Botryodiplodia theobromae and the ascomycete Sclerotinia sclerotiorum, are known to accumulate large amounts of storage protein. During spore formation in B. theobromae pycnidia, a new protein of 20 kd accounts for about 25% of the total spore protein (Van Etten et al., 1979). This protein disappears during germination and growth. During development of the sclerotium (a resistant mass of aggregated hyphae) in S. sclerotiorum, a 36-kd protein accumulates to a level equal to 35–40% of the total cellular protein (Russo et al., 1982).

The same protein could not be detected in hyphae produced from mature sclerotia but was still present when the sclerotia produced apothecia (fruiting structures). In both cases the protein seemed to be a storage form, and their stage-specific accumulation must require a high level of expression of the structural genes during development. Such storage proteins are rare in fungi but were also found in a few other species of *Sclerotinia*.

Ascomycetes

The sporulation systems of three ascomycetes, *A. nidulans, N. crassa,* and *S. Cerevisiae,* have been examined for differential gene expression. In the case of *A. nidulans*, the patterns of protein synthesis, using *in vivo* labeling or *in vitro* translation during sporulation, have not been reported (for a review see Timberlake et al., 1983). This fungus produces a stage-specific protein, laccase I, involved in production of the green conidial pigment [a similar, but distinct enzyme, laccase II, may be similarly specific to sexual development (Hermann et al., 1983)]. The sporulation laccase I, a *p*-diphenol oxidase coded by the *yA* locus and lacking in yellow *ya* mutants, is absent from wild-type mycelia but present in conidiating colonies (Clutterbuck, 1972, 1977). Using a semisynchronized sporulation system induced by surface aeration of pregrown mycelial mats, Law and Timberlake (1980) demonstrated that the laccase protein was produced *de novo* and that its appearance and increased amounts and activity were closely correlated with the onset of conidiation. Enzyme production began with the appearance of maturing conidia and peaked during the period of maximal conidial production. RNA and protein synthesis were required to maintain the level of enzyme, suggesting that it was continually being turned over throughout sporulation.

In a study on differential gene expression in the *Aspergillus* system, Timberlake (1980) isolated enriched cDNA preparations by means of "cascade hybridization" beginning with sporulation-phase and conidial poly(A) RNA. These enriched cDNAs were then used for hybridization to estimate the number of different poly(A) RNA sequences present in vegetative hyphae, midsporulation cultures, and mature conidia. Vegetative cells expressed 26–27% of the total genomic sequence complexity (5600–6000 diverse, average-sized poly(A) RNA molecules). Sporulating cultures contained these sequences and an additional 1050 mRNAs that were absent from the growing hyphae or conidia, expressing 30.7% of the genome. The conidia were found to contain another 300 sequences unique to that stage (the total conidial poly(A) RNA complexity $\cong$ 30.1% of the genome). Thus about 1350 new genes representing 6.2% of the total genomic sequence complexity were estimated as developmentally specific with this approach.

Three stages of *Neurospora* have been examined by hybridization: the vegetative mycelia, conidia, and early growth phase (Dutta and Chaudhuri, 1975). Hybridizations with single-copy tracer DNA and total RNA from each stage indicated that log-phase vegetative cells expressed 34% of the genomic single-copy sequence complexity (ca. 7344 average-sized mRNAs), while early-growth-phase hyphae expressed 25% (ca. 5400 average-sized mRNAs). The conidia contained only 15% of the complexity (ca. 3240 mRNAs), or about half the amount estimated above for *A. nidulans* conidia. The lower figure for *N. crassa* may reflect poor recoveries of undegraded RNA as a result of the method of extraction used.

Sporulation in the yeast *S. cerevisiae* has been a subject of interest for many years, and the reader is refered to reviews for detailed discussions of the subject (Esposito and Klapholz, 1981; Berry, 1983; Dawes, 1983; Klar et al., 1984). Yeast sporulation has virtues and drawbacks as a system for developmental analysis. Its main advantages are its well-defined genetic system with many mapped genes (true also of *Aspergillus* and *Neurospora*), simple morphology, and ease of manipulation. A disadvantage is that the shift to an unbalanced acetate medium without a nitrogen source, necessary for induction, is essentially a starvation induction system that results in significant metabolic adjustment and turnover (also true of aquatics). Also, two overlapping processes are involved, meiosis and spore formation. Finally, the synchrony of sporulation is at best not very high, with rarely more than 80% of the cells actually completing the process. The results of the metabolic shifts due to the induction method may be assessed by the use of homozygous mating type diploids (α/α, a/a) as controls. These cells do not sporulate but show most of the same early metabolic changes found for the sporulation-competent a/α diploids. About 18 hr is required for the formation of mature spores; premeiotic DNA synthesis and the meiotic divisions occur between $T_{1.5}$ and $T_{9.5}$ after induction, followed by formation of the prespore units and spore wall formation. Total RNA and protein levels increase from T_0 to T_4 and then decrease, while glycogen accumulates from T_0 to $T_{7.5}$ and is then reutilized during spore formation. Extensive RNA and protein turnover occur throughout most of sporulation and, although the poly(A) RNA content as a percentage of the total RNA remains the same, the translation activity decreases 30% by T_4 and 47% by T_8 (Kraig and Haber, 1980; Weir-Thompson and Dawes, 1984).

Several studies have been made on sporulation protein synthesis in yeast using either *in vivo* labeling and gel electrophoresis or *in vitro* translation followed by electrophoresis. The results have been somewhat contradictory and confusing because of the different methods used for labeling the proteins. Wright et al. (1981) used $^{35}SO_4$ labeling during presporulation growth and reported changes in 45 proteins during sporulation. Of these, 21 were observed only in a/α diploids; but modifications could not be eliminated as a cause of some changes rather than *de novo* synthesis because of the prelabeling procedure. Kraig and Haber (1980), using pulse labeling with [^{35}S]methionine during sporulation, observed changes in the amounts of some proteins but concluded that no new proteins could be detected among the 400 proteins compared on two-dimensional gels. Weir-Thompson and Dawes (1984) translated the total RNA from various stages in a reticulocyte lysate system and found a number of differences when the a/α diploids were compared with the a/a and α/α controls. Eighty proteins changed in intensity, with 43 unique to the a/α cells, and four new proteins were produced only in translations from the a/α cells (five others were found in all cells). Most of the changes were observed between T_4 and T_{12}. Finally, Kurtz and Lindquist (1984) also translated total RNA from sporulating cultures, but used a wheat germ system for translation. From T_4 to T_6 a number of vegetative proteins diminished or were absent from two-dimensional gels, and two (26 and 84 kd) proteins that comigrated with heat shock proteins also appeared. After T_6 eight new proteins appeared, reached a maximum by T_{8-14}, and gradually disappeared, and two additional proteins appeared after T_{16}. The only new proteins shared by the homozygous mating type controls were the two heat shock proteins. Comparable results were reported for several different a/α diploids. These authors suggested that the lower number of unique proteins detected by Weir-Thompson and Dawes (1984) was due to the use of a reticulocyte system in which signal recognition peptides may have blocked the translation of the secretory proteins with leader sequences (e.g., proteins that might be secreted into the area of wall formation lying between membrane layers). As in the higher fungi examined, relatively few new high-abundance mRNAs are produced during yeast sporulation. To date only three stage-specific proteins clearly associated with sporulation have been identified, an α-glucosidase involved in glycogen degradation (Colonna and Magee, 1978; Clancy et al., 1982), a 1, 3-β-glucanase (Del Rey et al., 1980), and a spore surface antigen (Dawes et al., 1983). A detailed study on the number of unique mRNA sequences expressed, using the hybridization method, has not been reported to the authors' knowledge, although Esposito and Klapholz (1981) quote a preliminary report of Mills that ca. 7% of the cDNA from ascospores failed to hybridize with vegetative or T_{12} sporulation-phase poly(A) RNA. Vegetative cells contain poly(A) RNA sequences representing 40% of the genomic sequence complexity (ca. 4000 average-sized mRNAs) (Hereford and Rosbash, 1977), and a level of stage-specific expression typical of other fungi might be expected.

Asexual reproduction is relatively rare in higher basidiomycetes that depend mainly on sexually produced basidiospores for dispersal. This means that studies comparable to those for sporulation in other fungi are not available. However, Wessels and his co-workers have used a similar approach to study early stages in the sexual process of the mushroom *S*.

commune. In this fungus, the beginning of the sexual cycle is initiated by the fusion of two compatible monokaryons (mycelia with haploid uninucleate cells) to produce a dikaryon with special characteristics (a specialized heterokaryon with one nucleus from each parental mono-karyon, conjugate nuclear division, and clamp connections around the cell septa). This dikaryon can grow indefinitely but can be induced to begin fruiting rather synchronously under appropriate conditions (for details and reviews see Wessels et al., 1981, 1985; Raper, 1983). Strains isogenic except for the mating type loci can be used to produce dikaryons and serve as controls in studies on differential gene expression, since they do not express developmentally specific genes for fruiting.

Hoge et al. (1982a) found no significant differences between mono-karyons and dikaryons when either the total RNA or poly(A) RNA was hybridized with tracer DNA. Both cell types contain about 10,000 dif-ferent rare mRNAs and 500 abundant mRNA species. Two-dimensional polyacrylamide gel analysis of *in vitro* translation products also failed to reveal any differences among the 500 proteins analyzed. In a second study (Hoge et al., 1982b) dikaryons that had begun to form many mushroom primordia (small aggregates of hyphae) were examined by the same methods. In this case cDNA prepared from the dikaryon with primordia contained additional sequences (5%, or equivalent to ca. 12 abundant and 610 rare mRNA sequences) that were not present in con-trol monokaryons or the preprimordial dikaryotic mycelia. However, few if any of the vegetative sequences were absent from the early-stage fruit-ing dikaryon. Two-dimensional gel studies on the proteins synthesized with RNA from the same monokaryons was translated. Finally, when De Vries and Wessels (1984) examined the proteins labeled *in vivo* on two-dimensional gels, 22 new proteins were found for the preprimordial vege-tative dikaryons and an additional 15 in the primordia. Interestingly, 7 of these new proteins were excreted, and 2 were specifically associated with the cell wall fraction. Although the functions of none of the dikaryon- or primordia-specific proteins are known, a possible role in hyphal adhesion to form mycelial aggregates was suggested for the wall-associated pro-teins.

An Assessment of Developmentally Regulated Gene Expression

Although the data on sporulation mRNA and protein synthesis are in-complete, it is possible to draw some tentative conclusions. The results of both hybridization and two-dimensional gel experiments indicate that a large number of genes are expressed throughout sporulation, and proba-bly throughout other developmental processes. Most of these are for growth-phase gene products, since a surprisingly small number of new abundant-class mRNAs are transcribed and translated during the devel-opmental processes examined. This observation is consistent with the suggestion made that most of the metabolic and biosynthetic activities needed for development and differentiation are also utilized for growth, with many of the important changes more likely to be associated with regulatory controls. A quite different picture emerges with regard to unique or low-abundance mRNAs whose products cannot be detected by two-dimensional gel systems or individually by hybridization. Hybridization experiments clearly point to the developmental expression of a significant number of genes whose transcriptional products are present at low concentrations and can only be detected individually by more sensitive techniques. The numbers of new mRNAs estimated from hybridization experiments are certainly impressive, ranging from 1300 to 7000 different sequences. It is difficult to surmise what roles so many new proteins might play in fungal development. It appears that, if the numbers reflect functional gene products, the effects of many of these genes must be rather subtle, since so few mutants have been identified by standard genetic techniques. This might also imply that many do not have central or essential functions, because they would be expected to show up as conditional mutants.

IDENTIFICATION OF DEVELOPMENTALLY EXPRESSED GENES BY CLONING PROCEDURES

Cloning Strategies

It is evident, from the work described in the preceding sections, that surprisingly few genes or gene products expressed uniquely in fungal development have been identified. This might mean that only a small number of unique gene products are required or that they are mostly genes whose mRNAs and protein products are produced at very low concentrations in the cells. At this point it is probable that relatively few new products are produced in significant amounts (e.g., pigments, wall components, storage materials, and any new enzymes needed for their synthesis), while a much larger number of gene products may be required for more subtle changes in regulating metabolism and biosynthesis. The relatively abundant new products are most likely those detected with the two-dimensional protein gels of *in vivo* or *in vitro* labeled proteins. The presence of rare gene products is nevertheless suggested by hybridization data.

To examine the problem of low-abundance mRNAs in development, more sensitive methods are necessary and have been provided by the rapid advances in molecular genetic technology—cloning procedures. This approach has advanced most rapidly in the analysis of cell structure and function in the single-celled yeast *S. cerevisiae*; however, relatively little attention has been directed to the use of cloning techniques to exam-

ine yeast sporulation. The molecular genetics of filamentous fungi such as
A. nidulans and *N. crassa*, which also have excellent haploid genetic systems, has developed more slowly because of additional difficulties. With
the recent reports of successful transformation with these species,
progress should be much more rapid.

A general cloning approach is to prepare either a genomic library
(from random fragments generated with restriction enzymes) or a cDNA
library prepared from a developmental-stage poly(A) RNA prepara-
tion(s) in various lambda phage or bacterial plasmid vectors. After the
host bacterium has been transformed, the cells are plated on selective
media and the small colonies are transferred to replica nitrocellulose
filters and probed by hybridization with ^{32}p-labeled cDNAs. Probes
prepared by the transcription of poly(A) RNA from vegetative cells and
developmental stages are used to detect colonies (clones) that carry vectors with DNA inserts that give a differential hybridization signal (the
presence or absence of label or a significant difference in intensity).
After plaque purification of such clones and retesting to eliminate false
positives, the isolated clones can be used in a number of ways. Usually
they are first tested by dot blot hybridization with RNA from different
stages to show the level of expression during growth and development.
Positive clones are then further analyzed by Northern blots in which the
labeled sequence (or often the entire vector DNA) is hybridized with
filters to which the RNA has been transferred after separation of the
RNAs by polyacrylamide gel electrophoresis. This can detect the presence of more than one mRNA transcript with a sequence complementary to a particular cloned sequence of genomic DNA and, if RNAs of
different stages are used, whether all the contiguous genes on the cloned
DNA are expressed at the same time(s) in the growth and developmental cycle. If the small fraction of developmental genes were distributed randomly throughout the genome, most isolated genomic clones of
moderate-size ($\pm$ 10,000 NTP) would be expected to contain vegetative
sequences as well as the developmental one used for selection. If developmentally specific clones from a cDNA library are identified, these can
in turn be used to probe genomic libraries in order to isolate the actual
gene sequences. Clones of sequences at or near the end of a gene or
group of genes can also be used as probes to isolate adjacent DNA
sequences via chromosome "walking."

Isolation of Developmentally Regulated Genes

The procedures just described are being used for the isolation of developmentally expressed genes of fungi. Dons et al. (1984a) cloned cDNAs
prepared from primordia-stage dikaryons of *S. commune* in the plasmid

pBR327 and screened the library with cDNA probes from the dikaryon and from isogenic monokaryons. Although a number of clones showed a decrease in mRNAs during the dikaryon stage, no dikaryon-specific clones were identified in the initial screening step. Nevertheless, one sequence was subsequently identified (clone 1G2) that was exclusively expressed in the dikaryon; the amount of 1G2 mRNA increased ca. 20-fold during establishment of the fruiting primordia. This sequence was then used to isolate a 9-kb genomic sequence (Dons et al., 1984b) that codes for a 13-kd protein and contains three small introns. The function of the protein is not known, though its small size, the relative abundance of its mRNA in dikaryons (11,000 copies per cell), and time of appearance suggest a relationship to the proteins secreted during the primordial stage (De Vries and Wessels, 1984).

Cloned genomic DNA libraries have been used to isolate sporulation-specific gene sequences from yeast. Clancy et al. (1983) cloned Mbo-restricted DNA fragments in lambda Charon 28 bacteriophage and screened replicate filters from plated colonies using cDNAs prepared from RNA of 7-hr sporulating a/α diploids and α/α nonsporulating diploids incubated in a sporulating medium. Six cloned fragments were isolated and examined further. Restriction analysis of these clones showed four distinct cloned sequences and of these, three carried sequences for two different sporulation mRNAs when examined by Northern hybridization with sporulation-stage poly(A) RNA. Percival-Smith and Segal (1984) used pBR322 to isolate 38 clones from a/α- α/α-, and a/a-derived cDNAs. Based on differential hybridization and restriction analysis, these fell into 15 groups (each with a common subfragment in restriction digests). Fourteen were spore-specific, 3 carried two spore-specific sequences, and 3 were expressed in the vegetative cells of all diploids but only in the a/α diploid under sporulation conditions. Using an actin probe as a control for the levels of expression, these authors estimated that the genes they had cloned produced a moderately abundant class of mRNAs. In both studies the transcripts of the cloned sequences appeared about 5–7 hr after induction (the period of meiosis). None of the functions are yet known, but the close linkage of developmentally expressed genes is of considerable interest. Although the multiple band of different molecular weights that were detected by Northern blots could represent precursors and products of the same mRNAs, the lack of hnRNA or significant processing (see the section on transcription and processing) suggests that this is an unlikely explanation.

In the case of *Blastocladiella*, we have taken a somewhat different approach and begun experiments to isolate the gene sequences for three known stage-specific and organelle-specific proteins, the two major gamma particle proteins and the nuclear cap-associated protein of the zoospore. A genomic library has been prepared in a lambda gt11 and replica filters probed with antisera against the purified proteins (J. Horak, T. M. Hohn, and J. S. Lovett, unpublished observations). Two different sequences have been cloned for the gamma particle proteins (based on dissimilar restriction patterns), and two for the 57-kd nuclear cap protein. The clones have yet to be analyzed by hybridization against RNA from different developmental stages to assess the specificity of expression during sporulation or possible linkage with other developmentally expressed genes.

Sporulation-specific genes have been examined most extensively by Timberlake and co-workers using the semisynchronous conidiation system of *A. nidulans*. Zimmerman et al. (1980) prepared a genomic library in lambda Charon 4A and screened it with cDNA probes prepared from sporulation-phase poly(A) RNA. The poly(A) RNA was greatly enriched for sporulation sequences by cascade hybridization which removed most of the poly(A) RNA species also present in vegetative mycelia (Timberlake, 1980). Of the 400 clones isolated, 350 contained sequences selectively accumulated or uniquely expressed in conidiating cultures. Four were examined in detail and found to contain more than one developmentally expressed sequence using Northern blot hybridization. Orr and Timberlake (1982) assessed random and nonrandom (clustered) distributions of sporulation-specific genes in genomic DNA and found the distribution to be significantly nonrandom; i.e., 80% of the selectively isolated sporulation-specific sequences were found in close association with one or more additional sporulation genes. Thus in *A. nidulans* a significant degree of clustering occurs for developmentally expressed genes.

One *Aspergillus* gene cluster, *AnSpoCl*, coded for five complete and one partial sequence of mRNAs found only in mature conidia (sizes 2.5, 2.3, 1.6, 0.86, 0.75 and 0.59 kb) (Timberlake and Barnard, 1981). This study was extended by chromosome walking in which the ends of existing probes were used to select for overlapping cloned sequences in the library. Five such overlapping clones permitted analysis of 53 kb of linear sequence (Gwynne et al., 1984) which codes for 19 poly(A) RNAs. A major point of interest is that the location and polarity of the sequences were established and 14 of the 19 genes were differentially expressed during sporulation. Thirteen produced poly(A) RNAs found only in conidia and one found only during sporulation. The remaining genes were expressed at all stages including the growth phase. The actual cluster encompasses 37 kb, with each end delimited by a 1.1-kb repeat sequence. The position within the cluster is not directly correlated with the time of expression (i.e., the earliest expression occurs with the sporulation gene located in the center of the group of conidium-specific genes). It is not intuitively evident how such a clustering could lead to coordinate regulation of transcription by a simple chromatin-level control mechanism (for discussion see Gwynne et al., 1984), although some sort of coordinate regulation is an attractive possibility.

It is perhaps important to emphasize that, with the exception of the *yA* gene product, laccase I, little or nothing is known concerning the functions of proteins encoded in the sequences isolated from the various fungi just discussed. It seems reasonable to conclude that significant numbers of developmentally expressed genes do exist and can be identified. However, the numbers are still small relative to the predictions derived from hybridization experiments. The higher yield of such clones isolated from *A. nidulans* by Timberlake and co-workers was probably due to the use of probes from enriched stage-specific poly(A) RNAs that greatly increased the concentrations of lower-abundance species. This encourages optimism about the future prospects of isolating many more developmental genes.

CONCLUDING REMARKS

This discussion of molecular approaches to fungal development in a quantitative sense might suggest that only modest progress has been made toward the ultimate goal of understanding the mechanisms of regulation at the cellular and subcellular levels. On the other hand, important progress has been made which supports an optimistic view for much faster progress in the future. The data now available provide rather convincing evidence that differential gene expression has a significant role in the development and differentiation of fungi. Until relatively recently, saying this would have been more an act of faith than a reading of the evidence. It also seems probable that the numbers of such genes, as suggested by hybridization studies, may indeed be larger than the earlier estimates based on the results of transmission genetics. How much larger a cohort of genes is involved and the roles of these genes remain to be established. Still there seems little reason to discard the useful hierarchical levels of developmental gene action proposed by Clutterbuck (1978) for *A. nidulans* based on mutant studies (i.e., tactical, strategic, auxiliary, and support loci).

The genes coding for synthetic enzymes for new pathways such as pigment formation or other unique products (auxiliary) should be the most amenable to analysis. Others with regulatory functions may be harder to detect and analyze because of their possibly subtler effects in a complex system. During fungal sporulation considerable translocation of material from adjacent hyphal cells must occur to support the development of aerial structures. This could require significant turnover even in cases of surface sporulation on growth-supporting media. Turnover is obviously an important phenomenon in aquatic fungi that "remodel" a preexisting cell into a sporangium in which zoospore differentiation occurs. The production of new pathways for secondary metabolite synthesis is also a concomitant of sporulation to some fungi (Campbell, 1983; Shepherd and Carels, 1983). The spores finally produced by fungi vary considerably in their degree of dormancy, but even the least dormant must have mechanisms for modulating or repressing many metabolic and biosynthetic pathways until germination begins and then reversing this regulation. This may involve the production and release or destruction of self-inhibitors, but little is known about how regulation is achieved in most fungal spores (Dahlberg and Van Etten, 1982). It certainly seems possible that the control and coordination of so many systems involves a significant number of specific gene products. Some of these could have roles in the immediate responses to changes in the environment that initiate sporulation, without being necessarily developmentally specific. Others could be involved in subsequent developmental responses triggered by the change (e.g., turnover, allosteric effectors, new enzymes, new protein products, inhibitors, and a protein modification such as phosphorylation-dephosphorylation).

The existence of a few genes with essential functions in switching from growth to a developmental pathway (e.g., sporulation competence; Yager et al., 1982; Butnick et al., 1984a, b) and genes that may operate at specific stages in development to modify the next type of cell(s) to be produced (Clutterbuck, 1977) seems evident from mutant studies. Since these genes are fundamental for the completion of development, their developmental analysis represents an important objective. The development of useful systems for the transformation of filamentous fungi (Ballance et al., 1983; Case et al., 1979; Kinsey and Rambosek, 1984; Stohl and Lambowitz, 1983; Tilburn et al., 1983; Yelton et al., 1984b), as well as those already established for yeast, including shuttle vectors, opens the way for the isolation of these kinds of regulatory genes by gene replacement techniques (B. Miller et al., personal communication; Yelton and Timberlake, 1985). This has already been demonstrated for several structural genes from yeast, e.g., actin (Ng and Abelson, 1980), cell cycle genes (Nasmyth and Reed, 1980) and RNA-processing genes (Soltyk et al., 1984). Once the regulatory genes have been isolated, it will be possible with available technology to isolate their mRNAs and the proteins translated from them and to establish the protein amino acid sequences and examine their expression, and so on, without knowledge of their specific functions.

It is unlikely, of course, that the identification of a polypeptide and a description of its temporal pattern of synthesis will be sufficient to explain its function. Rather, it will be necessary at a minimum to use a variety of biochemical and cell biological techniques to establish its location in cells, the system or systems with which it is associated, and how these systems are related to other pathways and systems. Because of the complex interrelationships that must exist between the large numbers of pathways, structures, and molecules involved in the development and maintenance of cell structure, even such a joint effort is likely to proceed slowly. However, it is possible that the most rapid advances in fungal developmental biology will come when the powerful techniques and methodologies of molecular biology are used in collaboration with these other more conventional approaches to cell function.

REFERENCES

BALLANCE, D. J., BUXTON, F. P., AND TURNER, G. 1983. Transformation of *Aspergillus nidulans* by the orotidine-5'-phosphate decarboxylase gene of *Neurospora crassa, Biochem. Biophys. Res. Commun.* **112**:284–289.

BARSTOW W. E., AND LOVETT, J. S. 1975. Formation of gamma particles during zoosporogenesis in *Blastocladiella emersonii, Mycologia* **67**:518–529.

BARTNICKI-GARCIA, S., RUIZ-HERRERA, J., AND BRACKER, C. E. 1979. Chitosomes and chitin synthesis, in: *Fungal Walls and Hyphal Growth* (J. H. Burnett and A. P. J. Trinci, eds.), pp. 149–168, Cambridge University Press, Oxford.

BEEVER, R. E., RIDGWELL, R. J., AND DEMPSEY, G. P. 1979. Purification and chemical characterization of the rodlet layer of *Neurospora crassa* conidia, *J. Bacteriol.* **140**:1063–1070.

BERRY, D. R. 1983. Ascospore formation in yeast, in: *Fungal Differentiation: A Contemporary Synthesis* (J. E. Smith, ed.), pp. 147–173, Marcel Dekker, New York.

BOLLEN, G. H. P. M., MOLENAAR, C. M. T., COHEN, L. H., VAN RAAMSDONK-DUIN, M. M. C., MAGER, W. H., AND PLANTA, R. J. 1982. Ribosomal protein genes of yeast contain intervening sequences, *Gene* **18**:29–37.

BRAMBL, R., DUNKLE, L. D., AND VAN ETTEN, J. L. 1978. Nucleic acid and protein synthesis during fungal spore germination, in: *The Filamentous Fungi: Vol. III, Developmental Mycology* (J. E. Smith and D. R. Berry, eds.), pp. 94–118, John Wiley, New York.

BRODY, S. 1981. Genetic and biochemical studies on *Neurospora* conidia germination and formation, in: *The Fungal Spore: Morphogenetic Controls* (G. Turian and H. R. Hohl, eds.), pp. 605–626, Academic Press, London.

BUSCH, H., REDDY, R., ROTHBLUM, L., AND CHOI, Y. C. 1982. SnRNAs, snRNPs, and RNA processing, *Annu. Rev. Biochem.* **51**:617–654.

BUTNICK, N. Z., YAGER, L. N., HERMANN, T. E., KURTZ, M. B., AND CHAMPE, S. P. 1984a. Mutants of *Aspergillus nidulans* blocked at an early stage of sporulation secrete an unusual metabolite, *J. Bacteriol.* **160**:533–540.

BUTNICK, N. Z., YAGER, L. N., KURTZ, M. B., AND CHAMPE, S. P. 1984b. Genetic analysis of mutants of *Aspergillus nidulans* blocked at an early stage of sporulation, *J. Bacteriol.* **160**:541–545.

CABIB, E. 1981. Chitin: Structure, metabolism and regulation of biosynthesis, in: *Plant Carbohydrates II: Extracellular Carbohydrates* (W. Tanner and F. A. Loewus, eds.), pp. 395–415, Encyclopedia of Plant Physiology, Vol. 13B, Springer-Verlag, Berlin.

CABIB, E., BOWERS, B., AND ROBERTS, R. L. 1983. Vectorial synthesis of a polysaccharide by isolated plasma membranes, *Proc. Natl. Acad. Sci. USA* **80:**3318–3321.

CAMPBELL, I. M. 1983. Correlation of secondary metabolism and differentiation, in: *Secondary Metabolism and Differentiation in Fungi* (J. W. Bennett and A. Ciegler, eds.), pp. 55–72, Marcel Dekker, New York.

CANTINO, E. C., AND MILLS, G. L. 1979. The gamma particle in *Blastocladiella emersonii:* What is it? in: *Viruses and Plasmids in Fungi* (P. Lemke, ed.), pp. 441–484, Marcel Dekker, New York.

CANTINO, E. C., AND MILLS, G. L. 1983. The Blastocladialean gamma particle: Once viral endosymbiant (?), now "chitosome" progenitor, in *Fungal Differentiation: A Contemporary Synthesis* (J. E. Smith, ed.), pp. 175–210, Marcel Dekker, New York.

CASE, M. E., SCHWEIZER, M., KUSHNER, S. R., AND GILES, N. H. 1979. Efficient transformation of *Neurospora crassa* by utilizing hybrid plasmid DNA, *Proc. Natl. Acad. Sci. USA* **76:**5259–5263.

CLANCY, M. J., SMITH, L. M., AND MAGEE, P. T. 1982. Developmental regulation of a sporulation-specific enzyme activity in *Saccharomyces cerevisiae, Mol. Cell. Biol.* **2:**171–178.

CLANCY, M. J., BUTEN-MAGEE, B. STRAIGHT, D. J., KENNEDY, A. L., PARTRIDGE, R. M., AND MAGEE, P. T. 1983. Isolation of genes expressed preferentially during sporulation in the yeast *Saccharomyces cerevisiae., Proc. Natl. Acad. Sci., USA* **80:**3000–3004.

CLUTTERBUCK, A. J. 1972. Absence of laccase from yellow-spored mutants of *Aspergillus nidulans, J. Gen. Microbiol.* **70:**423–435.

CLUTTERBUCK, A. J. 1977. The genetics of conidiation of *Aspergillus nidulans,* in: *The Physiology and Genetics of Aspergillus* (J. E. Smith and J. A. Pateman, eds.), pp. 305–317, Academic Press, New York.

CLUTTERBUCK, A. J., 1978, Genetics of vegetative growth and asexual reproduction, in: *The Filamentous Fungi,* Vol. III, *Developmental Mycology* (J. E. Smith and D. R. Berry, eds.), pp. 240–256, John Wiley, New York.

COLE, G. T., AND POPE, L. M. 1981. Surface wall components of *Aspergillus niger* conidia, in: *The Fungal Spore: Morphogenetic Controls* (G. Turian and H. R. Hohl, eds.), pp. 195–215, Academic Press, London.

COLONNA, W. J., AND MAGEE, P. T. 1978. Gluconeogenic enzymes in sporulating yeast, *J. Bacteriol.* **134:**844–853.

CORREA, J. U., LEMOS, E. M., AND LODI, W. R. 1978. Inhibition of sporulation in the water mold *Blastocladiella emersonii, Dev. Biol.* **66:**470–479.

DAHLBERG, K. R., AND VAN ETTEN, J. L. 1982. Physiology and biochemistry of fungal sporulation. *Ann. Rev. Phytopathol.* **20:**281–301.

DALLEY, N. E., AND SONNEBORN, D. R. 1982. Evidence that *Blastocladiella emersonii* zoospore chitin synthetase is located in the plasma membrane, *Biochim. Biophys. Acta* **686**:65–76.

DAVIDSON, E. H., AND BRITTEN, R. J. 1983. Organization, transcription, and regulation in the animal genome, *Rev. Biol.* **48**:565–613.

DAWES, I. W. 1983. Genetic control and gene expression during meiosis and sporulation in *Saccharomyces cerevisiae,* in: *Yeast Genetics: Fundamental and Applied Aspects* (J. F. T. Spencer, D. M. Spencer, and A. R. W. Smith, eds.), pp. 29–64, Springer-Verlag, New York.

DAWES, I. W., DONALDSON, S., EDWARDS, R., AND DAWES, J. 1983. Synthesis of a spore-specific surface antigen during sporulation of *Saccharomyces cerevisiae, J. Gen. Microbiol.* **129**:1103–1108.

DEL REY, F., SANTOS, T., GARCIA-ACHA, I., AND NOMBELA, C. 1980. Synthesis of β-glucanases during sporulation in *Saccharomyces cerevisiae:* Formation of a new sporulation-specific 1,3-β-glucanase, *J. Bacteriol.* **143**:621–627.

DE VRIES, O. M. H., AND WESSELS, J. G. H. 1984. Patterns of polypeptide synthesis in non-fruiting monokaryons and a fruiting dikaryon of *Schizophyllum commune, J. Gen. Microbiol.* **130**:145–154.

DONS, J. J. M., DE VRIES, O. M. H., AND WESSELS, J. G. H. 1979. Characterization of the genome of the basidiomycete *Schizophyllum commune, Biochim. Biophys. Acta* **563**:100–102.

DONS, J. J. M., SPRINGER, J., DEVRIES, S. C., AND WESSELS, J. G. H. 1984a. Molecular cloning of a gene abundantly expressed during fruiting body initiation in *Schizophyllum commune. J. Bacteriol.* **157**:802–808.

DONS, J. J. M., MULDER, G. H., ROUWENDAL, G. J. A., SPRINGER, J., BREMER, W., AND WESSELS, J. G. H. 1984b. Sequence analysis of a split gene involved in fruiting from the fungus *Schizophyllum commune, EMBO J.* **3**:2101–2106.

DUTTA, S. K., AND CHAUDHURI, R. K. 1975. Differential transcription of nonrepeated DNA during development of *Neurospora crassa, Dev. Biol.* **43**:35–41.

ESPOSITO, R. E., AND KLAPHOLZ, S. 1981. Meiosis and ascospore development, in: *The Molecular Biology of the Yeast Saccharomyces: Life Cycle and Inheritance* (J. N. Strathern, E. W. Jones, and J. R. Broach, eds.), pp. 27–58, Cold Spring Harbor Laboratory, Cold Spring Harbor, New York.

ESPOSITO, R. E., FRINK, N., BERNSTEIN, P., AND ESPOSITO, M. S. 1972. The genetic control of sporulation in *Saccharomyces.* II. Dominance and complementation of mutants of meiosis and spore formation, *Mol. Gen. Genet.* **114**:241–248.

FIRTEL, R. A., KINDLE, K., AND HUXLEY, M. P. 1976. Structural organization and processing of the genetic transcript in the cellular slime mold *Dictyostelium discoideum, Fed. Proc.* **35**:13–22.

FOUQUET, H., BIERWEILER, B., AND SAUER, H. W. 1974. Reassociation kinetics of nuclear DNA from *Physarum polycephalum, Eur. J. Biochem.* **44**:407–410.

GOODAY, G. W. 1981. Biogenesis of sporopollenin in fungal spore walls, in: *The Fungal Spore: Morphogenetic Controls* (G. Turian and H. R. Hohl, eds.), pp. 487–505, Academic Press, London.

GOODAY, G. W. 1983. Hormones and sexuality in fungi, in: *Secondary Metabolism and Differentiation in Fungi* (J. W. Bennett and A. Ciegler, eds.), pp. 239–266, Marcel Dekker, New York.

GRESSEL, J., AND RAU, W. 1983. Photocontrol of fungal development, in: *Encyclopedia of Plant Physiology*, Vol. 16B, (W. Ruhland et al., eds.) Springer-Verlag, New York, pp. 603–639.

GRIERSON, D. 1977. The nucleus and the organization and transcription of nuclear DNA, in: *The Molecular Biology of Plant Cells* (H. Smith, ed.), pp. 213–255, Blackwell, London.

GRIFFIN, D. H., AND BREUKER, C. 1969. Ribonucleic acid synthesis during the differentiation of sporangia in the water mold *Achlya, J. Bacteriol.* **98:**689–696.

GWYNNE, D. I., AND BRANDHORST, B. P. 1980. Antheridiole-induced differentiation of *Achlya* in the absence of detectable synthesis of new proteins, *Exp. Mycol.* **4:**251–259.

GWYNNE, D. I., AND BRANDHORST, B. P. 1982. Changes in gene expression during sporangium formation in *Achlya ambisexualis, Dev. Biol.* **91:**263–277.

GWYNNE, D. I., MILLER, B. L., MILLER, K. Y., AND TIMBERLAKE, W. E. 1984. Structure and regulated expression of the SpoC1 gene cluster from *Aspergillus nidulans, J. Mol. Biol.* **180:**91–110.

HAMES, B. D., AND ASHWORTH, J. M. 1974. The metabolism of macromolecules during the differentiation of myxamoebae of the cellular slime mould *Dictyostelium discoideum* containing different amounts of glycogen, *Biochem. J.* **142:**301–315.

HARSHEY, R. M., JAYARAM, M., AND CHAMBERLIN, M. E. 1979. DNA sequence organization in *Phycomyces blakesleeanus. Chromosoma* **73:**143–151.

HEREFORD, L. M., AND ROSBASH, M. 1977. Number and distribution of polyadenylated sequences in yeast, *Cell* **10:**453–462.

HERMANN, T. E., KURTZ, M. B., AND CHAMPE, S. P. 1983. Laccase localized in hull cells and cleistothecial primordia of *Aspergillus nidulans, J. Bacteriol.* **154:**955–964.

HINNEBUSCH, A. G., AND FINK, G. R. 1983. Repeated DNA sequences upstream from HIS 1 also occur at several other co-regulated genes in *Saccharomyces cerevisiae, J. Biol. Chem.* **258:**5238–5247.

HOGE, J. H. C., SPRINGER, J., ZANTINGE, J., AND WESSELS, J. G. H. 1982a. Absences of differences in polysomal RNAs from vegetative monokaryons and dikaryotic cells of the fungus *Schizophyllum commune, Exp. Mycol.* **6:**225–232.

HOGE, J. H. C., SPRINGER, J., AND WESSELS, J. G. H. 1982b. Changes in complex RNA during fruit-body initiation in the fungus *Schizophyllum commune, Exp. Mycol.* **6:**233–243.

HOHN, T. M., LOVETT, J. S., AND BRACKER, C. E. 1984. Characterization of the major proteins in gamma particles, cytoplasmic organelles in *Blastocladiella emersonii* zoospores, *J. Bacteriol.* **158:**253–263.

HOPPER, A. K. 1984. Genetic and biochemical studies of RNA processing in yeast, in: *Processing of RNA* (D. Apirion, ed.), pp. 91–117, CRC Press, Boca Raton, Florida.

HORGEN, P. A., AND O'DAY, D. H. 1975. The developmental patterns of lysosomal enzyme activities during Ca^{++}-induced sporangium formation in *Achlya bisexualis*. II. α-Mannosidase, *Arch. Microbiol.* 102:9–12.

HORGEN, P. A., SMITH, R., SILVER, J. C., AND CRAIG, G. 1975. Hormonal stimulation of ribosomal RNA synthesis in *Achlya ambisexualis*. *Can. J. Biochem.* 53:1341–1345.

HORGEN, P. A., MEYER, R. J., FRANKLIN, A. L., ANDERSON, J. B., AND FILION, W. G. 1985. Motile spores from resistant sporangia of *Blastocladiella emersonii* possess one-half the DNA of spores from ordinary colorless sporangia, *Exp. Mycol.* 9:70–73.

HUDSPETH, M. E. S., TIMBERLAKE, W. E., AND GOLDBERG, R. B. 1977. DNA sequence organization in the water mold *Achlya*, *Proc, Natl. Acad. Sci. USA* 74:4332–4336.

JAWORSKI, A. J., AND THOMPSON, K. 1980. A temporal analysis of the synthesis of the mRNA sequestered in zoospores of *Blastocladiella emersonii*, *Dev. Biol.* 75:343–357.

JOHNSON, S. A., AND LOVETT, J. S. 1984a. Base sequence complexity of the polyadenylated and nonpolyadenylated stored RNA in *Blastocladiella* zoospores, *Exp. Mycol.* 8:117–131.

JOHNSON, S. A., AND LOVETT, J. S. 1984b. Gene expression during development of *Blastocladiella emersonii*, *Exp. Mycol.* 8:132–145.

KINNAIRD, J. H., AND FINCHAM, J. R. S. 1983. The complete nucleotide sequence of the *Neurospora crassa am* (NADP-specific glutamate dehydrogenase) gene, *Gene* 26:253–260.

KINSEY, J. A., AND RAMOSEK, J. A. 1984. Transformation of *Neurospora crassa* with the cloned *am* (glutamate dehydrogenase) gene, *Mol. Cell. Biol.* 4:117–122.

KLAR, A. J. S., STRATHERN, J. N., AND HICKS, J. 1984. Developmental pathways in yeast, in: *Microbial Development* (R. Losick and L. Shapiro, eds.). pp. 151–195, Cold Spring Harbor Laboratory, Cold Spring Harbor, New York.

KRAIG, E., AND HABER, J. E. 1980. Messenger ribonucleic acid and protein metabolism during sporulation of *Saccharomyces cerevisiae*, *J. Bacteriol.* 144:1098–1112.

KRUMLAUF, R., AND MARZLUF, G. A. 1979. Characterization of the sequence complexity and organization of the *Neurospora crassa* genome, *Biochemistry* 18:3705–3713.

KURTZ, M. B., AND CHAMPE, S. P. 1982. Purification and characterization of the conidial laccase of *Aspergillus nidulans*, *J. Bacteriol.* 151:1338–1345.

KURTZ, S., AND LINDQUIST, S. 1984. Changing patterns of gene expression during sporulation in yeast. *Proc. Natl. Acad. Sci. USA* 81:7323–7327.

LANGFORD, C. J., AND GALLWITZ, D. 1983. Evidence for an intron-contained sequence required for the splicing of yeast RNA polymerase II transcripts, *Cell* 33:519–527.

LANGFORD, C. J., KLINZ, F. -J., DONATH, C., AND GALLWITZ, D. 1984. Point mutations identify the conserved, intron-contained TACTAAC box as an essential splicing signal sequence in yeast, *Cell* **36**:645–653.

LAUER, G. D., ROBERTS, T. M., AND KLOTZ, L. C. 1977. Determination of the nuclear DNA content of *Saccharomyces cerevisiae* and implications for the organization of DNA in yeast chromosomes, *J. Mol. Biol.* **114**:507–526.

LAW, D. J., AND TIMBERLAKE, W. E. 1980. Developmental regulation of laccase levels in *Aspergillus nidulans, J. Bacteriol.* **144**:509–517.

LODI, W. R., AND SONNEBORN, D. R. 1974. Protein degradation and protease activity during the life cycle of *Blastocladiella emersonii, J. Bacteriol.* **117**:1035–1042.

LOVETT, J. S. 1975. Growth and differentiation of the water mold *Blastocladiella emersonii*: Cytodifferentiation and the role of ribonucleic acid and protein synthesis, *Bacteriol. Rev.* **39**:345–404.

LOVETT, J. S. 1983. Macromolecular synthesis in *Blastocladiella,* in: *Fungal Differentiation: A Contemporary Synthesis,* (J. E. Smith, ed.), pp. 211–234, Marcel Dekker, New York.

MACKO, V. 1981. Inhibitors and stimulants of spore germination and infection structure formation, in *The Fungal Spore: Morphogenetic Controls* (G. Turian and H. R. Hohl, eds.), pp. 565–584, Academic Press, London.

MARTINELLI, S. D. 1979. Phenotypes of double conidiation mutants of *Aspergillus nidulans, J. Gen. Microbiol.* **114**:277–287.

MARZLUF, G. A. 1981. Regulation of nitrogen metabolism and gene expression in fungi, *Microbiol. Rev.* **45**:437–461.

MATSUYAMA, S. S., NELSON, R. E., AND SIEGAL, R. W. 1974. Mutations specifically blocking differentiation of macroconidia in *Neurospora crassa, Dev. Biol.* **41**:278–287.

METZENBERG, R. L. 1979. Implications of some genetic control mechanisms in *Neurospora, Microbiol. Rev.* **43**:361–383.

MICHALSKI, C. J. 1978. Protein synthesis during hormone stimulation in the aquatic fungus *Achlya, Biochim. Biophys. Acta* **84**:417–427.

MISHRA, N. C. 1977. Genetics and biochemistry of morphogenesis in *Neurospora, Adv. Genet.* **19**:341–405.

MYERS, R. B., AND CANTINO, E. C. 1974. The gamma particle, *Monogr. Dev. Biol.* **8**:1–117.

MYTH, K. A., AND REED, S. I. 1980. Isolation of genes by Complementation in yeasts: Molecular cloning of a cell cycle gene. *Proc. Natl. Acad. Sci.* **77**:2119–2123.

NG, R., AND ABELSON, J. 1980. Isolation and sequence of the gene for actin in *Saccharomyces cerevisiae, Proc. Natl. Acad. Sci. USA* **77**:3912–3916.

OJHA, M., TURLER, H., AND TURIAN, G. 1977. Characterization of *Allomyces* genome, *Biochim. Biophys. Acta* **478**:377–391.

ORR, W. C., AND TIMBERLAKE, W. E. 1982. Clustering of spore-specific genes in *Aspergillus nidulans. Proc. Natl. Acad. Sci. USA* **79**:5976–5980.

PERCIVAL-SMITH, A., AND SEGAL, J. 1984. Isolation of DNA sequences preferentially expressed during sporulation in *Saccharomyces cerevisiae, Mol. Cell. Biol.* **4**:142–150.

PERKINS, D. D., RADFORD, A., NEWMEYER, D., AND BJORKMAN, M. 1982. Chromosomal loci of *Neurospora crassa, Microbiol. Rev.* **46**:426–570.

PIKIELNY, C. W., TEEM, J. L., AND ROSBASH, M. 1983. Evidence for the biochemical role of an internal sequence for yeast nuclear mRNA introns: Implications for U1 RNA and metazoan mRNA splicing, *Cell* **34**:395–403.

RAPER, C. A. 1983. Controls for development and differentiation of the dikaryon in basidiomycetes, in: *Secondary Metabolism and Differentiation in Fungi* (J. W. Bennett, and A. Ciegler, eds.), pp. 195–236, Marcel Dekker, New York.

RAST, D. M., STUSSI, H., HEGNAUER, H., AND NYHLEN, L. E. 1981. Melanins, in: *The Fungal Spore: Morphogenetic Controls* (G. Turian and H. R. Hohl, eds.), pp. 507–531, Academic Press, London.

ROSBASH, M., HARRIS, P. K. W., WOOLFORD, J. L., AND TEEM, J. L. 1981. The effect of temperature-sensitive RNA mutants on the transcription products from cloned ribosomal protein genes of yeast, *Cell* **24**:679–686.

ROZAK, C. E., AND TIMBERLAKE, W. E. 1980. Absence of evidence for changes in messenger RNA populations during steroid hormone-induced cell differentiation in *Achlya, Exp. Mycol.* **4**:33–47.

RUSSO, G. M., DAHLBERG, K. R., AND VAN ETTEN, J. L. 1982. Identification of a development-specific protein in sclerotia of *Sclerotinia sclerotiorum, Exp. Mycol.* **6**:259–267.

SAN-BLAS, G., AND SAN-BLAS, F. 1984. Molecular aspects of fungal dimorphism, *CRC Critical Rev. Microbiol.* **11**:101–127.

SCHMIT, J. C., AND BRODY, S. 1976. Biochemical genetics of *Neurospora crassa* conidial germination, *Bacteriol. Rev.* **40**:1–41.

SERLUPI-CRESCENZI, O., KURTZ, M. B., AND CHAMPE, S. P. 1983. Developmental defects resulting from arginine auxotrophy in *Aspergillus nidulans, J. Gen. Microbiol.* **129**:3535–3544.

SHEPHERD, D., AND CARELS, M. 1983. Production formation and differentiation in fungi, in: *Fungal Differentiation: A Contemporary Synthesis* (J. E. Smith, ed.), pp. 515-535, Marcel Dekker, New York.

SHOEMAKER, S., SCHWEICKART, V., LADNER, M., GELFAND, D., KWOK, S., MYAMBO, K., AND INNIS, M. ET AL. Molecular cloning of exocellobiohydrolase 1 derived from *Trichoderma reesii* strain L27. *Biotechnology* **1**:691–696.

SILVER, J. C., AND HORGEN, P. A. 1974. Hormonal regulation of presumptive mRNA in the fungus *Achlya ambisexualis, Nature* **249**:252–254.

SMITH, J. E., AND BERRY, D. R. 1974. *Biochemistry of Fungal Development*, Academic Press, London.

SOLTYK, A., TROPAK, M., AND FRIESEN, J. D. 1984. Isolation and characterization of the RNA^{2+}, RNA^{4+}, and RNA^{+} genes of *Saccharomyces cerevisiae. J. Bacteriol.* **160**:1093–1100.

STEWART, P. R., AND ROGERS, P. J. 1983. Fungal dimorphism, in: *Fungal Differentiation: A Contemporary Synthesis* (J. E. Smith, ed.), pp. 267–313, Marcel Dekker, New York.

STOHL, L. L., AND LAMBOWITZ, A. M. 1983. Construction of a shuttle vector for the filamentous fungus *Neurospora crassa*. *Proc. Natl. Acad. Sci. USA* **80**:1058–1062.

SUTHERLAND, R. B., AND SCHUERCH, B. M., BALL S. F., AND HORGEN, P. A. 1976. The developmental patterns of lysosomal enzyme activities during Ca^{++}-induced sprangium formation in *Achlya bisexualis*. III. Ribonucleases, *Arch. Microbiol.* **109**:289–294.

SYPHERD, P. S., BORGIA, P. T., AND PAZNOKAS, J. L. 1978. Biochemistry of dimorphism in the fungus *Mucor, Adv. Microbial. Physiol.* **18**:67–104.

TILBURN, J., SCAZZACHIO, C., TAYLOR, G. G., ZABICKY-ZISSMAN, J. H., LOCKINGTON, R. A., AND DAVIES, R. W. 1983. Transformation by integration in *Aspergillus nidulans, Gene* **26**:205–221.

TIMBERLAKE, W. E. 1976. Alterations in RNA and protein synthesis associated with steroid hormone-induced sexual morphogenesis in the water mold *Achlya, Dev. Biol.* **51**:202–214.

TIMBERLAKE, W. E. 1977. Relationship between nuclear and polysomal RNA populations of *Achlya:* A simple eukaryotic system, *Cell* **10**:623–632.

TIMBERLAKE, W. E. 1978. Low repetitive DNA content of *Aspergillus nidulans, Science* **202**:773–775.

TIMBERLAKE, W. E. 1980. Developmental gene regulation in *Aspergillus nidulans, Dev. Biol.* **78**:497–510.

TIMBERLAKE, W. E., BARNARD, E. C. 1981. Organization of a gene cluster expressed specifically in the asexual spores of *A. nidulans, Cell* **26**:29–37.

TIMBERLAKE, W. E., AND ORR, W. C. 1984. Steroid hormone regulation of sexual reproduction in *Achlya,* in: Biological Regulation and Development, Vol. 3B, Hormone Action (R. F. Goldberger and K. R. Yamamoto, eds), pp. 255–283, Plenum Press, New York.

TIMBERLAKE, W. E., MCDOWELL, L., CHENEY, J., AND GRIFFIN, D. H. 1973. Protein synthesis during the differentiation of sporangia in the water mold *Achlya, J. Bacteriol.* **116**:67–73.

TIMBERLAKE, W. E., GWYNNE, D. I., HAMER, J. E., MILLER, B. L., MILLER, K. Y., MULLANEY, E. M., DE SOUZA, E. R., YELTON, M. M., AND ZIMMERMAN, C. R. 1983. Gene regulation during conidiation in *Aspergillus nidulans,* in: *Plant Molecular Biology* (R. Goldberg, Ed.), pp. 179–199, Alan R. Liss, New York.

TOLLERVEY, D., WISE, J. A., AND GUTHRIE, C. 1983. A U4-like small nuclear RNA is dispensable in yeast. *Cell* **75**:753–762.

ULLRICH, R. C., KOHORN, B. D., AND SPECHT, C. A. 1980. Absence of short-period repetitive-sequence interspersion in the basidiomy cete *Schizophyllum commune, Chromosoma* **81**:371–378.

VAN DEN ENDE, H. 1983. Fungal pheromones, in: *Fungal Differentiation: A Contemporary Synthesis* (J. E. Smith, ed.), pp. 449–479, Marcel Dekker, New York.

VAN ETTEN, J. L., FREER, S. N., AND MCCUNE, B. K. 1979. Presence of a major (storage?) protein in dormant spores of the fungus *Botryodiplodia theobromae*, *J. Bacteriol.* **138**:650–652.

VAN ETTEN, J. L., DAHLBERG, K. R., AND RUSSO, G. M. 1981. Nucleic acids, in: *The Fungal Spore: Morphogenetic Controls* (G. Turian and H. R. Hohl, eds.), pp. 277–299, Academic Press, London.

WEIR-THOMPSON, E. M., AND DAWES, I. W. 1984. Developmental changes in translatable RNA species associated with meiosis and spore formation in *Saccharomyces cerevisiae*, *Mol. Cell. Biol.* **4**:695–702.

WESSELS, J. G. H., AND SIETSMA, J. H. 1981, Fungal cell walls: A survey, in: *Plant Carbohydrates II: Extracellular Carbohydrates* (W. Tanner and F. A. Loewus, eds.) pp. 352–389, Encyclopedia of Plant Physiology, Vol. 13B, Springer-Verlag, Berlin.

WESSELS, J. G. H., DONS, J. J. M., HOGE, J. H. C., SPRINGER, J., DE VRIES, O. M. H., AND ZANTINGE, A. 1981. Genetic regulation of RNA and protein patterns in the monokaryon-dikaryon transition, in: *The Fungal Nucleus* (K. Gull and S. G. Oliver, eds.), pp. 295–314, Cambridge University Press.

WESSELS, J. G. H., DONS, J. J. M., AND DE VRIES, O. M. H. 1985. Molecular biology of fruit body formation in *Schizophyllum commune*, in: *Developmental Biology of Agarics* (D. Moore, L. A. Casselton, and D. A. Wood, eds.), Cambridge University Press, in press.

WISE, J. A., TOLLERVEY, D., MALONEY, D., SWERDLOW, H., DUNN, E. J., AND GUTHRIE, C. 1983. Yeast contains small nuclear RNAs encoded by single copy genes, *Cell* **35**:743–751.

WOOLFORD, J. L., HEREFORD, L. M., AND ROSBASH, M. 1979. Isolation of cloned sequences containing ribosomal protein genes from *Saccharomyces cerevisiae*, *Cell* **18**:1247–1259.

WOUDT, L. P., PASTINK, A., KEMPERS-VEENSTRA, A. E., JANSEN, A. E. M., MAGER, W. H., AND PLANTA, R. J. 1983. The genes coding for histone H3 and H4 in *Neurospora crassa* are unique and contain intervening sequences, *Nucleic Acids Res.* **11**:5347–5360.

WRIGHT, B. E., AND EMYANITOFF, R. 1983. Metabolic organization during differentiation in the slime mold, in: *Fungal Differentiation: A Contemporary Synthesis* (J. E. Smith, ed.), pp. 19–41, Marcel Dekker, New York.

WRIGHT, J. F., AJAM, N., AND DAWES, I. W. 1981. Nature and timing of some sporulation-specific changes in *Saccharomyces cerevisiae*, *Mol. Cell. Biol.* **1**:910–918.

YAGER, L. N., KURTZ, M. B., AND CHAMPE, S. P. 1982. Temperature-shift analysis of conidial development in *Aspergillus nidulans*, *Dev. Biol.* **93**:92–103.

YELTON, M. M., AND TIMBERLAKE, W. E. 1985. A cosmid for selecting genes by complementation in *Aspergillus nidulans*: Selection of the developmentally regulated *yA* locus, *Proc. Natl. Acad. Sci. USA*, in press.

YELTON, M. M., HAMER, J. E., DE SOUZA, E. R., MULLANEY, E. J., AND TIMBERLAKE, W. E. 1984a. Developmental regulation of the *Aspergillus nidulans* trpC gene, *Proc. Natl. Acad. Sci. USA* **80**:7576–7580.

YELTON, M. M. HAMER, J. E., AND TIMBERLAKE, W. E. 1984b. Transformation of *Aspergillus nidulans* by using a *trpC* plasmid, *Proc. Natl. Acad. Sci. USA* **81**:1470–1474.

ZANTINGE, B., HOGE, J. H. C., AND WESSELS, J. G. H. 1981. Frequency and diversity of RNA sequences in different cell types of the fungus *Schizophyllum commune, Eur. J. Biochem.* **113**:381–389.

ZIMMERMAN, C. R., ORR, W. C., LECLERC, R. F., BARNARD, E. C., AND TIMBERLAKE, W. E. 1980. Molecular cloning and selection of genes regulated in *Aspergillus* development, *Cell* **21**:709–715.

SECTION II
Nitrogen Fixation

Chapter 4

Molecular Aspects of the Physiology of Symbiotic Nitrogen Fixation in Legumes

Mary Lou Guerinot and Barry K. Chelm

BACTERIA OF THE GENERA *Rhizobium* and *Bradyrhizobium* are able to form nitrogen-fixing associations with leguminous plants. These associations, which link the photosynthetic capability of plants to the ability of bacteria to reduce atmospheric dinitrogen, are the culmination of a complex developmental process which requires the coordination of plant and bacterial gene expression. Establishment of the symbiosis begins with recognition and invasion of the appropriate legume host by rhizobia, followed by proliferation and differentiation of both host and bacterial cells to form highly specialized structures called root nodules. It is within these root nodules that conditions are established for bacterial nitrogen fixation and for host plant assimilation of the fixed nitrogen.

To better understand the *Rhizobium*-legume symbiosis, it is helpful to examine the interactions between the two organisms from several points of view. Recent reviews have approached the symbiosis from the host (Verma and Nadler, 1984) and the bacterial perspectives (Rolfe and

Shine, 1984), as a beneficial plant disease (Vance, 1983), and within the context of the endosymbiont theory considering rhizobia intracellular organelles analogous to chloroplasts and mitochondria (Verma and Long, 1983). The goal of this chapter is to discuss the molecular aspects of the symbiosis in the context of the specialized physiology of root nodules. This broad topic includes regulation of the basic metabolic processes of each of the partners, as well as the metabolism unique to the symbiosis. The current advances in the molecular biology of this system can be viewed in this light, as the essence of the symbiosis lies in the metabolic communication between the two organisms. The host plant provides the bacterial symbiont with carbon substrates and is in turn provided with fixed nitrogen. It is the tight coordination of this exchange of metabolites which we wish to emphasize in this chapter. In addition, we will present some of the available data regarding effector mechanisms for the integration of plant and bacterial metabolism.

To minimize any possible confusion we would like to describe the bacterial nomenclature used in this field. In the endosymbiotic state, differentiated rhizobial cells are known as bacteroids. Rhizobial cells grown in axenic culture, that is *ex planta,* are often referred to as free-living bacteria. Depending on their growth properties in culture, members of the genus *Rhizobium* have traditionally been divided into two working groups, fast growers and slow growers. Fast growers have mean generation times of less than 6 hr, are nutritionally more versatile than slow growers, and include *Rhixobium meliloti* and *R. leguminosarum* which nodulate alfalfa and peas, respectively. Slow growers, with mean generation times of greater than 6 hr, include *Bradyrhizobium japonicum* which nodulates soybeans and cowpea rhizobia which nodulate a variety of tropical legumes and some nonleguminous tropical plants. There are many metabolic differences between fast and slow growers, and in recognition of these differences a new genus, *Bradyrhizobium,* was created for slow-growing rhizobia (Jordan, 1982; Krieg, 1984). To simplify matters, we will use the term rhizobia to refer to both fast and slow growers. However, as the groups are physiologically quite distinct, it is important to keep in mind that results obtained with one group may not necessarily apply to the other. We will point out known differences as they arise in the discussion.

CARBON METABOLISM

Carbon metabolism serves two main functions in the nodule. It provides energy and reducing power to both bacterial and host cells, and it provides the carbon skeletons needed to synthesize the transport compounds which carry the nitrogen reduced in the nodule to the rest of the plant. The nature of the carbon compounds involved in these processes, their metabolism, and the partitioning of metabolites among the various sites of utilization are incompletely understood (for a review see Rawthorne et al., 1980). As it has been assumed that there must be a correlation between the carbon substrates which can be provided by the host plant and those which can be utilized by the bacteria, researchers have examined both of these facets in order to address nodule carbon metabolism.

Carbon Substrates Provided by the Host Plant

Photosynthesis is the ultimate source of all the carbon compounds utilized by nitrogen-fixing root nodules. Indeed, the availability of photosynthate is a major factor limiting nitrogen fixation by legumes (Hardy and Havelka, 1976; Havelka and Hardy, 1976). Factors which increase the photosynthate supply, such as increased illumination, CO_2 enrichment, and the grafting of additional shoots onto a nodulated root system, also increase nitrogen fixation, while those which decrease photosynthate supply, such as defoliation and darkness, decrease nitrogen fixation. This is substantiated by studies which have documented that a large proportion of the photosynthate produced in legumes passes through the nodules. For example, Minchin and Pate (1973) reported that as much as 32% of the total photosynthate in *Pisum* was translocated to the nodule where 5% was used for nodule growth, 12% for respiration, and 15% returned to the shoot via the xylem as nitrogenous transport compounds. This subject is discussed in greater detail in Chapter 4.

Photosynthate is known to be transported in the phloem as sucrose in many plant species. From studies employing radiolabeled CO_2 and analyses of phloem (for a sap review see Rawthorne et al., 1980), sucrose is believed to be the major form of carbon translocated in the phloem to the nodules. Legume nodules contain large quantities of sucrose and lower concentrations of glucose and fructose (Streeter, 1980). As alkaline invertase activity is higher in nodule tissue than in the surrounding roots (Robertson and Taylor, 1973), the sucrose is thought to be rapidly converted into glucose and fructose. These observations have led many researchers to suggest that sugars are the main substrate for bacteroids within the nodule cells. However, nodules also contain organic acids (Stumpf and Burris, 1979), implying that C_4 dicarboxylates may serve as substrates for bacterial catabolism. If bacteroids do utilize organic acids from the plant tricarboxylic acid (TCA) cycle, then they must be replenished, most probably via the phosphoenolpyruvate (PEP) carboxylase reaction. The PEP carboxylase activity of *Vicia faba* nodules has been reported to be 50-fold higher than that of roots (Lawrie and

Wheeler, 1975), and the increased activity of PEP carboxylase in the nodule cytosol is correlated with the onset of nitrogen fixation (Christeller et al., 1977).

Carbon Substrates Utilized by *Rhizobium*

Symbiotic Properties of Bacterial Mutants

Genetic and biochemical analyses of carbon utilization in *Rhizobium* have begun to yield an overview of the nutritional requirements and preferences of these bacteria, both as free-living cells and as nitrogen-fixing bacteroids (for a review see Elkan and Kuykendall, 1982). The results of such studies are helping to reveal which plant-supplied substrates are important for supporting the nitrogen fixation process.

Ronson and Primrose (1979) showed that mutants of *Rhizobium trifolii* defective in sucrose, glucose, or fructose catabolism were able to nodulate and fix nitrogen, indicating that sugars, despite their prevalence in nodules, are probably not the crucial source of energy used by bacteroids to support nitrogen fixation. Using Tn5 mutagenesis, Glenn et al. (1984) also showed that the capability to utilize particular C_6 and C_{12} sugars was apparently not essential for bacteroid development or the establishment of effective nitrogen fixation in *Rhizobium leguminosarum*. In contrast, there is one report that a Tn5-induced fructokinase mutant of *R. meliloti* formed nodules but was unable to fix nitrogen (Duncan, 1981). These conflicting results may be due to metabolic differences among rhizobial species; molecular analysis of the mutant alleles used in these studies might clarify this point. However, the recent report that glucose-6-phosphate dehydrogenase mutants of *R. meliloti* which cannot grow on a variety of sugars, including fructose, and which have normal symbiotic properties agrees with the previous observations on *R. trifolii* and *R. leguminosarum* (Cervenansky and Arias, 1948).

Further evidence that the metabolism of sugars is not essential for nitrogen fixation comes from studies on isolated bacteroids. Sugar transport was not detected for bacteroids of either *Rhizobium* or *Bradyrhizobium* (Hudman and Glenn, 1980; Glenn and Dilworth, 1981; deVries et al., 1982). However, transport of sugars was observed when the same strains were grown in culture. C_4 dicarboxylates have been shown to repress glucose transport when rhizobia are grown in culture (Stowers and Elkan, 1983; deVries et al., 1982), leading to the suggestion that glucose transport may be repressed by the C_4 dicarboxylates present in nodules. As bacteroids are unable to transport sugars, it is not surprising that glucose and sucrose do not stimulate N_2 reduction by isolated bacteroids (Bergersen and Turner, 1967). There is one report that both sucrose and glucose can support N_2 reduction by bacteroids isolated from French beans under conditions of low oxygen tension (Trinchant et al., 1981), but this was not observed for bacteroids isolated from peas and examined under a variety of oxygen conditions (deVries et al., 1982).

Similar approaches have been taken to determine whether the metabolism of C_4 dicarboxylates is required for nitrogen fixation. Cultured cells of *B. japonicum, R. leguminosarum,* and *R. trifolii* have all been shown to possess a C_4 dicarboxylate transport system which mediates the transport of succinate, fumarate, and malate into the cell (Finan et al., 1981; Glenn et al., 1980; Ronson et al., 1981; McAllister and Lepo, 1983). This system is also operative in isolated bacteroids (Finan et al., 1983; Reibach and Streeter, 1984; San Francisco and Jacobson, 1985). Furthermore, N_2 reduction by isolated bacteroids is greatly enhanced by succinate and fumarate (Bergersen and Tuner, 1967). Bacteroids from soybean have been shown to possess a functional TCA cycle (Stovall and Cole, 1978). The bacterial TCA cycle appears to be important in the symbiosis, as mutants of *R. meliloti*, one defective in α-ketoglutarate dehydrogenase (Duncan and Fraenkel, 1979) and another defective in succinate dehydrogenase (Gardiol et al., 1982), formed ineffective nodules on alfalfa ("ineffective" refers to nodules which do not fix nitrogen). Further evidence suggesting a role for C_4 dicarboxylates in supporting nitrogen fixation comes from studies on mutants specifically impaired in the transport of these compounds. Mutants of both *R. trifolii* and *R. leguminosarum* which are unable to transport and thus utilize succinate, fumarate, and malate form ineffective nodules on clover and peas, respectively (Ronson et al., 1981; Glenn and Brewin, 1981; Finan et al., 1983). These ineffective nodules lack leghemoglobin, which may be the result of a requirement for an exogenous supply of C_4 dicarboxylates as precursors in the synthesis of bacterially produced heme, since succinyl-CoA is an essential precursor in the bacterial synthesis of heme. While the bacterial synthesis of heme via succinyl-CoA has been shown to be required for the accumulation of leghemoglobin in *R. meliloti*-induced nodules on alfalfa (Leong et al., 1982), it is not essential for leghemoglobin formation in *B. japonicum*-induced nodules on soybeans (Guerinot and Chelm, 1985) (see the section on heme). C_4 dicarboxylates may also be required as energy substrates. As the nodules incited by these transport mutants contain what appear to be highly differentiated bacteroids, it seems that the rhizobia can utilize carbon sources other than C_4 dicarboxylates to fuel their multiplication and development into bacteroids. Another possibility is that the mutants use a low-affinity transport system for dicarboxylates which can provide enough carbon for differentiation but not enough to meet the high energy requirements of nitrogen fixation. There is one report of dual uptake mechanisms for dicarboxylates in bacteroids isolated from soybean nodules (Reibach and Streeter, 1984). However, as the soybeans had been inoculated with a commercial mix containing four different *B. japonicum* strains, it is possible that the differences in uptake were strain specific rather than components of a single system.

To facilitate study of the regulation of the rhizobial genes involved in C_4 dicarboxylate transport, Ronson et al. (1984) have identified three *dct* loci in *R. leguminosarum* by a combination of molecular cloning, genetic complementation, and transposon mutagenesis. Cosmids containing *dct* genes were identified from a gene bank of *R. leguminosarum* DNA made in the broad host range vector pLAFRl by the ability of a cosmid to complement *R. trifolii dct* mutants. The three loci identified are arranged within a 5.5-kilobase (kb) region of DNA in the order *dctA-dctB-dctC*. The results suggest that *dctA* encodes a structural component necessary for C_4 dicarboxylate transport, whereas *dctB* and *dctC* appear to encode positive regulatory elements.

The role of bacterial CO_2 assimilation in supporting nitrogen fixation has also been examined. *Bradyrhizobium japonicum* can fix CO_2 with energy derived from either H_2 metabolism, a process termed chemoautotrophic growth (Simpson et al., 1979; Lepo et al., 1980), or formate metabolism (Manian and O'Gara, 1982) via the enzyme ribulose bisphosphate carboxylase. Both formate metabolism and chemoautotrophic growth can support nitrogenase activity in free-living cultures of *B. japonicum* (Manian et al., 1982; Graham et al., 1984). However, *B. japonicum* strains which are deficient in ribulose bisphosphate carboxylase activity are able to form fully effective nodules on plants (Maier, 1981; Manian et al., 1984). This implies that bacterial CO_2 fixation via ribulose biophosphate carboxylase activity is not necessary for symbiotic nitrogen fixation. Whether the ability to utilize CO_2 is important for survival and competitiveness of the bacteria in the soil has not been examined.

Studies With Free-living Nitrogen-Fixing Cultures

Most of the studies described above employed either isolated bacteroids or cultured cells of rhizobia which were not actively reducing N_2. The ability to induce detectable levels of nitrogenase activity in free-living cultures of certain strains of rhizobia (McComb et al., 1975; Kurz and LaRue, 1975; Pagan et al., 1975; Stam et al., 1983) has provided another means of examining the nature of the carbon sources required for nitrogen fixation. Formate, sugars, sugar derivatives, and organic acids have all been used to support nitrogen fixation by free-living cultures. Although strains which can be induced differ with regard to the exact requirements for the expression of nitrogenase activity, a number of studies have indicated that maximal rates of N_2 reduction are supported by a mixture of two carbon sources, a sugar and a TCA intermediate (Pankhurst,

1981; Mohapatra and Greshoff, 1984). Mohapatra and Gresshoff (1984) have speculated that this dual requirement is due to a synergism between these two metabolic routes, for example, through maintenance of a balanced pH (Ronson et al., 1981) or through modulation of uptake and/or metabolism of one by the other (McAllister and Lepo, 1983). Studies with each carbon source alone suggest that it is the TCA cycle intermediate which is used as the energy source for N_2 reduction. This is consistent with data supporting a crucial role for C_4 dicarboxylates in symbiotic nitrogen fixation.

Bacterial Production of Exopolysaccharides and Polyhydroxybutyrate

Wild-type rhizobia produce large quantities of exopolysaccharide (EPS) under a variety of culture conditions. It has been suggested that, in the soil, this EPS is required for bacterial adhesion to root surfaces (Dazzo and Brill, 1978) and for the determination of host specificity (Dudman, 1977). Both of these topics have been extensively reviewed (e.g., see Dazzo and Truchet, 1983). Briefly, the results of numerous experiments have been interpreted to imply that the host plant can recognize bacterial polysaccharides and that this interaction can be host-specific and lead to specific attachment. As with other facets of carbon metabolism, mutants in EPS production have been sought in order to determine whether EPS production plays a role in the symbiosis. In general, EPS-deficient (Exo⁻) rhizobial mutants are incapable of forming effective nodules (Sanders et al., 1978; Law et al., 1982; Chakravorty et al., 1982; Finan et al., 1985; Leigh et al., 1985). Chakravorty et al. (1982) have reported on the complementation of a Tn5-induced R. trifolii mutant in which no water-soluble EPS could be detected. The acquisition of the wild-type DNA sequence resulted in restoration of the mutant's ability to synthesize normal levels of EPS and to form effective nodules on clover. Leigh et al. (1985) isolated 26 independent Tn5 mutants of R. meliloti which were deficient in the production of EPS. These mutant strains fall into six groups based on the ability of their Exo⁻ phenotype to be complemented by different recombinant plasmids from a R. meliloti gene bank. The ineffective nodules formed by these mutants were found to be "empty," containing no bacteroids, and appeared to form by an abnormal pathway which did not involve shepherd's crooks or infection threads. These nodules are similar to those induced by certain Agrobacterium tumefaciens hybrids carrying R. meliloti nodulation genes (Wong et al., 1983; Truchet et al., 1984; Hirsh et al., 1984, 1985). These results indicate that nodule formation can occur without normal invasion by bacteria and suggest that the R. meliloti exopolysaccharide is somehow involved in the initial invasion process. The polysaccharide might serve as a signal for the induction of an early step in the invasion process or it may be a necessary structural component of the nodule, possibly the infection thread matrix (Newcomb, 1980).

Conspicuous crystals of polyhydroxybutyrate (PHB) are stored in the nodule bacteroids of many legumes. Indeed, this polymer can account for up to 10% of the total dry weight of soybean nodules or up to 50% of the dry weight of *B. japonicum* bacteroids (Klucas and Evans, 1968; Wong and Evans, 1971). It has been suggested that PHB functions as an insoluble carbon store which may be used to support nitrogen fixation during periods of darkness when photosynthate is less readily available. While Wong and Evans (1971) found no evidence that PHB was utilized in the dark by soybean bacteroids, other workers found that the PHB content of lupin bacteroids decreased during periods of very active nitrogen fixation (Kretovich et al., 1977, Romanov et al., 1980).

If PHB does function as a bacteroid reserve material during photosynthate deprivation, it presumably provides reducing power through the β-hydroxybutyrate dehydrogenase reaction (Klucas and Evans, 1968) and carbon and energy from the glyoxalate cycle (Johnson et al., 1966). Wong and Evans (1971) reported a threefold increase in PHB depolymerase activity during nodule development, which was associated with the bacteroid fraction, and β-hydroxybutyrate dehydrogenase has been observed in bacteroids from several legume species (Gerson et al., 1978). *Bradyrhizobium japonicum* mutants have now been isolated which are deficient in β-hydroxybutyrate dehydrogenase activity (Emerich, 1985). These mutants form ineffective nodules on soybean, which implies that PHB metabolism plays an important role in symbiotic nitrogen fixation.

HYDROGEN METABOLISM

Nitrogenases from all known sources catalyze an ATP-dependent reduction of protons to form H_2 (Zumft and Mortensen, 1975). This evolution of hydrogen gas appears to be an inherent characteristic of the nitrogenase reaction (Rivera-Ortiz and Burris, 1975) which consumes a minimum of 25% of the energy flux through nitrogenase (for a review of the energetics of nitrogen fixation see Chapter 5). A limited number of rhizobial strains, primarily bradyrhizobia, possess an oxygen-dependent, unidirectional enzyme system termed the uptake hydrogenase (Hup) system, which allows oxidation of the H_2 released during the reduction of N_2 and thereby the regeneration of energy. The proposed beneficial effects of such recycling and the experimental evidence supporting these effects have been recently reviewed (Eisbrenner and Evans, 1983; Brewin, 1984). Briefly, hydrogen oxidation can regenerate chemical energy in the form of ATP or reducing power and, in addition, both H_2 and O_2 can be removed from the active site of nitrogenase where they act, respectively, as reversible and irreversible inhibitors of N_2 fixation (Dixon, 1972); Ruiz-Argueso et al., 1979; Emerich et al., 1979). Although the majority of the evidence supporting these proposals has been obtained from studies on free-living nitrogen-fixing cultures and isolated bacteroids, field studies also indicate that H_2 may act as an energy source in nodules and thus create less demand for potentially limiting respiratory substrates from the plant (Drevon et al., 1982; Rainbird et al., 1983). In some Hup-positive strains, sufficient H_2-oxidizing activity is present to recycle all the H_2 produced during symbiotic nitrogen fixation. In view of the importance of the Hup system to the energy efficiency of nitrogen fixation, there is currently great interest in introducing the hydrogen oxidation system into rhizobial strains lacking this trait. Such an introduction could theoretically achieve a 10–20% stimulation of plant growth which is dependent on symbiotic nitrogen fixation. It is important to remember, however, that although the genetic information for the synthesis of hydrogenase is bacterially encoded, the legume host has been shown to affect expression of the Hup system (Dixon, 1972; Keyser et al., 1982; Lopez et al., 1983; Bedmar et al., 1983; Garg et al., 1985; Phillips et al., 1985). Different levels of hydrogenase activity have been observed in nodules produced by the same strain of rhizobia when in association with different host plant cultivars. Therefore any attempts to optimize the efficiency of the symbiosis must also consider the role of the host plant.

If the hydrogenase system is to be manipulated, it is essential to know how many components are specifically involved in the oxidation of hydrogen and how the hydrogen oxidation system is regulated. Data have been forthcoming from both biochemical and genetic analyses. The membrane-bound hydrogenase from *B. japonicum* was originally isolated and characterized as an iron-sulfur protein composed of a single subunit (Arp and Burris, 1979). In addition, it has been shown to be a nickel-containing metalloprotein (Klucas et al., 1983; Stults et al., 1984). There are now conflicting reports as to whether hydrogenase is made up of one or two polypeptides (Stults et al., 1984; Harker et al., 1984). From data obtained using membranes prepared from free-living cells which had been derepressed for hydrogenase activity, H_2 oxidation was shown to involve cytochromes o and a-a_3 (O'Brian and Maier, 1982). Bacteroid membranes, however, contain cytochromes o and c-552 in addition to an unusual flavoprotein (O'Brian and Maier, 1983). Studies on bacteroid suspensions by Eisbrenner and Evans (1982a,b) implicated ubiquinone in the O_2-dependent uptake of H_2 and identified a b-type cytochrome which appeared to be unique to the pathway and which functioned in transporting electrons from H_2 to ubiquinone. O'Brian and Maier (1985a) have also presented data supporting a role for ubiquinone, but they have not found a b-type cytochrome involved in electron transfer for H_2 to ubiquinone. These authors have also studied the cytochrome composition of mutants which are constitutive for hydrogenase activity and present further evidence that there is no b-type cytochrome unique to the H_2 oxidation system (O'Brian and Maier, 1985b). They also present evidence that the originally identified b-type cytochrome involved in H_2 oxidation is indeed cytochrome o, which appears to be the primary oxidase involved in substrate oxidation under low pO_2 conditions (O'Brian and Maier, 1985b).

Analysis of the hydrogen uptake system has been greatly facilitated by the ability to induce hydrogenase activity in free-living cultures of certain strains of rhizobia (Maier et al., 1978a). From the information obtained from free-living *B. japonicum* cultures, it has been postulated that the oxygen tension of nodules, the quantity and nature of the carbon substrate provided by the plant host, and the amount of H_2 evolved by the nitrogenase enzyme complex are the primary factors that influence the derepression of hydrogenase. This is based on observations that certain carbon substrates and high oxygen concentrations prevent formation of the H_2 uptake system, whereas H_2 and CO_2 stimulate hydrogenase activity (Maier et al., 1978a; Maier et al., 1979). Rhizobial mutants have been obtained which are hypersensitive to oxygen-mediated repression of hydrogen oxidation (Maier and Merberg, 1982), as well as mutants that express hydrogenase activity in the presence of high concentrations of oxygen (Merberg and Maier, 1983). Other revertible mutants of *B. japonicum* that are defective in hydrogenase activity have been isolated either by monitoring the H_2-dependent reduction of the dye methylene blue (Maier et al., 1978b; Haugland et al., 1983) or by screening for the inability to carry out H_2-dependent chemoautotrophic growth (Maier, 1981; Lepo et al., 1981). As these Hup mutants fall into several classes, they support the view that multiple genes are required for a functional hydrogenase system.

At least some of the determinants for the Hup phenotype have been shown to be plasmid-encoded in *R. leguminosarum* (Brewin et al., 1980). There is no evidence for plasmid localization of these functions in *B. japonicum*. Cantrell et al. (1982) analyzed a series of Hup$^+$ and Hup$^-$ natural isolates of *B. japonicum* to determine whether a correlation existed between plasmid content and the presence of an H_2 uptake system. None of the strains with high H_2 uptake activities had discernible-plasmids. Using a gene bank of DNA from a Hup$^+$ strain, Cantrell et al. (1983) isolated several cosmids which could complement a Hup$^-$ derivative. Further analyses of these cosmids has indicated that Hup-specific sequences span a region of approximately 15 kb and appear to be organized in at least two and probably three transcription units (Haugland et al., 1984). Work by Hom et al. (1985) also suggests that *B. japonicum* contains a cluster of genes involved in H_2 uptake activity and that this cluster also contains a gene involved in the expression of both nitrogenase and hydrogenase activities. They identified a cosmid which complemented a Nif⁻Hup⁻ strain having a single revertible mutation (Moshiri et al., 1983). This cosmid also complements several other Nif⁺Hup⁻ mutants.

The isolation of the Nif⁻Hup⁻ mutant described above indicates that common biochemical and/or regulatory components are involved in both nitrogen fixation and hydrogen oxidation. Nitrogenase and uptake hydrogenase activities appear concurrently in nodules, and the expression of both activities by free-living cultures requires a microaerophilic environment such as that found in nodules. However, studies on free-living rhizobia have indicated that expression of nitrogenase activity requires significant levels of carbon substrates, while hydrogenase expression is substantially repressed by carbon substrates (Maier et al., 1978a; Hanus et al., 1979; Simpson et al., 1979; Merberg et al., 1983); this seems paradoxical in light of the fact that bacteroids are being supplied with carbon by the plant. By first screening for nitrogenase activity and then checking for hydrogenase activity, Graham et al. (1984) demonstrated high levels of both nitrogenase activity and hydrogenase activity with free-living cells grown on any of six different carbon sources, four of which were TCA cycle intermediates. They concluded that, under conditions where nitrogenase is expressed, hydrogenase is also expressed and that this indicates a coordinate relationship between the induction of nitrogenase and hydrogenase activities. There are, however, also rhizobial mutants which express up to nine times more hydrogenase activity than the wild type but express normal levels of nitrogenase (Graham et al., 1984), indicating that the control of these two systems can be uncoupled.

NITROGEN METABOLISM

Nitrogen metabolism in nodules is clearly a joint endeavor. The prokaryotic partner contains the machinery for nitrogen fixation, and the eukaryotic partner assimilates the ammonia produced into an organic form which is then used for the nutrition of the whole plant and also of the prokaryote. For both partners, this involves the induction of symbiotic-specific pathways of nitrogen metabolism. Nitrogenase, which can constitute 10–12% of the bacteroid protein (Whiting and Dilworth, 1974), is not found in rhizobial cells under normal free-living growth conditions. In the plant, synthesis of a nodule-specific glutamine synthetase for assimilation of the bacterially produced ammonia (Lara et al., 1983), as well as synthesis of enzymes such as uricase which are needed to form the nitrogenous transport compounds (Bergmann et al., 1983), are induced. In addition, the bacterial cells repress their ammonia-assimilating pathways (Brown and

Dilworth, 1975; Upchurch and Elkan, 1978a). The key enzymic steps in the nitrogen metabolism of root nodules have been the subject of a recent review (Miflin and Cullimore, 1984).

Nitrogen Metabolism in the Bacterial Symbiont

Early studies in which $^{15}N_2$ was fed to detached nodules confirmed that nodules indeed reduced atmospheric nitrogen to ammonia (Aprison et al., 1954). When nitrogenase activity was demonstrated for both isolated bacteroids and cell-free extracts of bacteroids (Bergersen and Turner, 1967; Koch et al., 1967), it seemed certain that the bacteria encoded the genetic information necessary for nitrogen fixation. Further proof came from isolation of the nitrogenase enzyme complex from bacteroids (Whiting and Dilworth, 1974) and the discovery that certain strains of rhizobia could fix N_2 in the absence of the host plant under the appropriate culture conditions (Kurz and LaRue, 1975; McComb et al., 1975; Pagan et al., 1975; Stam et al., 1983).

The nitrogenase enzyme complex has now been purified from several different strains of rhizobia, and it is biochemically similar to all other nitrogenases which have been characterized (for a review of the structure and function of nitrogenase see Mortensen and Thorneley, 1979). The complex is made up of two multisubunit components. Component 1 (MoFe protein or dinitrogenase) is a tetramer of two α subunits and two β subunits which are encoded by the *nifD* and *nifK* genes, respectively. Component 2 (Fe protein or dinitrogenase reductase) is a dimer of two identical subunits, and is encoded by *nifH*. Component 2 functions in passing electrons to component 1 which in turn reduces N_2 to ammonia. Because the amino acid sequence of nitrogenase has been highly conserved in evolution (Ruvkun and Ausubel, 1980), the genes encoding the protein components of this enzyme have been identified from several *Rhizobium* and *Bradyrhizobium* strains using cloned *Klebsiella pneumoniae nif* genes as hybridization probes. In this manner, the genes encoding the three nitrogenase polypeptides have been isolated from *R. meliloti* (Ruvkun and Ausubel, 1980; Ditta et al., 1980), *R. trifolii* (Scott et al., 1983a), *R. leguminosarum* (Ma et al., 1982; Schetgens et al., 1984), *R. phaseoli* (Quinto et al., 1982), *B. japonicum* (Hennecke, 1981; Adams and Chelm, 1984), and several strains of cowpea rhizobia (Scott et al., 1983b; Weinman et al., 1984). The products of several other genes are required for the synthesis of cofactors, assembly, and activation of this enzyme complex (Roberts and Brill, 1981). The analysis of these other *nif* genes has just been initiated for rhizobia.

Although there is a high degree of nucleotide sequence homology among the various *nif, H, D* and *K* genes which have been isolated and sequenced, the location and organization of these genes show striking differences. In fast-growing rhizobia, these genes are plasmid-encoded (Banfalvi et al., 1981) and occur within a cluster of genes required for nodulation and symbiotic nitrogen fixation (Long et al., 1982; Corbin et al., 1983; Downie et al., 1983; Schofield et al., 1983; Kondorosi et al., 1984). However, in slow-growing rhizobia, there is no evidence that the genes encoding the nitrogenase polypetides reside on plasmids, and they are presumed to be located on the chromosome as they are in free-living diazotrophs. Multiple copies of *nif* sequences have been found in a number of strains of *R. phaseoli* (Quinto et al., 1982, 1985) and *R. fredii* (Prakash and Atherly, 1984); there is at present no evidence for this type of reiteration in other fast-growing species or in slow growers. The role of extra *nif* gene copies is not clear, though other nitrogen-fixing organisms including the cyanobacteria *Anabaena* (Rice et al., 1982) and *Calothrix* (Kallas et al., 1983) and the purple nonsulfur bacterium *Rhodopseudomonas capsulata* (Scolnik and Haselkorn, 1984) have been shown to have multiple copies of *nif* genes. The existence of multigene families has been well documented in eukaryotic organisms, in contrast to prokaryotes where there are few examples.

In addition to the differences in copy number and location, differences in the operon structure of the genes encoding the nitrogenase polypeptides have also been observed. In *Rhizobium*, the genes encoding the three nitrogenase polypeptides are linked in a single operon, *nifHDK*, and are transcribed from a single promoter adjacent to *nifH* (Ruvkun et al., 1982; Corbin et al., 1983; Scetgens et al., 1984), as they are in *K. pneumoniae* (as reviewed in Kennedy et al., 1981). In *Bradyrhizobium*, these three genes are divided into two transcriptional units, *nifH* and *nifDK* (Kaluza et al., 1983; Scott et al., 1983b; Adams and Chelm, 1984; Adams et al., 1984), and both units are transcribed from their own promoters (Adams et al., 1984; Kaluza and Hennecke, 1984; Yun and Szalay, 1984).

Rhizobia normally fix nitrogen only as bacteroids, i.e., in a highly differentiated state within the roots of leguminous plants. Ausubel (1984), in a recent review of the regulation of nitrogen fixation genes, discussed what little is known about the control of *nif* gene expression in *Rhizobium*. He points out that what is known has largely been learned by comparisons to and extrapolations from the detailed understanding of the regulation of *nif* genes in *K. pneumoniae*. First, there is significant similarity among the various *Rhizobium nif* promoters and the *K. pneumoniae nif* promoters which have been examined. They have consensus sequences of 5'-TTGCA-3' at −15 to −10 and of 5'-CTGG-3' at −26 to −23, where the numbers refer to the nucleotide position relative to the start of transcription (Ow et al., 1983; Sundaresan et al., 1983b; Beynon et al., 1983). This suggests that rhizobial *nif* genes may be regulated in a manner analogous to *K. pneumoniae nif* genes. Briefly, in *K. pneumoniae,* *nif* genes are subject to two levels of regulation in response to ammonia and oxygen (for a recent study see Cannon et al., 1985). They are under the control of a general nitrogen regulatory system where three regulatory proteins, *ntrA, B,* and *C,* work together to control the expression of numerous genes involved in nitrogen assimilation (for a review of *ntr* see Magasanik, 1982) They are also controlled by a *nif*-specific regulatory system. Under conditions of nitrogen starvation, the *ntrC* and *ntrA* products activate operons involved in nitrogen assimilation such as *hut* (histidine utilization) and *put* (proline utilization) in addition to the *nifLA* operon whose products then mediate *nif*-specif regulation. The *nifA* product is a transcriptional activator required for the expression of all *nif* operons except its own and functions by a mechanism which also requires *ntrA* (Buchanan-Wollaston et al., 1981). The *nifL* gene product acts to repress *nif* transcription as a consequence of rising levels of ammonia and oxygen (Hill et al., 1981). In addition, when fixed nitrogen is in excess, the *ntrB* and *ntrC* products combine to eliminate *nif* gene expression entirely through the repression of the *nifLA* operon.

Using a *R. meliloti nifH-lacZ* fusion, Sundaresan et al. (1983a,b) showed that the *R. meliloti nifHDK* promoter could be activated in *E. coli* by either the *K. pneumoniae ntrC* or *nifA* products when *ntrA* was present. This suggests that *nif* and *ntr* promoters and the genes that regulate them have been conserved between *K. pneumoniae* and *R. meliloti.* In *R. meliloti,* a *nif*-specific regulatory gene has recently been identified that is essential for activation of the *nifHDK* operon and which shows homology to both *K. pneumoniae nifA* and *ntrC* (Zimmerman et al., 1983; Szeto et al., 1984; F. Ausubel, personal communication). Sequences which are homologous to the *K. pneumoniae nifA* gene have also been identified in *R. leguminosarum* (Rossen et al., 1984) and *B. japonicum* (Adams et al., 1984). Whether these *nifA*-like genes function in a manner analogous to the *nifA* gene in *K. pneumoniae* is yet to be determined. As *Rhizobium* normally fixes nitrogen only in the symbiotic state and rhizobial symbiotic nitrogen fixation is not subject to repression by ammonium, it seems likely that *Rhizobium nifA*-like genes will be found to be activated independently of a centralized nitrogen control system. This suggests that the molecular details of symbiotic *nif* regulation will differ from those found in free-living diazotrophs.

In free-living rhizobia, ammonia is assimilated by the coordinate activity of glutamine synthetase (GS) and glutamate synthase (GOGAT) (Brown and Dilworth, 1975; Kondorosi et al., 1977; Vairinhos et al., 1983). However, in bacteroids, GS activity is repressed in concert with the depression of nitrogenase activity (Brown and Dilworth, 1975; Upchurch and Elkan, 1978a). The regulation of GS then is of special interest, as its repression apparently leads to export of the newly fixed nitrogen to the plant cytoplasm. Usually cultures of free-living N_2-fixing *B. japonicum*, O'Gara and Shanmugam (1976) showed that as much as 94% of the fixed nitrogen was recovered as NH_4^+ from the culture supernatant. They also provided evidence that several of the key enzymes involved in ammonium assimilation, including GS, were repressed in free-living nitrogen-fixing cultures, in agreement with previous observations on bacteroid cells. They proposed that the rhizobial ammonium assimilatory activities may be repressed by amino acids supplied by the host plant; however, this has yet to be tested.

The study of GS regulation in rhizobia is complicated by the presence of two GS species, designated GSI and GSII (Darrow and Knotts, 1977) (Table 1). GSI is very similar to the single GS enzyme found in most other gram-negative bacteria and is similarly regulated by a reversible adenylylation cascade system (Darrow, 1980). In contrast, GSII is not known to be modified after translation (Darrow, 1980). These proteins are the products of different genes (Darrow, 1980; Somerville and Kahn, 1983) and are differentially regulated in response to changes in nitrogen source (Ludwig, 1980), carbon source (Darrow et al., 1981), and oxygen concentration (Rao et al., 1978; Darrow et al., 1981).

The mechanisms by which rhizobia regulate GSI and GSII activity are currently being investigated. The gene encoding GSI has now been cloned from *R. meliloti* (Somerville and Kahn, 1983) and *B. japonicum* (Carlson et al., 1983). Carlson et al. (1985) showed that the GSI gene was transcribed from a single promoter and that there was no dramatic transcriptional control of this gene. They concluded that GSI activity is primarily regulated at the posttranslational level, in agreement with the data on adenylylation of this enzyme. Using the isolated *R. meliloti* GSI gene to create a GSI mutant via marker exchange mutagenesis, Somer-

Table 1 Rhizobial Glutamine Synthetases.

Characteristic	GS I	GS II
Structure	Homododecamer, 59,000-dalton subunit, similar to all other prokaryotic GS	Homooctomer, 36,000-dalton subunit, unique to Rhizobiaceae
Biochemical properties	Posttranslational reversible inhibition by adenylylation	No known posttranslational control
	DNA binding activity	No DNA binding activity
Gene expression	Constitutive transcription	Highly regulated transcription

ville and Kahn (1983) showed that GSI was not necessary for the establishment of symbiotic nitrogen fixation. The gene encoding GSII, *glnII*, has recently been isolated from *B. japonicum* (T. Carlson and B. Chelm, unpublished results). The *glnII* gene is transcribed at high levels in free-living cultures grown with glutamate as a nitrogen source but at low levels when cells are grown in the presence of ammonia or at low oxygen concentrations. There is also decreased transcription of *glnII* in bacteroids.

Most of the work on the bacterial GS-GOGAT system has focused on GS. Glutamate synthase mutants (Asm$^-$) of *R. meliloti* have been reported to be capable of expressing nitrogenase activity (Kondorosi et al., 1977). In contrast, a report describing the isolation and characterization of an Asm$^-$ mutant of *B. japonicum* stated that the mutant strain failed to fix nitrogen in symbiotic association in soybean nodules although it was capable of inducing nitrogenase activity *ex planta* (O'Gara et al., 1984). Molecular analyses of GOGAT mutants should shed light on whether the difference in phenotype between the *R. meliloti* mutant and the *B. japonicum* mutant reflects a difference between fast growers and slow growers or merely reflects the nature of the specific alleles tested.

Plant Nitrogen Metabolism

Information from numerous physiological studies suggests that assimilation of ammonia occurs in the plant cytosol via the GS-GOCAT pathway rather than via glutamate dehydrogenase (GDH). Over 95% of the total nodule GS activity is present in the plant cytosol, and this enzyme may constitute up to 2% of the total soluble protein of nodules (McParland et al., 1976). The activity of GS in nodules may be 10- to 500-fold higher than in nonnodulated roots (Robertson et al., 1975; Sen and Schulman, 1980; Cullimore et al., 1982). Furthermore, both GS and GOGAT levels rise in parallel with nitrogenase activity and leghemoglobin content during nodule development (Robertson et al., 1975; Sen and Schulman, 1980), whereas GDH activity is not correlated with either (Sen and Schulman, 1980; Groat and Vance, 1981). In *Phaseolus*, the increase in GS activity is due to the production of a nodule-specific form of the enzyme (GS$_{n1}$) which can account for over 85% of the total nodule GS activity (Lara et al., 1983). Cullimore and Miflin (1983) have shown that production of this form of the enzyme appears to be due to changes occurring at the transcriptional level, as the amount of GS mRNA is substantially greater in nodules than in roots. Purification of GS mRNA by immunoprecipitation of nodule polysomes has allowed the identification of a cDNA clone encoding GS$_{n1}$ (Cullimore et al., 1984). Using this clone they showed that

GS was encoded by a small, multigene family which showed organ-specific patterns of expression.

Once the bacterially produced ammonia has been assimilated into an organic form in the cells of the root nodule, it must be transported to other parts of the plant. The form in which the nitrogen is transported varies depending on the legume species. Many temperate legumes, including *Pisum, Lupinus,* and *Medicago,* transport nitrogen from the nodule as amides, while a second group comprised mainly of tropical legumes, transports ureides under nitrogen-fixing conditions (Miflin and Cullimore, 1984). Species which transport ureides include *Glycine, Phaseolus,* and *Vigna.*

The pathway of asparagine synthesis in higher plants involves aspartate aminotransferase and asparagine synthetase (for a review see Lea and Miflin, 1980). In lupin, the activities of both these enzymes have been shown to increase in parallel with the activities of GS, glutamate synthetase, and nitrogen fixation (Reynolds and Farnden, 1979; Scott et al., 1976). When the asparagine synthetase activity in nodules is compared in a variety of legume species, 7- to 10-fold higher activity is found in amide-transporting as opposed to ureide-transporting legumes (Reynolds et al., 1982a). Aspartate aminotransferase, on the other hand, functions in ureide-transporting legumes as well as in amide-transporting legumes. Two forms of aspartate aminotransferase have been found in the nodules of both amide- and ureide-transporting legumes, and in each type only one of the forms was believed to be involved in the synthesis of nitrogenous transport compounds (Reynolds and Farnden, 1979; Hanks et al., 1983).

Experiments by a number of authors have shown that ureides are synthesized in large quantities only in nodules and then only from recently fixed nitrogen (as reviewed in Miflin and Cullimore, 1984). The pathway of ureide synthesis involves the formation and degradation of purine bases (for a review see Miflin and Cullimore, 1984). The activity of several of the enzymes involved in ureide production, including aspartate aminotransferase, xanthine dehydrogenase, uricase, and allantoinase, have been measured in a number of legume species and have been found to increase substantially during nodule development (Schubert, 1981; Atkins et al., 1980; Reynolds et al., 1982b). Two of these enzymes, xanthine dehydrogenase and uricase, are generally not found in root tissue and are considered nodule-specific. Nodulin-35, previously identified only as a nodule-specific protein, has been shown to be a subunit of uricase (Bergmann et al., 1983).

A scheme for the localization of ureide synthesis has been proposed (Boland et al., 1982; Shelp et al., 1983), and Hanks et al. (1983) have suggested that the whole of purine synthesis and catabolism occurs predominantly in the uninfected cells of the nodule. Immunocytological methods have been used to show the localization of uricase to uninfected nodule cells (Bergmann et al., 1983). The significance of this compartmentalization is not yet understood but suggests a fascinating separation of function within the nodule.

EFFECTORS FOR THE INTEGRATION OF PLANT AND BACTERIAL METABOLISM

The establishment of a successful symbiosis depends on the differential expression of both legume and rhizobial genes at successive stages of nodule development. Regulation of this process is likely to involve the exchange of specific chemical signals between the symbiotic partners, possibly as part of the induced changes in the physiological environment. In this section, we will briefly review and discuss the available information on the involvement of some of the obvious possible effectors which could mediate the integration of plant and bacterial metabolism.

Phytohormones

Many features of nodule development suggest the involvement of phytohormones. Root hair curling, the induction of cell division in the cortex, and the increase in the ploidy level of nodule cells are all examples of processes which appear to be likely candidates for hormonal control (for a review see Newcomb, 1980). To date, however, there is no direct and conclusive evidence of a specific role for phytohormones in nodule morphogenesis. There is evidence that diffusible factors play a role in nodule development, as normal nodule morphogenesis can be triggered at a distance by infecting rhizobia (Libbenga et al., 1973; Newcomb et al., 1979; Truchet et al., 1980). The nature of this diffusible factor is unknown, although exogenous indoleacetic acid (IAA) and cytokinin can induce cortical proliferation in pea root explants (Libbenga et al., 1973). There are now well-studied precedents for the involvement of microbially produced phytohormones in the establishment of plant-bacterial interactions, most notably in the *Agrobacterium*-induced crown gall and in the *Pseudomonas syringae* pv. *savastoni*-induced tumor systems (for a review see Chapter 2). However, these interactions lead to uncontrolled plant growth, hence to a disease condition, while the *Rhizobium*-legume system leads to a tightly controlled perturbation of normal plant growth which has a beneficial rather than a deleterious effect.

Rhizobia have been shown to produce both IAA and cytokinins. IAA has unequivocally been identified in culture supernatants of *R. leguminosarum* and *R. trifolii* strains by gas chromatography–mass spectrometry in studies directed at determining whether there was any correlation between the ability to produce IAA and the ability to nodulate (Badenoch-Jones et al., 1982; Wang et al., 1982a). The amount of IAA produced was measured for a number of nodulating and nonnodulating strains of *Rhizobium* in both the presence and absence of plant tissue. Similar concentrations were found for all strains examined, with no correlation between IAA production and exposure to plant tissue. There was also no correlation between the ability to produce IAA and the presence of the large nodulation plasmid in *R. leguminosarum* (Wang et al., 1982a).

Most of the reports on cytokinin production by *Rhizobium* have relied on bioassays alone (as reviewed in Greene, 1980). Using gas chromatography–mass spectrometry, Wang et al. (1982b) detected very low levels of isopentenyladenine in the culture supernatant of *R. leguminosarum*, as well as in pea nodules themselves. More importantly, the levels detected in root nodules were the same order of magnitude as those detected for root tissue. This was not in agreement with earlier reports which had stated that root nodules had elevated levels of cytokinin (Henson and Wheeler, 1976; Syono and Torrey, 1976).

Root nodules have also been reported to have elevated levels of IAA (Dullart, 1967). It has been proposed that the level of IAA in the nodule may be controlled by the pseudoperoxidase activity of leghemoglobin (Puppo and Rigaud, 1975) as well, as by nitrite reductase activity (Arp and Zumft, 1983). Verma et al. (1978) have suggested that IAA may be involved in the induction of hydrolytic enzymes such as cellulase; such an effect was demonstrated when IAA was applied to pea epicotyls (Verma et al., 1975). There is preliminary evidence correlating IAA application and *Rhizobium* infection with respect to the appearance of host gene products. Verma et al. (1983) have reported preliminary data suggesting there is a set of low-molecular-weight proteins which can be induced either by application of IAA or by exposure of plant roots to *Rhizobium*, implying that rhizobially produced IAA may be involved in the induction of nodulins.

Cyclic Nucleotides

Cyclic AMP (cAMP) is an important regulator of gene expression in both prokaryotes and eukaryotes and as such is a candidate for mediator of regulator cross talk in the symbiosis. To date, there is no definitive evidence that cAMP functions in the control of either rhizobial or legume physiology or that it plays a role in the symbiosis. Indeed, the very existence of cAMP in plants was questioned for many years (for a review see Amrhein, 1977), but the analysis of plant tissues by a highly sensitive gas chromatography—mass spectrometry procedure has finally provided the first reliable evidence that plants contain cAMP (Johnson et al., 1981). What is needed now are studies to determine whether cAMP functions as a secondary messenger in plants as it does in animals, where it has been shown to mediate the effects of hormones. Cyclic AMP has been detected in *Rhizobium*, and evidence is accumulating from studies on rhizobial cultures which supports a role for this molecule in regulating rhizobial metabolism. For example, it has been implicated in the control of ammonia assimilation enzymes in *B. japonicum* (Upchurch and Elkan, 1978b) and in control of the hydrogenase uptake system (Lim and Shanmugam, 1979). There is at present only one study reported in the literature in which investigators attempted to look at the effect of exogenous cAMP on the symbiosis (Moustafa and Hastings, 1973). In this study legume seedlings were inoculated with rhizobia in the presence or absence of various forms of the nucleotide; dibutyryl cAMP decreased the resultant levels of nitrogen fixation, whereas free cyclic AMP and AMP increased the resultant levels of nitrogen fixation. These results have yet to be reproduced.

In *Rhizobium* and *Bradyrhizobium*, the levels of cAMP seem to be correlated with the quality of the carbon source on which the rhizobia are growing. When rhizobia are grown on a source of carbon which allows for good rates of growth, such as malate, they have lower levels of cAMP than when they are grown on a carbon source which does not support good rates of growth, such as glutamate (Lim and Shanmugam, 1979) or formate (C. R. McClung, M. L. Guerinot, and B. K. Chelm, unpublished results). This is consistent with what is known from *E. coli* where cAMP levels change in response to the quality of the carbon or energy sources available for growth. In *Escherichia coli*, cAMP, complexed with its receptor protein, binds to DNA and exerts transcriptional control (Botsford, 1981; Ulmann and Danchin, 1983). This cAMP-based control system in *E. coli* is termed catabolite repression because good carbon sources repress the synthesis of enzymes needed to catabolize poorer carbon sources (Magasanik, 1961).

Data on the preferential use of carbon substrates from mixed carbon sources by rhizobia are relatively scarce and contradictory. Glucose does not prevent lactose utilization by *R. trifolii* (DeHollaender and Stouthamer, 1979) or β-galactosidase activity in *R. meliloti* (Ucker and Signer, 1978). In *B. japonicum*, glucose does not repress the induction of mannitol dehydrogenase (Kuykendall and Elkan, 1977), whereas it has been reported to be repressive in *R. meliloti* (Martinez de Drets et al., 1970). One of the best examples of a catabolite repression-like phenomenom has been reported for *R. meliloti* where succinate, which allows the highest observed rate of growth, caused an immediate reduction in β-galactosidase activity when added to cells growing in lactose (Ucker and Signer, 1978). However, unlike catabolite repression in *E. coli*, exogenous cAMP could not overcome this effect. Other workers have also reported catabolite repression-like phenomena, but the repression was not as extensive as that observed by Ucker and Signer (1978). When grown in the presence of two carbon sources, *R. leguminosarum* utilizes both substrates simultaneously, but the consumption of substrates which support slower growth rates is substantially reduced in comparison with that of substrates which support faster growth rates such as glucose and succinate (Dilworth et al., 1983).

Organic acids may function as catabolite repressors in *Rhizobium*, as glucose does in *E. coli*. Lim and Shanmugam (1979) showed that H_2 utilization by *B. japonicum* was significantly reduced when malate was present and demonstrated that this inhibition could be overcome by the addition of cAMP. This is consistent with the fact that growth on malate results in low levels of cAMP and suggests that some component of the hydrogen oxidation system requires cAMP to function. Organic acids have also been shown to exert catabolite repression in *Pseudomonas* which, like *Rhizobium*, utilizes organic acids in preference to glucose and other carbohydrates (Siegal et al., 1977). However, in *Pseudomonas*, this catabolite repression does not appear to be mediated through cAMP (Siegal et al., 1977).

As mentioned previously, growth on glutamate results in high levels of cAMP (Lim and Shanmugam, 1979). A number of workers have demonstrated that L-glutamate inhibits ammonia assimilation, and Upchurch and Elkan (1978b) showed that the addition of glutamate to an actively growing culture of *B. japonicum* drastically reduced growth and repressed GS activity. In addition, they demonstrated that cAMP also repressed ammonia assimilation. The suggested that control of ammonia assimilation in the nodule by glutamate is probable and that cAMP may be the messenger through which generalized ammonia assimilatory repression is mediated.

To further characterize the role of cAMP in rhizobial metabolism and to define its role in the symbiosis, genes encoding adenylate cyclase (*cya*), which catalyzes the formation of cAMP from ATP, have been cloned from *R. meliloti* (Kiely and O'Gara, 1983) and *B. japonicum* (Guerinot and Chelm, 1984). In both cases, the gene was identified by complementation of an *E. coli* adenylate cyclase mutant. The isolation of the *B. japonicum* adenylate cyclase gene in this manner is the first report of a gene from a slow-growing rhizobia being isolated by direct complementation in *E. coli*. The isolated *B. japonicum* DNA could complement at least eight different phenotypes in two independently derived *E. coli cya* mutants, and the cloned region gave rise to adenylate cyclase activity in *E. coli* and caused accumulation of significant levels of cAMP. The complementation of the *cya* phenotype with the *B. japonicum* gene was more complete than that seen with the *R. meliloti cya* region. This may be due to the difference in vectors used in the two studies; the plasmid vector pRK290 used in the *R. meliloti* study is maintained in *E. coli* at a lower copy number than pBR322 which was used in the *B. japonicum* study.

A *B. japonicum cya* mutant has been constructed in which the entire region which can complement *E. coli cya* mutants has been deleted (M. L. Guerinot and B. K. Chelm, unpublished results). Surprisingly, this mutant has wild-type levels of cAMP when cells are grown under a variety of conditions, and it forms fully effective nodules on soybeans. This has led us to propose that *B. japonicum* has isozymes of adenylate cyclase. This is not without precedent in the bacterial literature. *Myxococcus xanthus* has two isozymes of adenylate cyclase, only one of which has been implicated in the developmental process of fruiting body formation (Devi and McCurdy, 1984). Biochemical characterization of the adenylate cyclase enzyme(s) in *B. japonicum* should resolve this question.

The evidence that cyclic GMP (cGMP) may have a regulatory function in *Rhizobium* is the observation that the intracellular levels of this compound respond to the physiological changes imposed on cultures of free-living rhizobia and that these changing levels can be correlated with concomitant changes in a number of enzyme activities (Lim et al., 1979). In addition, exogenous cGMP has been shown to inhibit the expression of nitrogenase, nitrate reductase, and hydrogenase in free-living cultures of *B. japonicum* (Lim et al., 1979). Microaerophilic-aerobic shift experiments have shown that cellular cGMP concentrations increase from 0.25 to 2.6 pmol/mg cell protein upon the shift to air. Thus it is possible that the levels of cGMP may be under redox control such that a decrease in oxygen concentration lowers the cGMP levels, which in turn allows the expression of enzymes systems that have been found to be active under microaerophilic conditions. The levels of cGMP may therefore be involved in sensing O_2 through redox control of the enzyme responsible for its synthesis, as has been suggested for mammalian guanylate cyclase (Bradham and Cheung, 1982).

Oxygen

Many features of bacteroid metabolism might be regarded as adaptations to the environment in which bacteroids function. One aspect of this environment, low oxygen tension, may signal initiation of the bacteroid-specific changes known to occur in *Rhizobium*. We will first discuss how the nodule oxygen environment is maintained at low pO_2 and then present some of the evidence supporting a role for oxygen-mediated regulation of gene expression in the nodule.

The concentration of oxygen critically affects the symbiotic bacterial nitrogenase activity in root nodules. Oxygen is needed to support the respiratory metabolism of the nitrogen-fixing bacteroids. However, oxygen irreversibly inhibits nitrogenase and represses synthesis of the enzyme. The balance between these two conflicting demands is maintained by leghemoglobin, a myoglobin-like hemoprotein which regulates the oxygen tension in the nodule. Its presence allows the delivery of a high flux of O_2, albeit at a very low oxygen tension. Leghemoglobin and rhizobial respiration are the subject of a recent review (Appleby, 1984).

Leghemoglobin can constitute up to 40% of the total soluble protein in the nodule. Apoleghemoglobin (Lb) is a plant gene product (Balcombe and Verma, 1978; Sidloi-Lumbrosco et al., 1978), whereas the heme prosthetic group of leghemoglobin is thought to be provided by the bacteria (Cutting and Schulman, 1969; Godfrey and Dilworth, 1971; Nadler and Avissar, 1977). Apoleghemoglobin is also a symbiotic protein in that it is not made in uninfected plant roots but is synthesized exclusively in the root nodules, as are GS $_{nl}$ and uricase. Soybean nodules contain four major species of Lb. They are encoded by a small gene family which consists of four functional genes, one pseudo gene, and at least three truncated genes (Hyldig-Nielson et al., 1982; Wiborg et al., 1982; Brisson and Verma, 1982). The primary structure of Lb genes is similar to that of mammalian globin genes with respect to the position of two introns common to all globin genes, as well as the presence of several presumptive regulatory sequences on the 5' end of these genes (Mauro et al., 1985). Soybean Lb genes all contain a third, central intron (Brisson and Verma, 1982; Jensen et al., 1981; Wiborg et al., 1982) which is missing from animal globins but is in the exact position predicted from a computer analysis of animal globin protein structure, which suggests that plant globin genes are primitive. Six of the Lb genes are arranged in two independent clusters in the soybean genome. One cluster contains four Lb genes in the order 5'-Lb_a-Lb_{c1}-ψLb-Lb_{c3}-3' with the same transcriptional polarity, while the other cluster contains two genes in the order 5'-ψLb-Lb_{c2}-3' (Bojsen et al., 1983; Lee et al., 1983). About 7–8 days after infection the Lb genes are activated sequentially during a period of a few days in the order opposite that in which they are arranged in the soybean genome (Marcker et al., 1984). Then, on about day 12, there is a dramatic increase in the transcription rates of all the Lb genes except the Lb_{c2} gene. These observations are in complete accord with earlier work on the rate of appearance of different Lb proteins during the early stages of soybean nodule development and the abundance of the different Lb proteins within a fully developed nodule (Fuschman and Appleby, 1979).

As mentioned above, rhizobia have been thought to produce the heme moeity of leghemoglobin. Early studies employing radioactive precursors of heme showed that bacteroids could produce labeled heme, while the plant fraction was unable to do so (Cutting and Schulman, 1969). Further evidence for the bacterial synthesis of heme came from enzyme studies demonstrating that the first and last enzymes of heme biosynthesis, δ-aminolevulinic acid synthase (ALAS) (Nadler and Avissar, 1977) and ferrochelatase (Porra, 1975), were both associated predominantly with the bacterial fraction in root nodule extracts. In addition, the activity of the bacteroid ALAS increased in parallel with that of the total heme in nodules during nodule development (Nadler and Avissar, 1977). While these studies support the idea that bacteroids can be the site of heme formation, a role for the plant has not been rigorously eliminated. For example, the enzyme which catalyzes the second step in heme biosynthesis, δ-aminolevulinic acid dehydrase (ALAD), has been found in both the plant and bacteroid fractions of nodules. Godfrey et al. (1971) reported that the major portion of nodule ALAD activity was in the plant cytosol fraction and suggested that in the nodule there is a coordinated effort to produce heme, just as there is in animal cells where different intermediates in heme biosynthesis are produced in either the cytoplasm or the mitochondria.

Experiments on *Rhizobium* strains with mutations in the heme biosynthetic pathway should help answer the questions concerning heme formation in nodules. At least seven genes are required for the production of heme. While there are many rhizobial mutants which form white nodules, i.e., nodules which lack heme (e.g, see Noel et al., 1982), few of these mutant strains have been characterized as to the nature of their defect and most were selected in a manner which precluded the identification of auxotrophs. Leong et al. (1982) isolated and characterized a mutant of *R. meliloti* which was deficient in ALAS and identified a recombinant plasmid from their *R. meliloti* gene bank which could complement this mutant. The ALAS mutants formed white, ineffective nodules on alfalfa, and partial complementation was observed when plants were grown in the presence of δ-aminolevulinic acid. The gene encoding ALAS (*hemA*) has also been identified from a *B. japonicum* cosmid library using a fragment of the *R. meliloti* gene as a hybridization probe (Guerinot and Chelm, 1986). In order to verify that the region identified by homology with the *R. meliloti hemA* gene indeed encoded an ALAS, a *B. japonicum* mutant was constructed in which most of the coding sequence was deleted. As expected, this mutant had no ALAS activity and was unable to grow on minimal medium in the absence of exogenous ALA. However, unlike the *R. meliloti hemA* mutant, the *B. japonicum* mutant formed effective nodules containing leghemoglobin. Evidently, the soybean plant is able to rescue the *B. japonicum* ALAS-deficient strain, suggesting that regulation of the synthesis of the heme moiety might be more complex in soybean nodules than previously believed. The difference in symbiotic phenotypes between the *R. meliloti* and *B. japonicum* ALAS mutants may reflect inherent differences between the two bacterial species or between the different host plants.

While little is known about how heme biosynthesis is controlled during nodule development, it has been estimated that there would need to be a 10- to 20-fold increase in heme synthesis by the bacteria to accommodate all the apoleghemoglobin produced by the plant. There is evidence that reduced oxygen tension enhances the levels of some of the enzymes of heme biosynthesis and results in a 10-fold increase in cellular heme content in cultured cells of *B. japonicum* (Avissar and Nadler, 1978). With the isolated genes, it should now be possible to look at the effect of oxygen limitation on the transcription of *hemA*.

We have discussed heme as it pertains to leghemoglobin. It is possible that the heme molecule itself is involved in the regulation of gene expression. In yeast, it appears that there is a family of coordinately regulated genes controlled by heme. It has been proposed that coordination of expression is achieved by a regulatory protein that binds to activation sequences upstream of these genes. An attractive possibility is that such a protein requires a heme cofactor for its synthesis or binding. Regulation by heme is best studied with the *Sacchromyces cervisiae CYC1* gene which encodes iso-1-cytochrome c, whose transcription has been shown to be regulated by both catabolite repression and by the level of intracellular heme (Guarente and Mason, 1983). Since regulation by catabolite repression and regulation by heme have been shown to map to the same regulatory sites, Guarente et al. (1984) explored the possibility that both forms of control are mediated by a single protein-heme complex analogous to CPA-cAMP in *E. coli*. A locus has been identified which appears to encode a protein that mediates catabolite repression by responding to intracellular heme levels (Guarente et al., 1984).

A number of functions characteristic of bacteroids, such as the ability to fix nitrogen, can be induced under low pO_2 in laboratory culture. GSII activity disappears and GSI becomes adenylylated at approximately the same partial pressure of O_2 at which nitrogenase activity appears (Rao et al., 1978). This lack of GS activity is characteristic of bacteroids. Hydrogenase activity has also been shown to be dependent on low pO_2 levels (Maier et al., 1979). In addition, anaerobic growth of *B. japonicum* in the presence of nitrate as a terminal electron acceptor results in cells that have acquired some bacteroid characteristics. These include an altered pattern of heme proteins, nitrate reductase activity, and the production of several factors capable of transferring electrons to nitrogenase (Daniel and Appleby, 1972; Phillips et al., 1973). In a study on the effect of oxygen supply on nitrogenase levels in 39 strains of *B. japonicum*, Agarwal and Keister (1981) reported that oxygen was also involved in the regulation of EPS synthesis. Rhizobial cells grown under microaerophilic conditions or anaerobically on nitrate have very low levels of EPS. This correlates well with the apparent lack of EPS associated with bacteroid cells (R. Tully, personal communication).

Most of the work on oxygen effects in rhizobia has been done on the nitrogenase system. When microaerophilic suspensions of isolated bacteroids which are actively reducing N_2 are suddenly exposed to air, the rate of synthesis of the two nitrogenase components declines differen-

tially, as measured by pulse labeling followed by two-dimensional electrophoresis (Shaw, 1983). Shaw observed that the rate of synthesis of
component I declined much slower than that of component II. It appears
from a number of studies that the synthesis of components I and II is not
strictly coordinated, with the synthesis of component I preceding that of
component II in nodule development and the synthesis of the two components being differentially affected by oxygen. For example, under waterlogging conditions, nitrogenase activity by nodulated legumes is known
to decrease, presumably because of a decreased oxygen supply. Under
such conditions, the synthesis of component II, but not that of component
I, is repressed (Bisseling et al., 1980). Scott et al. (1979) reported that
component I was synthesized when *B. japonicum* cells were grown
anaerobically; they did not look at component II synthesis. It will be interesting to see what effects oxygen limitation has on the expression of the
nitrogenase genes.

pH and Osmoticum

pH and osmoticum seem likely to be involved in regulation of the symbiosis, possibly at the level of changes in gene expression. As the
rhizobial cells leave the soil and invade the plant root, particularly as they
are released into the host cells, changes in either the pH or the osmoticum
could signal the bacteria to differentiate into the symbiotic bacteroid
form. Although there is no research to date that has directly addressed
these possibilities, there are a few studies on the response of rhizobia to
osmotic stress, which have been prompted by the detrimental effects of
salinity on nodulation, plant growth, and nitrogen fixation. These studies
have shown that various strains of rhizobia accumulate K^+ glutamate
when grown under conditions of high osmolarity (Hua et al., 1982; Yap
and Lim, 1983; Yelton et al., 1983; Botsford, 1984) and may be pertinent
if the osmoticum is involved in the symbiotic process. The increase in glutamate and K^+ is all that is presently known about osmoregulation in
rhizobia, but an increase in the amino acid pool is a fundamental response
of gram-negative bacteria to osmotic stress (Measures, 1975). Therefore
it is likely that the concept of osmosensory proteins linking environmental
changes in osmotic strength to changes in gene expression, as developed
in *E. coli*, may also apply to rhizobia (LeRudulier et al., 1984).

SUMMARY

This chapter serves to illustrate that it is difficult to separate the physiology of the host from the physiology of its microsymbiont when trying to

understand how the nodule functions. While we have attempted to point out the benefits of examining both sides of the symbiosis, bacterial mutants which uncouple plant and bacterial aspects of nodulation, such as those described by Finan et al. (1985), may prove to be a powerful tool in analyzing which bacterial factors are necessary to induce host cells to proliferate and form a nodule.

By drawing on what is known about nodule physiology, much progress has been made in identifying rhizobial and plant genes which are important for the symbiosis. It should now be possible to dissect the mechanisms regulating the expression of these genes. As the bacterial partner lends itself more readily to mutational analysis, it seems likely that there will continue to be rapid progress in the molecular analysis of rhizobial gene expression. Studies on the appearance and accumulation of nodule-specific products may yield clues to what is effecting the integration of the plant and bacterial physiologies. Using such an approach, Fuller and Verma (1984) have shown that the time course of the induction of nodule-specific plant mRNA is distinct for different rhizobial strains when each is inoculated into identical hosts. These time course analyses also indicate that nodulin activity is necessary for creating an environment within the nodule which allows induction and utilization of rhizobial nitrogen fixation gene products. No comparable study on the developmental accumulation of bacterial mRNAs in an effective symbiosis has been reported. By identifying classes of both plant and bacterial genes which appear to be regulated coordinately, it may prove possible to establish the sequence of regulatory events which occur during nodule development.

ACKNOWLEDGMENTS

We thank all the authors who generously provided reprints and preprints of their manuscripts, Tom Adams and Rob McClung for a critical reading of the chapter, and Andrew Hanson and all the members of our laboratory for stimulating discussions. The research from our laboratory was supported by grants from the U.S. Department of Energy, Division of Biological Energy Research, and the U.S. Department of Agriculture, Science and Education Division.

REFERENCES

ADAMS, T. H., AND CHELM, B. K. 1984. The *nif*H and *nif*DK promoter regions from *Rhizobium japonicum* share structural homologies with each other and with nitrogen-related promoters from other organisms, *J. Mol. Appl. Genet.* **2**:392–405.

ADAMS, T. H., McCLUNG, C. R., AND CHELM, B. K. 1984. Physical organization of the *Bradyrhizobium japonicum* nitrogenase gene region, *J. Bacteriol.* **159:**857–862.

AGARWAL, A. K., AND KEISTER, D. L. 1981. Physiology of *ex planta* nitrogenase activity in *Rhizobium japonicum, Appl. Environ. Microbiol.* **45:**1592–1601.

AMRHEIN, N. 1977. The current status of cyclic AMP in higher plants, *Annu. Rev. Plant Physiol.* **28:**123–132.

APRISON, M. H., MAGEE, W. E., AND BURRIS, R. H. 1954. Nitrogen fixation by excised soybean root nodules, *J. Biol. Chem.* **208:**29–39.

APPLEBY, C. A. 1984. Leghemoglobin and *Rhizobium* respiration, *Annu. Rev. Plant Physiol.* **35:**443–478.

ARP, D. J., AND BURRIS, R. H. 1979. Purification and properties of the particulate hydrogenase from the bacteroids of soybean root nodules, *Biochem. Biophys. Acta* **570:**221–230.

ARP, D. J., AND ZUMFT, W. G. 1983. Regulation and control of nitrogenase activity, in: *Nitrogen Fixation: The Chemical-Biochemical-Genetic Interface* (A. Miller and W. E. Newton, eds.), pp. 149–179, Plenum Press, New York.

ATKINS, C. A., RAINBIRD, R. M., AND PATE, J. S. 1980. Evidence for a purine pathway of ureide synthesis in N_2-fixing nodules of cowpea (*Vigna unguiculata* L. Walp.), *Z. Pflanzenphysiol.* **97:**249–260.

AUSUBEL, F. M. 1984. Regulation of nitrogen fixation genes, *Cell* **37:**5–6.

AVISSAR, Y. J., AND NADLER, K. D. 1978. Stimulation of tetrapyrrole formation in *Rhizobium japonicum* by restricted aeration, *J. Bacteriol.* **135:**782–789.

BADENOCH-JONES, J., SUMMONS, R. E., DJORDJEVIC, M. J., SHINE, J., LETHAM, D. S. AND ROLFE, B. G. 1982. Mass spectrometric quantification of Indole-3-acetic acid in *Rhizobium* culture supernatants: Relation to root hair curling and nodule initiation, *Appl. Environ. Microbiol.* **44:**275–280.

BALCOMBE, D., AND VERMA, D. P. S. 1978. Preparation of a complementary DNA for leghemoglobin and direct demonstration that leghemoglobin is encoded by the soybean genome, *Nucleic Acids Res.* **5:**4141–4153.

BANFALVI, A., SAKANYAN, V., KONEZ, C., KISS, A., DUSHA, I., AND KONDOROSI, A. 1981. Location of nodulation and nitrogen fixation genes on a high molecular weight plasmid of *R. meliloti, Mol. Gen. Genet.* **184:**318–325.

BEDMAR, E. J., EDIE, S. A., AND PHILLIPS, D. A. 1983. Host plant cultivar effects on hydrogen evolution by *Rhizobium leguminosarum, Plant Physiol.* **72:**1011–1015.

BERGERSEN, F. J., AND TURNER, G. L. 1967. Nitrogen fixation by the bacteroid fraction of breis of soybean root nodules, *Biochem. Biophys. Acta* **141:**507–515.

BERGMANN, H., PREDDIE, E., AND VERMA, D. P. S. 1983. Nodulin-35: A subunit of specific uricase (uricase II) induced and localized in the uninfected cells of soybean nodules, *EMBO J.* **2:**2333–2339.

BEYNON, J., CANNON, M., BUCHANAN-WOLLASTON, V., AND CANNON, F. 1983. The *nif* promoters of *Klebsiella pneumoniae* have a characteristic primary structure, *Cell* **34:**673–682.

BISSELING, T., VAN STAVEREN, W., AND VAN KAMMEN, A. 1980. The effect of waterlogging on the synthesis of the nitrogenase components in bacteroids of *Rhizobium leguminosarum* in root nodules of *Pisum sativum, Biochem. Biophys. Res. Commun.* **93**:687–693.

BOJSEN, K., ABILDSTEN, D., JENSEN, E. O., PALUDAN, K., AND MARCKER, K. A. 1983. The chromosomal arrangement of six soybean leghemoglobin genes, *EMBO J.* **2**:1165–1168.

BOLAND, M. J., HANKS, J. F., REYNOLDS, P. H. S., BLEVINS, D. G., TOLBERT, N. E., AND SCHUBERT, K. R. 1982. Subcellular organization of ureide biogenesis from glycolytic intermediates and ammonium in nitrogen-fixing soybean nodules, *Planta* **155**:45–51.

BOTSFORD, J. L. 1981. Cyclic nucleotides in prokaryotes, *Microbiol. Rev.* **45**:620–642.

BOTSFORD, J. L. 1984. Osmoregulation in *Rhizobium meliloti:* Inhibition of growth by salts, *Arch. Microbiol.* **137**:124–127.

BRADHAM, L. S., AND CHEUNG, W. Y. 1982. Nucleotide cyclases, *Prog. Nucleic Acid Res. Mol. Biol.* **27**:189–231.

BREWIN, N. J. 1984. Hydrogenase and energy efficiency in nitrogen-fixing symbiosis, in: *Plant Gene Research: Genes Involved in Microbe-Plant Interactions* (D. P. S. Verma and T. Hohn, eds.), pp. 179–203, Springer-Verlag, New York.

BREWIN, N. J., DEJONG, T. M., PHILLIPS, D. A., AND JOHNSTON, A. W. B. 1980. Co-transfer of determinants for hydrogenase activity and nodulation ability in *Rhizobium leguminosarum, Nature* **288**:77–79.

BRISSON, N., AND VERMA, D. P. S. 1982. Soybean leghemoglobin gene family: Normal, pseudo and truncated genes, *Proc. Natl. Acad. Sci. USA* **79**:4055–4059.

BROWN, C. M. AND DILWORTH, M. J. 1975. Ammonia assimilation by *Rhizobium* cultures and bacteroids, *J. Gen. Microbiol.* **86**:39–48.

BUCHANAN-WOLLASTON, V., CANNON, M. C., BEYNON, J. L., AND CANNON, F. C. 1981. Role of the *nif*A gene product in the regulation of *nif* expression in *Klebsiella pneumoniae, Nature* **294**:776–778.

CANNON, M., HIU, S., KAVANAUGH, E. AND CANNON, F. 1985. A molecular genetic study of *nif* expression in *Klebsiella pneumoniae* at the level of transcription, translation and nitrogenase activity, *Mol. Gen. Genet.* **198**:198–206.

CANTRELL, M. A., HICKOK, R. E., AND EVANS, H. J. 1982. Identification and characterization of plasmids in hydrogen uptake positive and hydrogen uptake negative strains of *Rhizobium japonicum, Arch. Microbiol.* **131**:102–106.

CANTRELL, M. A., HAUGHLAND, R. A., AND EVANS, H. J. 1983. Construction of a *Rhizobium japonicum* gene bank and use in isolation of a hydrogen uptake gene, *Proc. Natl. Acad. Sci. USA* **80**:181–185.

CARLSON, T. A., GUERINOT, M. L., AND CHELM, B. K. 1983. Isolation of *Rhizobium japonicum* glutamine synthetase genes, in: *Plant Molecular Biology: UCLA Symposia on Molecular Biology,* New Series, Vol. 12 (R. B. Goldberg, ed.), pp. 291–302, Alan R. Liss, New York.

CARLSON, T. A., GUERINOT, M. L., AND CHELM, B. K. 1985. Characterization of the gene encoding glutamine synthase I (*gln* A) from *Bradyrhizobium japonicum*, *J. Bacteriol.* **162**:698–703.

CERVENANSKY, C., AND ARIAS, A. 1984. Glucose-6-phosphate dehydrogenase deficiency in pleiotropic carbohydrate-negative mutant strains of *Rhizobium meliloti*, *J. Bacteriol.* **160**:1027–1030.

CHAKRAVORTY, A. K., ZURKOWSKI, W., SHINE, J., AND ROLFE, B. J. 1982. Symbiotic nitrogen fixation: Molecular cloning of *Rhizobium* genes involved in exopolysaccharide synthesis and effective nodulation, *J. Mol. Appl. Genet.* **1**:585–596.

CHRISTELLER, J. T., LAING, W. A., AND SUTTON, W. D. 1977. Carbon dioxide fixation by lupin root nodules. I. Characterization, association with phosphoenolpyruvate carboxylase, and correlation with nitrogen fixation during nodule development, *Plant Physiol.* **60**:47–50.

CORBIN, D., BARREN, L., AND DITTA, G. 1983. Organization and expression of *Rhizobium meliloti* nitrogen fixing genes, *Proc. Natl. Acad. Sci. USA* **80**:3005–3009.

CULLIMORE, J. V., AND MIFLIN, B. J. 1983. Glutamine synthetase from the plant fraction of *Phaseolus* root nodules: Purification of the mRNA and *in vitro* synthesis of the enzyme, *FEBS Lett.* **158**:107–112.

CULLIMORE, J. V., LEA, P. J., MIFLIN, B. J. 1982. Multiple forms of glutamine synthetase in the plant fraction of *Phaseolus* root nodules, *Isr. J. Bot.* **31**:151–162.

CULLIMORE, J. V., GEBHARDT, C., SAARELAINEN, R., MIFLIN, B. J., IDLER, K. B., AND BARKER, R. F. 1984. Glutamine synthetase of *Phaseolus vulgaris* L.: Organ-specific expression of a multigene family, *J. Mol. Appl. Genet.* **2**:589–599.

CUTTING, J. A., AND SCHULMAN, H. M. 1969. The site of heme synthesis in soybean root nodules, *Biochim. Biophys. Acta* **192**:486–493.

DANIEL, R. M., AND APPLEBY, C. A. 1972. Anaerobic nitrate, symbiotic and aerobic growth of *Rhizobium japonicum:* Effects on cytochrome P-450, other haemoproteins, nitrate and nitrite reductases, *Biochim. Biophys. Acta* **275**:347–354.

DARROW, R. A. 1980. Role of glutamine synthetase in nitrogen fixation, in: *Glutamine Synthetase: Metabolism, Enzymology and Regulation* (J. Mora and R. Palacios, eds.), pp. 139–166, Academic Press, New York.

DARROW, R. A., AND KNOTTS, R. R. 1977. Two forms of glutamine synthetase in free-living root-nodule bacteria, *Biochem. Biophys. Res. Commun.* **78**:554–559.

DARROW, R. A., CRIST, D., EVANS, W. R., JONES, B. L., DEISTER, D. L., AND KNOTTS, R. R. 1981. Biochemical and physiological studies on the two glutamine synthetases of *Rhizobium*, in: *Current Perspectives in Nitrogen Fixation* (A. H. Gibson and W. E. Newton, eds.), pp. 182–185, Australian Academy of Science, Canberra.

DAZZO, F. B., AND BRILL, W. J. 1978. Bacterial polysaccharide which binds *Rhizobium trifolii* to clover root hairs, *J. Bacteriol.* **137**:1362–1373.

DAZZO, F. B., AND TRUCHET, G. L. 1983. Interactions of lectins and their saccharide receptors in the *Rhizobium*-legume symbiosis, *J. Membrane Biol.* **73**:1–16.

DEHOLLAENDER, J. A., AND STOUTHAMER, A. H. 1979. Multicarbon substrate growth of *Rhizobium trifolii*, *FEMS Microbiol. Lett.* **6**:57–59.

DEVI, A. L., AND MCCURDY, H. D. 1984. Adenylate cyclase and guanylate cyclase in *Myxococcus xanthus*, *J. Gen. Microbiol.* **130**:1851–1856.

DEVRIES, G. E., VAN BRUSSEL, A. A. N., AND QUISPEL, A. 1982. Mechanism and regulation of glucose transport in *Rhizobium leguminosarum*, *J. Bacteriol.* **149**:872–879.

DILWORTH, M. J., MCKAY, I., FRANKLIN, M., AND GLENN, A. R. 1983. Catabolite effects on enzyme induction and substrate utilization in *Rhizobium leguminosarum*, *J. Gen. Microbiol.* **129**:359–366.

DITTA, G., STANFIELD, S., CORBIN, D., AND HELINSKI, D. R. 1980. Broad host range DNA cloning system for gram-negtive bacteria: Construction of a gene bank of *Rhizobium meliloti*. *Proc. Natl. Acad. Sci. USA* **77**:7347–7351.

DIXON, R. O. D. 1972. Hydrogenase in legume root nodule bacteroids: Occurrence and properties, *Arch. Microbiol.* **85**:193–201.

DOWNIE, J. A., MA, O. -S., KNIGHT, C. D., HOMBRECHER, G., AND JOHNSTON, A. W. B. 1983. Cloning of the symbiotic region of *Rhizobium leguminosarum:* The nodulation genes are between the nitrogenase genes and a *nif*A-like gene, *EMBO J.* **2**:947–952.

DREVON, J. J., FRAZIER, L., RUSSELL, S. A., AND EVANS, H. J. 1982. Respiratory and nitrogenase activities of soybean nodules formed by hydrogen uptake negative (Hup⁻) mutant and reverant strains of *Rhizobium japonicum* characterized by protein patterns, *Plant Physiol.* **70**:1341–1346.

DUDMAN, W. F. 1977. The role of surface polysaccharides in natural environments, in: *Surface carbohydrates of the Prokaryotic Cell* (I. Sutherland, ed.), pp. 357–414, Academic Press, New York.

DULLART, J. 1967. Quantitative estimation of indoleacetic acid and indolecarboxylic acid in root nodules and roots of *Lupinus luteus* L., *Acta Bot. Neerl.* **16**:222–230.

DUNCAN, M. J. 1981. Properties of Tn5-induced carbohydrate mutants in *Rhizobium meliloti*, *J. Gen. Microbiol.* **122**:61–67.

DUNCAN, M. J., AND FRAENKEL, D. G. 1979. α-Ketoglutarate dehydrogenase mutant of *Rhizobium meliloti*, *J. Bacteriol.* **137**:415–419.

EISBRENNER, G., AND EVANS, H. J. 1982a. Spectral evidence for a component involved in hydrogen metabolism of soybean root nodules, *Plant Physiol.* **70**:1667–1672.

EISBRENNER, G., AND EVANS, H. J. 1982b. Carriers in electron transport from molecular hydrogen to oxygen in *Rhizobium japonicum* bacteroids, *J. Bacteriol.* **149**:1005–1012.

EISBRENNER, G., AND EVANS, H. J. 1983. Aspects of hydrogen metabolism in nitrogen-fixing legumes and other plant-microbe associations, *Annu. Rev. Plant Physiol.* **34**:105–136.

ELKAN, G. H., AND KUYKENDALL, L. D. 1982. Carbohydrate metabolism, in: *Nitrogen fixation*, Vol. 2, *Rhizobium* (W. J. Broughton, ed.), pp. 147–166, Clarendon Press, Oxford.

EMERICH, D. W. 1985. Characterization of carbon metabolism in *Rhizobium japonicum* bacteroids, in: *Nitrogen Fixation and CO_2 Metabolism*, (P. W. Ludden and J. E. Burris, eds.), pp. 21–30, Elsevier, New York.

EMERICH, D. W., RUIZ-ARGUESO, T., CHING, T. M., AND EVANS, H. J. 1979. Hydrogen dependent nitrogenase activity and ATP formation in *Rhizobium japonicum* bacteroids, *J. Bacteriol.* **137**:153–160.

FINAN, T. M., WOOD, J. M., AND JORDAN, D. C. 1981. Succinate transport in *Rhizobium leguminosarum*, *J. Bacteriol.* **148**:193–202.

FINAN, T. M., WOOD, J. M., AND JORDAN, D. C. 1983. Symbiotic properties of C_4-dicarboxylic acid transport mutants of *Rhizobium leguminosarum*. *J. Bacteriol.* **154**:1403–1413.

FINAN, T. M., HIRSCH, A. M., LEIGH, J. A., JOHANSEN, E., KULDAU, G. A., DEEGAN, S., WALKER, G. C., AND SIGNER, E. R. 1985. Symbiotic mutants of *Rhizobium meliloti* that uncouple plant from bacterial differentiation, *Cell* **40**:869–877.

FUCHSMAN, W. H., AND APPLEBY, C. A. 1979. Separation and determination of the relative concentrations of the homogeneous components of soybean leghemoglobin by isoelectric focussing, *Biochim. Biophys. Acta* **579**:314–324.

FULLER, F., AND VERMA, D. P. S. 1984. Appearance and accumulation of nodulin mRNAs and their relationship to the effectiveness of root nodules. *Plant Mol. Biol.* **3**:21–28.

GARDIOL, A., ARIAS, A., CERVENANSKY, C., AND MARTINEZ-DRETS, G. 1982. Succinate dehydrogenase mutant of *Rhizobium meliloti*. *J. Bacteriol.* **151**:1621–1623.

GARG, R. C., GARG, R. P., KUKREJA, K., SINDHU, S. S., AND TAURO, P. 1985. Host-dependent expression of uptake hydrogenase in cowpea rhizobia, *J. Gen. Microbiol.* **131**:93–96.

GERSON, T., PATEL, J. J., AND WONG, M. N. 1978. The effects of age, darkness, and nitrate on poly-β-hydroxybutyrate levels and nitrogen-fixing ability of *Rhizobium* in *Lupinus angustifolius*, *Physiol. Plant* **42**:420–424.

GLENN, A. R., AND BREWIN, N. J. 1981. Succinate-resistant mutants of *Rhizobium leguminosarum*, *J. Gen. Microbiol.* **126**:237–241.

GLENN, A. R., AND DILWORTH, M. J. 1981. The uptake and hydrolysis of dissaccharides by fast and slow-growing species of *Rhizobium*, *Arch. Microbiol.* **129**:233–239.

GLENN, A. R., POOLE, P. S., AND HUDMAN, J. F. 1980. Succinate uptake by free-living and bacteroid forms of *Rhizobium leguminosarum*, *J. Gen. Microbiol.* **119**:367–371.

GLENN, A. R., MCKAY, I. A., ARWAS, R., AND DILWORTH, M. J. 1984. Sugar metabolism and the symbiotic properties of carbohydrate mutants of *Rhizobium leguminosarum*, *J. Gen. Microbiol.* **130**:239–245.

GODFREY, C. A., AND DILWORTH, M. J., 1971. Haem biosynthesis from [^{14}C]-δaminolaevulinic acid in laboratory grown and root nodule *Rhizobium lupini*, *J. Gen. Microbiol.* **69**:385–390.

GRAHAM, L. A., STULTS, L. W., AND MAIER, R. J. 1984. Nitrogenase-hydrogenase relationships in *Rhizobium japonicum*, *Arch. Microbiol.* **140**:243–246.

GREENE, E. M. 1980. Cytokinin production by microorganisms, *Bot. Rev.* **46**:25–74.

GROAT, R. G., AND VANCE, C. P. 1981. Root nodule enzymes of ammonia assimilation in alfalfa (*Medicago sativa* L.). Developmental patterns and response to applied nitrogen, *Plant Physiol.* **67**:1198–1203.

GUARENTE, L., AND MASON, T. 1983. Heme regulates transcription of the CYC1 gene of *S. cerevisiae* via an upstream activation site, *Cell* **32**:1279–1286.

GUARENTE, L., LALONDE, B., GIFFORD, P., AND ALANI, E. 1984. Distinctly regulated tandem upstream activation sites mediate catabolite repression of the CYC1 gene of *S. cerevisiae*, *Cell* **36**:503–511.

GUERINOT, M. L., AND CHELM, B. K. 1984. Isolation and expression of the *Bradyrhizobium japonicum* adenylate cyclase gene (*cya*) in *Escherichia coli*, *J. Bacteriol.* **159**:1068–1071.

GUERINOT, M. L., AND CHELM, B. K. 1986. Leghemoglobin formulation in the soybean/*Bradyrhizobium japonicum* symbiosis: Bacterial δ-aminolevulinic acid synthetase activity is not essential, *Proc. Natl. Acad. Sci. USA* in press.

HANKS, J. F., SCHUBERT, K., AND TOLBERT, N. E. 1983. Isolation and characterization of infected and uninfected cells from soybean nodules, *Plant Physiol.* **71**:869–873.

HANUS, F. J., MAIER, R. J., AND EVANS, H. J. 1979. Autotrophic growth of H_2-uptake-positive strains of *Rhizobium japonicum* in an atmosphere supplied with hydrogen gas, *Proc. Natl. Acad. Sci. USA* **76**:1788–1792.

HARDY, R. W. F., AND HAVELKA, U. D. 1976. Photosynthate as a major limitation to N_2 fixation by field grown legumes with an emphasis on soybeans, in: *International Biological Programme: Symbiotic Nitrogen Fixation in Plants* (P. S. Nutnam, ed.), pp. 421–439, Cambridge University Press, New York.

HARKER, A. R., XU, L. -S., HANUS, F. J., AND EVANS, H. J. 1984. Some properties of the nickel-containing hydrogenase of chemolithotrophically grown *Rhizobium japonicum*, *J. Bacteriol.* **159**:850–856.

HAUGLAND, R. A., HANUS, F. J., CANTRELL, M. A., AND EVANS, H. J. 1983. Rapid colony screening method for identifying hydrogenase activity in *Rhizobium japonicum*, *Appl. Environ. Microbiol.* **45**:892–897.

HAUGLAND, R. A., CANTRELL, M. A., BEATY, J. S., HANUS, F. J., RUSSELL, S. A., AND EVANS, H. J. 1984. Characterization of *Rhizobium japonicum* hydrogen uptake genes, *J. Bacteriol.* **159**:1006–1012.

HAVELKA, U. D., AND HARDY, R. W. F. 1976. Legume N_2 fixation as a problem of carbon nutrition, in: *Proceedings of the 1st International Symposium on Nitrogen Fixation* (W. E. Newton and C. Y. Nyman, eds.), pp. 456–475, Washington State University Press, Pullman.

HENNECKE, H. 1981. Recombinant plasmids carrying nitrogen fixation genes from *Rhizobium japonicum*, *Nature* **291**:354–355.

HENSEN, I. E., AND WHEELER, C. T. 1976. Hormones in plants bearing nitrogen-fixing root nodules: The distribution of cytokinins in *Vicia faba* L., *New Phytol.* **76**:433–439.

HILL, S., KENNEDY, C., AND KAVANAUGH, E. 1981. Nitrogen fixation gene (*nif*L) involved in oxygen regulation of nitrogenase synthesis in *Klebsiella pneumoniae, Nature* **290**:424–426.

HIRSCH, A. M., WILSON, K. J., JONES, J. D. G., BANG, M., WALKER, V. V., AND AUSUBEL, F. M. 1984. *Rhizobium meliloti* genes allow *Agrobacterium tumefaciens* and *Escherichia coli* to form pseudonodules on alfalfa, *J. Bacteriol.* **158**:1133–1143.

HIRSCH, A. M., DRAKE, D., JACOBS, T. W., AND LONG, S. R. 1985. Nodules are induced on alfalfa roots by *Agrobacterium tumefaciens* containing small segments of the *Rhizobium meliloti* nodulation region, *J. Bacteriol.* **161**:223–230.

HOM, S. S. M., GRAHAM, L. A., AND MAIER, R. J. 1985. Isolation of genes (*nif*/*hup* cosmids) involved in hydrogenase and nitrogenase activities in *Rhizobium japonicum, J. Bacteriol.* **161**:822–887.

HUA, S. S. T., TASI, V. Y., LICHENS, G. M., AND NOMA, A. T. 1982. Accumulation of amino acids in *Rhizobium* sp. strain WR1001 in response to sodium chloride salinity, *Appl. Environ. Microbiol.* **44**:135–140.

HUDMAN, J. F., AND GLENN, A. R. 1980. Glucose uptake by free-living and bacteroid forms of *Rhizobium leguminosarum, Arch. Microbiol.* **128**:72–78.

HYLDIG-NIELSEN, J. J., JENSEN, E. O., PALUDAN, K., WIBORG, O., GARRETT, R., JORGENSEN, O. P., AND MARCKER, K. A. 1982. The primary structures of two leghemoglobin genes from soybean, *Nucleic Acids Res.* **10**:689–701.

JENSEN, E. O., PALUDAN, K., HYLDIG-NIELSEN, J. J., JORGENSEN, P., AND MARCKER, K. A. 1981. The structure of chromosomal leghemoglobin gene from soybean. *Nature* **291**:677–679.

JOHNSON, G. V., EVANS, H. J., AND CHING, T. M. 1966. Enzymes of the glyoxylate cycle in rhizobia and nodules of legumes, *Plant Physiol.* **41**:1330–1336.

JOHNSON, L. P., MACLEOD, J. L., PARKER, C. W., LETHAM, D. S., AND HUNT, N. H. 1981. Identification and quantitation of adenosine-3′:5′-cyclic monophosphate in plants using gas chromatography-mass spectrometry and high-performance liquid chromatography, *Planta* **152**:195–201.

JORDAN, D. C. 1982. Transfer of *Rhizobium japonicum* Buchanan 1980 to *Bradyrhizobium* gen. nov., a genus of slow-growing, root nodule bacteria from leguminous plants, *Int. J. Syst. Bacteriol.* **32**:136–139.

KALLAS, T., REBIERE, M. C., RIPPKA, R., AND TANDEAU DE MARSAC, N. 1983. The structural *nif* genes of the cyanobacteria *Gloeothece* sp. and *Calothrix* sp. share homology with those of *Anabaena* sp. but the *Gloeothece* genes have a different arrangement, *J. Bacteriol.* **155**:427–431.

KALUZA, K., AND HENNECKE, H. 1984. Fine structure analysis of the *nif*DK operon encoding the α and β subunits of dinitrogenase from *Rhizobium japonicum, Mol. Gen. Genet.* **196**:35–42.

KALUZA, K., FUHRMANN, M., HAHN, M., REGENSBURGER, B., AND HENNECKE, H. 1983. In *Rhizobium japonicum* the nitrogenase genes *nif*H and *nif*DK are separated, *J. Bacteriol.* **155**:915–918.

KENNEDY, C., CANNON, F., CANNON, M., DIXON, R., HIU, S., JENSEN, J., KUMAR, S., MCLEAN, P., MERRICK, M., ROBSON, R., AND POSTGATE, J. 1981. Recent advances in the genetics and regulation of nitrogen fixation, in: *Current Perspectives in Nitrogen Fixation* (A. H. Gibson and W. E. Newton, eds.), pp. 146–156, Australian Academy of Sciences, Canberra.

KEYSER, H. H., VANBERKUM, P., AND WEBER, D. F. 1982. A comparative study of the physiology of symbioses formed by *Rhizobium japonicum* with *Glycine max*, *Vigna ungiculata* and *Macroptilium atropurpurem*, *Plant Physiol.* **70**:1626–1630.

KIELY, B., AND O'GARA, F. 1983. Cyclic 3′5′-adenosine monophosphate synthesis in *Rhizobium:* Identification of a cloned sequence from *Rhizobium meliloti* coding for adenyl cyclase, *Mol. Gen. Genet.* **192**:230–234.

KLUCAS, R. V., AND EVANS, H. J. 1968. An electron donor system for nitrogenase-dependent acetylene reduction by extracts of soybean nodules. *Plant Physiol.* **43**:1458–1460.

KLUCAS, R. V., HANUS, F. J., RUSSELL, S. A., AND EVANS, H. J. 1983. Nickel: A micronutrient element for hydrogen-dependent growth of *Rhizobium japonicum* and for expression of urease activity in soybean leaves. *Proc. Natl. Acad. Sci. USA* **80**:2253–2257.

KOCH, B., EVANS, H. J., AND RUSSELL, S. 1967. Reduction of acetylene and nitrogen gas by breis and cell-free extracts of soybean root nodules, *Plant Physiol.* **42**:466–468.

KONDOROSI, A., SVAB, Z., KISS, G. B., AND DIXON, R. A. 1977. Ammonia assimilation and nitrogen fixation in *Rhizobium meliloti*, *Mol. Gen. Genet.* **151**:221–226.

KONDOROSI, E., BANFALVI, A., AND KONDOROSI, A. 1984. Physical and genetic analysis of a symbiotic region of *Rhizobium meliloti:* Identification of nodulation genes, *Mol. Gen. Genet.* **193**:445–452.

KRETOVICH, W. L., ROMANOV, V. I., YUSHKOVA, L. A., SHRAMKO, V. I., AND FEDULOVA, N. G. 1977. Nitrogen fixation and poly-β-hydroxybutyric acid content in bacteroids of *Rhizobium lupini* and *Rhizobium leguminosarum*, *Plant Soil* **48**:291–302.

KRIEG, N. R. 1984. *Bergey's Manual of Systematic Bacteriology*, Williams and Wilkins, Baltimore, pp. 234–244.

KURZ, W. G. W., AND LARUE, T. A. 1975. Nitrogenase activity in rhizobia in absence of plant host. *Nature* **256**:407–408.

KUYKENDALL, L. D., AND ELKAN, G. H. 1977. Some features of mannitol metabolism in *Rhizobium japonicum*, *J. Gen. Microbiol.* **98**:291–295.

LARA, M., CULLIMORE, J. V., LEA, P. J., MIFLIN, B. J., JOHNSTON, A. W. B., LAMB, J. W. 1983. Appearance of a novel form of plant glutamine synthetase during nodule development in *Phaseolus vulgaris* L., *Planta* **157**:254–258.

LAW, I. J., YAMAMOTO, Y., AND BAUER, W. D. 1982. Nodulation of soybean by *Rhizobium japonicum* mutants with altered capsule synthesis. *Planta* **154**:100–109.

LAWRIE, A. C., AND WHEELER, C. T. 1975. Nitrogen fixation in the root nodules of *Vicia faba* L. in relation to the assimilation of carbon. II. The dark fixation of carbon dioxide, *New Phytol.* **74**:437–445.

LEA, P. J., AND MIFLIN, B. J. 1980. Transport and metabolism of asparagine and other nitrogen compounds within the plant, in: *The Biochemistry of Plants* (B. J. Milfin, ed.), pp. 569–607, Vol. 5, Academic Press, New York.

LEE, J. S., BROWN, G. C., VERMA, D. P. S. 1983. Chromosomal arrangements of leghemoglobin genes in soybean, *Nucleic Acids Res.* **11**:5541–5553.

LEIGH, J. A., SIGNER, E. R., AND WALKER, G. S. 1985. Exopolysaccharide-deficient mutants of *Rhizobium meliloti* that form ineffective nodules. *Proc. Natl. Acad. Sci. USA*, **82**:6231–6235, 1985.

LEONG, S. A., DITTA, G. S., AND HELINSKI, D. R. 1982. Heme biosynthesis in *Rhizobium:* Identification of a cloned gene coding for δ-aminolevulinic acid synthetase from *Rhizobium meliloti*, *J. Biol. Chem.* **257**:8724–8730.

LEPO, J. E., HANUS, F. J., AND EVANS, H. J. 1980. Chemautotrophic growth of hydrogen-uptake-positive strains of *Rhizobium japonicum*, *J. Bacteriol.* **141**:664–670.

LEPO, J. E., HICKOK, R. E., CANTRELL, M. A., RUSSELL, S. A., AND EVANS, H. J. 1981. Revertible hydrogen uptake-deficient mutants of *Rhizobium japonicum*, *J. Bacteriol.* **146**:614–620.

LERUDULIER, D., STROM, A. R., DANDEKAR, A. M., SMITH, L. T., AND VALENTINE, R. C. 1984. Moleculer biology of osmoregulation, *Science* **224**: 1064–1068.

LIBBENGA, K. R., VAN IREN, R., BOGERS, R. J., AND SHRAAG-LAMERS, M. F. 1973. The role of hormones and gradients in the initiation of cortex proliferation and nodule formation in *Pisum sativum* L., *Planta* **114**:29–39.

LIM, S. T. AND SHANMUGAM, K. T. 1979. Regulation of hydrogen utilization in *Rhizobium japonicum* by cAMP, *Biochim. Biophys. Acta* **584**:479–492.

LIM, S. T., HENNECKE, H., AND SCOTT, D. B. 1979. Effect of cyclic guanosine 3',5'-monophosphate on nitrogen fixation in *Rhizobium japonicum*, *J. Bacteriol.* **139**:256–263.

LONG, S. R., BUIKEMA, W. J., AND AUSBEL, F. M. 1982. Cloning of *R. meliloti* nodulation genes by direct complementation of Nod⁻ mutants, *Nature* **298**:485–488.

LOPEZ, M., CARBONERO, V., CABRERA, E., AND RUIZ-ARGUESO, T. 1983. Effects of host on the expression of the H₂-uptake hydrogenase of *Rhizobium* in legume nodules, *Plant Sci. Lett.* **29**:191–199.

LUDWIG, R. A. 1980. Physiological roles of glutamine synthetases I and II in ammonium assimilation in *Rhizobium* sp. 32H1, *J. Bacteriol.* **141**:1209–1216.

MA, Q. -S, JOHNSTON, A. W. B., HOMBRECHER, G., AND DOWNIE, J. A. 1982. Molecular genetics of mutants of *Rhizobium leguminosarum* which fail to fix nitrogen, *Mol. Gen. Genet.* **187**:166–171.

MAGASANIK, B. 1961. Catabolite-repression, *Cold Spring Harbor Symp. Quant. Biol.* **26**:244–256.

MAGASANIK, B. 1982. Genetic control of nitrogen assimilation in bacteria, *Annu. Rev. Genet.* **16**:135–168.

MAIER, R. J. 1981. *Rhizobium japonicum* mutant strains unable to grow chemoautotrophically with H₂, *J. Bacteriol.* **145**:533–540.

MAIER, R. J., AND MERBERG, D. M. 1982. *Rhizobium japonicum* mutants that are hypersensitive to repression of H_2 uptake by oxygen, *J. Bacteriol.* **150**:161–167.

MAIER, R. J., CAMPBELL, N. E. R., HANUS, F. J., SIMPSON, F. B., RUSSELL, S. A., AND EVANS, H. J. 1978a. Expression of hydrogenase activity in free-living *Rhizobium japonicum, Proc. Natl. Acad. Sci. USA* **75**:3258–3262.

MAIER, R. J., POSTGATE, J. R., AND EVANS, H. J. 1978b. *Rhizobium japonicum* mutants unable to use H_2, *Nature* **276**:494–495.

MAIER, R. J., HANUS, F. J., AND EVANS, H. J. 1979. Regulation of hydrogenase in *Rhizobium japonicum, J. Bacteriol.* **137**:824–829.

MANIAN, S. S., AND O'GARA, F. 1982. Induction and regulation of ribulose bisphosphate carboxylase activity in *Rhizobium japonicum* during formate-dependent growth, *Arch. Microbiol.* **131**:51–54.

MANIAN, S. S., GUMBLETON, R., AND O'GARA, F. 1982. The role of formate metabolism in nitrogen fixation in *Rhizobium* spp., *Arch. Microbiol.* **133**:312–317.

MANIAN, S. S., GUMBLETON, R., BUCKLEY, A. M., AND O'GARA, F. 1984. Nitrogen fixation and carbon dioxide assimilation in *Rhizobium japonicum, Appl. Environ. Microbiol.* **48**:276–279.

MARCKER, A., LUND, M., JENSEN, E. O., AND MARCKER, K. A. 1984. Transcription of the soybean leghemoglobin genes during nodule development. *EMBO J.* **3**:1691–1695.

MARTINEZ DE DRETS, G., AND ARIAS, A. 1970. Metabolism of some polyols by *Rhizobium meliloti, J. Bacteriol.* **103**:97–103.

MAURO, V. P., NGUYEN, T., KATINAKIS, P., AND VERMA, D. P. S. 1985. Primary structure of the soybean nodulin-23 gene and potential regulatory elements in the 5'-flanking regions of nodulin and leghemoglobin genes, *Nucleic Acids Res.* **13**:239–260.

McALLISTER, C. F., AND LEPO, J. E. 1983. Succinate transport by free-living forms of *Rhizobium japonicum, J. Bacteriol.* **153**:1155–1162.

McCOMB, J. A., ELLIOTT, J., AND DILWORTH, M. J. 1975. Acetylene reduction by *Rhizobium* in pure culture, *Nature* **256**:409–410.

McPARLAND, R. H., GUEVARA, J. G., BECKER, R. R., AND EVANS, H. J. 1976. The purification of the glutamine synthetase from the cytosol of soya-bean root nodules, *Biochem. J.* **153**:597–606.

MEASURES, J. C. 1975. Role of amino acids in osmoregulation of nonhalophilic bacteria, *Nature* **257**:398–400.

MERBERG, D., AND MAIER, R. J. 1983. Mutants of *Rhizobium japonicum* with increased hydrogenase activity, *Nature* **220**:1064–1065.

MERBERG, D., O'HARA, E. B., AND MAIER, R. J. 1983. Regulation of hydrogenase in *Rhizobium japonicum:* An analysis of mutants altered in regulation by carbon substrates and oxygen, *J. Bacteriol.* **156**:1236–1242.

MIFLIN, B. J., AND CULLIMORE, J. V. 1984. Nitrogen assimilation in the legume-*Rhizobium* symbiosis: A joint endeavor, in: *Plant Gene Research: Genes Involved in Microbe-Plant Interactions* (D. ᛆ. S. Verma and T. Hohn, eds.) pp. 58–93, Springer-Verlag, New York.

MINCHIN, F. R., AND PATE, J. S. 1973. The carbon balance of a legume and the functional economy of its root nodules, *J. Exp. Bot.* **24:**259–271.

MOHAPTRA, S. S., AND GRESSHOFF, P. M. 1984. Carbon-nitrogen requirements for the expression of nitrogenase activity in cultures *Parasponia-Rhizobium* strain ANU289, *Arch. Microbiol.* **137:**58–62.

MORTENSEN, L. E., AND THORNELEY, R. N. F. 1979. Structure and function of nitrogenase, *Annu. Rev. Biochem.* **48:**387–418.

MOSHIRI, F., STULTS, L., NOVAK, P., AND MAIER, R. J. 1983. Nif⁻ Hup⁻ mutants of *Rhizobium japonicum, J. Bacteriol.* **155:**926–929.

MOUSTAFA, E., AND HASTINGS, A. 1973. Effect of dibutyryl cyclic AMP and other nucleotides on nitrogen fixation by legume root nodules, *Nature* **244:**461–462.

NADLER, K. D., AND AVISSAR, Y. J. 1977. Heme synthesis in soybean root nodules. I. On the role of bacteroid δ-aminolevulinic acid synthase and δ-aminolevulinic acid dehydrase in the synthesis of the heme of leghemoglobin, *Plant Physiol.* **60:**433–436.

NEWCOMB, W. 1980. Control of morphogenesis and differentiation of pea root nodules, in: *Nitrogen Fixation,* Vol. II, *Symbiotic associations and Cyanobacteria* (W. E. Newton and W. H. Orme-Johnson, eds.), pp. 87–102, University Park Press, Baltimore, Maryland.

NEWCOMB, W., SIPPEL, D., AND PETERSON, R. L. 1979. The early morphogenesis of *glycine max* and *Pisum sativum* root nodules, *Can. J. Bot.* **57:**2603–2616.

NOEL, K. D., STACEY, G., TANDON, S. R., SILVER, L. E., AND BRILL, W. J. 1982. *Rhizobium japonicum* nutants defective in symbiotic nitrogen fixation, *J. Bacteriol.* **152:**485–494.

O'BRIAN, M. R., AND MAIER, R. J. 1982. Electron transport components involved in hydrogen oxidation in free-living *Rhizobium japonicum, J. Bacteriol.* **152:**422–430.

O'BRIAN, M. R., AND MAIER, R. J. 1983. Involvement of cytochromes and a flavoprotein in hydrogen oxidation in *Rhizobium japonicum* bacteroids, *J. Bacteriol.* **155:**481–487.

O'BRIAN, M. R., AND MAIER, R. J. 1985a. Role of ubiquinone in hydrogen-dependent electron transport in *Rhizobium japonicum, J. Bacteriol.* **161:**775–777.

O'BRIAN, M. R., AND MAIER, R. J. 1985b. Expression of cytochrome o in hydrogen-uptake constitutive mutants of *Rhizobium japonicum, J. Bacteriol.* **161:**507–514.

O'GARA, F., AND SHANMUGAM, K. T. 1976. Regulation of nitrogen fixation by rhizobia: Export of fixed N_2 as NH_4. *Biochim. Biophys. Acta* **437:**313–321.

O'GARA, R., MANIAN, S., AND MEADE, J. 1984. Isolation of an Asm⁻ mutant of *Rhizobium japonicum* defective in symbiotic N_2 fixation, *FEMS Microbiol. Lett.* **24:**241–245.

OW, D. W., SUNDARESAN, V., ROTHSTEIN, D. M., BROWN, S. E., AND AUSUBEL, F. M. 1983. Promoters regulated by *glnG* (*ntrC*) and *nifA* gene products share a heptameric consensus sequence in the −15 region, *Proc. Natl. Acad. Sci. USA* **80:**2524–2528.

PAGAN, J. D., CHILD, J. J. SCOWCROFT, W. R., AND GIBSON, A. H. 1975. Nitrogen fixation by *Rhizobium* cultured on a defined medium, *Nature* **256**:406–407.

PANKHURST, C. E. 1981. Nutritional requirement for the expression of nitrogenase activity by *Rhizobium* sp. in agar culture, *J. Appl. Bacteriol.* **50**:45–54.

PHILLIPS, D. A., DANIEL, R. M., APPLEBY, C. A., AND EVANS, H. J. 1973. Isolation from *Rhizobium* of factors which transfer electrons to soybean nitrogenase, *Plant Physiol.* **51**:136–138.

PHILLIPS, D. A., BEDMAR, E. J., WAULSET, C. O., AND TEUBER, L. R. 1985. Host legume control of *Rhizobium* function, in: *Nitrogen Fixation and CO₂ Metabolism* (P. W. Ludden and J. E. Burris, eds.), pp. 203–212, Elsevier, New York.

PRAKASH, R. K., AND ATHERLY, A. G. 1984. Reiteration of genes involved in symbiotic nitrogen fixation by fast growing *Rhizobium japonicum*, *J. Bacteriol.* **160**:758–787.

ORRA, R. J. 1975. A rapid spectrophotometric assay for ferrochelatase activity in preparations containing much endogenous hemoglobin and its application to soybean root-nodule preparations. *Anal. Biochem.* **68**:289–298.

PUPPO, A., AND RIGAUD, J. 1975. Indole-3-acetic acid oxidation by leghemoglobin from soybean nodules, *Physiol. Plant.* **35**:181–185.

QUINTO, C., DE LA VEGA, H., FLORES, M., FERNANDEZ, L., BALLADO, T., SOBERON, G., AND PALACIOS, R. 1982. Reiteration of nitrogen fixation gene sequences in *Rhizobium phaseoli*, *Nature* **200**:724–726.

QUINTO, C., DE LA VEGA, H., FLORES, M., LEEMANS, J., CEVALLOS, M. A., PARDO, M. A., AZPIROZ, R., DE LOURDES GIRARD, M., CARVA, E., AND PALACIOS, R. 1985. Nitrogenase reductase: A functional multigene family in *Rhizobium phaseoli*, *Proc. Natl. Acad. Sci. USA* **82**:1170–1174.

RAINBIRD, R. M., ATKINS, C. A., PATE, J. S., AND SANFORD, P. 1983. Significance of hydrogen evolution in the carbon and nitrogen economy of nodulated cowpea, *Plant Physiol.* **71**:122–127.

RAO, V. R., DARROW, R. A., AND KEISTER, D. L. 1978. Effect of oxygen tension on nitrogenase and on glutamine synthetases I and II in *Rhizobium japonicum* 61A76, *Biochem. Biophys. Res. Commun.* **81**:224–231.

RAWTHORNE, S., MINCHIN, F. R., SUMMERVIELD, R. J., COOKSON, C., AND COOMBS, J. 1980. Carbon and nitrogen metabolism in legume root nodules, *Phytochemistry* **19**:341–355.

REIBACH, P. H., AND STREETER, J. G. 1984. Evaluation of active versus passive uptake of metabolites by *Rhizobium japonicum* bacteroids, *J. Bacteriol.* **159**:47–52.

REYNOLDS, P. H. S., AND FARNDEN, K. J. F. 1979. The involvement of aspartate aminotransferases in ammonium assimilation in lupin nodules, *Phytochemistry* **18**:1625–1630.

REYNOLDS, P. H. S., BLEVINS, D. G., BOLAND, M. J., SCHUBERT, K. R., AND RANDALL, D. D. 1982a. Enzymes of ammonia assimilation in legume nodules: A comparison between ureido- and amide-transporting plants, *Physiol. Plant.* **55**:255–260.

REYNOLDS, P. H. S., BOLAND, M. J., BLEVINS, D. G., SCHUBERT, K. R., AND RANDALL, D. D. 1982b. Enzymes of amide and ureide biogenesis in developing soybean nodules, *Plant Physiol.* **69**:1334–1338.

RICE, D., MAZUR, B. J., AND HASELKORN, R. J. 1982. Isolation and physical mapping of nitrogen fixation genes from the cyanobacterium *Anabaena* 7120, *J. Biol. Chem.* **257**:13157–13163.

RIVERA-ORTIZ, J. M., AND BURRIS, R. H. 1975. Interactions among substrates and inhibitors of nitrogenase, *J. Bacteriol.* **123**:537–545.

ROBERTS, G. P., AND BRILL, W. J. 1981. Genetics and regulation of nitrogen fixation, *Annu. Rev. Microbiol.* **35**:207–235.

ROBERTSON, J. G., AND TAYLOR, M. R. 1973. Acid and alkaline invertase in roots and nodules of *Lupinus angustifolia* infected with *Rhizobium lupini*, *Planta* **112**:1–6.

ROBERTSON, J. G., FARNDEN, K. J. F., WARBURTON, M. P., BANKS, J. M. 1975. Induction of glutamine synthetase during nodule development in lupin, *Aust. J. Plant Physiol.* **2**:265–272.

ROLFE, B., AND SHINE, J. 1984. *Rhizobium*-leguminosae symbiosis: The bacterial point of view, in: *Plant Gene Research: Genes Involved in Microbe-Plant Interactions* (D. P. S. Verma, T. Hohn, eds.), pp. 95–128, Springer-Verlag, New York.

ROMANOV, V. I., FEDULOVA, N. G., TCHERMENSKAYA, I. E., SHRAMKO, V. I., MOLCHANOV, M. I., AND KRETOVICH, W. L. 1980. Metabolism of poly-β-hydroxybutyric acid in bacteroids of *Rhizobium lupini* in connection with nitrogen fixation and photosynthesis, *Plant Soil* **56**:379–390.

RONSON, C. W., AND PRIMROSE, S. B. 1979. Carbohydrate metabolism in *Rhizobium trifolii:* Identification and symbiotic properties of mutants, *J. Gen. Microbiol.* **122**:77–88.

RONSON, C. W., LYTTLETON, P., AND ROBERTSON, J. G. 1981. C_4-dicarboxylate transport mutants of *Rhizobium trifolii* form ineffective nodules on *Trifolium repens*, *Proc. Natl. Acad. Sci. USA* **78**:4284–4288.

RONSON, C. W., ASTWOOD, P. M., AND DOWNIE, J. A. 1984. Molecular cloning and genetic organization of C_4-dicarboxylate transport genes from *Rhizobium leguminosarum*, *J. Bacteriol.* **160**:903–909.

ROSSEN, L., MA, Q. -S., MUDD, E. A., JOHNSTON, A. W. B., AND DOWNIE, J. A. 1984. Identification and DNA sequence of *fixZ*, a *nif*B-like gene from *Rhizobium leguminosarum*, *Nucleic Acids Res.* **12**:7123–7134.

RUIZ-ARGUESO, T., EMERICH, D. W., AND EVANS, H. J. 1979. Characteristics of the H_2-oxidizing hydrogenase system in soybean nodule bacteroids, *Arch. Microbiol.* **121**:199–206.

RUVKUN, G. B., AND AUSUBEL, F. M. 1980. Interspecies homology of nitrogenase genes, *Proc. Natl. Acad. Sci. USA* **77**:191–195.

RUVKUN, G. B., SUNDARESAN, K., AND AUSUBEL, F. M. 1982. Directed transposon Tn5 mutagenesis and complementation analysis of *Rhizobium meliloti* symbiotic nitrogen fixation genes, *Cell* **29**:551–559.

SANDERS, R. E., CARLSON, R. W., ALBERSHEIM, P. 1978. *Rhizobium* mutant incapable of nodulation and normal polysaccharide secretion, *Nature* **271**:240–242.

SAN FRANCISCO, M. J. D., AND JACOBSON, G. R. 1985. Uptake of succinate and malate in cultured cells of bacteroids of two slow-growing species of *Rhizobium, J. Gen. Microbiol.* **131**:765–773.

SCHETGENS, T. M. P., BAKKEREN, G., VANDUN, C., HONTELEZ, J. G. J., VANDENBOS, R. C., AND VANKAMMEN, A. 1984. Molecular cloning and functional characterization of *Rhizobium leguminosarum* structural *nif*-genes by site-directed transposon mutagenesis and expression in *Escherichia coli* minicells, *J. Mol. Appl. Genet.* **2**:406–421.

SCHOFIELD, P. R., DJORDJEVIC, M. A., ROLFE, B. G., SHINE, J., AND WATSON, J. M. 1983. A molecular linkage map of nitrogenase and nodulation genes in *Rhizobium trifolii, Mol. Gen. Genet.* **192**:459–465.

SCHUBERT, K. R. 1981. Enzymes of purine biosynthesis and catabolism in *Glycine max.* I. Comparisons of activities with N_2 fixation and composition of xylem exudate during nodule development, *Plant Physiol.* **68**:1115–1122.

SCOLNIK, P. A., AND HASELKORN, R. 1984. Activation of extra copies of genes coding for nitrogenase in *Rhodopseudomonas* capsulata, *Nature* **307**:289–292.

SCOTT, B. D., ROBERTSON, J. G., AND FARNDEN, K. J. F. 1976. Ammonia assimilation in lupin nodules, *Nature* **263**:703–708.

SCOTT, B. D., HENNECKE, H., AND LIM, S. T. 1979. The biosynthesis of nitrogenase MoFe protein polypeptides in free-living cultures of *Rhizobium japonicum, Biochim. Biophys. Acta* **565**:365–378.

SCOTT, K. F., ROLFE, B. G., AND SHINE, J. 1983a. Biological nitrogen fixation: Primary structure of the *Rhizobium trifolii* iron protein gene, *DNA* **2**:149–155.

SCOTT, K. F., ROLFE, B. G., AND SHINE, J. 1983b. Nitrogenase structural genes are unlinked in the nonlegume symbiont *Parasponia rhizobium, DNA* **2**:141–148.

SEN, D., AND SCHULMAN, H. M. 1980. Enzymes of ammonia assimilation in the cytosol of developing soybean root nodules, *New Phytol.* **85**:243–250.

SHAW, B. D. 1983. Non-coordinate regulation of *Rhizobium* nitrogenase synthesis by oxygen: Studies with bacteroids from nodulated *Lupinus angustifolius, J. Gen. Microbiol.* **129**:849–857.

SHELP, B. J., ATKINS, C. A., STORER, P. J., AND CANVIN, D. T. 1983. Cellular and subcellular organization of pathways of ammonia assimilation and ureide synthesis in nodules of cowpea (*Vigna unguiculata* L. Walp.),*Arch. Biochem. Biphys.* **224**:429–441.

SIDLOI-LUMBROSCO, R., KLEINMAN, L., AND SCHULMAN, H. M. 1978. Biochemical evidence that leghemoglobin genes are present in the soybean but not in the *Rhizobium* genome, *Nature* **273**:558–560.

SIEGEL, L. S., HYLEMON, P. B., AND PHIBBS, P. V. 1977. Cyclic adenosine 3', 5'-monophosphate levels and activities of adenylate cyclase and cyclic adenosine 3', 5'-monophosphate phosphodiesterase in *Pseudomonas* and *Bacteriodes, J. Bacteriol.* **129**:87–96.

SIMPSON, F. B., MAIER, R. J., AND EVANS, H. J. 1979. Hydrogen-stimulated CO_2 fixation and coordinate induction of hydrogen and ribulose bisphosphate carboxylase in a H_2-uptake positive strain of *Rhizobium japonicum, Arch. Microbiol.* **123**:1–8.

SOMERVILLE, J. E., AND KAHN, M. L. 1983. Cloning of the glutamine synthetase I gene from *Rhizobium meliloti, J. Bacteriol.* **156**:168–176.

STAM, H., VAN VERSEVELD, H. W., AND STOUTHAMER, A. H. 1983. Depression of nitrogenase in chemostat cultures of the fast growing *Rhizobium leguminosarum, Arch. Microbiol.* **135**:199–204.

STOVALL, I., AND COLE, M. 1978. Organic acid metabolism by isolated *Rhizobium japonicum* bacteroids, *Plant Physiol.* **61**:787–790.

STOWERS, M. D., AND ELKAN, G. H. 1983. The transport and metabolism of glucose in cowpea rhizobia, *Can. J. Microbiol.* **29**:398–406.

STREETER, J. G. 1980. Carbohydrates in soybean nodules. II. Distribution of compounds in seedlings during the onset of nitrogen fixation, *Plant Physiol.* **66**:471–476.

STULTS, L. W., O'HARA, E. B., AND MAIER, R. J. 1984. Nickel is a component of hydrogenase in *Rhizobium japonicum, J. Bacteriol.* **159**:153–158.

STUMPF, D. K., AND BURRIS, R. H. 1979. A micromethod for the purification and quantification of organic acids of the tricarboxylic acid cycle in plant tissues, *Anal. Biochem.* **95**:311–315.

SUNDARESAN, V., JONES, J. D. G., OW, D. W., AND AUSUBEL, F. M. 1983a. *Klebsiella pneumoniae nif*A product activates the *Rhizobium meliloti* nitrogenase promoter, *Nature* **301**:728–732.

SUNDARESAN, V., OW, D. W., AND AUSUBEL, F. M. 1983b. Activation of *Klebsiella pneumoniae* and *Rhizobium meliloti* nitrogenase promoters by *gln (ntr)* regulatory proteins. *Proc. Natl. Acad. Sci. USA* **80**:4030–4034.

SYONO, K., AND TORREY, J. G. 1976. Identification of cytokinins of root nodules of the garden pea, *Pisum sativum* L., *Plant Physiol.* **57**:602–606.

SZETO, W. W., ZIMMERMAN, J. L., SUNDARESAN, V., AND AUSUBEL, F. M. 1984. A *Rhizobium meliloti* symbiotic regulatory gene, *Cell* **36**:1035–1043.

TRINCHANT, J. C., BIROT, A. M., AND RIGAUD, J. 1981. Oxygen supply and energy-yielding substrates for nitrogen fixation (acetylene reduction) by bacteroid preparations, *J. Gen. Microbiol.* **125**:159–165.

TRUCHET, G., MICHEL, M. DENARIE, J. 1980. Sequential analysis of the organogenesis of lucerne (*Medicago sativa*) root nodules using symbiotically-defective mutants of *Rhizobium meliloti, Differentiation* **16**:163–172.

TRUCHET, G., ROSENBERG, C., VASSE, J., JUILLOT, J. -S. CAMUT, S., AND DENARIE, J. 1984. Transfer of *Rhizobium meliloti* pSym genes into *Agrobacterium tumefaciens:* Host-specific nodulation by atypical infection, *J. Bacteriol.* **158**:134–142.

UCKER, D. S., AND SIGNER, E. R. 1978. Catabolite-repression-like phenomenon in *Rhizobium meliloti, J. Bacteriol.* **136:**1197–1200.

ULLMAN, A., AND DANCHIN, A. 1983. Role of cyclic AMP in bacteria. *Adv. Cyclic Nucleotide Res.* **15:**1–53.

UPCHURCH, R. G., AND ELKAN, G. H. 1978a. Ammonia assimilation in *Rhizobium japonicum* colonial derivatives differing in nitrogen-fixing efficiency, *J. Gen. Microbiol.* **104:**219–225.

UPCHURCH, R. G., AND ELKAN, G. H. 1978b. The role of ammonia, l-glutamate and cyclic adenosine 3′, 5′-monophosphate in the regulation of ammonia assimilation in *Rhizobium japonicum, Biochem. Biophys. Acta* **538:**244–248.

VAIRINHOS, F., BHANDARI, B., AND NICHOLAS, D. J. D. 1983. Glutamine synthase, glutamate synthase and glutamate dehydrogenase in *Rhizobium japonicum* strains grown in cultures and in bacteroids from root nodules of *Glycine max, Planta* **159:**207–215.

VANCE, C. P. 1983. *Rhizobium* infection and nodulation: A beneficial plant disease? *Annu. Rev. Microbiol.* **37:**399–424.

VERMA, D. P. S., AND NADLER, K. D. 1984. Legume-*Rhizobium*-symbiosis: Host's point of view, in: *Plant Gene Research: Genes involved in Microbe-Plant Interactions* (D. P. S. Verma and T. Hohn, eds.), pp. 58–93, Springer-Verlag, New York.

VERMA, D. P. S., AND NADLER, K. D. 1984. Legume-*Rhizobium*-symbiosis: legume symbiosis, in: *Int. Rev. Cytol.* Suppl 14, pp. 211–245.

VERMA, D. P. S., MACLACHLAN, G. A., BYRNE, H., AND EWINGS, D. 1975. Regulation and *in vitro* translation of messenger ribonucleic acid for cellulase from auxin-treated pea epicotyls, *J. Biol. Chem.* **250:**1019–1026.

VERMA, D. P. S., HUNTER, N., AND BAL, A. K. 1978. Asymbiotic association of *Rhizobium* with pea epicotyls treated with a plant hormone. *Planta* **138:**107–110.

VERMA, D. P. S., BEWLEY, J. D., AUGER, S., FULLER, F., PUROHIT, J. AND KUNSTNER, P. 1983. Host genes involved in symbiosis with *Rhizobium*, in: *Genetic Engineering: Application to Agriculture* (L. D. Owens, ed.), pp. 236–245, Romman and Allanhead, New Jersey.

WANG, T. L., WOOD, E. A., BREWIN, N. J. 1982a. Growth regulators, *Rhizobium* and nodulation in peas: Indole-3-acetic acid from the culture medium of nodulating and non-nodulating strains of *R. leguminosarum, Planta* **155:**345–349.

WANG, T. L., WOOD, E. A., AND BREWIN, N. J. 1982b. Growth regulators, *Rhizobium* and nodulation in peas: The cytokinin content of a wild-type and a Ti-plasmid-containing strain of *R. leguminosarum, Planta* **155:**350–355.

WEINMAN, J. J., FELLOWS, F. F., GRESSHOFF, P. M., SHINE, J., AND SCOTT, K. F. 1984. Structural analysis of the genes encoding the molybdenum-iron protein of nitrogenase in the *Parasponia rhizobium* strain ANU289, *Nucleic Acids Res.* **12:**8329–8344.

WHITING, M. J., AND DILWORTH, M. J. 1974. Legume root nodule nitrogenase: Purification, properties and studies on its genetic control. *Biochim. Biophys. Acta* **371:**337–351.

WIBORG, O., HYLDIG-NILESON, J. J., JENSEN, E. O., PALUDAN, K., AND MARCKER, K. A. 1982. The nucleotide sequences of two leghemoglobin genes from soybean, *Nucleic Acids Res.* **10**:3487–3494.

WONG, C. H., PANKHURST, C. E., KONDOROSI, A., AND BROUGHTON, W. J. 1983. Morphology of root nodules and nodule-like structures formed by *Rhizobium* and *Agrobacterium* strains containing a *Rhizobium meliloti* megaplasmid, *J. Cell Biol.* **97**:787–794.

WONG, P. P., AND EVANS, H. J. 1971. Poly-β-hydroxybutyrate utilization by soybean (*Glycine max* Merr.) nodules and assessment of its role in maintenance of nitrogenase activity, *Plant Physiol.* **47**:750–755.

YAP, S. F., AND LIM, S. T. 1983. Response of *Rhizobium* sp. UMK20 to sodium chloride stress, *Arch. Microbiol.* **135**:224–228.

YELTON, M. M., YANG, S. S., EDIE, S. A., AND S. T. LIM. 1983. Characterization of an effective salt tolerant, fast-growing strain of *Rhizobium japonicum, J. Gen. Microbiol.* **129**:1537–1547.

YUN, A. C., AND SZALAY, A. A. 1984. Structural genes of dinitrogenase and dinitrogenase reductase are transcribed from two separate promoters in the broad host range cowpea *Rhizobium* strain IRC78. *Proc. Natl. Acad. Sci. USA* **81**:7358–7362.

ZIMMERMAN, J. L., SZETO, W. W., AND AUSBEL, F. M. 1983. Molecular characterization of Tn5 induced, symbiotic (Fix) mutants of *Rhizobium meliloti, J. Bacteriol.* **156**:1025–1034.

ZUMFT, W. G., AND MORTENSEN, L. E. 1975. Nitrogen-fixing complex of bacteria, *Biochim Biophys. Acta* **416**:1–52.

Chapter 5

The Energetics and Energy Costs of Symbiotic Nitrogen Fixation

Leonard L. Saari and Paul W. Ludden

BIOLOGICAL NITROGEN FIXATION accounts for the conversion of molecular dinitrogen to approximately 200×10^6 tons of ammonia per year (Evans, 1975; Hardy and Havelka, 1975a). Of this, about 12×10^6 tons is fixed by crop plants in the United States. On a per-hectare basis, crop plants in association with bacteria can produce 100–400 kg/ha/year. Soil bacteria in rhizosphere associations with plants are estimated to contribute far less nitrogen to the soil, approximately 8–15 kg/ha/year.

The fixation of molecular nitrogen by nitrogenase on earth is approximately equaled by the global rate of denitrification. Thus there is a continuing need for fixed nitrogen. Crop plants which do not fix nitrogen must use residual soil nitrogen or rely on nitrogen addition in the form of animal waste products or industrially produced fertilizer. Industrial production of nitrogen via the Haber-Bosch process is ultimately dependent on fossil energy for reductant and for energy needed to overcome the high energy of activation in breaking the dinitrogen triple bond.

That nitrogen fixation is an energetically expensive process for organisms in satisfying their need for nitrogen is well established from theoretical calculations, from observed growth rates and yields of free-living organisms which fix nitrogen, and from knowledge of the requirements of the enzyme nitrogenase. In symbiotic systems, the energy costs of using molecular nitrogen versus fixed-nitrogen sources such as ammonia and nitrate are also considerable; there are costs associated with the "housing" or "maintenance" of the symbiosis in addition to those associated with the conversion of N_2 to ammonia. In this chapter, we will attempt to define the maintenance costs of the symbiosis and arrive at an estimate of the energy cost to the plant for nitrogen fixation. This cost will be compared to the cost for utilization of a fixed-nitrogen source such as nitrate. As models for more complex symbiotic systems, the energy costs of nitrogen fixation to free-living organisms will be discussed.

THE SYMBIOTIC SYSTEM

The site of nitrogen fixation in legumes is the root nodule which forms as a consequence of a *Rhizobium* sp. invading a legume root hair (see Bauer, 1981). A legume is infected by a species of *Rhizobium* specific for that legume. For instance, *Rhizobium japonicum* infects and nodulates roots of *Glycine max* but not those of *Phaseolus vulgaris*. The mechanism for the recognition between rhizobium and a root hair is unknown but possibly involves a rhizobial surface polysaccharide (Dazzo and Hubbell, 1975; Bauer, 1981) and/or a protein factor from the root (Halverson and Stacey, 1984).

Two types of nodules, determinant and indeterminant, are produced depending on the host plant (Newcomb, 1976; Newcomb et al., 1979). Determinant nodules ultimately become limited in growth, whereas indeterminant nodules grow much longer because of the presence of an apical meristem. Furthermore, the major forms of fixed nitrogen exported from determinant nodules are ureides, whereas amides are the export products of indeterminant nodules (Sprent, 1980; Schubert and Boland, 1984).

Upon infection of the host cell, rhizobia differentiate into bacteroids which are able to fix nitrogen. The bacteroids are enclosed in a peribacteroid membrane derived from the host cell plasma membrane (Verma et al., 1978). Not all of the cells in the nodule become infected (Dart, 1977); the uninfected cells, at least in ureide exporters, participate in producing the export products of nitrogen fixation (Newcomb et al., 1985).

The symbiosis between rhizobia and legumes results in the synthesis of a number of nodule-specific host proteins called nodulins (Auger and Verma, 1981). One of the most prominent of these is leghemoglobin, a monomeric oxygen-binding hemoprotein. The functions of leghemoglobin are to provide an increased flux of oxygen to the respiring bacteroid and to serve as an oxygen buffer (Appleby, 1984).

The commitment of the plant and the bacterium to nitrogen fixation is metabolically expensive. Up to 29% of the total bacteroid protein (Haaker and Wassink, 1984) and 40% of the total soluble nodule protein (Nash and Schulman, 1976) are nitrogenase and leghemoglobin, respectively. These two proteins, as well as other nodulins, must be synthesized and their functions supported. From the discussion in the following section, it will be obvious that nitrogenase function is energy-expensive.

THE NITROGENASE SYSTEM

There are several properties of the nitrogenase enzyme system that merit special attention when considering the costs to the plant. The nitrogenase enzyme system is composed of two proteins, and the ratio of the proteins can affect the energetics of the reaction (Hageman and Burris, 1978). Nitrogenase unavoidably carries out a wasteful side reaction, the reduction of protons to H_2 gas (Bulen et al., 1965; Simpson and Burris, 1984). The ability of the organism to minimize this reaction affects the energetics of the system. The components of nitrogenase are extremely oxygen-labile, and yet the endophytes of symbioses with legumes and nonlegume systems are obligate aerobes. Thus protecting nitrogenase from oxygen while allowing oxidative metabolism is a challenge to the organisms of the symbiosis.

Properties of Nitrogenase

The properties of nitrogenase are described in detail in several reviews (Mortenson and Thorneley, 1979; Emerich and Burris, 1978b; Burgess, 1984) and will be outlined briefly here. The nitrogenase enzyme system consists of two oxygen-labile proteins. The smaller of the two, dinitrogenase reductase (also called component II or the Fe protein), has a molecular weight of 60,000–65,000, depending on the source, and has a single, 4Fe-4S sulfur cubane center prosthetic group. It can be oxidized and reduced by the loss or addition of a single electron and, in the reduced form, exhibits an electron paramagnetic resonance (EPR) signal in the $g = 1.94$ region; the oxidized form is EPR-silent. Dinitrogenase reductase is a dimer with identical subunits and binds two molecules of Mg-ATP per protein molecule (Tso and Burris, 1973); the dinitrogenase reductase from *Clostridium pasteurianum* binds a single molecule of Mg-ADP, and the enzyme from *Azotobacter vinelandii* binds one or more molecules of Mg-ADP (Tso and Burris, 1973; Mortenson and Upchurch, 1981).

Dinitrogenase, also referred to as the MoFe protein or component I, is a tetramer with an $\alpha_2\beta_2$-subunit structure with an estimated molecular weight of 210,000–250,000, depending on the source of the enzyme. It has a complex set of metal prosthetic groups: four 4Fe-4S clusters, referred to as P clusters; two molecules of the iron-molybdenum cofactor (FeMo-co), each of which contains molybdenum, iron, and sulfur in a ratio of 1:6–8:4–9 (Shah and Brill, 1977; Burgess et al., 1980; Nelson et al., 1983). Possibly, a third type of Fe center exists, which is referred to as the S center (Münck et al., 1976). The FeMo-co groups of the proteins are responsible for the characteristic EPR signal of dinitrogenase; this signal has lines at $g = 3.65$ and 2.01. The P clusters of the native protein are EPR-silent but can be observed with Mössbauer spectroscopy. The P clusters exhibit an EPR signal when the protein is partially denatured with urea or dimethyl sulfoxide (DMSO), while the S center is observed only with Mössbauer spectroscopy. Although the redox states of the protein have not been completely defined, the dinitrogenase molecule must be capable of a range of redox states spanning eight electrons in order to carry out the reduction of N_2 and H_2.

Dinitrogenase has been reported to have weak binding sites for four Mg-ATP molecules and to carry out the exchange of the gamma P of ATP (Miller et al., 1980). The enzyme is thought to be the site of N_2 reduction based on the presence of Mo in this protein and on the sequence of electron transfer reactions established by EPR studies on the reaction.

The nitrogenase system reduces a number of triple-bonded molecules (Rivera-Ortiz and Burris, 1975), the strained double-bonded molecule cyclopropene (McKenna et al., 1979), and protons, In this chapter, the most significant substrates for discussion are N_2, the biologically important substrate, protons which are reduced to H_2 gas, and acetylene which is reduced to ethylene. Although two-electron reduction of acetylene is not physiologically significant, this compound is nevertheless the most widely used substrate for measurements of nitrogenase activity *in vitro* and *in vivo* (Stewart et al., 1967; Koch and Evans, 1966).

Conservation of the properties of the nitrogenase proteins is quite marked among the more extensively studied systems in *A. vinelandii*, *Klebsiella pneumoniae*, *C. pasteurianum*, and probably *Rhizobium* sp. This is seen in the ability of the dinitrogenase subunits from one organism to form active complexes with the subunits of dinitrogenase reductase from a different organism (Emerich and Burris, 1978a). Some of the complexes are inactive because the two nitrogenase proteins form a tight complex which is incapable of turnover. The amino acid sequences of a number of dinitrogenase reductases have been determined and show high degrees of homology (Hausinger and Howard, 1982) and high degrees of amino acid sequence conservation; the amino acid sequences of the dinitrogenases of several species also show high degrees of amino acid sequence homology and conservation. However, dinitrogenases and dinitrogenase reductases share no significant amino acid sequence homology. (Lammers and Haselkorn, 1983).

Characteristics of the Nitrogenase Reaction

The physiological electron donor to nitrogenase is a ferredoxin or flavodoxin, both of which are found in most nitrogen-fixing organisms. In most cases it is not known which protein(s) actually is the physiological electron donor. In *K. pneumoniae,* mutants lacking flavodoxin (nifF$^-$; see below) do not exhibit nitrogenase activity *in vivo*, although nitrogenase activity can be detected in cell-free extracts in complete assay mixtures. Thus, in this organism, it is clear that there is a requirement for a specific electron donor to nitrogenase (Nieva-Gomez et al., 1980; Shah et al., 1983).

The path of electrons to nitrogen in the nitrogenase reaction is shown in Fig. 1. The source of reducing power is reduced ferredoxin or flavodoxin which donates electrons to dinitrogenase reductase. Because dinitrogenase reductase carries a single electron, eight cycles of association and dissociation must occur in order for the dinitrogenase molecule to accumulate enough electrons to accomplish the reduction of N_2 to ammonia ($6e-$) and $2H^+$ to H_2 ($2e-$). There appear to be two binding sites for the reductase on dinitrogenase (Emerich and Burris, 1976) and two molecules of the FeMo-co center per dinitrogenase molecule. The rate-limiting step in the dinitrogenase reductase cycle is thought to be release of the oxidized form of Mg-ADP-dinitrogenase reductase from the nitrogenase complex, which may also limit the overall rate of nitrogen reduction (Lowe and Thorneley, 1984). The details of the mechanism by which the reduced dinitrogenase molecule carries out substrate reduction are not yet completely known.

A significant feature of the mechanism is the apparent obligate reduction of two protons for every N_2 reduced (Simpson and Burris, 1984). Thus the overall reaction is an eight-electron reduction, with two of the electrons used to generate H_2. The ability of the organism to minimize the flow of electrons to H^+ and to recapture the energy lost in the production of H_2 is crucial to the overall efficient use of energy for nitrogen reduction (Fig. 1). Hydrogen is both a product and an inhibitor of nitrogenase activi-

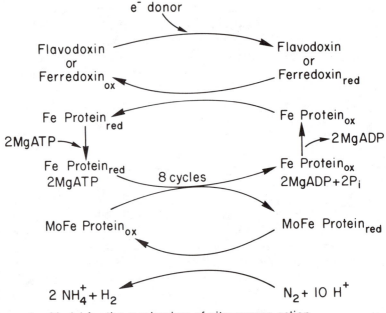

Figure 1. Model for the mechanism of nitrogenase action.

ty; hydrogen specifically inhibits the reduction of N_2 and not other substrates (Rivera-Ortiz and Burris, 1975). Thus, in the presence of hydrogen, the reduction of protons to produce more hydrogen continues and is in fact enhanced because the flow of electrons is diverted from N_2 to protons. Because the reduction of H^+ to H_2 appears to be an intrinsic part of the mechanism of nitrogenase, no organism has a nitrogenase system with less hydrogen production. However, many organisms possess an uptake hydrogenase (Dixon, 1972) which appears to scavenge hydrogen and thereby removes the inhibiting product; part of the energy utilized in the production of H_2 by nitrogenase is also recovered. The effect of these uptake hydrogenases on yield is discussed below.

Factors Affecting the Electron Flux through Nitrogenase

The total flow of electrons through nitrogenase is referred to as the electron flux. The partitioning of the total flow to various substrates is known as electron partitioning or electron allocation and is important in determining the efficiency of nitrogen reduction. Because the enzyme is always measured in aqueous solutions, it is not possible to eliminate protons as a substrate. Thus, when N_2 or acetylene reductions are measured, the transfer of electrons to protons must be taken into account. As discussed above, while the reduction of N_2 occurs with concomitant production of one H_2 per each N_2 reduced, reduction of acetylene occurs with little loss of electrons for H_2 production. Estimates of total nitrogen fixation based on acetylene reduction measurements therefore must take this difference into account.

The ratio of the amounts of the two components of nitrogenase affects the partitioning of electrons and the overall rate of nitrogen production. The optimal dinitrogenase reductase/dinitrogenase ratio is about 2:1. At lower ratios, the reductase becomes limiting and the electron flow per dinitrogenase molecule is decreased. In such cases, the likelihood of the enzyme reducing protons to H_2 (a two-electron process) increases because a sufficient supply of electrons to reduce N_2 (a six-electron process) is not available. At extremely low reductase/dinitrogenase ratios, a lag in enzyme turnover is observed (Hageman and Burris, 1978). This is thought to be the time necessary for the enzyme to accumulate one, and then a second, electron necessary for proton reduction. It should be noted that ATP hydrolysis shows no lag, indicating that the lag affects substrate reduction, not the dinitrogenase reductase cycle.

Effect of ATP Concentration on Nitrogenase Efficiency

The transfer of each electron from dinitrogenase reductase to dinitrogenase requires the hydrolysis of 2Mg-ATPs. The K_m of nitrogenase for Mg-ATP is 20 μM. Mg-ADP is a potent inhibitor of nitrogenase and has a K_i of 10 μM with nitrogenase from *C. pasteurianum*. The concentrations of Mg-ATP found *in vivo* are well in excess of the K_m of nitrogenase for Mg-ATP. However, the energy charge (energy charge = ATP + $\frac{1}{2}$ADP/ATP + ADP + AMP) of cells grown with N_2 as the nitrogen source is significantly less than the energy charge of cells grown with good nitrogen sources such as ammonia (Upchurch and Mortenson, 1980; Privalle and Burris, 1983; Paul and Ludden, 1984). Cells grown on ammonia have energy charges of 0.7–0.8; cells grown on N_2 or limiting nitrogen have energy charges of 0.6–0.65. Most of the differences arise from the decrease in ATP and increase in ADP concentrations. Theoretically, at the concentrations of ADP found *in vivo*, nitrogenase should be significantly inhibited. However, for unknown reasons, the expected inhibition is not observed. It is possible that the [Mg^{2+}] potentiates the inhibitory effect (Davis and Kotake, 1980). If Mg^{2+} is limiting, the apparent ATP/ADP ratio will not reflect the *in vivo* Mg-ATP/Mg-ADP ratio which would shift in favor of Mg-ATP because of the higher affinity of ATP for Mg^{2+}.

Comparison of *In Vivo* and *In Vitro* Activities

Although it is useful to extrapolate from the known properties of isolated enzymes to determine the requirements of the enzyme *in vivo*, it is important to remember that the catalyst may behave differently *in vivo* than *in vitro*. *In vivo*, the enzyme is very concentrated and is in the presence of many cellular constitutents including its natural electron donor; it may have specific associations with other proteins, small molecules, or the cell membrane that are disrupted on cell lysis. In the case of nitrogenase, there is reason to suspect that there are factors which affect or control the *in vivo* activity that we do not yet understand. If the whole-cell activity of nitrogenase is compared to the total extractable activity, it can be calculated that the whole-cell activity is four- to fivefold higher than the total extractable activity.

One advantage of the nitrogenase enzyme system is that it can be measured *in vivo* without disrupting the cell. This can be accomplished by the acetylene reduction assay, by manometry, or by feeding $^{15}N_2$ to the cells. Thus it is possible to make an estimate of total activity *in vivo* and compare the activity to total extractable activity.

The Cost of Nitrogenase Synthesis

Because nitrogenase is an inherently slow enzyme with a turnover number of about 10 electron pairs per second, a large amount of the enzyme must be produced to supply the cell with sufficient nitrogen. Nitrogenase is estimated to comprise from 5 to 10% of the total cell protein in free-living bacteria, and therefore the cost of synthesis of nitrogenase is itself significant. In addition, there are a number of gene products produced beyond the structural genes for dinitrogenase and dinitrogenase reductase that are required for operation of the system *in vivo* and also represent a cost to the cell (Brill, 1980). For example, several of the *nif* gene products are involved in the synthesis of the Fe-, and S-, and Mo-containing cofactors of nitrogenase (QBNEV and perhaps M); since these cofactors must be produced in large amounts, the costs in cell energy would be high. Included in these costs are the increased uptake of Fe and Mo. In the case of *K. pneumoniae*, the *nif* F and J proteins are a flavodoxin and pyruvate:flavodoxin oxidoreductase, respectively. These constitute the physiological source of electrons for dinitrogenase reductase *in vivo*. The cost of the synthesis of these products is in addition to that for those needed for ammonia assimilation when N_2 is the nitrogen source. The cost of their syntheses cannot be measured directly but is included as part of the maintenance cost of the legume nodule. In the case of symbiotic systems, there are high costs for synthesis and maintenance of leghemoglobin, which may represent as much as 40% of the cytosolic protein of the infected nodule cell (Nash and Schulman, 1976). The cost of synthesis of *nif* proteins is partially minimized by the stability of mRNAs and the gene products in the cell. During conditions conducive to nitrogen fixation, *nif* mRNA is significantly stable in *K. pneumoniae*, the only case where message stability has been extensively studied. However, accelerated turnover of mRNA occurs when *K. pneumoniae* is exposed to oxygen. Because the nitrogenase proteins are extremely oxygen-labile, continued synthesis of the proteins would be wasteful in an aerobic environment. Likewise, the nitrogenase components themselves are quite stable in the cell. When fixed nitrogen is added to a culture of nitrogen-fixing *A. vinelandii* or *K. pneumoniae,* synthesis of new nitrogenase stops quickly, but the proteins that are already present are lost primarily by dilution, not by accelerated degradation (Tubb and Postgate, 1973; Shah et al., 1972). Such studies are yet to be made on *Rhizobium* sp. and other nitrogen-fixing species; the modes of regulation of synthesis and turnover may vary among the many nitrogen-fixing organisms and deserve much more study.

In some nitrogen-fixing organisms, the activity of nitrogenase is regulated by posttranslational modification of dinitrogenase reductase (Kanemoto and Ludden, 1984). When cells of *Rhizobium rubrum* are exposed to a source of fixed nitrogen such as ammonia or glutamine, the dinitrogenase reductase protein is modified by the addition of adenosine diphosphate ribose to a specific arginyl residue (Pope et al., 1985). The modified protein cannot transfer electrons to dinitrogenase. A number of other effectors also cause modification of dinitrogenase reductase. Upon removal of fixed nitrogen, the modifying group is removed, and the protein becomes activated. This posttranslational system undoubtedly operates to save energy for the cell, but the system is not observed in all nitrogen-fixing systems.

ENERGY COSTS OF NITROGEN FIXATION
IN LEGUMES

Studies performed before 1981 on the energy costs of nitrogen fixation in legumes have been reviewed (Pate et al., 1981; Phillips, 1980; Schubert and Ryle, 1980; Minchin et al., 1981; Schubert, 1982; Hardy et al., 1983; Salsac et al., 1984). Three reviews, by Phillips (1980), Minchin et al. (1981), and Schubert (1982), are particularly useful in understanding the methodology and results of previous studies. Our aim in this section is to emphasize more recent methodologies and investigations from the perspective of information already assimilated. Thus most research before 1981 will be discussed in a more general manner (as condensed by the aforementioned reviews), and the emphasis will be on more recent work.

Methods of Estimating Energy Costs of Nitrogen Fixation

Determination of Carbon Costs

Investigators of the energetics of nitrogen fixation in legumes have used one of three approaches to estimate carbon costs (Schubert, 1982). One method compares the carbon contents and growth rates of plants grown either on fixed-nitrogen or on nitrogen-free media (symbiotically). Differences in the carbon accumulated between legumes grown symbiotically and nonsymbiotically are assumed to reflect differences in the energy required for nitrogen fixation. A second method estimates the cost of nitrogen fixation by comparing yields of plants grown either on fixed nitrogen or symbiotically. Both of these approaches are noninvasive and thereby avoid problems associated with manipulations such as denodulation. Yield estimates have the potential advantage of determining the agronomic significance of nitrogen fixation but have not yet delineated the agronomic cost between nitrogen fixation and fixed-nitrogen utilization. Some legume species grown on fixed nitrogen have higher growth rates (Mahon and Child, 1979) and better carbon utilization efficiencies (Finke et al., 1982) than symbiotically grown plants. Evidence of this type leads Minchin et al. (1981) to suggest that symbiotic nitrogen fixation is an additional energy burden to the plant.

A third approach is to compare the rates of respiration of plants or plant parts to the rates of nitrogen fixation. This last approach has been the predominant method of study recently because the respiratory costs associated with nitrogen fixation can be separated from the growth and maintenance costs of the roots and nodules. The remainder of the chapter will focus on studies in which respiration measurements were used to estimate carbon costs of nitrogen fixation.

Most estimates of nodule or root respiration are performed by measuring the efflux of CO_2 from these tissues using an infrared gas analyzer; this assumes that the CO_2 efflux is a function of the energy used for nitrogen fixation. Oxygen uptake is not used as a method to estimate the respiration of roots and nodules because nitrogen fixation and ammonia assimilation use energy equivalents both in the form of ATP and reductant. Oxygen uptake measurements would measure only the photosynthate oxidized for phosphorylation and would not reflect the reductive costs

Schubert (1982). Thus CO_2 efflux measurements may yield more accurate estimates of the reductant and ATP requirements of nitrogen fixation than oxygen consumption.

Measurements of CO_2 evolution from nodulated roots, denodulated roots, and detached nodules are used to separate energy costs associated with roots plus nodules versus nodules alone. However, nodule removal may cause wound respiration which would alter the CO_2 efflux and the subsequent estimation of energy cost (Schubert, 1982). Also, varying amounts of CO_2 may be refixed by carboxylation (Minchin et al., 1981; Pate et al., 1981; Schubert, 1982); estimates of the amount of CO_2 reassimilated in soybean nodules range from 9 to 30% (Coker and Schubert, 1981). The CO_2 concentration in the measuring apparatus also must be considered, because high concentrations of CO_2 inhibit pea nodule respiration (Mahon, 1977a), whereas very low CO_2 concentrations alter markedly the metabolic conditions encountered *in situ* (Schubert, 1982).

Methods have been developed to measure the CO_2 efflux and C_2H_2 reduction of attached root nodules enclosed in flow-through gastight cuvets (Layzell et al., 1979; Winship and Tjepkema, 1982), which avoids toxic gas accumulation. While this method avoids the errors introduced by manipulations such as denodulation, respiration measurements will include a small component due to root respiration. Measurements on individual nodules require extrapolation to the total nodule population, which may increase the error of determination (Minchin et al., 1981). Nodule respiration alone was estimated by subtracting the denodulated root respiration from the nodulated root respiration (Minchin and Pate, 1973; Layzell et al., 1979) but, as mentioned previously, wound respiration may alter the results.

Although respiration (CO_2 efflux) is commonly measured with detached nodules (Minchin and Pate, 1973; Herridge and Pate, 1977; Pate and Herridge, 1978; Wittey et al., 1983), removal of nodules from pea roots (Witty et al., 1983) and soybean roots (Ralston and Imsande, 1982) resulted in decreased nodule respiration and nitrogen fixation. Because the stoichiometry (i.e., mol CO_2 respired per minute/mol C_2H_2 reduced per minute) remains unchanged after detachment (Witty et al., 1983), it is argued that detached nodules can be used to estimate metabolic costs associated with nitrogen fixation. It was also found that the sum of the respiratory components of detached nodules and denodulated roots was nearly equal to the respiration of the nodulated roots of cowpea (Herridge and Pate, 1977) and pea (Minchin and Pate, 1973). This type of evidence suggests that denodulation has little effect on respiration and indicates that detached nodule respiration is useful in determining energy costs of nitrogen fixation. However, researchers must be aware that nodule detachment decreases dark CO_2 fixation (Coker and Schubert, 1981; Pate et al., 1981), affects the normal flow of metabolites (Minchin et al., 1981), and possibly decreases the permeability to oxygen (Ralston and Imsande, 1982).

CO_2 efflux from nodulated roots was also determined by exposing the aerial portions of soybean plants to $^{14}CO_2$ for 24 hr, followed by exposure to $^{12}CO_2$ (Warembourg, 1983). Simultaneously, the root systems were exposed to $^{15}N_2$. The $^{14}CO_2$ evolution by root systems was a measure of respiration, whereas ^{15}N incorporation into cellular metabolites was a measure of nitrogen fixation. Nonnodulated plants served as controls to determine the respiratory costs associated with the roots alone. Although this method is advantageous in that plant manipulation is minimal and overall cost analysis is more direct, comparisons between nodulated versus nonnodulated plants may not be valid in that the two plant types differ both physiologically and morphologically.

Heytler and Hardy (1984) designed a calorimeter that measures simultaneously the heat release and gas exchange of detached root nodules under a controlled gas stream. The rise in temperature due to the turnover of gaseous substrates by nitrogenase in the detached nodule was converted into a ΔH value. The calculated ΔH value was converted into a carbohydrate cost by assuming that glucose was the nodule energy source and that glucose oxidation to CO_2 yielded -673 kcal/mol. The CO_2 efflux of the detached nodules was measured in addition to the heat release, and estimates of the metabolic costs obtained by CO_2 measurements were similar to those obtained by ΔH measurements. Thus it appears that the two independent methods measure the same respiratory costs.

Determination of Nitrogen Fixation Rates

Nitrogen fixation rates associated with respiratory costs in root nodules are usually measured by one or more of the following methods: total nitrogen accumulation, acetylene reduction, ^{15}N incorporation, or H_2 evolution. However, in measurements of the rate of nitrogen fixation by nitrogenase, protons are concomitantly reduced along with nitrogen. Simpson and Burris (1984) determined that a minimum of 27% of the electron flux through nitrogenase resulted in H_2 production even at 50 atm of N_2. The electron allocation coefficient (EAC) describes the relative electron flux through nitrogenase to reducible substrate:

$$\text{EAC} = \text{electrons to exogenous reduced substrate/electrons to } H_3O^+ \text{ and exogenous reducible substrate}$$

The EAC for nitrogen decreases as the electron flux decreases (Hageman and Burris, 1980). Thus, to measure nitrogenase activity accurately, the EAC values for nitrogen and hydrogen must be considered.

The EAC for nitrogen is determined by estimating separately the nitrogen fixed and the total electron flux through nitrogenase. The latter determination is performed by assaying nitrogenase under Ar/O_2 (80/20, v/v) (Schubert and Evans, 1976) so that the total electron flux goes into hydrogen production (i.e., the EAC for protons is 1.0). Alternatively, reduction of acetylene yields the total electron flux through nitrogenase because no proton reduction occurs under saturating concentrations of acetylene. The nitrogen fixation rate corrected for proton reduction can be calculated from hydrogen evolution in the presence and absence of nitrogen:

$$N_2 \text{ fixed} = [(H_2 \text{ evolved in } Ar/O_2) - (H_2 \text{ evolved in air})]/3$$

Since nitrogen reduction is a six-electron process and hydrogen production is a two-electron process, division by 3 converts the hydrogen produced to the nitrogen fixed.

Symbiotic systems that possess an uptake hydrogenase present difficulties in the measurement of hydrogen produced by nitrogenase, since nodules formed with Hup^+ *Rhizobium* strains can reoxidize H_2 evolved by nitrogenase. Schubert and Evans (1976) defined the relative efficiency (RE) of electron transfer to nitrogen by nitrogenase as:

$$RE = 1 - (H_2 \text{ evolved in air})/(H_2 \text{ evolved in } Ar/CO_2/O_2)$$

or

$$RE = 1 - (H_2 \text{ evolved in air})/(C_2H_2 \text{ reduced in } C_2H_2/air)$$

The RE *in vivo* is dependent on the EAC for N_2 and the ability of the uptake hydrogenase to reoxidize the evolved H_2. In the absence of hydrogenase activity (Hup^-), the RE is theoretically equal to the EAC. Thus the estimation of nitrogenase activity and the EAC to N_2 by hydrogen production (e.g., Rainbird et al., 1984a) requires that the endophyte be Hup^-.

The EAC for nitrogen is also determined by measurements of $^{15}N_2$ incorporation before measuring acetylene reduction (Witty et al., 1983). These measurements yield a ratio of C_2H_2 reduced to N_2 reduced. The method assumes that both substrates are reduced at the same rate and do not perturb the nodules differently. A disadvantage of $^{15}N_2$ measurements is the expense of the isotope and the requirement for a mass spectrometer. Acetylene reduction assays (Koch and Evans, 1966) are simple, inexpensive, and very sensitive, but alone do not take into account the EAC for protons when nitrogen is the substrate. However, problems can arise because C_2H_2 reduction activity in some symbioses is inhibited after a few minutes' incubation in a 10% C_2H_2/21% O_2/69% N_2 mixture (Minchin et al., 1983). Thus long incubations in the presence of C_2H_2 may underestimate N_2 fixation rates. Cessation of ammonia production may be responsible for the decline in nitrogenase activity, because Ar/O_2 atmospheres mimic the effects observed with atmospheres containing acetylene (Minchin et al., 1983).

Acetylene reduction activity can be converted to nitrogen fixation activity by using the theoretical estimate of 4 mol C_2H_2 reduced per 1 mol N_2 reduced. Implicit is the assumption that 75% of the electrons go into nitrogen reduction and 25% into proton reduction. Since the EAC for nitrogen varies with the electron flux through nitrogenase (Hageman and Burris, 1980), this conversion factor must be used cautiously.

The methods mentioned above estimate nitrogen fixation over a relatively short span of time. While this is advantageous for most experiments, care must be taken to avoid short-term environmental effects on nitrogen fixation such as diurnal variation (Rainbird et al., 1983a) (see section on the effect of diurnal variations). A more integrated view of nitrogen fixation over a season can be obtained by determining the total nitrogen accumulated, but this method has the disadvantage of being subject to limiting-growth conditions (Schubert, 1982).

Separation of Nodule Maintenance and Growth from Nitrogen Fixation

Techniques are now available that allow separation of the costs of nitrogen fixation from those of growth and maintenance. With this approach, nitrogen fixation and respiration associated with nitrogen fixation are manipulated by various experimental conditions without significantly affecting the respiration associated with growth and maintenance. Thus metabolic costs due to nitrogen fixation can be calculated by the difference observed in respiration rates as the nitrogen fixation rate changes.

Total respiration of a nodulated root system can be described by the equation:

$$R = R_M W + R_G \frac{dW}{dt} + R_F(N_2\text{ase})$$

where R = total respiration
W = root plus nodule dry weight
t = time
N_2ase = nitrogenase activity
R_M, R_G, R_F = maintenance, growth, and fixation coefficients, respectively
(Mahon, 1977a)

When respiration measurements are performed before exposure of plants to acetylene or Ar/O_2, then ammonia is produced and respiration costs of nitrogen fixation include costs for ammonia assimilation. When respiration measurements are performed simultaneously with acetylene reduction measurements, ammonia is not formed.

Experimentally, if the respiration rates associated with growth and maintenance are shown to be constant for a given manipulation (e.g., addition of NO_3^-), any variation in the observed respiration rate will be directly related to the respiration associated with nitrogen fixation. Thus, on a graph of the observed respiration rate versus the nitrogen fixation rate, the slope gives the metabolic costs of nitrogen fixation and the intercept value is the respiratory cost of growth and maintenance. A limiting case occurs when mature nodules are studied (Rainbird et al., 1984a); then the term for respiration associated with nodule growth approaches zero, since little or no growth occurs ($dW/dt = 0$). Thus the respiration costs associated with nodule maintenance alone are equal to the value of the y intercept of a graph of the observed respiration rate versus the nitrogen fixation rate.

The rate of nitrogen fixation and its attendant respiration in the root nodule has been varied by four methods: (1) addition of fixed nitrogen to nodulated roots, (2) exposure of plants to dark-light regimes, (3) addition of acetylene and variation in oxygen concentration to roots and nodules, and (4) exposure of nodules to 1.0 atm oxygen. Treatment of plants with fixed nitrogen or short dark treatments (Mahon, 1977a) decrease nitrogen fixation rates, possibly by affecting the availability of photosynthate (Hardy and Havelka, 1975b; Noel et al., 1982). However, the addition of fixed nitrogen may cause increased nitrate reductase activity or senescence (Minchin et al., 1981). Linear relationships were demonstrated between nodule respiration and nitrogen fixation by the addition of fixed nitrogen (Patterson and LaRue, 1983) and the use of varying dark-light dark periods (Patterson and LaRue, 1983; Rainbird et al., 1984a). In the latter study, respiration measurements on denodulated roots were constant, supporting the conclusion that the observed reduction in respiration with longer dark periods was due to decreased respiration associated with nitrogen fixation and not decreased root growth or maintenance respiration (nodule growth was assumed to be zero, and nodule maintenance was assumed to be constant).

In another method for varying the rate of nitrogen fixation and the associated respiration, nodulated roots are exposed to acetylene and varying concentrations of oxygen (Witty et al., 1983). Acetylene inhibits nodule respiration within a few minutes (Winship and Tjepkema, 1982) concurrently with nitrogen fixation (Minchin et al., 1983), perhaps as a result of cessation of ammonia production (Minchin et al., 1983) or decreased oxygen diffusion (Witty et al., 1983) in root nodules. Oxygen, at both subambient (Criswell et al., 1976; Witty et al., 1983; Ryle et al., 1984) and supraambient (Witty et al., 1983; Patterson et al., 1983; Patterson and LaRue, 1983) levels, decreases nitrogen fixation and the associated respiration of either intact roots or detached nodules. Why oxygen has an effect on the nitrogen fixation and respiration rates of nodules is unclear.

Possible causes of this include loss of reductant (Bergersen, 1962) and changes in the permeability of the tissues to oxygen (Sheehy et al., 1983). Acetylene, subambient oxygen levels, and supraambient oxygen levels have been used to vary the nitrogen fixation rate (Witty et al., 1983). Respiration from the detached nodules formed by ineffective strains of rhizobia was not affected by either acetylene or altered O_2 levels. Also, respiration by denodulated roots under 3% oxygen remained fairly stable (Ryle et al., 1984). These controls allowed separation of nodule and root growth and maintenance respiration.

Nodulated versus nonnodulated isolines of soybeans have been used to establish the magnitude of nodule respiration (Heytler and Hardy, 1979; Hardy et al., 1983; Heytler et al., 1985). In addition, the nitrogen fixation cost can be separated from the nodule maintenance cost by comparing the respiration of nodules produced with either effective (Nif$^+$) or ineffective (Nif$^-$) *Rhizobium* strains. Different metabolic paths may be induced in nonnodulated versus nodulated roots resulting in different rates of CO_2 efflux. However, the use of isogenic plant lines minimizes differences between the comparisons.

Factors Influencing the Metabolic Costs of Nitrogen Fixation

A number of factors can affect the efficiency of nitrogen fixation in legume root nodules. For instance, the presence of an uptake hydrogenase, which catalyzes H_2 oxidation, or phosphoenolpyruvate carboxylase, which catalyzes dark CO_2 fixation in legume nodules, may significantly change the costs (or at least affect the measurement). Once symbiosis is established, a number of other processes such as ammonia assimilation and fixed-nitrogen transport add to the overall energy cost; these are discussed below.

Effect of Uptake Hydrogenase

The production of hydrogen by nitrogenase appears to be obligatory, because *Rhizobium* strains have evolved uptake hydrogenase systems rather than a loss of H_2-generating capabilities (Phillips, 1980), and hydrogen evolution by nitrogenase is not eliminated by a very high pN$_2$ *in vitro* (Simpson and Burris, 1984).

Hydrogen evolution by soybean nodules was initially reported by Hoch et al. (1957). The demonstration that pea root nodules could take up hydrogen (Phelps and Wilson, 1941; Dixon, 1967) led Dixon (1972) to propose that *Rhizobium* uptake hydrogenase has three functions: (1) to scavenge oxygen to protect the oxygen-labile nitrogenase; (2) to use up hydrogen and prevent nitrogenase inhibition by hydrogen; and (3) to recover energy by oxidation of H_2 gas. In addition, uptake hydrogenase may decrease the total energy costs by providing reducing power for nitrogenase function (Salminen and Nelson, 1984). For detailed discussions on the advantages of hydrogen-utilizing systems in the legume-*Rhizobium* symbiosis, the reader is referred to the reviews of Eisenbrenner and Evans (1983) and Evans et al. (1985). The relative importance of each of the proposed functions of uptake hydrogenase needs to be documented and may vary in plant species.

Hydrogen oxidation both protects nitrogenase from oxygen and increases the ATP concentration in *R. japonicum* bacteroids (Emerich et al., 1979). The protection of nitrogenase in *Rhizobium leguminosarum* bacteroids (Nelson and Salminen, 1982). From model studies, Dixon et al. (1981) calculated that hydrogen concentrations would be inhibitory and proposed that the main function of uptake hydrogenase is hydrogen removal.

The benefits of uptake hydrogenase in reducing energy costs was demonstrated with nodules infected with Hup⁺ or Hup⁻ bacteroids; CO_2 evolution was less from Hup⁺ nodules than from Hup⁻ controls (Drevon et al., 1982; Rainbird, et al., 1983c). Thus less carbon is used for respiration in symbioses containing Hup⁺ *Rhizobium*.

In trials with Hup⁺ and Hup⁻ strains of *Rhizobium*, 10% increases in dry matter and total nitrogen were reported for legumes inoculated with Hup⁺ strains (Eisenbrenner and Evans, 1983). In a recent study on *Rhizobium* strains isogenic except for the Hup determinant, total plant nitrogen was 11% higher in plants inoculated with Hup⁺ strains (Evans et al., 1985). Plants must be grown to maturity in order to realize the full benefits of Hup⁺ strains on yield and total nitrogen (Evans et al., 1985). The supply of photosynthate must be limiting, otherwise hydrogen reoxidation will not exert such a significant effect on nitrogen fixation and growth (Evans et al., 1985).

To explain the effects on yield, it is useful to calculate the metabolic costs of nitrogen fixation in legumes as a function of (1) varying electron flow through nitrogenase to hydrogen production and (2) the percentage of evolved hydrogen reoxidized by uptake hydrogenase (Pate et al., 1981). The extremes of these two parameters can be used to demonstrate the variability of costs; in the absence of an uptake hydrogenase, the metabolic cost is between 1.9 (the EAC for hydrogen $=0.25$) and 3.5 (the EAC for hydrogen $=0.60$) g C/g N fixed. *In vivo*, the EAC for hydrogen probably varies between 0.25 and 0.60 (Pate et al., 1981). If an uptake hydrogenase were present and recycled all of the evolved hydrogen, the metabolic costs at 25 and 60% of the electron flux to proton reduction would be 1.7 and 2.6 g C/g N fixed, respectively. In arriving at these figures, Pate et al. (1981) assumed that the reduction of 1 mol of nitrogen requires 0.55 mol glucose, 38 ATP are formed per mol of glucose oxidized, the respiration of 1 mol of glucose results in 6 mol of CO_2, and 3

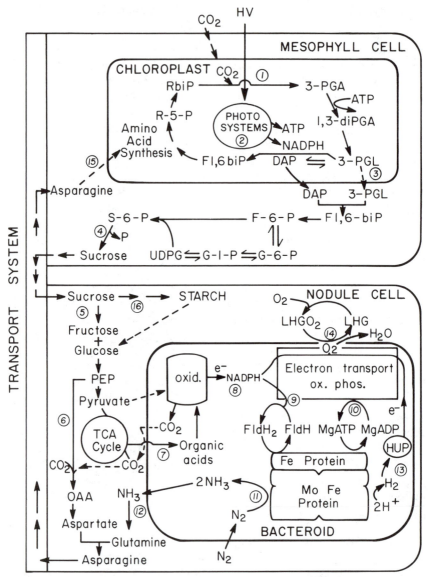

Figure 2. Relationships among photoassimilation, photoassimilate transport, nonphotosynthetic CO_2 fixation, and nitrogen fixation in nodulated legumes. Key reactions of photosynthesis in the plant include photoassimilation by (1) ribulose 1,5 bisphosphate carboxylase and photophosphorylation and NADPH production by photosystems I and II (2). The triose phosphates, DAP (dihydroxyacetone phosphate) and 3-PGL (glyceraldehyde-3-phosphate) are exported from the chloroplast to the cytoplasm via the phosphate translocator system (3); they are converted to sucrose via F 1,6-biP (fructose 1,6 bisphosphate), F-6-P (fructose-6-phosphate), G-6-P (glucose-6-phosphate), G-1-P (glucose-1-phosphate) and UDPG (uridine diphosphate glucose). Sucrose 6-phosphate synthase catalyzes the synthesis of sucrose 6-phosphate (S-6-P) from F-6-P and UDPG; S-6-P is converted to sucrose by S-6-P phosphohydrolase (4). Sucrose is transported to nodule

mol ATP is recovered on oxidation of 1 mol of hydrogen by hydrogenase. The presence of a fully efficient uptake hydrogenase reduces the carbon cost of nitrogen fixation by 11% at 25% electron flux to H^+ and by 26% at 60% electron flux to H^+. Thus an uptake hydrogenase has a greater impact in reducing the carbon costs of nitrogen fixation if hydrogen evolution via nitrogenase is large. However, the ATP recovered by reoxidation of the evolved hydrogen does not equal the ATP consumed in the production of hydrogen by nitrogenase. For this reason, the ideal symbiosis in terms of energy usage would consist of a nitrogenase that allows the minimum electron flux (25%) to H^+ and a fully efficient uptake hydrogenase; this model provides a goal for those seeking to design a "perfect" *Rhizobium* for further improving the yields of legumes.

Role of CO_2 Fixation

Fixation of CO_2 by legume roots and nodules directly affects estimates of the metabolic costs of nitrogen fixation by influencing net root and nodule respiration, supplying carbon for C_4 acid biosynthesis, and providing substrates for energy generation.

The enzymes for CO_2 fixation in the legume nodule include phosphoenopyruvate (PEP) carboxylase and NADP-malic enzyme (Rawsthorne et al., 1980). The contribution of NADP-malic enzyme to nodular CO_2 fixation probably is minimal, since alfalfa nodule extracts fail to catalyze significantly the reaction between CO_2 and pyruvate (Duke and Henson, 1985). On the other hand, PEP carboxylase (Fig. 2) provides a

---

cells and is hydrolyzed via invertase (5) to yield glucose and fructose. The latter are utilized via glycolysis for energy generation and carbon intermediates for nodule maintenance and ammonia assimilation. PEP carboxylase (6) catalyzes nonphotosynthetic carbon dioxide fixation to yield oxalacetate (OAA) for asparagine synthesis via aspartate. Nodule respiration yields organic acids (7) which are taken up by bacteroids; organic acids are consumed in the bacteroid to yield reducing power (NADPH) (8). Bacteroid oxidative phosphorylation and electron transport (9) convert flavodoxin quinone (F1dH) to flavodoxin semi quinone ($F1dH_2$) which is utilized for nitrogen reduction via nitrogenase (11); bacteroid oxidative phosphorylation (10) yields ATP for nitrogen reduction (11). Ammonia assimilation (12) occurs by reactions diagrammed in Fig. 3. Hydrogen uptake system (13) utilizes hydrogen generated from nitrogen reduction. Leghemoglobin scavanges oxygen in nodule cell and transfers it to the bacteroid oxidative phosphorylation system. Asparagine is transported from nodule cells to mesophyll cells and into the chloroplast for amino acid synthesis (15). Sucrose may be converted to starch (16) in the nodule cell; starch is a temporary carbon storage form and is readily reutilized for energy generation and carbon intermediates by the nodule cell.

major route of CO_2 fixation and comprises approximately 1–2% of the total soluble protein of alfalfa nodules (Vance and Stade, 1984); other properties of PEP-carboxylase are summarized by Duke and Henson (1985).

Without appropriate correction, the refixation of CO_2 during measurements of CO_2 efflux from roots and/or nodules will result in an inaccurate estimation of the respiration. For instance, Coker and Schubert (1981) es-

Table 1 Carbon Cost of Nitrogen Fixation in Legume Root Nodules.

| | | METHOD FOR | | |
| | | ESTIMATING C COST | ESTIMATING N_2 FIXATION | SEPARATING COSTS[a] |
SPECIES	TISSUE(S)			
Pea	Detached nodules	CO_2 efflux	C_2H_2 reduction	C_2H_2 and O_2
Lucerne	Nodulated roots	CO_2 efflux	C_2H_2 reduction	C_2H_2 and O_2
Cowpea	Nodulated roots vs. denodulated roots	CO_2 efflux	H_2 evolution	
Soybean	Nodulated roots	CO_2 efflux	Total N	
Soybean	Nodulated roots	CO_2 efflux	C_2H_2 reduction	Supraambient O_2
Soybean	Nodulated roots (effective-ineffective) vs. nonnodulated roots	CO_2 efflux	H_2 evolution or C_2H_2 reduction	Effective vs. ineffective nodules
Soybean	Nodulated roots vs. nonnodulated roots	$^{14}CO_2$ efflux	$^{15}N_2$ uptake	
Soybean	Nodulated roots versus denodulated roots	CO_2 efflux	H_2 evolution	Extended dark period
Soybean	Detached nodules	Calorimetry	H_2 evolution	Supraambient O_2
Soybean	Detached nodules	CO_2 efflux	H_2 evolution	Supraambient O_2
Soybean	Nodulated vs. denodulated nodulated roots	CO_2 efflux	C_2H_2 reduction	Nodule detachment and 20% or 3% O_2 levels

[a]That is, separating N_2 fixation costs from nodule growth and maintenance costs.
[b]Does not separate costs of nodule maintenance from nitrogen fixation.
[c]Ratio of reported CO_2 evolved/C_2H_4 (grams carbon respired/gram of nitrogen fixed) produced converted to g C/g N by assuming EAC for N_2 = 0.75.
[d]EAC for N_2 assumed to be 0.75.

timated that $9-31\%$ of the respired carbon was refixed by soybean nodules; alfalfa nodules reassimilated 25% of respired CO_2 (Vance et al., 1983). Rainbird et al. (1984a) determined that 32% of the CO_2 effluxed was reassimilated in soybean nodules by subtracting the observed rate of CO_2 fixation from the expected CO_2 efflux.

Some measurements on the carbon costs of nitrogen fixation reported in Table 1 include corrections for CO_2 fixation. For example, Christeller

COST CORRECTED FOR			
DARK CO_2 FIXATION?	EAC?	COST (g C RESPIRED/ g N FIXED)	REFERENCE
No	Yes	5.4	Witty et al. (1983)
No	Yes	4.6	Witty et al. (1983)
No	Yes	4.2 (day)[b] 2.4 (night)[b]	Rainbird et al. (1983a)
No	Not applicable	5.8[b]	Patterson and LaRue (1983)
No	No	3.6[c]	Patterson and LaRue (1983)
No	No	2.4	Hardy et al. (1983)
No	Not applicable	2.4−7.0	Warembourg (1983)
No	Yes	2.6	Rainbird et al. (1984a)
Not applicable	Yes	3.8	Heytler amd Hardy (1984)
Yes	Yes	3.6	Heytler and Hardy (1984)
Yes	No[d]	3.3−3.7[d]	Ryle et al. (1984)

et al. (1977) reported that approximately one CO_2 was fixed per N_2 fixed in lupine nodules. This ratio has been used to estimate the amount of CO_2 fixation by nodules in some studies (e.g., Heytler and Hardy, 1984; Ryle et al, 1984).

Nodule CO_2 fixation also provides carbon skeletons for nitrogen assimilation and substrates for energy production. In lupine nodules (Christeller et al., 1977) and alfalfa nodules (Vance et al., 1983), a close correlation was observed between rates of CO_2 fixation and nitrogen fixation. Thus a primary function of CO_2 fixation in lupine nodules is to provide carbon skeletons for asparagine and aspartate synthesis (Fig. 2) (Christeller et al., 1977). According to Coker and Schubert (1981), CO_2 fixation in soybean roots and nodules fulfills a dual role of providing carbon skeletons for ammonia assimilation and carbon substrates for energy metabolism. The transport form of fixed nitrogen may determine whether CO_2 fixation and nitrogen fixation are correlated; such correlations are seen in legumes such as alfalfa that export amides and can more directly use the products of PEP carboxylase (i.e., oxaloacetic acid) and nitrogenase (i.e., ammonia) for the synthesis of amides (Christeller et al., 1977; Vance et al., 1983). These correlations are not as apparent for ureide producers (Coker and Schubert, 1981), possibly because ureide synthesis is less dependent on oxaloacetic acid.

Cultivar and Strain Effects

As discussed earlier, symbioses derived from legumes inoculated with Hup[+] strains of *Rhizobium* performed better in terms of total dry matter and nitrogen accumulation (Evans et al., 1985). Both the bacterial and host genotypes have a great influence on the expression of Hup genes. In surveys of *R. leguminosarum* strains (Ruiz-Argueso et al., 1979; Nelson and Salminen, 1982), only a fraction of the isolates displayed uptake hydrogenase activity when tested against a single pea cultivar. Keyser et al. (1982) showed that two *R. japonicum* strains displayed a Hup[+] phenotype on a cowpea cultivar, but a Hup[-] phenotype with a soybean cultivar. The biochemical cause of the difference in the expression of Hup is unknown (Phillips et al., 1985).

The legume host may affect the RE and the EAC of a symbiosis. For example, when pea plants infected with Hup[-] strains of *Rhizobium* were exposed to extended dark periods, the RE increased and the nodule soluble sugar content decreased (Edie and Phillips, 1983). It was hypothesized that the decreased carbohydrate supply (a host effect) resulted in the altered RE and presumably the altered EAC of nitrogenase, since the RE should equal the EAC in Hup[-] strains. Different uptake hydrogenase activities may occur in the two symbioses; one nodulated cultivar of pea evolved 24 times more H_2 than another pea cultivar infected with the same *R. leguminosarum* strain (Bedmar et al., 1983). The increased Hup[+]

phenotype may result from the action of a transmissible factor from shoots of the pea plant (Bedmar and Phillips, 1984). In addition, the EAC of nitrogenase may change as a result of plant shoot factors (Bedmar and Phillips, 1984). It will be interesting to determine whether these plant host effects translate into increased yields.

Cultivar and *Rhizobium* strain differences also affect the metabolic costs of nitrogen fixation in root nodules. Minchin et al. (1981), however, observed no differences in the grams of carbon respired per grams of nitrogen fixed in nodules infected with different strains of rhizobia. Similarly, Patterson et al. (1983) detected little difference in the respiration rates associated with nitrogen fixation in 13 soybean cultivars, but Witty et al. (1983) reported strain-dependent variations of amount of CO_2 respired per amount of C_2H_2 reduced within a single pea species and species differences in CO_2/C_2H_2 values using the same *Rhizobium* strain. With the present methods of separating nitrogen fixation costs from maintenance costs, it may be possible to determine more precisely the differences in metabolic costs due to strains and cultivars.

Ontogenetic Effects

The relationship between nitrogen fixation and photosynthesis throughout the life cycle of the legume plant has been intensively investigated and is summarized by Minchin et al. (1981). In general, the rates of nitrogen fixation are relatively higher during vegetative growth and decline during periods of reproductive development. Photosynthetic rates are relatively lower during vegetative growth than at the later stages of development.

Electron flux through nitrogenase changes with plant development (Bethlenfalvay and Phillips, 1977; Edie, 1983), in two symbioses of cowpea infected with either a Hup$^+$ or a Hup$^-$ strain of *Rhizobium*, the RE changed during plant development, suggesting that the metabolic cost of nitrogen fixation also varied with ontogeny (Rainbird et al., 1983c). Although the two symbioses did not differ in total dry matter or nitrogen accumulation at maturity, plants with the hydrogen-evolving *Rhizobium* strain lost more CO_2 from their nodules than plant nodules with the Hup$^+$ strain. Thus the carbon economy was superior, at least for part of the plant's lifetime, in symbiosis possessing a hydrogen uptake system.

When costs of nitrogen fixation were separated from nodule maintenance costs, carbon consumed per nitrogen fixed was found to vary with plant age; thus soybean nodules utilized 15–9 g CH_2O/g N fixed (6.0–3.6 g C/g N) in young and older plants, respectively (Heytler and Hardy, 1979). In another study with soybeans, 2.5 g C/g N was consumed within 69 days after sowing and increased to 7–8 g C/g N thereafter (Warembourg, 1983). Ryle et al. (1984) reported the cost of nitrogen fixation in soybean as varying from 3.3–3.7 g C/g N from early reproductive growth to later pod development. In contrast, Witty et al. (1983) found that the respiratory costs (measured as CO_2 evolved/C_2H_2 reduced) associated with nitrogen fixation were relatively constant with plant age for detached pea nodules, nodulated roots of lucerne, and nodulated roots of field bean. Because no attempt was made to determine the RE of the symbiosis (Witty et al., 1983), differences in the respiratory costs of nitrogen fixation possibly were not realized.

Effect of Diurnal Variations

Since fixation of nitrogen in the bacteroid is dependent on photosynthate, nitrogen fixation rates generally peak during the day and decline at night (Minchin et al., 1981). These rates can be related to diurnal variations in photosynthate export, root respiration, and carbohydrate content in plants (Minchin et al., 1981). In estimating the nitrogen fixation of legumes using acetylene reduction rates, care should be taken with regard to the time of sampling, because acetylene reduction rates may vary diurnally (e.g., Minchin and Pate, 1974).

The metabolic costs of nitrogen fixation appear to vary diurnally as well. In cowpea (Rainbird et al., 1983a) and soybean (Rainbird et al., 1984a) nodules, increased RE values were observed during darkness and in cowpea were related to a smaller electron flux to proton reduction, whereas the nitrogen fixation rate remained relatively constant (Rainbird et al., 1983a). The respiratory costs varied from 9.74 (day) to 5.70 (night) mol CO_2 evolved/mol N_2 fixed.

The variation in RE with day or night regimes may be due to fluctuations in temperature (Rainbird et al., 1983b). Cowpea plants infected with Hup⁻ strains of *Rhizobium* showed diurnal variations in RE, but these variations in RE were abolished in plants maintained under the same diurnal regimes but at constant temperature (Rainbird et al., 1983b). The temperature was related linearly to the electron flux to H^+ between 15 and 47°C. In contrast, the metabolic costs of nitrogen fixation in soybean nodules infected with Hup⁻ strains of *Rhizobium* were identical (2 mol CO_2 evolved/mol H_2 evolved) at 22°C and 30°C, but the nodule maintenance costs varied with temperature (Rainbird et al., 1984a).

Total Metabolic Costs of Nitrogen Fixation in Nodulated Roots

The respiratory costs measured on nodulated roots include costs associated with root growth and maintenance, nodule growth and maintenance, nitrogen fixation, ammonia assimilation and transport, and CO_2 fixation (Fig. 2). Minchin et al. (1981) list the respiratory costs of nitrogen fixation in nodulated roots systems during a period of increasing nitrogen fixation from previous studies. The values ranged from 4.5 g C/g N fixed for cowpea (Herridge and Pate, 1977) to 14 g C/g N for pea (Mahon, 1977a,b; Mahon and Child, 1979). Patterson and LaRue (1983) estimated that 5.8 g C/gN fixed in soybean roots over the lifetime of a plant. Possible reasons for the wide variation in values are diurinal fluctuations in the C/N ratios and ontogenetic changes (Minchin et al., 1981).

Separated Metabolic Costs of Root Maintenance, Nodule Maintenance, and Nitrogen Fixation

The costs of nitrogen fixation in nodulated roots include the costs of nitrogen fixation alone (Table 1) and nodule maintenance and growth (Table 2). The carbon cost for nitrogen fixation alone in legume nodules varies from 2.4 to 7.0 g C/g N (Table 1). Because ammonia is not produced by nitrogenase in an atmosphere of either acetylene or Ar/O_2, the carbon consumed also includes the cost of ammonia assimilation and transport when the respiration rate was measured before establishing the rate of nitrogen fixation by acetylene reduction or hydrogen production. The estimated costs of nodule maintenance (Table 2) range from 1.32 to 2.28 μmol CO_2/min/g dry weight of nodule.

The calculations of Witty et al. (1983) (Table 1) of 5.4 and 4.6 g C/g N for detached pea nodules and nodulated roots of lucerne, respectively, were determined by analyzing regression gradients of CO_2 efflux versus acetylene reduction. Factors relating $^{15}N_2/C_2H_2$ and total N/C_2H_2 were used for the conversion of acetylene reduced to nitrogen fixed for detached pea nodules and lucerne roots, respectively. Either long-term (nitrogen accumulation) or short-term (^{15}N accumulation) conversion factors result in similar carbon costs. Maintenance costs of 1.34 and 1.45 μmol CO_2/min/g dry weight nodules for pea and French bean, respectively, were estimated from regression analysis of respiration versus nitrogen fixation for detached nodules. The contribution of Hup[+] phenotypes was not considered.

Cowpea nodules formed by Hup[-] rhizobia were used to study diurnal variations with respect to nitrogen fixation costs (Table 1) (Rainbird et al., 1983a). A higher cost during the day (4.2 g C/g N) than at night (2.4 g C/g N) was attributed to increased hydrogen evolution during the day. However, the reported metabolic costs were not separated into those associated with nitrogen fixation or nodule maintenance.

CO_2 efflux and acetylene reduction by nodulated and denodulated soybean roots at 20 and 3% oxygen were used to calculate by difference the nodule growth and maintenance costs versus the nitrogen fixation costs (Ryle et al., 1984). The difference in respiration rate between

Table 2 Growth and Maintenance Costs of Root Nodules.

Species	Age[a,b] (days)	Tissue(s)	Nodule Growth and Maintenance Cost (µmol CO_2/min/g dry weight nodule)[b]	Percentage of Total Nodule Respiration[b]	Temperature (C°)[b]	Reference
Pea	NA	Detached nodules	1.34	NA	NA	Witty et al. (1983)
French bean	NA	Detached nodules	1.45	NA	NA	Witty et al. (1983)
Soybean		nif+ nodulated roots, nonnodulated roots, and nif- nodulated roots	NA	25	NA	Hardy et al. (1983)
Soybean	20–30	Detached nodules	1.83	25–30	23.5[c]	Heytler and Hardy (1983)
Soybean	30–40	Nodulated and denodulated roots	1.32	22	22	Rainbird et al. (1984a)
Soybean	30–40	Nodulated and denodulated roots	2.28	—	30	Rainbird et al. (1984a)
Soybean	35–40	Nodulated and denodulated roots	1.59[d]	20	20–21	Ryle et al. (1984)

[a]After sowing.
[b]NA, Not available.
[c]Initial temperature
[d]Reported as 20% of total nodule respiration which was 7.95 µmol CO_2/min/g dry weight nodule.

nodulated and denodulated roots was assumed to equal the total nodule respiration at 20% O_2 (growth, maintenance, and nitrogen fixation) and to equal nodule growth and maintenance respiration at 3% O_2. At 3% O_2, the nitrogen fixation rate of nodulated roots was nearly zero; thus respiration associated with nitrogen fixation was assumed to be absent. Nodule growth and maintenance respiration were subtracted from total nodule respiration to calculate the respiration due to nitrogen fixation and ammonia assimilation. The average total respiration cost for the nodule was 13.2 g CO_2/g N (g CO_2 efflux per g N fixed) until the end of plant development based on the assumption that the ratio of acetylene reduced per nitrogen reduced was 4. Twenty percent of the total nodule respiration was used for growth and maintenance. The cost of nitrogen fixation and ammonia metabolism ranged from 11 to 12.5 g CO_2/g N. The latter values were converted to 3.3–3.7 g C/g N assuming that the respiration was underestimated by 10% as a result of CO_2 fixation.

Regression analysis of respiration and nitrogen fixation of nodulated roots using soybean plants exposed to varying lengths of darkness gave a value of 2.6 g C/g N for the cost of nitrogen fixation (Rainbird et al., 1984a). Denodulated root respiration was subtracted from the intercept value (which equaled that of the nodule plus root growth and maintenance) to obtain the nodule respiration associated with growth and maintenance. The maintenance cost was temperature-dependent; at 22 and 30°C, the growth and maintenance costs for the nodules were 1.32 and 2.28 μmol CO_2/min/g dry weight of nodule, respectively. Nodule maintenance accounted for 22% and nitrogenase consumed 52% of the total nodule respiration. However, these nitrogen fixation and maintenance costs may be underestimated, because Rainbird et al. (1984a) estimated that 32% of the CO_2 efflux was refixed but did not correct for the calculated CO_2 fixation. Interpretation of the results is difficult because the RE values exceeded 0.8 (0.86 in the light and 0.94 in the dark) although Hup$^-$ strains of *Rhizobium* were used. Theoretically, the EAC for nitrogen should not exceed 0.75 (see Simpson and Burris, 1984) and the RE should approach the EAC when Hup$^-$ strains are used.

Calorimetry was used to estimate detached nodule respiration (Heytler and Hardy, 1984), and a regression analysis of the heat evolved versus the hydrogen produced by detached soybean nodules infected with Hup$^-$ strains of *Rhizobium* yielded a cost for nitrogen fixation of 3.8 g C/g N fixed. Although CO_2 fixation is not a problem in this approach, the criticisms associated with detached nodules still apply. The RE of the Hup$^-$ *Rhizobium* strains was approximately 0.7. The maintenance costs were estimated at 25–30% of the total heat evolved by detached nodules. The carbon cost of nitrogen fixation in the nodule was also estimated by CO_2 efflux. On the assumption that CO_2 fixation and NH_3 production were correlated (see section on the role of CO_2 fixation), the estimated nitrogen fixation cost was 3.6 g C/g N. Thus the two independent methods of estimating energy costs resulted in similar values.

In studies where $^{14}CO_2$ efflux and ^{15}N fixed from roots were measured simultaneously (Warembourg, 1983), the estimate of the cost associated with nitrogen fixation was between 2.5 g C/g N and 7.0 g C/g N during periods of high nitrogen fixation rates. Estimates on the cost of nitrogen fixation assume that the slow rate of $^{14}CO_2$ efflux from the roots after reexposure of the plant tops to $^{12}CO_2$ was due to costs associated with maintenance; comparisons were made between nodulated and non-nodulated roots to determine nodule costs, and these comparisons may not be valid.

Nodulated and nonnodulated isolines of soybean along with effective and ineffective *Rhizobium* strains were used to estimate energy costs of 2.4 g C/g N (Hardy et al., 1983). Total root and nodule respiration in plants with high rates of nitrogen fixation was determined to consist of root respiration (25%), nodule maintenance respiration (25%), and respiration associated with nitrogen fixation and ammonia assimilation (50%).

Patterson and LaRue (1983) compared four methods of determining the respiration in the same soybean cultivar. The respiration associated with nitrogen fixation was 2.10, 2.90, 4.08, and 4.36 mole CO_2 evolved/mol C_2H_2 reduced when the respiration and nitrogen fixation were varied by exposing the roots to 1.0 atm oxygen, adding NO_3^- to the roots, partial denodulation, and dark-light manipulations, respectively. The EAC for neither nitrogen nor CO_2 fixation was estimated; thus no attempt was made to convert CO_2/C_2H_2 ratios to g C/g N estimates. For comparison, the respiration values obtained using supraambient levels of oxygen were converted to grams of carbon consumed per grams of nitrogen fixed (Table 1) assuming a 0.75 EAC for nitrogen. The relationship between acetylene reduction and CO_2 efflux probably reflects costs associated with nitrogen fixation, because supraambient levels of oxygen had no effect on the respiration of nonnodulated roots (Patterson et al., 1983).

The estimated cost of nitrogen fixation in nodules ranges from 1.1 to 7.6 g C respired/g N fixed (Minchin et al., 1981). However, most of these estimates do not separate nodule growth and maintenance costs from nitrogen fixation costs. The costs directly associated with nitrogen fixation vary from 2.4 to 7.0 g C/g N (Table 1), whereas the theoretical cost of nitrogen fixation in legumes is 2.5 g C/g N (Pate et al., 1981). Some estimates of costs associated with nitrogen fixation (Table 1) include costs relating to ammonia assimilation and transport. Thus actual costs are surprisingly close to theoretical estimates.

The costs associated with growth and maintenance as opposed to nitrogen fixation are listed in Table 2. Maintenance costs represent less than 30% of the energy costs associated with the nodule. While each method of study can be criticized, it should be noted that a number of independent methods result in maintenance values (1.3–2.3 μmol CO_2/min/g dry weight of nodule) that represent 22–30% of the total nodule respiration. It must be concluded that the majority of the energy utilization in the legume nodule is involved in the reduction of nitrogen.

ATP Costs of Nitrogen Fixation in Legumes

Determination of the metabolic costs of nitrogen fixation in terms of ATP equivalents requires many assumptions. First, suppositions as previously discussed are inherent in calculating the carbon respired per nitrogen fixed. In order to relate carbon consumption to ATP synthesis, a knowledge of the stoichiometry between the carbon respired and the ATP synthesized is required. However, while the stoichiometry should be constant with respect to a given source of carbon, it may not be constant *in planta* either between different legumes or even within a single plant. The variability is due to the differences in energy sources that are delivered to the bacteroids; depending on their physiological age, plant shoots may deliver different carbon substrates that are absorbed and utilized by the bacteroids with varying efficiency. Although some carbohydrate metabolism in the nodule may occur via fermentation under stress conditions (Tajima and LaRue, 1982), the synthesis of ATP in the bacteroid occurs mainly by oxidative phosphorylation (see Appleby, 1984) and is presumably a function of the proton motive force. ATP synthesis by the H^+-ATPase is driven by the translocation of protons across the cell membrane according to chemiosmotic theory (Nichols, 1982). The number of protons required for the net synthesis of 1 mol of ATP may vary with different metabolic conditions. For instance, the number of protons translocated by the H^+-ATPase per ATP synthesized in a cowpea *Rhizobium* strain varied from 1.8–2.3 to 3.7–3.8 when the cells were grown in 21 and 0.2% O_2, respectively (Gober and Kashket, 1984). Most reviews on the energetics of nitrogen fixation calculate the ATP cost by assuming a P/O or $P/2e^-$ ratio of 2 or 3 (Minchin et al., 1981; Pate et al., 1981; Schubert, 1982), but the stoichiometry between ADP phosphorylation and O_2 utilization or CO_2 efflux may be a noninteger value, since ATP synthesis occurs as a function of proton translocation by the H^+-ATPase. Ratcliffe et al. (1983) suggest that the P/O ratio of free-living *R. leguminosarum* is closer to 1 rather than 3.

In estimates of the ATP cost of nitrogen fixation in rhizobia, Stam et al. (1984) reported that *Rhizobium* ORS 571 from *Sesbania rostrata* grown *ex planta* under nitrogen utilized 42 mol ATP/mol N_2 fixed. The EAC for protons was approximately 0.70. The costs measured in this study presumably were those of nitrogen fixation and bacterial growth.

The ATP cost of nitrogen fixation calculated from growth yields of cells grown at 5% oxygen were quite high; possibly, the efficiency of respiration would have been higher in bacteroids bathed in oxygenated leghemoglobin. *Rhizobium japonicum* bacteroids (Appleby et al., 1975; Bergersen and Turner, 1975) respired more efficiently in terms of nitrogen fixation in the presence of oxygenated leghemoglobin. In studies on *R. leguminosarum* bacteroids, conditions that altered the electron flux through nitrogenase *in vitro* did not affect the electron flux *in vivo* (Haaker and Wassink, 1984). More experiments of this nature are required to appreciate the ATP costs of nitrogen fixation *in vivo*.

The substrate available for the bacteroids plays a major role in determining the energy costs of nitrogen fixation. Sucrose is the predominant form of carbohydrate transported to the nodules (Lawrie and Wheeler, 1975; Reibach and Streeter, 1983) (Fig. 2). To be utilized by the bacteroid, sources of carbon must transverse the nodule cell membrane, the peribacteroid membrane, and the bacteroid membrane. In general, organic acids and amino acids derived from the citric acid cycle are preferred substrates for bacteroids and for free-living rhizobia for nitrogen fixation. Organic acids (primarily malate) and amino acids (primarily glutamate) are the predominant products of sucrose metabolism in soybean nodules (Reibach and Streeter, 1983). Carbohydrates are relatively poor sources of carbon for bacteroids because they must be converted to organic acids before being used.

Other compounds may be important storage compounds for the bacteroid. Trehalose accumulates only in nodulated legumes as opposed to nonnodulated legumes (Philips et al., 1984) and also accumulates in *R. japonicum* bacteroids (Reibach and Streeter, 1983). The storage product of bacteroids, poly-β-hydroxybutyrate, may also be an important energy source in some symbioses (Appleby, 1984).

The details of carbon metabolism and ATP synthesis in legume nodules are still not understood well enough to confidently predict the ATP costs for nitrogen fixation. Based on experimental data, the average of the estimated costs of nitrogen fixation in nodules (Table 1) is approximately 4 g C/g N for studies which exclude nodule maintenance costs. If it is assumed that 4 g of carbon is equivalent to 10 g of glucose and that the oxidation of 1 mol of glucose yields 36 mol of ATP, 56 mol of ATP is utilized in the reduction of 1 mol of nitrogen.

The theoretical cost of reducing 1 mol of dinitrogen to 2 mol of ammonia is 28 ATP (ATP plus reductant costs) assuming that the EAC for nitrogen is 0.75, hydrogen oxidation does not occur, and the P/O ratio equals 3 (Minchin et al., 1981; Schubert, 1982). Depending on the type of export product, the total theoretical ATP costs of nitrogen fixation and ammonia assimilation vary between 25.5 and 49 mol ATP/mol nitrogen fixed for 0–50% electron flux to proton reduction (Minchin et al., 1981).

The reason for the discrepancy between theoretical and experimental energy estimates of nitrogen fixation is unknown. However, independent studies utilizing different methods of analysis have yielded reasonably similar carbon costs for nitrogen fixation in legume nodules. It is also possible that the assumptions regarding ATP costs are incorrect or that the process of nitrogen fixation possesses a certain degree of inefficiency.

INTERACTIONS OF NITROGEN FIXATION WITH HOST METABOLISM

Ammonia Assimilation and Transport of Fixed Nitrogen

Ammonia, the product of the nitrogenase-catalyzed reduction of nitrogen in the bacteroid, is exported to the plant cytosol (Fig. 2, 3). Both passive (Dilworth and Glenn, 1985) and active (Schubert, 1982) mechanisms of ammonia transport have been suggested. The energy costs would be minimal if the ammonia diffusion were passive; however, it has been suggested that during export from the bacteroid ammonia must traverse two membranes (bacteroid and host) at a cost of 1 mol ATP/mol ammonia transported per membrane crossed (Schubert, 1982). Ammonia is not assimilated in the bacteroid, because the activities of ammonia-assimilating enzymes are low in the bacteroid but high in the plant cytosol (Dunn and Klucas, 1973; Boland et al., 1980). Ammonia assimilation in the plant cytosol of amide exporters occurs via the action of glutamine synthetase, glutamate synthase, and asparagine synthase (Boland et al., 1980) at an apparent cost of 7 ATP/1 asparagine produced, for 2 mol ammonia assimilated per asparagine (Minchin et al., 1981; Schubert, 1982) (Fig. 3).

Legumes of tropical origin such as soybean (McClure and Israel, 1979; Streeter, 1979) and cowpea (Herridge et al., 1978) export mainly ureides, whereas other legumes of temperate origin such as pea export mainly the amides glutamine and asparagine (Pate and Wallace, 1964; Scott et al., 1976; Sprent, 1980; Schubert and Boland, 1984). The most probably pathway for ureide synthesis in the nodule is the *de novo* synthesis and subsequent degradation of purines (Atkins et al., 1982; Boland and Schubert, 1982; Christensen and Jochimsen, 1983; Reynolds et al., 1982) (Fig. 4). The ureide concentration in the shoot axis of soybeans is considered to be a quantitative indicator of nitrogen fixation (Herridge, 1982), but nitrate concentrations may also affect the ureide concentration in the xylem (McNeil and LaRue, 1984).

It has been suggested that export in the form of ureides decreases the amount of reduced carbon required for nitrogen transport to the shoots (Schubert, 1981). In calculating the metabolic costs of producing export

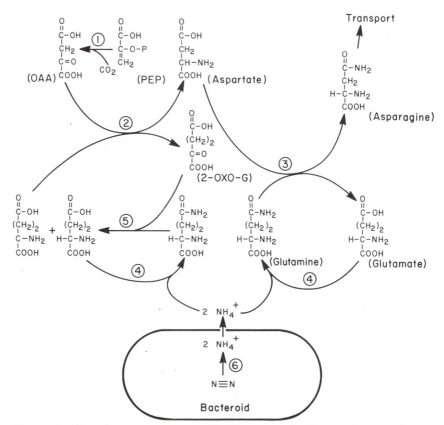

Figure 3. Reactions concerned with nitrogen reduction and ammonia assimilation in legume nodules. Enzymes and the reactions catalyzed are: (1) PEP carboxylase, synthesis of oxalacetate from phospoenol pyruvate (PEP) and CO_2; (2) amino transferase, synthesis of aspartate from oxalacetate and glutamate; (3) asparagine synthetase, synthesis of asparagine from aspartate, glutamine and ATP; (4) glutamine synthetase, synthesis of glutamine from glutamate and ATP; (5) glutamate synthase, production of 2 glutamates from 2-oxoglutarate (2-OXO-G), glutamine and NADPH; (6) nitrogenase, reduction of dinitrogen to yield ammonia.

products, Pate et al. (1981) suggest that the type of solute exported does not affect the overall carbon cost of nodule function significantly.

Photosynthesis and Nitrogen Fixation

Carbon substrates are required during symbiotic nitrogen fixation as sources of energy and as acceptor molecules for ammonia assimilation. In the legume nodule these substrates are derived from photosynthate transported from the leaves, dark CO_2 fixation in the nodule, and the metabolism of photosynthate stored as carbohydrate reserves in the nodule (Fig. 2). Dark CO_2 fixation by root nodules reassimilates 9–30% of the CO_2 released by soybean nodules (Coker and Schubert, 1981). Thus the main source of carbon for use in nitrogen fixation is photosynthesis. During periods of increasing nitrogen fixation in grain legumes, it is estimated that 40–50% of the total net photosynthate is transported to the nodulated roots and that 25–33% of the net photosynthate is required by the nodulated root system to support nitrogen fixation (Minchin et al., 1981). Sheehy (1983) calculated that 20% of the gross photosynthate is required for nitrogen fixation in the soybean using data obtained from plant growth studies. For a number of legumes studied, it is estimated that 5–13% of the net photosynthate is respired by the nodules alone (Minchin et al., 1981).

Factors that affect rates of photosynthesis directly affect rates of nitrogen fixation. Treatments that increase photosynthesis, such as CO_2 enrichment (Hardy and Havelka, 1975b; Quebedeaux et al., 1975) and increased light (Bethlenfalvay and Phillips, 1977), also increase nitrogen fixation, whereas treatments such as defoliation (Lawn and Brun, 1974) and darkness or shading (Lawn and Brun, 1974; Schweitzer and Harper, 1980) decrease both photosynthesis and nitrogen fixation.

Figure 4. The production of allantoin and allantoic acid from adenine (1). Deamination of adenine yields hypoxanthine (2); the latter is oxidized via xanthine oxidase to yield successively, xanthine (3) and uric acid (4). Uricase catalyzes the oxidation of uric acid to yield allantoin (5). Allantoin is converted to allantoic acid (6) by the action of allantoinase.

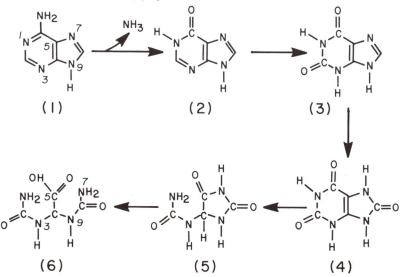

Although Hardy and Havelka (1975b) suggest that an increased photosynthate supply is responsible for the increased nitrogen fixation observed in nodulated soybean plants exposed to increased levels of CO_2, results of recent studies suggest that photosynthesis per se under normal conditions does not limit nitrogen fixation. Increases in acetylene reduction rates of nodulated soybean plants exposed to increased levels of CO_2 probably were due to increases in nodule mass (Finn and Brun, 1982; Williams et al., 1982). The fact that short-term enrichment with CO_2 did not affect acetylene reduction rates led to the conclusion that photosynthesis itself did not limit nitrogen fixation. However, the partitioning or utilization of photosynthate may ultimately limit nitrogen fixation.

Nitrogen Fixation versus Nitrate Utilization by Legumes

The energy requirements of nitrogen fixation in legumes include costs of nitrogen fixation (reductant and ATP costs), ammonia assimilation, and fixed-nitrogen transport. For the process of nitrogen fixation, assuming that 4 NADH and 16 ATP are utilized, an energy input equivalent to 271 kcal is required for the reduction for 1 mol of nitrogen (or 136 kcal/mol ammonia formed). This contrasts with energy requirements for nitrate reduction in nonnodulated legumes growing on nitrate; the reduction of nitrate to ammonia in plants requires two enzymes, nitrate reductase and nitrite reductase, which use reduced pyridine nucleotides and ferredoxin, respectively, as reductants. Reduction of 1 mol of nitrate to ammonia uses 1 mol NAD(P)H and 6 mol ferredoxin, an equivalent usage of 69 kcal/mol ammonia (Schubert, 1982).

One factor that affects the energy costs is the site of nitrate reduction; in roots, the reductant is generated from the oxidation of photosynthate. In shoots the reductant is generated by photosynthesis and is therefore abundant and "cost-free." Some authors (Rufty et al., 1984; Andrews et al., 1984; Timpo and Neyra, 1983) claim that the capacity to reduce nitrate in a number of legumes of both tropical and temperate origin resides primarily in the shoots. On the other hand, the root system of pea has an active nitrate reductase system (see Minchin et al., 1981), and nitrate reduction is proposed to occur in the roots of this species (Schubert, 1982).

The differences in energy costs of nitrogen fixation versus nitrate reduction have been calculated by comparing respiration rates, growth rates, and yields of legumes actively fixing and administered fixed nitrogen. In 16 studies reviewed by Schubert (1982), the relative energy cost of nitrogen fixation compared to nitrate reduction was greater in 8 studies, equal in 7 studies, and less in 1 study. Thus the theoretical calculations are not upheld by experimental data. Calculations of the specific root respiration (respiration rate per gram dry weight of root) are equally contradictory; it was found to be lower (see Minchin et al., 1981) and higher in nonnodulated legumes (Ryle et al., 1983) than in nodulated legumes. The basis of this contradiction is unknown. In the early stages of soybean plant development, nitrogen-fixing plants retained 8–12% less photosynthate than nitrate-dependent plants (Finke et al., 1982); this reflects the energy demands for nitrogen fixation. Environmental effects such as water stress (DeJong and Phillips, 1982) may affect the energetics of nitrogen fixation more than nitrate use.

The seemingly contradictory data indicate that it is difficult to compare directly the energy costs of nitrogen fixation versus nitrate assimilation. Since the morphology, physiology, and biochemistry of plants grown on the two nitrogen sources are different, a comparison of energy costs may not be valid. The biochemical differences are apparent from the changes observed upon nitrate addition to nodulated legume roots (Ursino et al., 1982). Xylem and phloem exudates as well as photosynthate transport differ substantially in NO_3^--grown versus nitrogen-fixing legumes (Ursino et al., 1982; Layzell and LaRue, 1982; McNeil and LaRue, 1984; Crafts-Brandner et al., 1984). In addition, CO_2 fixation by nodules is inhibited by the addition of nitrate to nodulated alfalfa roots (Vance et al., 1983). The relative costs of nitrogen fixation versus nitrate utilization by legumes may be variable depending on the legume species and environmental conditions.

GROWTH YIELDS OF FREE-LIVING, NITROGEN-FIXING ORGANISMS AS A MODEL SYSTEM FOR SYMBIOTIC SYSTEMS

The rate of growth and the growth yield of free-living microorganisms grown on N_2 or fixed nitrogen can serve as a model for the growth of symbiotic systems on N_2 versus growth of the plant on fixed nitrogen. Such comparisons must be made cautiously, since free-living microorganisms growing on N_2 are provided with other growth substrates in excess, while for the plant nitrogen may be only one substance in the complex of limiting nutrients. For example, if light is limiting, the plant may be limited for carbon, which ultimately provides energy for the reduction of N_2. Nevertheless, the examination of free-living systems is instructive in assessing the energy cost of nitrogen fixation.

The growth rates of free-living microorganisms including *C. pasteurianum, pneumoniae, Rhizobium ORS 571,* and *R. rubrum* with N_2 as the nitrogen source are significantly slower than their growth rates on fixed nitrogen (Daesch and Mortenson, 1967, 1972; Tubb and Postgate,

1973; Munson and Burris, 1969; Stam et al., 1984). This might be explained by nitrogenase being the rate-limiting reaction for cell growth; alternatively, the energy demands of nitrogenase could result in slower growth. In *Rhizobium* grown on N_2, a case most relevant to symbiotic systems, the molar growth yield on ATP for cells grown on NH_4^+ in a succinate-limited chemostat is approximately double that of cells grown on N_2 (Stam et al., 1984). Furthermore, Stam et al. (1984) estimate that 70% of the flux of the electrons through nitrogenase in succinate-limited *Rhizobium* goes into the production of H_2 and only 30% into the reduction of N_2 to NH_4^+. This indicates that an advantage exists for cells expressing an uptake hydrogenase to recycle some of the energy lost to H_2 production. The addition of H_2 to the gas phase of succinate-limited nitrogen-fixing chemostat cultures increased the growth yield. This concurs with the results of Evans and co-workers (1985), who found increased yields from soybeans infected with *Rhizobium japonicum* strains that express uptake hydrogenase (Hup$^+$).

SUMMARY

Nitrogen fixation in legumes occurs because of a symbiosis between plant and rhizobia. The site of nitrogen fixation is the root nodule which houses the nitrogen-fixing bacteroids. Along with nitrogenase, the enzyme responsible for nitrogen fixation, many tissues and molecules must be synthesized to support nitrogen fixation. From studies on free-living nitrogen-fixing microorganisms, it is known that the synthesis and functioning of nitrogenase is very energy-expensive.

Nitrogenase structure is well conserved among nitrogen-fixing organisms. It consists of a molydenum-iron protein tetramer and an iron-protein dimer. A minimum of 16 ATP is required for the reduction of 1 mol of dinitrogen to ammonia. At least 25% of the total electron flux through nitrogenase is devoted to hydrogen production, an apparently wasteful process. Many factors such as [ATP] and nitrogenase protein ratios affect the relative amount of dinitrogen versus number of protons reduced.

Because of the economic importance of legumes, the energy cost of nitrogen fixation in these plants is of interest. The total costs include those associated with nodule growth and maintenance, nitrogen fixation, and ammonia assimilation and transport. Separation of the respiratory costs of nitrogen fixation from those of nodule growth and maintenance is now possible using a variety of techniques. The majority of nodule respiration is used for nitrogen fixation. Each technique, however, is subject to criticism, and the resultant energy estimate questioned. Because a number of independent methods arrive at similar energy partitioning and costs in the legume nodule, it seems that the reported costs of 2.4–7.0 g C respired/g

N fixed are reliable. Approximately 22–30% of the total nodule respiration is used for nodule growth and maintenance. The energy costs of nitrogen fixation in terms of ATP are more difficult to predict because of variations in the substrates available to the bacteroid, ontogenetic effects, and lack of knowledge regarding the relationship between carbon respired and ATP synthesized. Based on a number of assumptions, the reduction of 1 mol of dinitrogen to ammonia in the nodule is estimated to require 56 mol of ATP.

Several factors influence the cost of nitrogen fixation in legumes. Uptake hydrogenase and phosphenolpyruvate carboxylase in legume nodules apparently recover some of the energy and carbon lost during nitrogen fixation and respiration. Both plant cultivar and rhizobial strain exert differences in the relative efficiency of nitrogen fixation. Also, diurnal and seasonal changes affect the cost of nitrogen fixation in legumes. Since nitrogen fixation is a complex process and is affected by many variables, the costs vary; thus it is difficult to make comparisons between nitrogen-fixing and nitrate-utilizing legumes.

ACKNOWLEDGMENTS

Preparation of this chapter was supported by the McKnight Foundation grant to the Center for the Study of Nitrogen Fixation, by USDA Competitive Grants Program grant 81-CRCR-1-703, and by NSF grant PCM8302630. We thank Robert Lowery, Roy Kanemoto, Timothy Hoover, and Dr. Stanley Duke for their suggestions regarding the manuscript.

REFERENCES

ANDREWS, M., SUTHERLAND, J. M., THOMAS, R. J., AND SPRENT, J. I. 1984. Distribution of nitrate reductase activity in six legumes: The importance of the stem, *New Phytol.* **98**:301–310.

APPLEBY, C. A. 1984. Leghemoglobin and *Rhizobium* respiration, *Annu. Rev. Plant Physiol.* **35**:443–478.

APPLEBY, C. A., TURNER, G. L., AND MACNICOL, P. K. 1975. Involvement of oxyleghemoglobin and cytochrome P-450 in an efficient oxidative phosphorylation pathway which supports nitrogen fixation in *Rhizobium, Biochim. Biophys. Acta* **387**:461–474.

ATKINS, C. A., PATE, J. S., RITCHIE, A., AND PEOPLES, M. B. 1982. Metabolism and translocation of allantoin in ureide-producing grain legumes, *Plant Physiol.* **70**:476–482.

AUGER, S., AND VERMA, D. P. S. 1981. Induction and expression of nodule-specific host genes in effective and ineffective root nodules of soybean, *Biochemistry* **20**:1300–1306.

BAUER, W. D. 1981. Infection of legumes by rhizobia, *Annu. Rev. Plant Physiol.* **32**:407–449.

BEDMAR, E. J., AND PHILLIPS, D. A. 1984. A transmissible plant shoot factor promotes uptake hydrogenase activity in *Rhizobium* symbionts, *Plant Physiol.* **75**:629–633.

BEDMAR, E. J., EDIE, S. A., AND PHILLIPS, D. A. 1983. Host plant cultivar effects on hydrogen evolution by *Rhizobium leguminosarum, Plant Physiol.* **72**:1011–1015.

BERGERSEN, F. J. 1962. The effects of partial pressure of oxygen upon respiration and nitrogen fixation by soybean root nodules, *J. Gen. Microbiol.* **29**:113–125.

BERGERSEN, F. J., AND TURNER, G. L. 1975. Leghaemoglobin and the supply of O_2 to nitrogen-fixing root nodule bacteroids: Presence of two oxidase systems and ATP production at low free O_2 concentration, *J. Gen. Microbiol.* **91**:345–354.

BETHLENFALVAY, G. J., AND PHILLIPS, D. A. 1977. Ontogenetic interactions between photosynthesis and symbiotic nitrogen fixation in legumes, *Plant Physiol.* **60**:419–421.

BOLAND, M. J., AND SCHUBERT, K. R. 1982. Biosynthesis of purines by a proplastid fraction from soybean nodules, *Arch. Biochem. Biophys.* **220**:179–187.

BOLAND, M. J., FARNDEN, K. J. F., AND ROBERTSON, J. G. 1980. Ammonia assimilation in nitrogen-fixing legume nodules, in: *Nitrogen Fixation,* Vol. II (W. E. Newton and W. H. Orme-Johnson, eds.), pp. 33–52, University Park Press, Baltimore, Maryland.

BRILL, W. J. 1980. Biochemical genetics of nitrogen fixation, *Microbiol. Rev.* **44**:449–467.

BULEN, W. A., LeCOMTE, J. R., BURNS, R. C., AND HINKSON, J. 1965. Nitrogen fixation studies with aerobic photosynthetic bacteria, in: *Non-heme Iron Proteins for Energy Conversion* (A. San Pietro, ed.), pp. 261–274, Antioch Press, Yellow Springs, Ohio.

BURGESS, B. K. 1984. Structure and reactivity of nitrogenase—An overview, in: *Advances in Nitrogen Fixation Research* (C. Veeger and W. E. Newton, eds.), pp. 103–115, Martinus Nijhoff/Dr. W. Junk, Pudoc, The Netherlands.

BURGESS, B. K., STIEFEL, E. I., AND NEWTON, W. E. 1980. Oxidation-reduction properties and complexation reactions of the iron-molybdenum cofactor of nitrogenase, *J. Biol. Chem.* **255**:353–356.

CHRISTELLER, J. T., LAING, W. A., AND SUTTON, W. D. 1977. Carbon dioxide fixation by lupin root nodules. I. Characterization, association with phosphoenolpyruvate carboxylase, and correlation with nitrogen fixation during nodule development, *Plant Physiol.* **60**:47–50.

CHRISTENSEN, T. M. I. E., AND JOCHIMSEN, U. 1983. Enzymes of ureide synthesis in pea and soybean, *Plant Physiol.* **72**:56–59.

COKER, G. T., III, AND SCHUBERT, K. R. 1981. Carbon dioxide fixation in soybean roots and nodules. I. Characterization and comparison with N_2 fixation and composition of xylem exudate during early nodule development, *Plant Physiol.* **67**:691–696.

CRAFTS-BRANDNER, S. J., BELOW, F. E., HARPER, J. E., AND HAGEMAN, R. H. 1984. Effect of nodulation on assimilate remobilization in soybean, *Plant Physiol.* **76**:452–455.

CRISWELL, J. G., HAVELKA, U. D., QUEBEDEAUX, B., AND HARDY, R. W. F. 1976. Adaptation of nitrogen fixation by intact soybean nodules to altered rhizosphere pO_2, *Plant Physiol.* **58**:622–625.

DAESCH, G., AND MORTENSON, L. E. 1967. Sucrose catabolism in *Clostridium pasteurianum* and its relation to N_2 fixation, *J. Bacteriol.* **96**:346–351.

DAESCH, G., AND MORTENSON, L. E. 1972. Effect of ammonia on the synthesis and function of the N_2-fixing enzyme system in *Clostridium pasteurianum, J. Bacteriol.* **110**:103–109.

DART, P. J. 1977. Infection and development of leguminous nodules, in: *A Treatise on Dinitrogen Fixation* (R. W. F. Hardy and W. S. Silver, eds.), pp. 367–372, John Wiley, New York.

DAVIS, L. C., AND KOTAKE, S. 1980. Regulation of nitrogenase activity in aerobes by Mg^{2+} availability: An hypothesis, *Biochem. Biophys. Res. Commun.* **93**:934–940.

DAZZO, F. B., AND HUBBELL, D. H. 1975. Cross-reactive antigens and lectins as determinants of symbiotic specificity in *Rhizobium trifolii*–clover association, *Appl. Microbiol.* **30**:1017–1033.

DEJONG, T. M., AND PHILLIPS, D. A. 1982. Water stress effects on nitrogen assimilation and growth of *Trifolium subterraneum* L. using dinitrogen or ammonium nitrate. *Plant Physiol.* **69**:416–420.

DILWORTH, M. J., AND GLENN, A. R. 1985. Transport in *Rhizobium* and its significance to the legume symbiosis, in: *Nitrogen Fixation and CO_2 Metabolism: A Steenbock Symposium in Honor of Robert H. Burris* (P. W. Ludden and J. E. Burris, eds.), pp. 53–61, Elsevier, New York.

DIXON, R. O. D. 1967. Hydrogen uptake and exchange by pea root nodules, *Ann. Bot.* **31**:179–188.

DIXON, R. O. D. 1972. Hydrogenase in legume root nodule bacteroids: Occurrence and properties, *Arch. Mikrobiol.* **85**:193–201.

DIXON, R. O. D., BLUNDEN, E. A. G., AND SEARL, J. W. 1981. Intercellular space and hydrogen diffusion in pea and lupin root nodules, *Plant Sci. Lett.* **23**:109–116.

DREVON, J. J., FRAZIER, L., RUSSELL, S. A., AND EVANS, H. J. 1982. Respiratory and nitrogenase activities of soybean nodules formed by hydrogen uptake negative (Hup⁻) mutant and revertant strains of *Rhizobium japonicum* characterized by protein patterns, *Plant Physiol.* **70**:1341–1346.

DUKE, S. H., AND HENSON, C. A. 1985. Legume nodule carbon utilization in the synthesis of organic acids for the production of transport amides and amino acids, in: *Nitrogen Fixation and CO_2 Metabolism: A Steenbock Symposium in Honor of Robert H. Burris* (P. W. Ludden and J. E. Burris, eds.), pp. 293–302, Elsevier, New York.

DUNN, S. D., AND KLUCAS, R. V. 1973. Studies on possible routes of ammonium assimilation in soybean root nodule bacteroids, *Can. J. Microbiol.* **19**:1493–1499.

EDIE, S. A. 1983. Acetylene reduction and hydrogen evolution by nitrogenase in a *Rhizobium*-legume symbiosis, *Can. J. Bot.* **61**:780–785.

EDIE, S. A., AND PHILLIPS, D. A. 1983. Effect of the host legume on acetylene reduction and hydrogen evolution by *Rhizobium* nitrogenase, *Plant Physiol.* **72**:156–160.

EISBRENNER, G., AND EVANS, H. J. 1983. Aspects of hydrogen metabolism in nitrogen-fixing legumes and other plant-microbe associations, *Annu. Rev. Plant Physiol.* **34**:105–136.

EMERICH, D. W., AND BURRIS, R. H. 1976. Interactions of heterologous nitrogenase components that generate catalytically inactive complexes, *Proc. Natl. Acad. Sci. USA* **73**:4369–4373.

EMERICH, D. W., AND BURRIS, R. H. 1978a. Complementary functioning of the component proteins of nitrogenase from several bacteria, *J. Bacteriol.* **134**:936–943.

EMERICH, D. W., AND BURRIS, R. H. 1978b. Preparation of nitrogenase, *Methods Enzymol.* **53**:314–329.

EMERICH, D. W., RUIZ-ARGÜESO, T., CHING, T. M., AND EVANS, H. J. 1979. Hydrogen-dependent nitrogenase activity and ATP formation in *Rhizobium japonicum* bacteroids, *J. Bacteriol.* **137**:153–160.

EVANS, H. J. 1975. *Enhancing Biological Nitrogen Fixation, Proceedings of a Workshop on Nitrogen Fixation*, National Science Foundation, Washington, D.C.

EVANS, H. J., HANUS, F. J., RUSSELL, S. A., HARKER, A. R., LAMBERT, G. R., AND DALTON, D. A. 1985. Biochemical characterization, evaluation, and genetics of H_2 recycling in *Rhizobium*, in: *Nitrogen Fixation and CO$_2$ Metabolism: A Steenbock Symposium in Honor of Robert H. Burris* (P. W. Ludden and J. E. Burris, eds.), pp. 3–11, Elsevier, New York.

FINKE, R. L., HARPER, J. E., AND HAGEMAN, R. H. 1982. Efficiency of nitrogen assimilation by N_2-fixing and nitrate-grown soybean plants (*Glycine max*) [L.] (Merr.), *Plant Physiol.* **70**:1178–1184.

FINN, G. A., AND BRUN, W. A. 1982. Effect of atmospheric CO_2 enrichment on growth, nonstructural carbohydrate content, and root nodule activity in soybean, *Plant Physiol.* **69**:327–331.

GOBER, J. W., AND KASHKET, E. R. 1984. H^+/ATP stoichiometry of cowpea *Rhizobium* sp. strain 32H1 cells grown under nitrogen-fixing and nitrogen-nonfixing conditions, *J. Bacteriol.* **160**:216–221.

HAAKER, H., AND WASSINK, H. 1984. Electron allocation to H^+ and N_2 by nitrogenase in *Rhizobium leguminosarum* bacteroids, *Eur. J. Biochem.* **142**:37–42.

HAGEMAN, R. V., AND BURRIS, R. H. 1978. Kinetic studies on electron transfer and interaction between nitrogenase components from *Azotobacter vinelandii*, *Biochemistry* **17**:4117–4124.

HAGEMAN, R. V., AND BURRIS, R. H. 1980. Electron allocation to alternative substrates of *Azotobacter* nitrogenase is controlled by the electron flux through dinitrogenase, *Biochim. Biophys. Acta* **591**:63–75.

HALVERSON, L. J., AND STACEY, G. 1984. Host recognition in the *Rhizobium*-soybean symbiosis, *Plant Physiol.* **74**:84–89.

HARDY, R. W. F., AND HAVELKA, U. D. 1975a. Nitrogen fixation research: A key to world food, *Science* **188**:633–643.

HARDY, R. W. F., AND HAVELKA, U. D. 1975b. Photosynthate as a major factor limiting nitrogen fixation by field grown legumes with emphasis on soybeans, in: *Symbiotic Nitrogen Fixation in Plants,* International Biology Program Series, Vol. 7 (P. S. Nutman, ed.), pp. 421–439, Cambridge University Press, London.

HARDY, R. W. F., HEYTLER, P. G., AND RAINBIRD, R. M. 1983. Status of new nitrogen imputs for crops, in: *Better Crops for Food*, Ciba Foundation Symposium 97 (J. Nugent and M. O'Connor, eds.), pp. 28–48, Pitman, London.

HAUSINGER, R. P., AND HOWARD, J. B. 1982. The amino acid sequence of the nitrogenase iron protein from *Azotobacter vinelandii*, *J. Biol. Chem.* **257**:2483–2490.

HERRIDGE, D. F. 1982. Relative abundance of ureides and nitrate in plant tissues of soybean as a quantitative assay of nitrogen fixation, *Plant Physiol.* **70**:1–6.

HERRIDGE, D. F., AND PATE, J. S. 1977. Utilization of net photosynthate for nitrogen fixation and protein production in an annual legume, *Plant Physiol.* **60**:759–764.

HERRIDGE, D. F., ATKINS, C. A., PATE, J. S., AND RAINBIRD, R. M. 1978. Allantoin and allantoic acid in the nitrogen economy of cowpea (*Vigna unguiculata* [L.] (Walp.), *Plant Physiol.* **62**:495–498.

HEYTLER, P. G., AND HARDY, R. W. F. 1979. Energy requirement for N-fixation by rhizobial nodules in soybeans, *Plant Physiol.* **63S**:84.

HEYTLER, P. G., AND HARDY, R. W. F. 1984. Calorimetry of nitrogenase-mediated reductions in detached soybean nodules, *Plant Physiol.* **75**:304–310.

HEYTLER, P. G., REDDY, G. S., AND HARDY, R. W. F. 1985. *In vivo* energetics of symbiotic nitrogen fixation in soybeans, in: *Nitrogen Fixation and CO₂ Metabolism: A Steenbock Symposium in Honor of Robert H. Burris* (P. W. Ludden and J. E. Burris, eds.), pp. 283–292, Elsevier, New York.

HOCH, G. E., LITTLE, H. N., AND BURRIS, R. H. 1957. Hydrogen evolution from soybean root nodules, *Nature* **179**:430–431.

KANEMOTO, R. H., AND LUDDEN, P. W. 1984. Effect of ammonia, darkness and phenazine methosulfate on whole-cell nitrogenase activity and Fe protein modification in *Rhodospirillum rubrum*, *J. Bacteriol.* **158**:713–720.

KEYSER, H. H., VAN BERKUM, P., AND WEBER, D. F. 1982. A comparative study of the physiology of symbioses formed by *Rhizobium japonicum* with *Glycine max, Vigna unguiculata,* and *Macroptilium atropurpurem, Plant Physiol.* **70**:1626–1630.

KOCH, B., AND EVANS, H. J. 1966. Reduction of acetylene to ethylene by soybean root nodules, *Plant Physiol.* **41**:1748–1750.

LAMMERS, P. J., AND HASELKORN, R. 1983. Sequence of the nifD gene coding for the β subunit of dinitrogenase from the cyanobacterium *Anabaena*, *Proc. Natl. Acad. Sci. USA* **80**:4723–4727.

LAWN, R. J., AND BRUN, W. A. 1974. Symbiotic nitrogen fixation in soybeans. I. Effect of photosynthetic source-sink manipulations, *Crop. Sci.* **14**:11–16.

LAWRIE, A. C., AND WHEELER, C. T. 1975. Nitrogen fixation in the root nodules of *Vicia faba* L. in relation to the assimilation of carbon. II. The dark fixation of carbon dioxide, *New Phytol.* **75**:437–445.

LAYZELL, D. B., AND LARUE, T. A. 1982. Modeling C and N transport to developing soybean fruits, *Plant Physiol.* **70**:1290–1298.

LAYZELL, D. B., RAINBIRD, R. M., ATKINS, C. A., AND PATE, J. S. 1979. Economy of photosynthate use in nitrogen-fixing legume nodules, *Plant Physiol.* **64**:888–891.

LOWE, D. J., AND THORNELEY, R. N. F. 1984. The mechanism of *Klebsiella pneumoniae* nitrogenase action, *Biochem. J.* **224**:877–886.

MAHON, J. D. 1977a. Root and nodule respiration in relation to acetylene reduction in intact nodulated peas, *Plant Physiol.* **60**:812–816.

MAHON, J. D. 1977b. Respiration and the energy requirement for nitrogen fixation in nodulated pea roots, *Plant Physiol.* **60**:817–821.

MAHON, J. D., AND CHILD, J. J. 1979. Growth response of inoculated peas (*Pisum sativum*) to combined nitrogen, *Can. J. Bot.* **57**:1687–1693.

MCCLURE, P. R., AND ISRAEL, D. W. 1979. Transport of nitrogen in the xylem of soybean plants, *Plant Physiol.* **64**:411–416.

MCKENNA, C. E., MCKENNA, M. -C., AND HUANG, C. W. 1979. Low stereoselectivity in methylacetylene and cylopropene reductions by nitrogenase, *Proc. Natl. Acad. Sci. USA* **76**:4773–4777.

MCNEIL, D. L., AND LARUE, T. A. 1984. Effect of nitrogen source on ureides in soybean, *Plant Physiol.* **74**:227–232.

MILLER, R. W., ROBSON, R. L., YATES, M. G., AND EADY, R. R. 1980. Catalysis of exchange of terminal phosphate groups of ATP and ADP by purified nitrogenase proteins, *Can. J. Biochem.* **58**:542–548.

MINCHIN, F. R., AND PATE, J. S. 1973. The carbon balance of a legume and the functional economy of its root nodules, *J. Exp. Bot.* **24**:259–271.

MINCHIN, F. R., AND PATE, J. S. 1974. Diurnal functioning of the legume root nodule, *J. Exp. Bot.* **25**:295–308.

MINCHIN, F. R., SUMMERFIELD, R. J., HADLEY, P., ROBERTS, E. H., AND RAWSTHORNE, S. 1981. Carbon and nitrogen nutrition of nodulated roots of grain legumes, *Plant Cell Environ.* **4**:5–26.

MINCHIN, F. R., WITTY, J. F., SHEEHY, J. E., AND MÜLLER, M. 1983. A major error in the acetylene reduction assay: Decreases in nodular nitrogenase activity under assay conditions, *J. Exp. Bot.* **34**:641–649.

MORTENSON, L. E., AND THORNELEY, R. N. F. 1979. Structure and function of nitrogenase, *Ann. Rev. Biochem.* **48**:387–418.

MORTENSON, L. E., AND UPCHURCH, R. G. 1981. Effect of adenylates on electron flow and efficiency of nitrogenase, in: *Current Perspectives in Nitrogen Fixation* (A. H. Gibson and W. E. Newton, eds.), pp. 75–78, Australian Academy of Science, Canberra.

MÜNCK, E., RHODES, H., ORME-JOHNSON, W. H., DAVIS, L. C., BRILL, W. J., AND SHAH, V. K. 1976. Nitrogenase. VIII. Mössbauer and EPR spectroscopy: The MoFe component from *Azotobacter vinelandii* OP, *Biochim. Biophys. Acta* **400**:32–44.

MUNSON, T. O., AND BURRIS, R. H. 1969. Nitrogen fixation by *Rhodospirillum rubrum* grown in nitrogen-limited continuous culture, *J. Bacteriol.* **97**:1093–1098.

NASH, D. T., AND SCHULMAN, H. M. 1976. Leghemoglobins and nitrogenase activity during soybean root nodule development, *Can. J. Bot.* **54**:2790–2797.

NELSON, L. M., AND SALMINEN, S. O. 1982. Uptake hydrogenase activity and ATP formation in *Rhizobium leguminosarum* bacteroids, *J. Bacteriol.* **151**:989–995.

NELSON, M. J., LEVY, M. A., AND ORME-JOHNSON, W. H. 1983. Metal and sulfur composition of iron-molybdenum cofactor of nitrogenase, *Proc. Natl. Acad. Sci. USA* **80**:147–150.

NEWCOMB, W. 1976. A correlated light and electron microscope study of symbiotic growth and differentiation in *Pisum sativum* root nodules, *Can. J. Bot.* **54**:2163–2186.

NEWCOMB, W., SIPPEL, D., AND PETERSON, R. L. 1979. The early morphogenesis of *Glycine max* and *Pisum sativum* nodules, *Can. J. Bot.* **57**:2603–2616.

NEWCOMB, E. H., SELKER, J. M. L., TANDON, S. R., MENG, F., AND KOWAL, R. R. 1985. Uninfected cells in ureide- and amide-exporting legume root nodules, in: *Nitrogen fixation and CO₂ Metabolism: A Steenbock Symposium in Honor of Robert H. Burris* (P. W. Ludden and J. E. Burris, eds.), pp. 31–40, Elsevier, New York.

NICHOLS, D. G. 1982. *Bioenergetics: An Introduction to the Chemiosmotic Theory*, Academic Press, London.

NIEVA-GOMEZ, D., ROBERTS, G. P., KLEVICKIS, S., AND BRILL, W. J. 1980. Electron transport to nitrogenase in *Klebsiella pneumoniae*, *Proc. Natl. Acad. Sci. USA* **77**:2555–2558.

NOEL, K. D., CARNEOL, M., AND BRILL, W. J. 1982. Nodule protein synthesis and nitrogenase activity of soybeans exposed to fixed nitrogen, *Plant Physiol.* **70**:1236–1241.

PATE, J. S., AND HERRIDGE, D. F. 1978. Partitioning and utilization of net photosynthate in a nodulated annual legume, *J. Exp. Bot.* **29**:401–412.

PATE, J. S., AND WALLACE, W. 1964. Movement of assimilated nitrogen from the root system of the field pea (*Pisum arvense* L.), *Ann. Bot.* **28**:83–99.

PATE, J. S., ATKINS, C. A., AND RAINBIRD, R. M. 1981. Theoretical and experimental costing of nitrogen fixation and related processes in nodules of legumes, in: *Current Perspectives in Nitrogen Fixation, Proceedings of the Fourth International Symposium on Nitrogen Fixation* (A. H. Gibson and W. E. Newton, eds.), pp. 105–116, Elsevier, New York.

PATTERSON, T. G., AND LARUE, T. A. 1983. Root respiration associated with nitrogenase activity (C_2H_2) of soybean, and a comparison of estimates, *Plant Physiol.* **72:**701–705.

PATTERSON, T. G., PETERSON, J. B., AND LARUE, T. A. 1983. Effect of supraambient oxygen on nitrogenase activity (C_2H_2) and root respiration of soybeans and isolated soybean bacteroids, *Plant Physiol.* **70:**695–700.

PAUL, T. D., AND LUDDEN, P. W. 1984. Adenine nucleotide levels in *Rhodospirillum rubrum* during switch-off of whole cell nitrogenase activity, *Biochem. J.* **224:**961–969.

PHELPS, A. S., AND WILSON, P. W. 1941. Occurrence of hydrogenase in nitrogen-fixing organisms, *Proc. Soc. Exp. Biol. Med.* **47:**473–476.

PHILLIPS, D. A. 1980. Efficiency of symbiotic nitrogen fixation in legumes, *Annu. Rev. Plant Physiol.* **31:**29–49.

PHILLIPS, D. A., BEDMAR, E. J., QUALSET, C. O., AND TEUBER, L. R. 1985. Host legume control of *Rhizobium* function, in: *Nitrogen Fixation and CO_2 Metabolism: A Steenbock Symposium in Honor of Robert H. Burris* (P. W. Ludden and J. E. Burris, eds.), pp. 203–212, Elsevier, New York.

PHILLIPS, D. V., WILSON, D. O., AND DOUGHERTY, D. E. 1984. Soluble carbohydrates in legumes and nodulated nonlegumes, *J. Agric. Food Chem.* **32:**1289–1291.

POPE, M. R., MURRELL, S. A., AND LUDDEN, P. W. 1986. Covalent modification of nitrogenase from *Rhodospirillum rubrum* by adenosine diphosphoribosylation of a specific arginyl residue, *Proc. Natl. Acad. Sci. USA,* **82:**3173–3177.

PRIVALLE, L. S., AND BURRIS, R. H. 1983. Adenine nucleotide levels in and nitrogen fixation by the cyanobacterium *Anabaena* sp. strain 7120, *J. Bacteriol.* **154:**351–355.

QUEBEDEAUX, B., HAVELKA, U. D., LIVAK, K. L., AND HARDY, R. W. F. 1975. Effect of altered pO_2 in the aerial part of soybean on symbiotic N_2 fixation, *Plant Physiol.* **56:**761–764.

RAINBIRD, R. M., ATKINS, C. A., AND PATE, J. S. 1983a. Diurnal variation in the functioning of cowpea nodules, *Plant Physiol.* **72:**308–312.

RAINBIRD, R. M., ATKINS, C. A., AND PATE, J. S. 1983b. Effect of temperature on nitrogenase functioning in cowpea nodules, *Plant Physiol.* **73:**392–394.

RAINBIRD, R. M., ATKINS, C. A., PATE, J. S., AND SANFORD, P. 1983c. Significance of hydrogen evolution in the carbon and nitrogen economy of nodulated cowpea, *Plant Physiol.* **71:**122–127.

RAINBIRD, R. M., HITZ, W. D., AND HARDY, R. W. F. 1984a. Experimental determination of the respiration associated with soybean/*Rhizobium* nitrogenase function, nodule maintenance, and total nodule nitrogen fixation, *Plant Physiol.* **75:**49–53.

RALSTON, E. J., AND IMSANDE, J. 1982. Entry of oxygen and nitrogen into intact soybean nodules, *J. Exp. Bot.* **33:**208–214.

RATCLIFFE, H. D., DROZD, J. W., AND BULL, A. T. 1983. Growth energetics of *Rhizobium leguminosarum* in chemostat culture, *J. Gen. Microbiol.* **129:**1697–1706.

RAWSTHORNE, S., MINCHIN, F. R., SUMMERFIELD, R. J., COOKSON, C., AND COOMBS, J. 1980. Carbon and nitrogen metabolism in legume root nodules, *Phytochemistry* **19**:341–355.

REIBACH, P. H., AND STREETER, J. G. 1983. Metabolism of ^{14}C-labeled photosynthate and distribution of enzymes of glucose metabolism in soybean nodules, *Plant Physiol.* **72**:634–640.

REYNOLDS, P. H. S., BOLAND, M. J., BLEVINS, D. G., SCHUBERT, K. R., AND RANDALL, D. D. 1982. Enzymes of amide and ureide biogenesis in developing soybean nodules, *Plant Physiol.* **69**:1334–1338.

RIVERA-ORTIZ, J. M., AND BURRIS, R. H. 1975. Interactions among substrates and inhibitors of nitrogenase, *J. Bacteriol.* **123**:537–545.

RUFTY, T. W., JR., ISRAEL, D. W., AND VOLK, R. J. 1984. Assimilation of $^{15}NO_3^-$ taken up by plants in the light and in the dark, *Plant Physiol.* **76**:769–775.

RUIZ-ARGÜESO, T., MAIER, R. J., AND EVANS, H. J. 1979. Hydrogen evolution from alfalfa and clover nodules and hydrogen uptake by free-living *Rhizobium meliloti*, *Appl. Environ. Microbiol.* **37**:582–587.

RYLE, G. J. A., ARNOTT, R. A., POWELL, C. E., AND GORDON, A. J. 1983. Comparisons of the respiratory effluxes of nodules and roots in six temperate legumes, *Ann. Bot.* **52**:469–477.

RYLE, G. J. A., ARNOTT, R. A., POWELL, C. E., AND GORDON, A. J. 1984. N_2 fixation and the respiratory costs of nodules, nitrogenase activity, and nodule growth and maintenance in Fiskeby soybean, *J. Exp. Bot.* **35**:1156–1165.

SALMINEN, S. O., AND NELSON, L. M. 1984. Role of uptake hydrogenase in providing reductant for nitrogenase in *Rhizobium leguminosarum* bacteroids, *Biochim. Biophys. Acta* **764**:132–137.

SALSAC, L., DREVON, J., ZENGBÉ, M., CLEYET-MAREL, J., AND OBATON, M. 1984. Energy requirement of symbiotic nitrogen fixation, *Physiol. Veg.* **22**:509–521.

SCHUBERT, K. R. 1981. Enzymes of purine biosynthesis and catabolism in *Glycine max*. I. Comparison of activities with N_2 fixation and composition of xylem exudate during nodule development, *Plant Physiol.* **68**:1115–1122.

SCHUBERT, K. R. (ED.) 1982. *The Energetics of Biological Nitrogen Fixation: Workshop Summaries—I*, American Society of Plant Physiologists, Rockville, Maryland.

SCHUBERT, K. R., AND BOLAND, M. J. 1984. The cellular and intracellular organization of the reactions of ureide biogenesis in nodules of tropical legumes, in: *Advances in Nitrogen Fixation Research, Proceedings of the 5th International Symposium on Nitrogen Fixation* (C. Veeger and W. E. Newton, eds.), pp. 445–451, Martinus Nijhoff/Dr W. Junk, Boston.

SCHUBERT, K. R., AND EVANS, H. J. 1976. Hydrogen evolution: A major factor affecting the efficiency of nitrogen fixation in nodulated symbionts, *Proc. Natl. Acad. Sci. USA* **73**:1207–1211.

SCHUBERT, K. R., AND RYLE, G. J. A. 1980. The energy requirements for nitrogen fixation in nodulated legumes, in: *Advances in Legume Science* (R. J.

Summerfield and A. H. Bunting, eds.), pp. 85–96, Royal Botanic Gardens, Kew, England.

SCHWEITZER, L. E., AND HARPER, J. E. 1980. Effect of light, dark, and temperature on root nodule activity (acetylene reduction) of soybeans, *Plant Physiol.* **65:**51–56.

SCOTT, D. B., ROBERTSON, J. G., AND FARNDEN, K. J. F. 1976. Ammonia assimilation in lupin nodules, *Nature* **263:**703–705.

SHAH, V. K., AND BRILL, W. J. 1977. Isolation of an iron-molybdenum cofactor (FeMo-co) from nitrogenase, *Proc. Natl. Acad. Sci. USA* **74:**3249–3253.

SHAH, V. K., DAVIS, L. C., AND BRILL, W. J. 1972. Nitrogenase. I. Repression and derepression of the iron-molybdenum and iron proteins of nitrogenase in *Azotobacter vinelandii, Biochim. Biophys. Acta* **256:**498–511.

SHAH, V. K., STACEY, G., AND BRILL, W. J. 1983. Electron transport to nitrogenase—Purification and characterization of pyruvate-flavodoxin oxidoreductase, the nifJ gene groduct, *J. Biol. Chem.* **258:**12064–12068.

SHEEHY, J. E. 1983. Relationships between senescence, photosynthesis, nitrogen fixation and seed filling in soya bean *Glycine max* (L.) Merr., *Ann. Bot.* **51:**679–682.

SHEEHY, J. E., MINCHIN, F. R., AND WITTY, J. F. 1983. Biological control of the resistance to oxygen flux in nodules, *Ann. Bot.* **52:**565–571.

SIMPSON, F. B., AND BURRIS, R. H. 1984. A nitrogen pressure of 50 atmospheres does not prevent evolution of hydrogen by nitrogenase, *Science* **224:**1095–1097.

SPRENT, J. I. 1980. Root nodule anatomy, type of export product and evolutionary origin in some Leguminosae, *Plant Cell Environ.* **3:**35–43.

STAM, H., VERSEVELD, H. W. VAN, VRIES, W. DE, AND STOUTHAMER, A. H. 1984. Hydrogen oxidation and efficiency of nitrogen fixation in succinate-limited chemostat cultures of *Rhizobium* ORS 571, *Arch. Microbiol.* **139:**53–60.

STEWART, W. D. P., FITZGERALD, G. P., AND BURRIS, R. H. 1967. *In situ* studies on N$_2$ fixation using the acetylene reduction technique, *Proc. Natl. Acad. Sci. USA* **58:**2071–2078.

STREETER, J. G. 1979. Allantoin and allantoic acid in tissues and stem exudate from field-grown soybean plants, *Plant Physiol.* **63:**478–480.

TAJIMA, S., AND LARUE, T. A. 1982. Enzymes for acetyldehyde and ethanol formation in legume nodules, *Plant Physiol.* **70:**388–392.

TIMPO, E. E., AND NEYRA, C. A. 1983. Expression of nitrate and nitrite reductase activities under various forms of nitrogen nutrition in *Phaseolus vulgaris* L., *Plant Physiol.* **72:**71–75.

TSO, M. -Y. W., AND BURRIS, R. H. 1973. The binding of ATP and ADP by nitrogenase components from *Clostridium pasteurianum, Biochim. Biophys. Acta* **309:**263–270.

TUBB, R. S., AND POSTGATE, J. R. 1973. Control of nitrogenase synthesis in *Klebsiella pneumoniae, J. Gen. Microbiol.* **79:**103–117.

UPCHURCH, R. G., AND MORTENSON, L. E. 1980. *In vivo* energetics and control of nitrogen fixation: Changes in the adenylate energy charge and adenosine-5'-diphosphate/adenosine-5'-triphosphate ratio of cells during growth on dinitrogen versus growth on ammonia, *J. Bacteriol.* **143**:274–284.

URSINO, D. J., HUNTER, D. M., LAING, R. D., AND KEIGHLEY, J. L. S. 1982. Nitrate modification of photosynthesis and photoassimilate export in young nodulated soybean plants, *Can. J. Bot.* **60**:2665–2670.

VANCE, C. P., AND STADE, S. 1984. Alfalfa root nodule carbon dioxide fixation, *Plant Physiol.* **75**:261–264.

VANCE, C. P., STADE, S., AND MAXWELL, C. A. 1983. Alfalfa root nodule carbon dioxide fixation. I. Association with nitrogen fixation and incorporation into amino acids, *Plant Physiol.* **72**:469–473.

VERMA, D. P. S., KAZAZIAN, V., ZOGBI, V., AND BAL, A. K. 1978. Isolation and characterization of the membrane envelope enclosing the bacteroids in soybean root nodules, *J. Cell Biol.* **78**:919–935.

WAREMBOURG, F. R. 1983. Estimating the true cost of dinitrogen fixation by nodulated plants in undisturbed conditions, *Can. J. Microbiol.* **29**:930–937.

WILLIAMS, L. E., DEJONG, T. M., AND PHILLIPS, D. A. 1982. Effect of changes in shoot carbon-exchange rate on soybean root nodule activity, *Plant Physiol.* **69**:432–436.

WINSHIP, L. J., AND TJEPKEMA J. D. 1982. Simultaneous measurement of acetylene reduction and respiratory gas exchange of attached root nodules, *Plant Physiol.* **70**:361–365.

WITTY, J. F., MINCHIN, F. R., AND SHEEHY, J. E. 1983. Carbon costs of nitrogenase activity in legume root nodules determined using acetylene and oxygen, *J. Exp. Bot.* **34**:951–963.

Chapter 6

Cellular Aspects of Root Nodule Establishment in *Frankia* Symbioses

Alison M. Berry

MEMBERS OF THE NITROGEN-FIXING actinomycetal genus *Frankia* can invade and inhabit the roots of over 200 species of woody plants. One can view the resulting actinorhizal root nodules as plant developmental systems specialized for symbiotic nitrogen fixation. Nodule development is regulated at least initially by some stimulus from the infective mocroorganism. At the same time, it is difficult to discuss the mechanisms of *Frankia* infection in a fruitful manner without also considering principles of host nodule tissue differentiation. The purpose of this chapter is to present current information on the related processes of infection and nodulation in the *Frankia*-nodulated symbiotic system.

ACTINORHIZAL HOSTS

The actinorhizal hosts described to date are shrubs and tree species from eight phylogenetically diverse angiosperm families (Table 1). The host genera are by and large pioneer species which tolerate nitrogen-deficient soil conditions. In other respects they occupy many different, often extreme, habitats.

While most of the nodulating species have been known for some years (Bond, 1976), new or incompletely documented species continue to be described regularly (Heisey et al., 1980, Becking, 1984; Newcomb, 1981). The monotypic family Datiscaceae was added to the list of actinorhizal hosts as recently as 1978 (Chaudhary, 1978). It is reasonable to expect that the list is still incomplete, at least at the generic level.

Utilization of these species for forestry-related purposes has been the subject of considerable research and discussion during the last decade. The current and projected uses for actinorhizal plants include use as fuelwood or timber, for biomass plantations, for erosion control, and as intercrops for forest species (Gordon et al., 1979, Gordon and Wheeler, 1983; Briggs et al., 1978; Akkermans et al., 1984; Midgley et al., 1983; National Research Council, 1980, 1984). Species of *Alnus* and *Casuarina,* for example, have exceptionally rapid early growth rates. A likely application for several species will continue to be in revegetation under arid and desert conditions (National Research Council, 1984). There are as yet no long-term breeding or selection programs for any of the actinorhizal species.

Table 1 Actinorhizal Host Plants.[a]

Family	Nodulated Genera	Ecological Sites
Betulaceae	*Alnus*	Poor soils, sand, till, mine dumps, gravel, bogs
Casuarinaceae	*Casuarina*	Very wide range: sand dunes, salt marshes, tropical forests, desert areas
Coriariaceae	*Cariaria*	Sandy, gravelly soils, lowland to subalpine
Datiscaceae	*Datisca*	Stream beds in chaparral
Elaeagnaceae	*Elaeagnus, Shepherdia*	Disturbed areas, sandy soils
Myricaceae	*Myrica*	Wetlands, sand dunes, mine wastes
	Comptonia	Sandy, gravelly sites
Rhamnaceae	*Ceanothus*	Chaparral, dry subalpine
Rosaceae	*Cercocarpus*	Dry subalpine
	Dryas	Sand, gravel; subarctic

[a]Modified from Torrey (1978).

FRANKIA

Frankia organisms are branched, septate, filamentous actinomycetes with a complex pattern of differentiation *in vitro* and in the nodule. For many years, *Frankia* was believed to be an obligate symbiont. Although there were reports of *Frankia* isolates during this time, plant inoculation tests and repeated subcultures were inconclusive. The first confirmed *Frankia* strain was isolated from nodules of *Comptonia* and reported in 1978 (Callaham et al., 1978).

Hundreds of strains have now been isolated from nodules of most recorded host species (Lechevalier, 1986). While taxonomic and physiological work is still in the early stages, it has already become evident that the genus *Frankia* includes a metabolically diverse group of organisms which nevertheless share certain developmental features.

A nitrogenase enzyme system coupled with uptake hydrogenase is responsible for nitrogen fixation in *Frankia* (Benson et al., 1979, 1980). Ruvkin et al. (1982) found DNA homology between portions of the *Frankia* genome and *Klebsiella* (nitrogen fixation) (*nif*) gene probes. Functional biochemical compatibility has also been reported to exist between portions of *Frankia* nitrogenase and that of *Azotobacter* (Benson et al., 1979).

The Vesicle

Frankia vesicles are differentiated branch hyphae with enlarged, spherical to club-shaped terminal cells. These structures appear to have important functional significance in terms of nitrogenase activity. In culture, vesicles are produced on the *Frankia* mycelium under conditions of log-phase growth and nitrogen starvation. *In vitro* and in the infected cells of the nodule, the timing of vesicle differentiation has been shown to be correlated with the onset of nitrogenase activity (Tjepkema et al., 1980; Schwintzer et al., 1982; Mian and Bond, 1978). Even small amounts of added nitrogen inhibit both nitrogenase activity and vesicle differentiation *in vitro* (Fontaine et al., 1984).

Additional lines of evidence have indicated that the endophyte vesicle is specialized for nitrogenase activity. *Frankia* is capable of fixing nitrogen *in vitro* at atmospheric oxygen tension, in sharp contrast to the conditions required by *Rhizobium* for nitrogenase activity (Tjepkema et al., 1980). In cell-free preparations, however, nitrogenase activity is oxygen-labile (Benson et al., 1979), as one would expect. Consequently, a structural or metabolic adaptation within the organism that restricts oxygen diffusion at the site of nitrogen fixation is indicated.

The vesicle wall is thick and multilamellar in electron micrographs of freeze-fracture preparations and has been hypothesized to serve as a gas diffusion barrier (Torrey and Callaham, 1982). The wall layers consisted of a material which was easily disrupted by organic solvents, but which was stabilized by osmium-thiocarbohydrazide treatment, and this suggested the possibility of a glycolipid, analogous to observations on the heterocyst wall of *Anabaena* (Torrey and Callaham, 1982). However, the occurrence of a glycolipid in vesicle-containing cultures of *Frankia* strain HFPArI3 was not demonstrated (Lopez et al., 1983).

High levels of respiratory activity accompany the process of nitrogen fixation in the *Frankia* vesicle (Akkermans, 1971), and respiration may provide an alternate or additional mechanism for creating a localized oxygen deficit. In addition, an iron-binding superoxide dismutase has been identified in several *Frankia* species (Puppo et al., 1985). The enzyme was inducible following depression of nitrogenase activity (Steele and Stowers, 1985).

Recently, a *Frankia* strain from *Casuarina* has been shown to fix nitrogen *in vitro* without forming vesicles (M. A. Murry and J. G. Torrey, personal communication). Vesicles have never been observed within *Casuarina* nodules. Thus oxygen protection in *Frankia* is not universally restricted to the vesicle structure, and at least in some strains of *Frankia* nitrogenase activity may also occur in the hyphae. The extent to which these hyphae are specialized in wall structure or in respiratory activity remains to be determined.

In the nodule, host cellular mechanisms exist for regulating oxygen distribution as well. Hemoglobin was reported to occur in *Myrica* and *Casuarina* nodules but not in *Ceanothus* or *Datisca* (Tjepkema, 1983); hemoglobin is also not detectable in *Alnus* nodule extracts (J. Tjepkema, personal communication). Suberin, or possibly lignin, appears to permeate the walls of infected cells, again notably in *Casuarina* (Berg, 1983) and *Myrica* (R. H. Berg, personal communication). In addition, tight packing of infected cortical cells in several species may contribute to creating or maintaining locally low levels of oxygen (Tjepkema, 1979).

Frankia differentiation, both structural and metabolic, is regulated by conditions of nitrogen starvation: Within 1 day of transfer to a nitrogen-free medium, vesicle differentiation is observable in culture (Fontaine et al., 1984). Within hours of transfer to a similar medium, changes in the pattern of ammonium assimilation have been detected (Mazzucco and Benson, 1984). Endogenous carbohydrates, particularly trehalose and glycogen, appear to function as energy reserves during rapid growth or differentiation of HFPArI3 following transfer to inductive media (Lopez et al., 1984).

Sporangia and Spores

In some but not all strains, sporangia can differentiate *in vitro* as a function of time after subculture. Sporangia also occur in nodules of certain hosts, but not all compatible *Frankia* strains produce sporangia in a given host (vanDijk and Merkus, 1976; vanDijk, 1978). The cytology of sporangial development in effective nodules apparently differs from that in ineffective nodules, where sporangia can form quite early in infected cell differentiation (VandenBosch and Torrey, 1983). Very little is known about the metabolic factors governing sporangial differentiation either *in vivo* or *in vitro*.

Sporangiospores form by repeated septation within portions of hyphae, followed by cell enlargement (Newcomb et al., 1978). The spores are thick-walled at maturity, and a distinctive triple-track membrane is present at the outer spore wall surface (Lechevalier and Lechevalier, 1979). *Frankia* spores are presumed to function as resting propagules, either in senescent nodules or in the soil. Spore germination *in vitro* has been difficult to achieve consistently, but spores should be an important means of obtaining single-celled isolates. Recently, Tisa and Ensign (1986) have reported the successful production of *Frankia* protoplasts from hyphal segments.

Carbon and Nitrogen Metabolism

Physiologically, at least in terms of carbon substrate utilization *in vitro*, there appears to be considerable diversity among *Frankia* strains. Substrates utilized can include organic acids such as propionic, succinic, and malic, as well as glucose, sucrose, arabinose, and trehalose, among other substances (Lechevalier and Ruan, 1984; Blom et al., 1980; Shipton and Burggraaf, 1982). The pathways of carbon assimilation in the nodule have not been definitely established. Vesicle clusters isolated from nodules of three different host genera respired succinate. Two of these also utilized malate and glutamate plus NAD (Akkermans et al., 1983). As mentioned in the previous section, glycogen and trehalose appear to function as carbon reserves.

Studies on ammonia assimilation in nodules using ^{13}N ammonium indicate that glutamine and, in lesser amounts, glutamate are major early products in the assimilatory pathway (Schubert et al., 1981). In fractionation studies, glutamine synthetase activity was restricted to the plant cytosol fraction and was not significant in isolated vesicle clusters. *In vitro,* however, *Frankia* has been shown to utilize ammonia, N_2, and a range of amino acids as a substrate for nitrogen metabolism, probably via glutamine synthetase (Akkermans et al., 1983).

Host Compatibility

Previously described host compatibility groupings (Becking, 1974) have been challenged recently by the finding that some isolates are capable of nodulating hosts from different compatibility groups (Lechevalier and Ruan, 1984). Other isolates are unable to reinfect the host from which they were derived. These results open the way for interesting research concerning the nature of host specificity.

INFECTION PROCESS

The actinorhizal root nodule is a perennial structure formed by repeated branchings of modified lateral roots termed nodule lobes (Fig. 1, part 5). The yearly production of new nodule lobes, initiated by the pericycle of higher-order lobes, contributes to the formation of a compact irregular or coralloid structure, the nodule. The nodule lobe has a central vascular cylinder, a cortical zone, and a periderm. Within this general pattern there is considerable variation in nodule morphology among actinorhizal hosts (Torrey, 1978).

Within the individual nodule lobe, *Frankia* passes from cell to cell in acropetal fashion, invading cells of the root cortex behind the nodule meristem, apparently during cell expansion. The cytology of *Frankia* differentiating within the developing nodule, treated briefly in the previous section, has been described in several papers (Newcomb et al., 1978; van-Dijk and Merkus, 1976; Schwintzer et al., 1982; Baker et al., 1980; VandenBosch and Torrey, 1983) (Fig. 1 part 3). The differentiation of vesicles correlates highly with the onset of nitrogenase function in the symbiotic tissue. There are varying degrees of compatibility between a given host and the strains of *Frankia* which can infect the host. When incompatible combinations occur, nodules are characteristically smaller and have fewer infected cells (Baker et al., 1980; VandenBosch and Torrey, 1983). Vesicles may not differentiate at all, and there may be low or no detectable nitrogenase activity. It is not known what factors interfere with or limit successful nodulation or whether they are host- or *Frankia*-determined.

Initial Infection Events

Two quite distinctive early infection sequences have been described which precede the induction of nodule lobes in actinorhizal species. In most of the cases reported thus far, initial entry by *Frankia* into the host root occurs through a highly deformed root hair (Callaham and Torrey, 1977; Callaham et al., 1979; Berry et al., 1986). The invasion of *Frankia* hyphae through the root hair and into the cells in the cortical zone stimulates cell expansion and other changes, producing a "prenodule." Subsequently, lateral root primordia initiated in the infection zone expand and become infected by *Frankia* from the prenodule tissue.

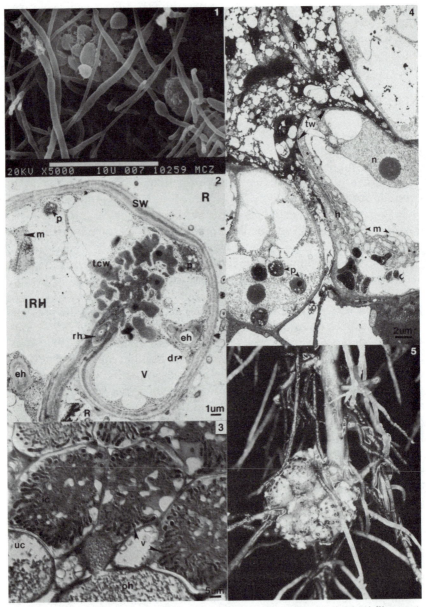

Figure 1. (1) *Frankia* strain HFPArI3 in culture, showing individual filaments and a single sporangial region with mature spores. Scanning electron micrograph. Copyright A. M. Berry, 1985. (2) Longitudinal section through a deformed root hair of *Alnus rubra*, infected by *Frankia*. A portion of *Frankia* hypha (rh) is seen in the rhizosphere (R), in the folded region between two portions of the infected hair (IRH), very near the site of passage through the host wall. Within the hair, *Frankia* hyphae are encapsulated by host-derived

In the other infection sequence, reported for *Elaeagnus* (Miller and Baker, 1985), nodulation takes place in regions of the root which often lack root hairs. Infection is apoplastic between epidermal cells and through the cortical tissue. When nodule lobe primordia are initiated, *Frankia* appears to penetrate into the host symplast near or within the primordia, and thence into the developing nodule tissue.

Infection sequences have not yet been reported for a number of actinorhizal genera. The alternate infection sequence bypassing prenodule events may not be restricted to *Elaeagnus*, and it is possible that nodulation in a given host can occur by either means. Both symplastic and apoplastic early infection have been reported for legumes as well (Turgeon, 1982; Lancelle and Torrey, 1984; Chandler et al., 1982).

A third likelihood is that stem nodules may form, as reported for several legume species (Dreyfus et al., 1984), especially under conditions which favor the expression of adventitious roots (e.g., wounds, girdling, or waterlogging). This type of infection pattern has not yet been reported for actinorhizal species. In the rhizobial stem nodule infection sequence, the endophyte infects intercellularly in stem tissue through the wound created by emergence of the adventitious root. Intracellular passage is effected through root (nodule lobe), as opposed to stem, tissue in a fashion similar to the apoplastic cortical infection described above.

---

wall material (eh). Note wall proliferation at the zone of infection, typical of transfer cell wall (tcw). Otherwise the inner wall of the infected hair forms layers (SW, secondary wall). Plastids (p) and especially mitochondria (m) are present in number in the infected hair cytoplasm. Reprinted by permission from Berry et al. (1986, in press). (3) Mature cortical cells in nodule of *Myrica gale*. Infected cells are much enlarged. Note central hyphal mass within the infected cells. Club-shaped vesicles (v) in the cell periphery terminate hyphae branching from the central mass. Hyphal strands also may be seen to traverse cell walls between infected cells. Host cells retain an active cytoplasm. Nearby uninfected cells (uc) often contain vacuolar phenolic accumulations (ph). Light micrograph of thick plastic section stained with toluidine blue and photographed with Nomarski optics. Reprinted with permission from Schwintzer et al. (1982). (4) Cells of the prenodule just below the infected root hair (*Alnus rubra*). Pathway of hyphal invasion (h) is from the upper, electron dense cell into the cell at the lower right. Note the thin-walled portion of the lower cell, which has expanded into the cell at the upper left. Hyphae may be seen growing into the lower cell through this wall region (h). Note numerous mitochondria in infected cells, in the vicinity of the hyphae (m), as well as large plastids lacking starch grains (p). Nuclei are enlarged (n) with prominent nucleoli. Copyright A. M. Berry, 1985. (5) Root nodule of *Alnus rubra* from aeroponics culture. Nodule lobes are approximately 5–10 mm in length. Copyright A. M. Berry, University Microfilm, 1983. Used by permission.

Adhesive Mechanisms of *Frankia*

There is no structural evidence of fimbriae or other specialized structures in the *Frankia* organism which might have an anchoring function at the root cell boundary. The organism grows by hyphal extension followed by septation and by hyphal branching. The branch hyphae initiated are often extremely fine (Newcomb et al., 1978), on the order of 0.1 μm in diameter. At the site of root hair penetration, numerous incipient branches occur (A. M. Berry et al., unpublished observations). Such evidence suggests that the growing hyphal tip is the initial site of *Frankia* contact with the host cell surface. As such, multiple branch hyphae may serve as loci of bacterial adhesion.

Adhesive Properties of the Root Hair Surface

The secretion of extracellular polysaccharide matrices by host plants plays a significant role in a number of plant-microbe interactions. For example, adhesion of *Agrobacterium* to plant tissue at wound sites appears to be mediated by pectic wall fractions, as reviewed by Lippincott et al. (1984). In certain host resistance situations, phytoalexin production is induced by the oligosaccharide fragments from epidermal cell walls generated by microbial degradative processes (Darvill and Albersheim, 1984).

Secretion of mucilage by root hairs has been demonstrated in many plant species, including *Alnus* (Vermeer and McCully, 1982; Berry et al., 1983). Mucilage secretion from root cap cells is mediated by Golgi-derived vesicles (Morré et al., 1967), and it is reasonable to anticipate a similar mechanism in root hair secretion. In the elongating *Alnus* root hair, Golgi secretion has been observed to occur just proximal to the growing tip (A. M. Berry, unpublished observations).

Rhizosphere bacteria can increase the amount of mucilage secreted by plant roots (Greaves and Darbyshire, 1972; Berry and Torrey, 1983). Structural evidence indicates that bacterial adherence to the root hair surface is mediated by a component of the root hair mucilage (Berry et al., 1983). Ultrastructurally, the mucilage component is a heteropolymer which forms flattened globular to chainlike strands connecting bacteria to the wall surface (Berry et al., 1984). This material is not specific to *Frankia* binding, occurring also in axenic conditions and in the presence of a pseudomonad inoculant.

Bacterial inoculation alters the histochemical properties of the *Alnus* root hair wall as compared with those of axenically grown root hairs (Berry and Torrey, 1983). The changes appear to be correlated with an increased susceptibility to acridine orange staining of polyanionic wall fractions, perhaps polyuronates. It is not known whether these observations are related to the secretory activity of *Frankia*-inoculated root hairs.

During successful *Frankia* infection, a smooth, electron-dense matrix may be observed at the infected root hair surface, which is distinct from the loosely fibrillar mucilage observed under other conditions. The significance of this matrix in the infection process is not clear.

It is worthy of note that in the *Elaeagnus* infection sequence, although there is no intracellular invasion of the host root cortex before nodule lobe initiation, there is production of an unusual extracellular matrix surrounding cortical cells presumably of host origin, within which *Frankia* may be observed (Miller and Baker, 1985).

Recognition

Since it was first proposed for legumes by Bohlool and Schmidt in 1973, the theory that lectin molecules on the surface of the root hair specifically bind sugar moieties at corresponding sites on the endosymbiont cell envelope has been the chief explanation of the observed host strain specificity in nodulation. Currently one can neither support not refute the possibility that a component of the root hair wall surface is a determinant in molecular recognition events involving the host and *Frankia*. The root hair mucilage component which binds bacteria, described in the previous section, is not specific to *Frankia* in its binding reaction. In instances where nodulation is initiated via an essentially intercellular infection mechanism, *Frankia* strains appear to bypass any surface recognition phenomena occurring on the outside of the root or root hair.

Root Hair Infection

As a result of bacterial inoculation, root hairs can become branched and lobed (Knowlton et al., 1980). Such deformation occurs as a function of modified cell elongation (tip growth) in the host root hair; hence the timing of deformation coincides with the timing of root hair expansion (Berry and Torrey, 1983). The nature of the bacterial stimulus for root hair deformation is currently unknown. Following compatible *Frankia* inoculation, branching and lobing of root hairs is pronounced. One or several of these hairs can become infected successfully by the actinomycete. The distal portions of the infected hairs become quite vacuolate, while the base of the hair remains densely cytoplasmic.

Ultrastructural details of root hair infection and early nodulation have recently been reported for *Alnus rubra* (Berry et al., 1986). Ultrastructural accounts of root hair infection in other actinorhizal genera are limited. However, a number of the features of the process can be observed in *Comptonia* (Callaham et al., 1979) and in *Alnus glutinosa* (Lalonde, 1977) and, using light microscopy, have been reported for other species as well (Callaham and Torrey, 1977; Callaham et al., 1979).

In the *A. rubra* infection sequence, a *Frankia* hyphal segment grows into the hair interior by traversing the cell wall from a zone outside the hair where the wall is folded and highly contorted (Fig. 1, part 2). The fibrillar component of the hair wall in the infection zone shows evidence of structural disorganization. Hydrolytic enzymes produced by *Frankia* may produce such a localized alteration of the root hair wall, although there is no definite break in the wall, as has been observed in the clover-*Rhizobium* infection sequence (Callaham and Torrey, 1981; Robertson et al., 1986).

Root hair infection appears also to involve altered patterns of root hair wall synthesis. At the infection site, the inner wall layers exhibit irregular wall proliferation characteristic of a transfer-cell wall (Fig. 1, part 2). In other regions of the infected hair, the secondary wall is highly organized and multilamellar. The formation of a secondary wall is in itself a feature of the infected root hair: Cell walls of uninfected hairs consist of a primary wall. Thus there is increased wall synthesis in response to *Frankia* infection, particularly in the actual infection zone. The new wall material is not callose (A. M. Berry and M. E. McCully, unpublished observations).

Within the root hair, and indeed throughout the subsequent nodule, *Frankia* is surrounded by a host-derived encapsulating wall, evidently polyanionic in nature (Lalonde and Knowles, 1975). In the infected hair, this encapsulation appears to be continuous with the transfer-cell wall region. Microtubular arrays can be observed running parallel to (or spirally around) the encapsulated hyphae and also aligned axially in the cytoplasm of the root hair base (Berry et al., 1986).

Like a transfer cell, the infected root hair appears to be specialized for solute transfer or accumulation at the interface between the host cytoplasm and the *Frankia* hypha. Wall synthesis, and indeed directed wall synthesis, is a major function of the host root hair during infection. The cytoplasm contains quantities of ribosomes, and organelles present include Golgi bodies, endoplasmic reticulum, mitochondria, and somewhat enlarged plastids. There is an extensive increase in the surface area of the plasmalemma (Berry et al., 1986). Solute transfer may be mediated by the polyanionic wall material or by constituents of the plasmalemma.

Prenodule Formation

Based on cytological observations, development of the prenodule may be divided into three stages: (1) localized root cortical mitotic activity, (2) cell expansion and *Frankia* infection, and (3) differentiation of infected host cells.

New cell divisions are initiated in the host cortical tissue in response to infection by *Frankia*. These divisions are at first localized at the bases of deformed or deforming (but not necessarily infected) root hairs. The precise timing of cell division in the infection sequence has not been established, but mitosis at least precedes the passage of *Frankia* into the cortical tissue. Calvert et al. (1983) found that cortical mitotic activity in soybean began before root hair infection.

In any case, the pathway of cortical infection by *Frankia* is through a population of newly divided cells which are enlarging at or near the time of *Frankia* penetration. The passage of *Frankia* from cell to cell appears to occur through thin-walled regions, either newly expanded or still expanding at the time *Frankia* grows through the wall (Fig. 1, part 4). Compared with uninfected cells, infected cortical cells subsequently become hypertrophied. The enlarged infected cells constitute the prenodule, which is in effect a redifferentiation of the cortex within a localized sector of the original root axis.

Cytology of Nodule Cortical Infection

Infected cortical cells exhibit distinctive structural features as compared with cortical cells of uninfected roots. The nuclei within the cells are large and have prominent nucleoli. In the genera *Coriaria* and *Datisca*, multiple nuclei arise within a single infected cell (Newcomb and Pankhurst, 1982). Kodama (1970) determined that the nodule meristem in three species of *Alnus* was diploid, however, with normal mitotic figures. DNA content has not been determined for any actinorhizal nodule tissue.

At maturity, infected cortical cells retain an active cytoplasm but are filled with a mass of *Frankia* hyphae and vesicles (Fig. 1, part 3). Numerous ribosomes and mitochondria are present in infected cortical cells in close association with hyphae or vesicles (Newcomb et al., 1978) (Fig. 1, part 4), indicating local metabolic activity. Plastids in infected cells do not usually contain starch grains and may appear enlarged and later degenerate (Newcomb et al., 1978) (Fig. 1, part 4). As mentioned, Golgi bodies are frequently observed during early infection in the root hair and prenodule stage (Berry et al., 1986).

Although nodule cortical cells are usually larger than root cortical tissues, uninfected nodule cells are often quite small (Fig. 1, part 3). In some host species, there are abundant starch grains in uninfected nodule tissue. Phenolic compounds accumulate in uninfected cells also, often conspicuously in the vacuoles (Newcomb et al., 1978; Schwintzer et al., 1982) (Fig. 1, part 3). Phenolics are not entirely absent from infected cells, are observable in the cell wall and, in some species, in the cytoplasm (Newcomb and Pankhurst, 1982).

REGULATION OF INFECTION

Frankia may be considered a rhizosphere inhabitant and an infectious agent in plants. It has often been surmised that the nodulating bacteria have developed adaptive mechanisms regulating the infection process, which may share common molecular features with adaptations of phytopathogens, rhizosphere dwellers, or both. Possibilities include regulation at the genetic level, production or regulation of plant growth substances by the invading organism, and endophyte involvment with processes of cell wall synthesis or degradation.

Cell Wall Interactions

At this stage, the only evidence of the production of wall hydrolytic enzymes by *Frankia* is cytological, i.e., the apparent disorganization of the host cell wall in the zone of infection, as described above. Such local wall degradation, or even partial wall degradation, could facilitate *Frankia* invasion. Wall hydrolytic enzymes might also act indirectly in releasing cell wall fragments which could act as signaling mechanisms, activating host wall synthesis and cell expansion, for example. Similar processes have been described for certain host-pathogen interactions (Darvill and Albersheim, 1984).

Patterns of host cell wall synthesis are also apparently affected by *Frankia* infection. If the wall structure or composition is altered, this might remove a mechanical barrier to invasion or open up the wall to direct hydrolytic activity. Alternatively, cell wall fragments might be released by endogenous enzyme activity (cell wall turnover) to serve as messenger molecules. There is as yet no experimental evidence concerning cell wall production and degradation in actinorhizal nodules.

Finally, *Frankia* may interact with the host cell wall in adhesion, as described in an earlier section. Cell wall fragments might function as information molecules in recognition as well, perhaps at the sites of adhesion.

Phytohormone Production

Many microorganisms which occupy plant surfaces or cortical tissues produce plant growth regulators. Such organisms are often implicated in enhancing or inhibiting plant growth processes. In recent years, it has become evident that microbial genes for auxin and cytokinin production are of central importance in regulating host cell division and expansion in response to several pathogens (Nester et al., 1984; Barry et al., 1984). Recently, indoleacetic acid (IAA) production *in vitro* has been demonstrated to occur at low levels in one strain of *Frankia* (Wheeler et al., 1984) in a medium supplemented with tryptophan. Further research needs to be done to elucidate the pathway of IAA synthesis in *Frankia* and to assess its significance in the nodulation process.

Nodule levels of IAA, zeatin and zeatin derivatives, and gibberellin activity have been demonstrated to be high relative to those in nonnodule host tissues (Henson and Wheeler, 1977; Wheeler et al., 1979) in several actinorhizal species. Cytokinin levels in *Alnus* nodules were highest at budbreak, before the onset of nitrogenase activity (Wheeler et al., 1979). Presumably the host tissue produces high levels of growth regulators in response to some influence by *Frankia*. Exogenously applied auxins can cause initiation of an increased number of nodule lobe primordia (i.e., lateral root primordia) in the vicinity of root hair infections (Angulo Carmona, 1974). There is also a report that kinetin application stimulated nodule-like structures in otherwise uninfected roots of *Alnus* (Rodriguez-Barrueco and Bermudez de Castro, 1973).

Genetic Aspects of Regulation

As mentioned earlier, research has only begun on the molecular genetic aspects of nodulation. Some degree of hybridization with *Frankia* DNA has been reported using a *nod c* probe from *Rhizobium meliloti* (Drake et al., 1985).

CONCLUDING REMARKS

Frankia symbioses, still relatively recently described, are of great interest and have great potential as differentiating biological systems for molecular genetic research and for field utilization.

There are striking parallels at least on the cytological level between the actinorhizal and the legume-*Rhizobium* infection sequences. In both symbioses, early invasion is initiated via root hair infection in many species but can also occur by an extracellular pathway through the root cortex. Thus, like the nitrogenase function, the early stages of nodule formation appear to be relatively strongly conserved from one symbiotic system to the other.

In actinorhizal species, subsequent nodule lobe development is generally comparable regardless of the type of early infection. At the cellular level, host cell expansion and cell-to-cell passage of *Frankia* are similar to that occurring in the prenodule and in the nodule lobe. Intracellular invasion appears to be related to host cell wall synthesis and new wall expansion. The cytoplasmic organelles of the infected host tissue are specialized for functions associated with transcription, energy production, wall synthesis and, in the infected root hair, secretion or solute transfer. Such structural evidence suggests that the host cell membrane and/or wall constituents may play a role in a number of infection-related processes. Phytohormones produced either by *Frankia* or by the host tissue may also serve a regulatory function.

The phylogenetic diversity of the actinorhizal host species suggests that the physiological processes underlying nodulation are common to a wide range of woody plants. If this is so, it may be possible to extend the range of hosts species susceptible to *Frankia* through molecular genetic techniques or through plant breeding. The challenge for the near future will be to identify the signaling molecules from *Frankia* which trigger nodule cell differentiation, to determine the corresponding host physiological processes regulating nodule formation, and to specify the cellular activities necessary for successful infection.

ACKNOWLEDGMENTS

Special thanks go to William Ormerod, Tammy Sage, and Ed Seling for technical assistance, and to Dee Drinks for manuscript preparation.

REFERENCES

AKKERMANS, A. D. L. 1971. Nitrogen fixation and nodulation of *Alnus* and *Hippophae* under natural conditions, Ph.D. Dissertation, University of Leiden, Netherlands.

AKKERMANS, A. D. L., BAKER, D., HUSS-DANELL, K., AND TJEPKEMA, J. D. (EDS.) 1984. *Frankia Symbioses,* Martinus Nijhoff/Dr. W. Junk, The Hague.

AKKERMANS, A. D. L., HUSS-DANELL, K., AND W. ROELOFSEN. 1981. Enzymes of the tricarboxylic acid cycle and the malate-aspartate shuttle in the N_2-fixing endophyte of *Alnus glutinosa. Physiol. Plant.* **53**:289–294.

AKKERMANS, A. D. L., ROELOFSEN, W., BLUM, J., HUSS-DANELL, K., AND HARKINK, R. 1983. Utilization of carbon and nitrogen compounds by *Frankia* in synthetic media and in root nodules of *Alnus glutinosa, Hippophae rhamnoides,* and *Datisca cannabina. Can. J. Bot.* **61**(11):2793–2800.

ANGULO CARMONA, A. F. 1974. La formation des nodules fixateurs d'azote chez *Alnus glutinosa* (L.) Vill. *Acta Bot. Neerl.* **23**:257–303.

BAKER, D., NEWCOMB, W., AND TORREY, J. G. 1980. Characterization of an ineffective actinorhizal microsymbiont, *Frankia* sp. EuI1 (Actinomycetales), *Can. J. Microbiol.* **26**:1072-1089.

BARRY, G. F., ROGERS, S. G., FRALEY, R. T., AND BRAND, L. 1984. Identification of a cloned cytokinin biosynthetic gene. *Proc. Natl. Acad. Sci. USA* **81**:4776-4780.

BECKING, J. H. 1974. Frankiaceae, in: *Bergey's Manual of Determinative Bacteriology*, 8th ed. (R. S. Buchanan and N. E. Gibbons, eds.), pp. 701-706, Williams and Wilkins, Baltimore, Maryland.

BECKING, J. H. 1984. Identification of the endophyte of *Dryas* and *Rubus* (Rosaceae), *Plant Soil* **76**:105-128.

BENSON, D. R., ARP, D. J., AND BURRIS, R. H. 1979. Cell-free nitrogenase and hydrogenase from actinorhizal nodules, *Science* **205**:688-689.

BENSON, D. R., ARP, D. J., AND BURRIS, R. H. 1980. Hydrogenase in actinorhizal root nodules and root nodule homogenates, *J. Bacteriol.* **142**:138-144.

BERG, R. H. 1983. Preliminary evidence for the involvement of suberization in infection of *Casuarina, Can. J. Bot.* **61**:2910-2918.

BERRY, A. M., AND TORREY, J. G. 1983. Root hair deformation in the infection process of *Alnus rubra, Can. J. Bot.* **61**:2863-2876.

BERRY, A. M., TORREY, J. G., AND McCULLY, M. E. 1983. The fine structure of the root hair wall and surface mucilage in the actinorhizal host, *Alnus rubra*, in: *Plant Molecular Biology* (R. Goldberg, ed.), pp. 319-327, Allan R. Liss, New York.

BERRY, A. M., McINTYRE, L., AND McCULLY, M. E. 1986. Fine structure of root hair infection leading to nodulation in the *Frankia-Alnus* symbiosis, *Can. J. Bot.*, in press.

BLOM, J., ROELOFSEN, W., AND AKKERMANS, A. D. L. 1980. Growth of *Frankia* AvcI1 on media containing Tween 80 as C-source, *FEMS Microbiol. Lett.* **11**:221-224.

BOHLOOL, B. B., AND SCHMIDT, E. L. 1974. Lectins: A possible basis for specificity in the *Rhizobium*-legume root nodule symbosis, *Science* **185**:269-271.

BOND, G. 1976. The results of the IBP survey of root-nodule formation in nonleguminous angiosperms, in: *Symbiotic Nitrogen Fixation in Plants* (P. S. Nutman, ed.), pp. 443-474, Cambridge University Press.

BRIGGS, D. G., DEBELL, D. S., AND ATKINSON, W. A. (EDS.) 1978. *Utilization and Management of Alder*, Gen. Tech. Rep. PNW-70, USDA Forestry Service.

CALLAHAM, D., AND TORREY, J. G. 1977. Prenodule formation and primary nodule development in roots of *Comptonia* (Myricaceae), *Can. J. Bot.* **55**:2306-2318.

CALLAHAM, D., AND TORREY, J. G. 1981. The structural basis for infection of root hairs of *Trifolium repens* by *Rhizobium, Can. J. Bot.* **59**:1647-1664.

CALLAHAM, D., DEL TREDICI, P., AND TORREY, J. G. 1978. Isolation and cultivation *in vitro* of the actinomycete causing root nodulation in *Comptonia, Science* **199**:899-902.

CALLAHAM, D., NEWCOMB, W., TORREY, J. G., AND PETERSON, R. L. 1979. Root hair infection in actinomycete-induced root nodule initiation in *Casuarina, Myrica* and *Comptonia, Bot. Gaz.* **140**(Suppl.):S51–S59.

CALVERT, H., PENCE, M. K., PIERCE, M., AND BAUER, W. D. 1983. The pattern of *Rhizobium* infection in soybean roots, 9th North American Rhizobium Conference, Ithica, New York, p. 15. Abstract.

CHANDLER, M. R., DATE, R. A., AND ROUGHLEY, R. J. 1982. Infection and root-nodule development in *Styolosanthes* species by *Rhizobium, J. Exp. Bot.* **33**(132):47–57.

CHAUDHARY, A. H., 1978. The discovery of root nodules on new species of non-leguminous angiosperms from Pakistan and their significance, in: *Limitations and Potentials for Biological Nitrogen Fixation in the Tropics* (J. Dobereiner, R. H. Burris, and A. Hollaender, eds.), p. 359, Plenum Press, New York.

DARVILL, A. G., AND ALBERSHEIM, P. 1984. Phytoalexins and their elicitors—A defense against microbial infection in plants, *Annu. Rev. Plant Physiol.* **35**:243–275.

DRAKE, D., LEONARD, J. T., AND HIRSCH, A. M. 1985. Symbiotic genes in *Frankia*. 6th International Symposium on Nitrogen Fixation, Corvallis, Oregon. Abstract.

DREYFUS, B. L., ALAZARD, D., AND DOMMERGUES, Y. R. 1984. Stem-nodulating bacteria, in: *Current Perspectives in Microbiol Ecology* (M. J. Klug and C. A. Reddy, eds.), pp. 161–169, American Society for Microbiology, Washington, D.C.

FONTAINE, M. S., LANCELLE, S., AND TORREY, J. G. 1984. Initiation and on-togeny of vesicles in cultured *Frankia* sp. strain HFPArI3, *J. Bacteriol.* 160, 921–927.

GORDON, J. C., AND WHEELER, C. T. (EDS.) 1983. *Biological Nitrogen Fixation in Forest Ecosystems: Foundations and Applications,* Martinus Nijhoff/Dr. W. Junk, The Hague.

GORDON, J. C., WHEELER, C. T., AND PERRY, D. A. (EDS.). 1979. *Symbiotic Nitrogen Fixation and the Management of Temeperate Forests,* Oregon State University, Corvallis.

GREAVES, M. P., AND DARBYSHIRE, J. F. 1972. The ultrastructure of the mucilaginous layer on plant roots, *Soil Biol. Biochem.* **4**:443–449.

HEISEY, R. M., DELWICHE, C. C., VIRGINIA, R. A., WRONA, A. F., AND BRYAN, B. A. 1980. A new nitrogen-fixing non-legume: *Chamaebatia foliosa* (Rosaceae), *Am. J. Bot.* **67**:429–431.

HENSON, I. E., AND WHEELER, C. T. 1977. Hormones in plants bearing nitrogen-fixing root nodules: Cytokinin levels in roots and root nodules of some non-leguminous plants, *Z. Pflanzenphysiol.* **84**:179–182.

KNOWLTON, S., BERRY, A., AND TORREY, J. G. 1980. Evidence that associated soil bacteria may influence root hair infection of actinorhizal plants by *Frankia, Can. J. Microbiol.* **26**:971–977.

KODAMA, A. 1970. Cytological and morphological studies on the plant tumors. II. Root nodules of three species of *Alnus, J. Sci. Hiroshima Univ., Ser. B, Div.* 2, **13**:261–264.

LALONDE, M. 1977. The infection process of *Alnus* root nodule symbiosis, in: *Recent Developments in Nitrogen Fixation* (J. R. Postgate and C. Rodriguez-Barrueco, eds.), pp. 569–589, Academic Press, London.

LALONDE, M., AND KNOWLES, R. 1975. Ultrastructure, composition and biogenesis of the encapsulation material surrounding the endophyte in *Alnus crispa* var. *mollis* root nodules, *Can. J. Bot.* **53**:1951–1971.

LANCELLE, S., AND TORREY, J. G. 1984. Early development of *Rhizobium*-induced root nodules of *Parasponia rigida*. I. Infection and early nodulation, *Protoplasma* **123**:26–37.

LECHEVALIER, M. P. 1986. (in press). *Catalog of Frankia strains: The Actinomycetales*, Waksman Institute of Microbiology, Piscataway, New Jersey.

LECHEVALIER, M. P., AND LECHEVALIER, H. 1979. The taxonomic position of the actinomycetic endophytes, in: *Symbiotic Nitrogen Fixation in the Management of Temperate Forests* (J. C. Gordon, C. T. Wheeler, and D. A. Perry, eds.), pp. 111–122. Oregon State University, Corvallis.

LECHEVALIER, M. P., AND RUAN, J. 1984. Physiology and chemical diversity of *Frankia* spp. isolated from nodules of *Comptonia peregrina* (L.) Coul and *Ceanothus americanus* L, in: *Frankia Symbioses* (A. D. L. Akkermans, D. Baker, K. Huss-Danell, and J. D. Tjepkema, eds.), pp. 15–22, Martinus Nijhoff/Dr. W. Junk, The Hague.

LIPPINCOTT, J. A., LIPPINCOTT, B. B., AND SCOTT, J. J. 1984. Adherence and host recognition in *Agrobacterium* infection, in: *Current Perspectives in Microbial Ecology* (M. J. Klug and C. A. Reddy, eds.), pp. 230–236, American Society for Microbiology, Washington, D.C.

LOPEZ, M., WHALING, C. S., AND TORREY, J. G. 1983. The polar lipids and free sugars of *Frankia* in culture, *Can. J. Bot.* **61**(11):2834–2842.

LOPEZ, M. F., FONTAINE, M. S., AND TORREY, J. G. 1984. Levels of trehalose and glycogen in *Frankia* sp. HFPArI3 (Actinomycetales). *Can. J. Microbiol.* **30**(6):746–752.

MAZZUCCO, C. E., AND BENSON, D. R. 1984. ^{14}C Methylammonium transport by *Frankia* sp. strain CpI1, *J. Bacteriol.* **160**:636–641.

MIAN, S., AND BOND, G. 1978. The onset of nitrogen fixation in young alder plants and its relation to differentiation in the nodular endophyte, *New Phytol.* **80**: 187–192.

MIDGLEY, S. J., TURNBULL, J. W., AND JOHNSTON, R. D. 1983. *Casuarina Ecology, Management and Utilization*, CSIRO, Melbourne.

MILLER, I. M., AND BAKER, D. D. 1985. The initiation, development and structure of root nodules in *Elaeagnus angustifolia* L (Elaegnaceae), *Protoplasma*, **128**:107–119.

MORRÉ, D. J., JONES, D. D., AND MOLLENHAUER, H. H. 1967. Golgi apparatus mediated polysaccharide secretion by outer root cap cells of *Zea mays*. 1. Kinetics and secretory pathway, *Planta* **74**:286–301.

MURRY, M. A., FONTAINE, M. S., AND TORREY, J. G. 1984. Growth kinetics and nitrogenase induction in *Frankia* sp. HFSrI3 grown in batch culture, *Plant Soil* **78**:61–79.

NATIONAL RESEARCH COUNCIL. 1980. *Firewood Crops,* National Academy of Sciences, Washington, D.C.

NATIONAL RESEARCH COUNCIL. 1984. *Casuarinas: Nitrogen-Fixing Trees for Adverse Sites,* National Academy Press, Washington, D.C.

NESTER, E. W., GORDON, M. P., AMASINO, R. M., AND YANOFSKY, M. F. 1984. Crown gall: A molecular and physiological analysis, *Annu. Rev. Plant Physiol.* **35:** 387–413.

NEWCOMB, W. 1981. Fine structure of the root nodules of *Dryas drummondii* Richards (Rosaceae), *Can. J. Bot.* **59:**2500–2514.

NEWCOMB, W., AND PANKHURST, C. E. 1982. Fine structure of actinorhizal root nodules of *Coriaria arborea* (Coriariaeae), *New Zealand J. Bot.,* **20:**93–103.

NEWCOMB, W., PETERSON, R. L., CALLAHAM, D., AND TORREY, J. G. 1978. Structure and host-actinomycete interactions in developing root nodules of *Comptonia peregrina, Can. J. Bot.* **56:**502–531.

PUPPO, A., DIMITRIJEVIC, L., DIEM, H., AND DOMMERGUES, Y. 1985. Superoxide dismutases in *Frankia* strains: Taxonomic importance, 6th International Symposium on Nitrogen Fixation, Corvallis, Oregon. Abstract.

ROBERTSON, J. G., WELLS, B., BREWIN, N. J., KNIGHT, C. D., WOOD, E. A., AND DOWNIE, J. A. 1986. Early events in nodulation of legumes, *J. Cell Sci.,* Suppl. 2, in press.

RODRIGUEZ-BARRUECO, C., AND BERMUDEZ DE CASTRO, F. 1973. Cytokinin-induced pseudonodules on *Alnus glutinosa, Physiol. Plant.* **29:**277–280.

RUVKIN, G. B., AND AUSUBEL, F. M. 1980. Interspecies homology of nitrogenase genes, *Proc. Natl. Acad. Sci. USA* **77**(7):191–195.

SCHUBERT, K. R., COKER, G. T., AND FIRESTONE, R. B. 1981. Ammonia assimilation in *Alnus glutinosa* and *Glycine max, Plant Physiol.* **67:**662–665.

SCHWINTZER, C. R., BERRY, A. M., AND DISNEY, L. D. 1982. Seasonal patterns of root nodule growth, endophyte morphology, nitrogenase activity, and shoot development in *Myrica gale, Can. J. Bot.* **60:**746–757.

SHIPTON, W. A., AND BURGGRAAF, A. J. P. 1982. A comparison of the requirements for various carbon and nitrogen sources and vitamins in some *Frankia* isolates, *Plant Soil* **69:**149–161.

STEELE, D. B., AND STOWERS, M. D. 1985. Enzymatic mechanisms for the protection of nitrogenase from oxygen in *Frankia.* 6th International Symposium on Nitrogen Fixation, Corvallis, Oregon. Abstract.

TISA, L. S., AND ENSIGN, J. C. 1986. Studies of vesicles and protoplasts of *Frankia. Plant Soil,* in press.

TJEPKEMA, J. 1979. Oxygen relations in leguminous and actinorhizal nodules, in: *Symbiotic Nitrogen Fixation in the Management of Temperate Forests* (J. C. Gordon, C. T. Wheeler, and D. A. Perry, eds.), pp. 175–186, Oregon State University, Corvallis.

TJEPKEMA, J. D. 1983. Hemoglobins in the nitrogen-fixing root nodules of actinorhizal plants, *Can. J. Bot.* **61:**2924–2929.

TJEPKEMA, J. D., ORMEROD, W., AND TORREY, J. G. 1980. Vesicle formation and acetylene reduction activity in *Frankia* sp. CpI1 cultured in defined media, *Nature* **287:**633–635.

TORREY, J. G. 1978. Nitrogen fixation by actinomycete-nodulated angiosperms, *Bioscience* **28**(9):586–592.

TORREY, J. G., AND CALLAHAM, D. 1982. Structural features of the vesicle of *Frankia* sp. CpI1 in culture, *Can. J. Microbiol.* **18**:749–757.

TURGEON, B. G. 1982. Ph.D. Dissertation, University of Dayton, Dayton, Ohio.

VANDENBOSCH, K. A., AND TORREY, J. G. 1983. Host-endophyte interactions in effective and ineffective nodules inducted by the endophyte of *Myrica gale*, *Can. J. Bot.* **61**:2898–2909.

VANDIJK, C. 1978. Spore formation and endophyte diversity in root nodules of *Alnus glutinosa* (L.) Vill., *New Phytol.* **81**: 601–615.

VANDIJK, C., AND MERKUS, E. 1976. A microscopic study of the development of a spore-like stage in the lite cycle of the root nodule endophyte of *Alnus glutinosa* (L.) Gaertn., *New Phytol.* **77**:73–90.

VERMEER, J., AND MCCULLY, M. E. 1981. Fucose in the surface deposits of axenic and field grown roots of *Zea mays* L., *Protoplasma* **109**:233–248.

WHEELER, C. T., HENSON, I. E., AND MCLAUGHLIN, M. E. 1979. Hormones in plants bearing actinomycete nodules, *Bot. Gaz.* **140**(Suppl):952–957.

WHEELER, C. T., CROZIER, A., AND SANDBERG, G. 1984. The biosynthesis of indole-3-acetic acid by *Frankia*, *Plant Soil* **78**:99–104.

SECTION III

Establishment of Microbes in Plants

Chapter 7

Recognition and Infection Processes in Plant Pathogen Interactions

Julie E. Ralton, Michael G. Smart, and Adrienne E. Clarke

PLANTS COME IN CONTACT with a variety of microorganisms during their lifetime. Some of them are symbiotic (e.g., mycorrhizae and *Rhizobium*), and others may be pathogenic. The outcome of contact between the plant and the microorganism, i.e., whether or not a successful symbiotic or pathogenic interaction is established, depends on the mutual perception, or recognition, between the interacting cells. Our understanding of the way in which such interactions are controlled is at present rudimentary but is currently being approached from a number of angles. Ultimately, the interactions are under genetic control, and the identification of the controlling genes and their products, using the techniques of molecular biology, is an important new approach. The information obtained will complement that obtained from studies at the biochemical and cellular levels. In this chapter, we discuss cell recognition between higher plants and their fungal pathogens, concentrating on what is known of the surfaces of the cells in contact at both the cellular and molecular levels and how this information relates to the recognition events. In doing this, we briefly review the principles of cell recognition between animal cells, which are understood in detail in some cases. We also consider the nature of the cell wall and the evidence that it plays an active role in recognition. We then examine four host-pathogen interactions which have been studied by classical microscopy and interpret these studies in terms of possible recognition events. Finally, we point out that studies using microscopy and molecular techniques can be interactive, the microscopy being used to define the cell types best selected for molecular studies and the molecular studies leading to specific probes which can be used in microscopy to define the role of specific components.

CELL RECOGNITION

Cell recognition has been defined as the initial event of cell-cell communication which elicits a defined biochemical, physiological, or morphological response (Clarke and Knox, 1978). However, as Daly (1984) points out, this definition leads to a requirement that the terms "initial event" and "communication" be defined precisely in molecular terms. This is not yet possible for any plant cell interaction, although some progress toward understanding the initial events is being made in a number of systems, notably the interaction of pollen (or pollen tubes) with the female pistil in systems where self-incompatibility genes operate to prevent inbreeding. Control of pollen tube growth through the female sexual tissues has a number of parallels with the control of fungal growth in host tissues (Hogenboom, 1983; Clarke and Gleeson, 1981). For both systems, control of the interaction (compatibility or incompatibility) is defined genetically, and incompatible interactions result in an arrest in growth of the pollen tube or fungal hypha. In both cases, a series of recognition events may be involved, and in both cases progress toward understanding the basis of these events is being made by application of molecular genetics to understanding the nature of the products of the R (resistance) and S (self-incompatibility) genes ultimately involved in the recognition events. Both interactions are also of great economic importance, as crop yields depend on efficient fertilization and control of disease.

In the pollen-pistil interaction, progress has depended on being able to identify glycoprotein components of the style which segregate with a particular genotype (Clarke et al., 1986). The cDNA encoding the protein component can then be used to identify products of the same allele in pollen, and a knowledge of the structures of the interacting components will lead to an understanding of how they interact. Host-pathogen interactions are more complex, and definition of the initial event is not as far advanced as for pollen-pistil recognition. A major complicating factor is that two quite different genomes are involved (host and pathogen), in contrast to the pollen-pistil interaction where the haploid and diploid genomes of the same plant determine the recognition events (for reviews see Heslop-Harrison, 1983; Harris et al., 1984). A second difficulty is that in no host-pathogen interaction has a protein product corresponding directly to expression of the resistance (R gene) been identified (for further discussion see Gilchrist and Yoder, 1984).

The second part of the definition of recognition requires an understanding of the term "communication." Again, we cannot offer a molecular explanation of this term for interacting plant cells, although (see section on application of animal cell recognition) some progress is being made in understanding the role protein kinases and Ca^{2+} channels, both important in animal cell-cell communication, may play in plant systems.

Sequiera (1978) has pointed out that in host-pathogen interactions the response elicited or triggered by the initial or early events may either facilitate or impede further growth of the pathogen. This point has direct parallels with pollen-pistil interactions in which recognition could theoretically lead either to active arrest of incompatible pollen tube growth (the oppositional hypothesis) or to active stimulation of the growth of compatible tubes (the complementary hypothesis). The body of opinion favors the outcome or recognition as being active arrest of incompatible pollen tube growth (Heslop-Harrison, 1983); Mulcahy and Mulcahy (1983) have listed evidence which may support the complementary hypothesis, although the reasoning behind this suggestion has recently been challenged by Lawrence et al. (1985). Undoubtedly, the uncertainties will eventually be resolved when complete molecular information is available. In the meantime, we are, as Daly (1984) points out, left with a somewhat imprecise definition of recognition but, as genes associated with specific expression of resistance are identified, the definition will be refined.

PRINCIPLES OF CELL-CELL RECOGNITION BETWEEN ANIMAL CELLS

As many of our approaches to understanding cell-cell recognition in plant systems are based on experience obtained from investigations of animal cell interactions, it is useful to review the principles of cell recognition involving animal cells. In animal systems, recognition involves specific binding of signals to complementary membrane-bound receptor proteins, and it is believed that this also applies to plant cell-cell recognition, notwithstanding the differences between plant and animal cells. The most studied systems are those in which one cell secretes a chemical signal which is received by a specific receptor on a target cell some distance away (Alberts et al., 1983). Examples are the action of water-soluble peptide growth hormones and neurotransmitters. The main points which have emerged from studies on such recognition systems are:

1. The secreted signal molecules bind specifically, reversibly, and saturably with high affinity to specific receptor proteins (association constant $K_a = 10^8 M^{-1}$) on the surface of target cells. This signal-receptor binding initiates intracellular signals (second messengers) which alter the behavior of the target cells.

2. Cyclic AMP (cAMP) is one of two known second messengers. The biological effect of specific binding of signal molecules to target cells may be achieved by regulation of the intracellular cyclic AMP concentration via activation of adenylate cyclase. Cyclic GMP (cGMP) can act in a similar way. Cyclic GMP produces its effect by activating specific cyclic AMP, Ca^{2+}-dependent protein kinases which in turn phosphorylate and thereby activate specific enzymic proteins. These activated enzymes often produce a biological effect directly, but in some cases they initiate another reaction which produces the biological effect.

3. Ca^{2+} is the other known second messenger. In this case the cell surface receptors are functionally coupled to Ca^{2+} channels. The binding of signal molecules to the receptors alters gated ion channels in the plasma membrane; this results in either an ion flux across the plasma membrane or an influx of Ca^{2+} into the cytosol. In both cases the change in Ca^{2+} concentration is transient. Calmodulin is a ubiquitous intracellular receptor for Ca^{2+}, and a number of cellular proteins are regulated by Ca^{2+}-calmodulin in a Ca^{2+}-dependent way. These proteins include protein kinases, adenylate cyclase, and cyclic nucleotide phosphodiesterase. There is overlap between the activities regulated by cyclic AMP and Ca^{2+}-calmodulin to the extent that the same protein kinase may be activated both by the cyclic AMP-dependent kinase and by the binding of Ca^{2+} to calmodulin (Nestler and Greengard, 1983; Alberts et al., 1983).

Another mechanism by which signal-receptor binding may initiate an intracellular signal is a receptor-mediated endocytosis: Receptor molecules with bound signal molecules move laterally through the membrane into a zone on the cell surface (a coated pit) which then undergoes endocytosis to form a vesicle. This vesicle then fuses with a lysosome to form a secondary lysosome in which various changes in the receptor molecule, the signal molecule, or both may occur. This receptor-mediated endocytosis may or may not occur in reactions involving enzyme activation (Brown et al., 1983; Hunter, 1984). An exciting recent finding in animal hormone biology, which may have parallels in plant systems, is that the epidermal growth factor (EGF) precursor may play a dual role as both a growth factor precursor and a cell surface receptor (Rall et al., 1985). Other structural homologies between secreted molecules and their membrane receptors have been reported for animal systems (Pfeffer and Ullrich, 1985). This would provide a model for recognition in the self-incompatibility system in which like genes carried by the pollen and style lead to an arrest of pollen tube growth. It is possible that different splicing patterns, intragenic duplication, and posttranslational modification events produce domains on the interacting molecules which allow specific and complementary interaction.

Recognition in fungal host-pathogen interactions is controlled by genes of the host and pathogen which are assumed to have different and complementary products. However, the experience with animal systems and the relatively simple self-incompatibility system may give us valuable insights into the design of experimental approaches to understanding recognition in the more complex host-pathogen interactions.

APPLICATION OF PRINCIPLES OF ANIMAL CELL-CELL RECOGNITION TO PLANTS

In considering the possible mechanisms for movement of signals between plant cells, we find a major difference between plants and animals, namely, the presence of a cell wall which encases and protects the protoplast. In some examples of host-pathogen interactions, the pathogen grows between the wall and the plasma membrane and is thus in direct contact with the plasma membrane. In other cases, growth occurs via the middle lamella so that contact is with the cell wall (see section on host-pathogen interaction, case 4). In cases where the contact is between the fungal cell wall and the middle lamella, signals may move directly or indirectly across the cell wall to the plasma membrane; how they are received and transduced to give intracellular cytoplasmic responses is not understood. Cyclic AMP, protein kinases, Ca^{2+} channels, and calmodulin have all been described in plant cells (for references see Polya et al., 1986), but their role, if any, in plant cell recognition is not yet defined. The principles of recognition of extracellular signals between animal and plant target cells are compared in Fig. 1. No plant plasma membrane receptor for a particular signal has yet been isolated and fully characterized, although there is evidence for the presence of plasma membrane-associated receptors for β-glucans (Peters et al., 1978; Yoshikawa et al., 1983), auxins (Jacobs and Gilberts, 1983), and abscisic acid (Hornberg and Weiler, 1984). Indeed, our understanding of the molecular structure of plant plasma membranes is poor compared with that of animal plasma

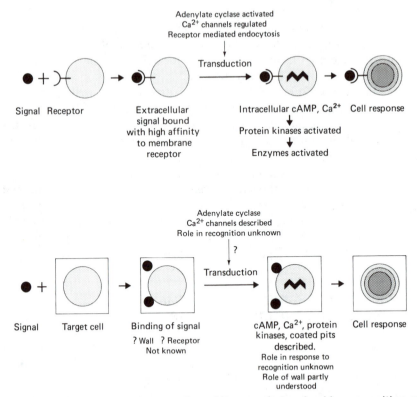

Figure 1. Schematic representation of the events involved in recognition of an extracellular signal by (A) an animal cell and (B) a plant cell. (From Clarke et al., 1986.)

membranes (Bowles, 1982). A significant difference between animal and plant plasma membranes is that enzymes involved in the synthesis of some cell wall components, e.g., cellulose synthase and $(1 \rightarrow 3)$-β-glucan synthase, are commonly associated with plant plasma membrane preparations. Chemical signals (e.g., auxins, cytokinins, gibberellins) are secreted in plants, but we have a poor understanding of the way in which these and other molecules move extracellularly.

There are several other interactions involving single plant cells, which are simpler and better defined than recognition in host-fungal pathogen interactions. For example, yeast mating is understood in some detail at the level of molecular genetics and the chemistry of the interacting surfaces; in algae *Chlamydomonas* mating interactions are also relatively simple and have been studied intensively (for list of interactions and details of references see Harris et al., 1984).

ROLE OF THE CELL WALL IN HOST-FUNGUS RECOGNITION

During penetration of plant tissues by fungi, the cell wall barrier is often breached and extensive intercellular and intracellular hyphal growth may occur. During interactions between fungal pathogens and their hosts, the major structures in contact are the fungal hyphae with the middle lamella or cell wall or (less frequently) the plasma membrane of the host cell. During intracellular growth, an extrahaustorial matrix, believed to be of fungal origin, makes direct contact with the host plasma membrane. (See section on interaction at tissue level for details.) Logically we would expect the cell walls of the host and the pathogen to play an important role in this interaction, and experimental evidence supports this view. However, this role may not be related directly to the products of the genes controlling the specificity of the interaction.

The Plant Cell Wall

Our understanding of the structure and organization of the individual wall components of plant cells is incomplete (McNeill et al., 1984). The proportion and detailed structures of the major groups of cell wall matrix polysaccharides, the pectic polymers and the other noncellulosic polysaccharides, the pectic polymers and the other noncellulosic polysaccharides, can differ markedly in wall preparations from the same cell type of different species and in wall preparations from different cell types in the same organ. The matrix polysaccharides and the wall-associated proteins, including the hydroxyproline-rich glycoprotein, have all been implicated in particular recognition phenomena (for reviews see West, 1981; Keen, 1982; Kosuge, 1981; Bacic and Clarke, 1983; Darvill and Albersheim, 1984; Sequiera, 1984), although there is as yet no way of relating these observations to the genetic control of any single host-pathogen interaction.

There are several ways in which the cell wall as a whole may act in the recognition process:

The Wall as a Molecular Sieve Restricting Access of Extracellular Material to Receptors Associated with the Plasma Membrane

The matrix of primary cell walls is effectively a negatively charged porous gel. The net charge may vary with the proportion of acidic polysaccharides and the degree of methylation of these components, as well as with the content of high-isoelectric-point proteins and glycoproteins. Both the charge and porosity of the gel matrix influence the movement of extracellular material through the wall to the plasma membrane; both these characteristics may differ for different cell types and for cells at different stages of growth. The effective porosity of the cell wall is controlled, at least in part, by the formation of junction zones between unsubstituted regions of different wall polymers (Rees, 1977). Estimates of the molecular weight of globular proteins which can move from the external medium through primary cell walls to the plasma membrane vary from 17 Kilodaltons (kd) (Carpita et al., 1979) to 60 kd (Tepfer and Taylor, 1981). Cell walls may also be penetrated by cross-channels which are seen in some freeze-fracture studies (e.g., Clarke et al., 1980). Interpretation of these attempts to describe the capacity of the cell wall to allow movement of molecules from the external medium to the plasma membrane is difficult because of the possibility of artifacts induced by the experimental techniques. Movement of material in the opposite direction, from the plasma membrane through the wall to the external medium, may involve a different pathway, as high-molecular-weight glycoconjugates ($>$100 kd) such as arabinogalactan-proteins are secreted by many cell types, e.g., suspension cultured cells (for a review see Fincher et al., 1983). Products of the host and fungal genes which are involved in determining the outcome of a host-pathogen interaction may move passively through the cell walls of the host and/or pathogen, although there is no precise information on this possibility as these gene products have not been identified.

The Wall as the Primary Target From Which Secondary Signals Are Generated

If, for instance, the primary signal from one cell is an enzyme such as a hydrolase or lyase, it may catalyze the release of fragments from the cell wall of the target cell. These wall fragments may become involved in the recognition process by interacting directly with a plasma membrane receptor. Alternatively, the primary signal (e.g., an enzyme), by modifying the number of junction zones between wall components, may change the gel properties and effective porosity of the wall and thereby alter the accessibility of the primary and/or secondary signals to membrane-associated receptors.

Studies on interactions between pathogenic fungi and their host plants indicate that host cell wall fragments released by fungal enzymes can elicit cytoplasmic changes (for reviews see Bell, 1981; Kosuge, 1981; West, 1981; Darvill and Albersheim, 1984). One well-studied mechanism of defense involves the accumulation of phytoalexins at the site of infection (Deverall 1977). West and co-workers (Lee and West, 1981; Bruce and West, 1982) isolated a heat-labile endopolygalacturonase from culture filtrates of *Rhizopus stolonifer* which could elicit accumulation of the phytoalexin casbene in castor bean (*Ricinus communis*). This en-

dopolygalacturonase releases heat-stable water-soluble cell wall fragments from the host plant. These wall fragments contain 70% of the
carbohydrate as galacturonic acid or its methyl ester, indicating their
derivation from pectic polysaccharides. Partial hydrolysis of polygalacturonic acid by the fungal enzyme also releases elicitor-active fragments
which require a minimum degree of polymerization of 10–11 units for activity (Jin and West, 1984). Oligosaccharides derived from pectic fragments are also active in inducing the production of proteinase inhibitors
(Ryan, 1984; Walker-Simmons et al., 1984). Presumably, these low-
molecular-weight cell wall fragments, which entered through pores in the
wall, interact with primary receptors at the plasma membrane to elicit the
cytoplasmic response. The nature of the presumptive receptors and the
precise structural requirements of the pectic fragments remain undefined.
Albersheim and co-workers have isolated pectic fragments which are
released from cell walls by acid hydrolysis; these fragments are active
elicitors of phytoalexin production in soybeans. The fragments, known as
endogenous elicitors are believed to contain dodeca-α-($1 \rightarrow 4$)-D-galac-
turonide as the active component (Darvill and Albersheim, 1984).

Wall-Associated Enzymes, Lectins, and Agglutinins Which Alter Extracellular Signals

Another potential mechanism for the modification of a primary extracellular signal is through interaction with wall-associated enzymes.
This could generate secondary signals which are accessible to putative
plasma membrane receptors. Alternatively, wall-bound lectins and agglutinins may bind, hence alter or immobilize extracellular signals. For instance, glycoproteins related to the hydroxyproline-rich cell wall glycoprotein, which are not lectins, have been isolated from extracts of potato
tubers (Leach et al., 1982), tobacco callus (Mellon and Helgeson, 1982),
and carrot (Van Holst and Varner, 1984). These glycoproteins are effective nonspecific bacterial agglutinins by virtue of their high pI. For
reviews of the role of cell wall components in microbial attachment see
Whatley and Sequeira (1981) and Lippincott and Lippincott (1984).
Another example of agglutination by cell wall components is the observation that a polygalacturonic acid-containing fraction from sweet potato
agglutinates germinated spores of *Ceratocystis fimbriata* with some
specificity in the presence of Ca^{2+} (Kojima et al., 1982).

The Fungal Cell Wall

Fungal cell walls generally are constructed with an inner part containing
either chitin or cellulose embedded in a matrix, and an outer alkali-soluble

Table 1 Various Polymers Occurring in the Cell Wall of Fungi [Wessels and Sietsma, 1981].

| Taxonomic Groups | CELL WALL POLYMERS | |
	Alakli-soluble	Alkali-insoluble
Basidiomycetes	Xylo-manno-protein	$(1 \rightarrow 3)$-β/$(1 \rightarrow)$-β-D-Glucan
	$(1 \rightarrow)$-α-D-glucan[a]	Chitin
Ascomycetes	(Galacto)-manno-protein	$(1 \rightarrow)$-β/$(1 \rightarrow 6)$-β-D-Glucan
	$(1 \rightarrow 3)$-α-D-glucan[a]	Chitin
Zygomycetes	Glucuronomanno-protein polyphosphate	Polyglucuronic acid Chitosan Chitin
Chytridiomycetes	Glucan[b]	Glucan[b] Chitin
Hyphochytridio- mycetes	Not determined	Chitin cellulose
Oomycetes	$(1 \rightarrow 3)$-β/$(1 \rightarrow 6)$-β-D-Glucan	$(1 \rightarrow 3)$-β/$(1 \rightarrow 6)$-β-D-Glucan cellulose

[a]In a number of cases the alkali-soluble fractions also contains part of the $(1 \rightarrow 3)$-β/$(1 \rightarrow 6)$-β-D-glucan.
[b]Incompletely characterized; probably $(1 \rightarrow)$-β and $(1 \rightarrow 6)$-β-linked.

layer. The cell wall compositions of the major taxonomic groups of fungi are summarized in Table 1. The common plant pathogens are members of the Basidiomycetes, Ascomycetes, and Oomycetes, all of which contain $(1 \rightarrow 3)$ - $(1 \rightarrow 6)$-β-glucans as major alkali-soluble wall components. These compounds are associated with chitin in the Basidiomycetes and Ascomycetes, and with cellulose in the Oomycetes. Differences which might be species-specific are found in the alkali-soluble glycoproteins (for a review see Wessels and Sietsma, 1981). In spite of the analytical information available, we know little about the relationship between the individual components or about their function in permeability, growth, and morphogenesis. Most analyses have been performed on wall material isolated from cultures of hyphae, and little is known regarding the wall components of the fungus when grown in a plant host or those of the growing tip of the fungus. Like the host cell wall, the fungal cell wall may act as a molecular sieve for extracellular signals originating from the plant host, or it may act as a primary target from which secondary signals are generated. Evidence that hyphal wall components may be involved in signaling a response in host cells comes from studies by Albersheim, Darvill, and co-workers at the University of Colorado on the structure of the fungal β-glucan elicitor of phytoalexin accumulation in soybean (for a review see Darvill and Albersheim, 1984). These neutral glucans are derived from hyphal cell walls and culture filtrates of the Oomycete *Phytophthora megasperma* f.sp. *glycinea*. The smallest elicitor-active oligosaccharide purified from a partial acid hydrolysate of hyphal cell walls is a hepta-β-glucoside alditol (Fig. 2). This oligoglucoside is effective in elicit-

Figure 2.   Structures of eight hepta-β-glucoside alditols purified from a partial acid hydrolysate of *P. megasperma* cell walls. Only one structure (*) is active in eliciting phytoalexin accumulation.

ACTIVE*

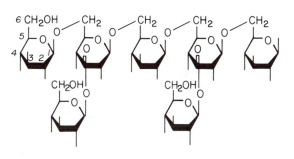

INACTIVE

(A)

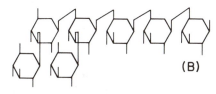
(B)

(C)

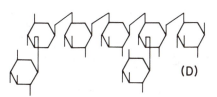
(D)

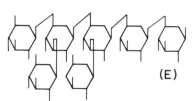
(E)

(F)

(G)

ing phytoalexin production at concentrations of 10^{-9}–10^{-10} M. Seven other hepta-β-glucoside alditols were inactive as elicitors of phytoalexins at concentrations up to 20 ng per soybean cotyledon. The minimum structure required for biological activity is a $(1 \rightarrow 6)$-β-glucosyl backbone with single glucosyl residues at the $C(0)$-3 position at two of the glucosyl backbone residues. The two substituted glucosyl backbone residues are separated by a single unbranched glucosyl residue. Darvill and Albersheim (1984) suggest that β-glucan hydrolases associated with soybean cell walls degrade the high-molecular-weight β-glucan elicitors to produce the most active elicitor, a hepta-β-glucoside.

The absolute requirement for a particular glucan structure emphasizes the informational potential of complex carbohydrates. It also implies that there must be complementary receptors for the active saccharide which, if protein in nature, would in effect be lectins. The importance of binding requirements of lectins for particular saccharide sequences (Kauss, 1985), which allows them to discriminate between closely related saccharide sequences, becomes apparent in this context. The host cell wall saccharides are involved in responses of the plant to pathogen attack and in a range of regulatory functions in growth and development (Darvill and Albersheim, 1984), but just how these phenomena are related to specific genetically controlled interactions is not known.

Another fungal wall polysaccharide which is an effective elicitor of phytoalexin production in peas is chitosan [polymeric $(1 \rightarrow 4)$-β-glucosamine, Kohle et al. (1984)]. In this case a heptamer is the smallest oligosaccharide active in eliciting the production of pisatin in peas (Kendra and Hadwiger, 1984).

An Extracellular Fungal Enzyme Involved in Pathogenesis

Fungal enzymes which facilitate penetration through the host tissues may also be involved in specific interactions. An example is cutinase produced by germinating fungal spores of *Fusarium solani* f.sp. *pisi*. Inhibition of cutinase prevents fungal penetration. This enzyme has recently been cloned and sequenced (Soliday et al., 1984), and the gene shown to be transcribed only in the presence of either intact or partially hydrolyzed cutin. This was the first fungal gene directly associated with pathogenicity to be cloned.

THE INVOLVEMENT OF THE CELL WALL IN RESPONSE TO A RECOGNITION EVENT

In plant cells the response to an extracellular signal is not restricted to the cytoplasm, as the cell wall itself may be dramatically altered, by the laying down of papillae, lignification, suberization, and increased production of the hydroxyproline-rich cell wall glycoprotein (for reviews see Aist, 1983; Heath, 1980). Papillae are wall modifications often containing callose, which is a glucan containing predominantly $(1 \rightarrow 3)$-β-glucosidic linkages and which gives a yellow fluorescence when treated with decolorized aniline blue. Studies on the production of papillae in response to contact with fungal hyphae show that for some, but not all, cell types papilla production involves proliferation of the plasma membrane followed by the laying down of a core of electron-dense material and an overlaying with electron-lucent material. The plasma membrane proliferation is preceded by changes in the host cell wall at the point of contact with the fungal hyphae (Fig. 3) (Zeyen and Bushnell, 1979; Hinch et al., 1985; Wetherbee et al., 1985). β-Glucan synthases involved in biosynthesis of callose are characteristically associated with the plasma membrane (Henry et al., 1983); it may be that proliferation of the plasma membrane observed is a mechanism for increasing the localized activity of these enzymes. In this regard, the observation that a Ca^{2+}-dependent $(1 \rightarrow 3)$-β-glucan synthase in soybean suspension cultured cells is activated by a Ca^{2+} influx, which can be induced nonspecifically by, for example, chitosan or polylysine, may be important (Kauss, 1985). This suggests that a signal may be transduced by the activation of Ca^{2+} channels, resulting in a change in the cell wall composition.

Possible pathways by which a signal with glycan hydrolase activity may affect the cell wall and the cytoplasm of a target cell are shown schematically in Fig. 4. The amount of cell wall hydroxyproline-rich glycoprotein (extensin) increases in cell walls in response to several types of fungal infection (Esquerre-Tugaye et al., 1979; Toppan et al., 1982; Hammerschmidt et al., 1984). A cDNA probe encoding a proline-rich protein has recently been isolated (J. Varner, unpublished observations) and is being used to study the control of the extensin gene expression by fungal elicitors (C. Lamb and A. S. Showalter, unpublished observations) in suspension cultures of *Phaseolus vulgaris*. This is an important new approach to understanding the role of cell wall components in infection processes.

ROLE OF THE PLASMA MEMBRANE IN RECOGNITION

It is believed, based on an analogy with animal cell recognition systems, that the plant plasma membrane plays a critical role in the recognition events between host and pathogen. There is however, only indirect evidence to support this theory because of the difficulty of obtaining purified plasma membrane preparations from plants. One approach is to use isolated protoplasts which are agglutinated by lectins (Larkin, 1981) and

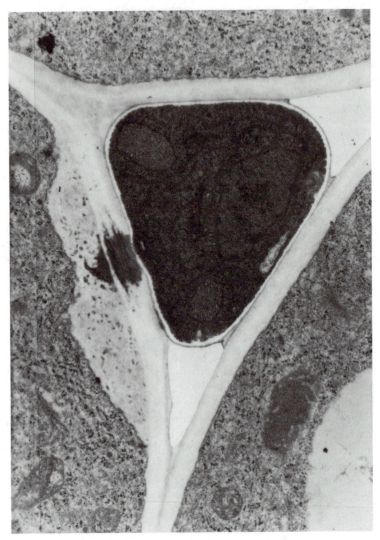

Figure 3. Electron micrograph of sections through *Zea mays* root cortex after inoculation with *Phytophthora cinnamoni* showing cross section of a hypha occupying a cortical air space. One of the host cells has responded to the presence of the hypha by producing an electron-lucent deposit (possibly callose). The region of initial response is visible at the center of the deposit. Deposition of the heterogeneous outer material continues, producing an extensive deposit that covers a greater area of the host surface than the actual interface with the hypha. Magnification: × 12,000. (From Wetherbee et al., 1985.)

by a fungal β-glucan preparation (Peters et al., 1978). Fungal cell wall preparations also bind to membrane fractions (Yoshikawa et al., 1983). Further evidence for the interaction of plant plasma membranes and fungal hyphae is the persistent adherence of plasma membranes to intracellular hyphae following plasmolysis of infected tissue (Nozue et al., 1979) and the hypersensitive-type response of protoplasts to hyphal wall components (Doke and Tomiyama, 1980).

OTHER FACTORS WHICH DETERMINE HOST-PATHOGEN INTERACTIONS

Apart from phytoalexin production, which is one of the most studied aspects of molecular plant pathology, there are a number of other responses which show specificity in some systems, e.g., production of host-specific toxins (Gilchrist, 1983; Yoder, 1980). In this case the specificity depends on the host having complementary receptors for the toxin. The host also

Figure 4.   Diagram showing possible routes by which extracellular enzymic signals may elicit a response in a plant cell. The enzyme releases fragments of the wall matrix components. These fragments may interact with other wall components (1) to modify existing junction zones and in so doing alter the effective porosity of the walls. Alternatively, the released fragments may move through the wall (2) to a plasma membrane receptor. This may induce a cytoplasmic response (3), possibly by enzymic or ion channel activation (as occurs for animal cells). The binding of wall fragments to a membrane receptor may initiate channels in the membrane itself (4), which in turn may initiate wall modifications. None of the stages described is understood in detail. (From Clarke et al., 1975.)

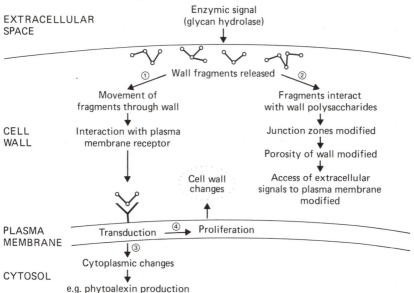

must lack the capacity to detoxify the toxin for a successful interaction to occur.

INTERACTION AT THE TISSUE LEVEL

Observations at the electron and light microscopic levels can provide valuable information on structures which may be involved in recognition events determining the progress of fungal growth in the plant. Although many host-pathogen systems have been examined microscopically, most published studies have been reported on the nature of host responses in particular tissues after inoculation under laboratory conditions. Questions relevant to recognition events, e.g., those concerning which surfaces of the host and pathogen are in contact at a particular stage of pathogenesis, have rarely been specifically addressed. Even where studies on these events are undertaken, there are very real limitations of interpretation both because of the nature of the technology and because of the physiology of the experimental systems. Most microscopic studies rely on fixed tissues and are, to a certain extent, selective. Some systems are amenable to *in vivo* observations which give another order of information, e.g., a demonstration of which plant responses occur at the right place and time to be effective in inhibiting pathogen growth (Aist, 1983). By and large, these studies are not quantitative, although some of their aspects can be quantified (Johnson et al., 1979, 1982; Aist and Israel, 1977b; Skou et al., 1984). Another difficulty is comparing different studies on the same system and assessing the significance of the differences recorded. This difficulty is compounded by the problem that, when studies on the same disease are carried out in different laboratories, it is rare for the same host cultivars and pathogen races to be used.

The host-pathogen systems themselves are quite diverse, and this diversity to some extent reflects the different nutritional patterns of the pathogens. An obligate parasite, such as a rust or a mildew, must establish haustorial connections for its nutrition and further growth in host tissue. On the other hand, facultative parasites such as *Phytophthora* spp. tend to grow intercellularly during the early stages of infection and subsequently ramify intracellularly. A further complexity is that the progress of disease in a particular host-pathogen pair may vary with such factors as inoculation procedure, maturity and general health of the plant, and environmental conditions at the time of inoculation.

In spite of all these caveats, microscopic studies can provide crucial information in planning molecular studies. Results of careful microscopy can suggest the best tissue to select and the timing of a particular interaction to give an optimum yield of protein or mRNA associated with particular recognition events. Furthermore, specific molecular probes (such as cDNA and monoclonal antibodies) produced by these studies can be used in microscopic studies (*in situ* hybridization or immunoelectron microscopy) to unite the molecular and microscopic approaches.

We now present some examples illustrating the diversity of host-pathogen interactions and the diversity of the structures in contact at the time recognition is presumed to occur. The recognition events may vary with the type of resistance expressed by the plant, i.e., nonhost or host resistance. Nonhost resistance is that expressed by a host species to all biotypes of a potential pathogen. Host resistance refers to that displayed by certain cultivars of a plant species, normally considered host to a particular pathogen, to particular biotype(s) of the pathogen (Heath, 1981). Host resistance may be governed by major, dominant resistance genes, as in gene-for-gene relationships (Ellingboe, 1981), or by the combined action of many genes, as in horizontal (or field) resistance (Van der Plank, 1982). Nonhost resistance and horizontal resistance are frequently more durable in the field than major gene host resistance, but whether this reflects fundamental differences in recognition and response mechanisms is not known.

Case 1: Bean Rust, Uromyces phaseoli var. vignae–Vigna sinensis

This is one of the few examples in which detailed studies on the interaction of both host and nonhost plants with the same fungus are available (Heath, 1974, 1977). Bean rust is an obligate biotrophic pathogen and is a member of the Basidiomycetes.

Invasion is initiated by contact between rust uredospores and leaf surfaces. The spores germinate equally well on leaves of hosts and of most nonhosts. The germ tubes grow across the leaf surface and form appressoria over stomates. At this stage some difference between the host and nonhost leaf surfaces is perceived by the pathogen, as functional appressorium formation is 20% less effective on nonhost leaves (Heath, 1974). Infection pegs then grow through the stomates, swelling to form substomatal vesicles from which infection hyphae grow intercellularly into the palisade mesophyll to form haustorial mother cells. At this point, there are clear differences between the growth patterns in hosts and nonhosts. In nonhosts there is little growth beyond the infection hypha stage, and often the infection hyphae die before the initiation of haustoria. When haustoria do start to form, their growth is inhibited. This inhibition is commonly associated with the deposition of silicon-containing material on or in plant cell walls adjacent to the haustorial mother cell (Heath, 1977, 1979) or poor adherence of the haustorial mother cell to the adjacent mesophyll cell wall (Heath, 1974). When haustoria do form in nonhosts, there is a rapid collapse of both the plant cell and the haustorium (Heath, 1974). This rapid cell necrosis associated with fungal penetration is commonly described as the hypersensitive response (Bushnell, 1982; Bailey, 1983). An individual plant species responds similarly to a range of rusts nonpathogenic to it, suggesting that these nonhost reactions are broadly based (Heath, 1977).

In contrast, haustoria formation is very successful in host cells, occurring in at least 80% of penetration sites 12–24 hr after inoculation in both compatible and incompatible combinations. In a highly resistant cultivar, haustoria are commonly encased in callose-containing material or may induce a rapid collapse of the host cell similar to that associated with nonhost reaction. Appositions containing callose also form in resistant host cell walls adjacent to infection hyphae or around points of cell penetration (Heath, 1971, 1974). Such deposits are rarely observed at nonhost penetration sites (Heath, 1974).

In susceptible hosts, haustorial mother cells, haustoria, and intercellular mycelia continue to develop until a small area of the leaf is packed with fungus and uredospores begin to develop beneath the epidermis.

This study illustrates that, in both nonhosts and highly resistant host cultivars, there are defined stages at which pathogenesis is arrested. Each stage may be described in terms of the interacting plant and fungal surfaces: the wall of the infection hypha or haustorial mother cell with the middle lamella or cell wall of the leaf mesophyll, or the plant plasma membrane with the developing haustorium. Interaction (recognition) at one or more of these stages leads to host cell wall modification, death of the infection hyphae, or collapse and death of both the host cell and the hyphae. We have little detailed molecular information on the nature of these surfaces; bean leaf mesophyll cell walls have not been isolated and analyzed, although cell wall analyses of parenchyma cells of runner bean pods are available (O'Neill and Selvendran, 1980). These analyses may or may not be similar to those of leaf mesophyll cell walls and give us only limited insight into the nature and organization of the molecular species present. The same is true of the components of the middle lamella which is difficult to analyze directly; most of our limited information comes from microscopic studies using specific cytochemical probes. From these studies, we know that the middle lamella in some tissues contains arabinogalactan-proteins (Clarke et al., 1979) and pectic polysaccharides (Roland and Vian, 1981), and there are probably a range of other components which are not detected by the probes available. Similar limitations apply to analyses of fungal cell walls. Although the major polymeric components of the various taxonomic groups of fungi are known (Table 1), the structures of minor components are poorly understood. There is limited analytical data comparing the wall composition at different stages in the life cycle of a particular fungus, and there is no clear picture of the molecular changes which occur during the transition from one stage to another (for a review see Wessels and Sietsma, 1981). One illustration that differences in surface components do occur at different stages is the study by Mendgen and colleagues (1985) who germinated spores on collodion membranes and demonstrated that the surfaces of the spores, appressoria, and infection hyphae differed in their patterns of lectin binding. This indicates that differences in the surface components of the saccharides are expressed at different stages of fungal growth *in vitro* (Fig. 5).

This type of host-pathogen interaction is further complicated by the fact that, although uredospores are responsible for secondary disease cycles during a particular growing season, basidiospores arise from overwintering spores in the spring and serve as the next source of initial inoculum. Hyphae from these basidiospores enter the host by a route different from that of uredospores. In the interaction of *P. vulgaris* with

Figure 5. Infection structures of *Puccinia coronata* spread on a membrane after incubation with FITC-WGA. Germ tube (gt) and appressorium (a) exhibit bright fluorescence. The substomatal vesicle (sv) and the infection hypha (ih) show no fluorescence. Magnification: × 350. (From Mendgen et al., 1985.)

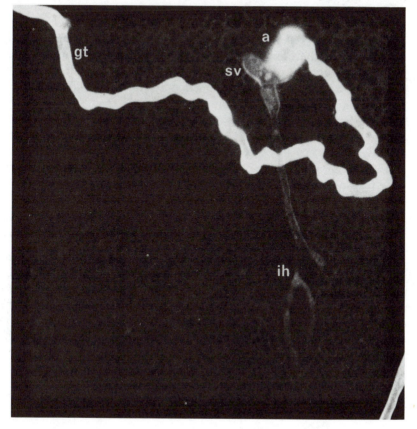

Uromyces appendiculatus var. *appendiculatus* (causing bean rust), the basidiospores germinate and penetrate the leaf epidermis directly rather than entering via the stomates (Agrios, 1978; Gold and Mendgen, 1984a,b). Hyphae grow intracellularly into other epidermal cells and intercellularly to invade the mesophyll cells. No haustorial mother cells, or infection hyphae as such, are formed. Papillae form in epidermal cells, but cell death usually does not occur in host tissues (Gold and Mendgen, 1984a,b), in keeping with the biotrophic nature of the fungus. There are no studies on the cytology of nonhost interactions with basidiospores or on interactions involving resistant hosts to compare with this study on a compatible interaction. However, the important point is that there may be different patterns of pathogen growth within the host plant for pathogens like rust fungi with complex life cycles. The different penetration pathways imply different recognition mechanisms.

From these studies it appears that:

1. Nonhost and host resistance are expressed at different stages of invasion. In nonhosts, fungal growth is arrested mainly at the infection hypha stage and few haustoria form; host resistance is expressed after successful haustoria formation.
2. Major points at which recognition events leading to the arrest of penetration are expressed may be defined microscopically. They involve contact of the host mesophyll cell wall, middle lamella, or plasma membrane with the infection hypha, haustorial mother cell, or haustorium of the pathogen.
3. We have little information on the chemistry of the surfaces in contact, or whether these are actively or passively involved in the recognition.
4. Different stages of the life cycle of the fungus may penetrate tissue by quite different routes.

Case 2: Barley Powdery Mildew Caused by Erysiphe graminis f.sp. hordei

Powdery mildew fungi (which are Ascomycetes) are also biotrophs but, unlike the situation in rusts (and most other pathogenic fungi), the main biomass of the fungus remains on the surface of diseased host plants (Agrios, 1978). Pathogenesis is initiated by adhesion of a conidium to the leaf surface. Up to four germ tubes can be produced on germination, and one of them usually becomes a primary germ tube which penetrates the epidermal cell wall about 3 hr after inoculation (Kunoh, 1981). The physiological function of this organ is not well understood; experimentally it translocates dyes, water, and metal ions from the host to the conidium before elongation of the appressorial germ tube, although primary germ tube formation may not be essential for appressorium formation (Kunoh, 1982). The host responds to intrusion by the primary germ tubes by forming cytoplasmic aggregates and papillae; in addition, some resistance to subsequent penetration of the epidermis by the appressorium may be induced (Woolacott and Archer, 1984). Recognition between the primary germ tube and the host epidermal cell wall presumably results in these responses.

By 10–14 hr after inoculation, one of the other germ tubes has elongated to form an appressorium, which is a thickened hyphal tip delimited by a septum at its base. Penetration of the host cell wall by the "hook" of appressorium is followed by expansion of the penetration peg to form a haustorium. The haustorium is surrounded by a membrane derived from, and continuous with, the host plasma membrane (Bushnell and Berquist, 1975). The fungal growth, as with rust fungi, is apoplastic. In contrast to the situation in rusts, however, in this fungus the extrahaustorial matrix is largely fluid (Bushnell, 1972). Secondary hyphae develop from the appressorium by 24 hr after inoculation and may form new appressoria and haustoria. In this way, an epicuticular mass develops which culminates in the production of conidiophores and conidia for the initiation of secondary disease cycles (Agrios, 1978; Spencer, 1978).

The genetics (Ellingboe and Slesinski, 1971), physiology (Gay and Manners, 1981), and cytology (Bushnell and Berquist, 1975; Aist and Israel, 1977a,b; Kunoh, 1981, 1982) of this interaction have been intensely studied. In contrast to the pathogenesis patterns of some fungi, such as the *Phytophthora* spp., the cytology of the diseases caused by powdery mildews is relatively simple and is a favored model system for host-pathogen studies.

We now discuss two examples in which the expression of resistance is genetically controlled, in one case by the *ML-O* locus and in the other by the *MLa* locus.

Nature of Resistance Governed by the ML-O Locus. Resistance conditioned by the *ML-O* locus is recessively inherited, is without race specificity (Jørgensen, 1977), and is expressed during the penetration phase (Jørgensen and Mortensen, 1977). Coincident with the expression of resistance at penetration is the formation of cytoplasmic aggregates and papillae (Skou et al., 1984). Papillae and cytoplasmic aggregates also form in the susceptible (ML-O) isolines, but there is evidence that they form later than in resistant (ml-o) lines (Skou et al., 1984). The papillae of ml-o lines are also larger (Skou et al., 1984; Stolzenburg, 1983) and stain more intensely with lacmoid (Skou et al., 1984.) These larger papillae formed in resistant isolines may be involved in the expression of resistance (Stolzenburg et al., 1984a,b). The evidence for this suggestion is that, when the formation of papillae and the accompanying cytoplasmic aggregates is prevented by centrifugation or heat shock treatment at the time fungal penetration is occurring, the penetration efficiency of the fungus into resistant (ml-o) isolines is significantly increased over control values. The corresponding penetration efficiency into susceptible (ML-O) isolines is not significantly altered (Stolzenburg, 1983; Stolzenburg et al., 1984b), indicating that in these plants the smaller, later-formed papillae do not affect fungal growth. These results imply that resistance is a function of the speed of expression of the host response mechanisms, dependent on contact and recognition, between the appressorium or penetration peg and the epidermal wall.

   Resistance Conditioned by the MLa Locus. In leaves attacked by *E. graminis* f.sp. *hordei*, the *MLa* locus confers resistance at two points in pathogenesis, the penetration phase and the stage of haustorium maturation of germlings which survive the penetration phase (Sherwood and Vance, 1982). In contrast, coleoptiles do not show such a pronounced two-point expression of resistance. Few germlings are stopped at the penetration stage; most penetrate successfully and induce hypersensitive cell collapse in tissues bearing the *MLa* (resistant) allele (Bushnell, 1982). Living, infected coleoptiles can be observed at the light microscope level, as partially dissected coleoptiles are optically clear. The critical host responses observed by this method are cessation of cytoplasmic streaming followed by disorganization and collapse of the coleoptile epidermal cells about 18 hr after inoculation (Bushnell, 1981; Edwards, 1975). Treatment of the tissues with cytochalasin B inhibits both cyclosis and cell death, as well as the expression of resistance (Hazen and Bushnell, 1983), suggesting the possibility of a casual relationship. Thus the critical recognition events leading to resistance in coleoptiles apparently occur at the time of haustorium formation.

   These studies illustrate:

1. The interactions of the plant and pathogen in powdery mildew diseases are cytologically relatively simple. The physiological role of the primary germ tube is not unequivocally established; it may function to signal maturation of the appressorial germ tube.
2. Genetic control of resistance may be expressed at different stages of pathogenesis in different tissues. Resistance controlled by the *ML-O* locus is expressed during the penetration of leaves, while that controlled by the *MLa* locus may be expressed mainly during the penetration of leaves or during haustorium formation in coleoptiles.
3. Recognition may be expressed differently in systems in which resistance is under the control of different loci. There may be differential timing of the same response in resistant and susceptible lines (*ML-O* locus) or differential host responses (*MLa* locus).

Case 3: Root Rot of Alfalfa. Phytophthora megasperma f.sp. medicaginis and P. megasperma f. sp. glycinea–Medicago sativa (Alfalfa)

The genus *Phytophthora* (from the Oomycetes) contains many economically important plant pathogens (Erwin et al., 1983) which, in contrast to rusts and mildews, are facultative pathogens. *Phytophora megasperma* f.sp. *medicaginis* (Pmm) is a root-rotting pathogen of alfalfa. The invasion of alfalfa, both by a pathogen (Pmm) and by a nonpathogen, *Phytophora megasperma* f.sp. *glycinea* (Pmg), have been studied (Miller and Maxwell, 1984a,b). Initial root penetration is similar in both the pathogen (Pmm) and the nonpathogen (Pmg). Zoospores encyst and germinate on the root surface, and germ tubes penetrate intercellularly through the epidermis.

Subsequent growth through the root cortex and stele is inter- and intracellular, and differences between growth of the pathogen and of the nonpathogen are apparent after 12 hr. Inter- and intracellular hyphae extensively colonize the roots of a susceptible host, but in resistant host and nonhost roots hyphae are confined to the epidermis and outer cortex and grow mainly intercellularly. In this example, host and nonhost resistance reactions seem to be qualitatively similar and involve plasmolysis and rapid necrosis of root cells. Nonhost resistance appears to act earlier, or be more effective than, host resistance, as colonization of root tissue by Pmg is less extensive (Miller and Maxwell, 1984a,b).

This study illustrates:

1. The infection pathway of a less specialized facultative pathogen and the complexity of the interactions leading to successful penetration. Zoospore motility, encystment, germination, and growth of hyphae are all essential to pathogenesis.   ,
2. Resistance is expressed as a restriction of intercellular growth in the epidermis and outer cortex.
3. The basis of nonhost resistance may be an earlier or more effective expression of similar mechanisms utilized in host resistance.

Case 4: Other Phytophthora-Plant Interactions

There are a number of other detailed studies on genetically defined resistance to *Phytophthora* spp., e.g., studies comparing the reactions of potato leaf and tuber tissue to the facultative biotroph *Phytophthora infestans* when resistance (and susceptibility) are governed by different genetic systems within the host (Hachler and Hohl, 1984; Hohl and Stossel, 1976; Hohl and Suter, 1976; Allen and Friend, 1983; Coffey and Wilson, 1983). The tuber studies are primarily directed toward examining host responses to fungal attack and generally involve wounded tissue so that the host-pathogen surfaces in contact do not reflect the natural penetration pathway (Allen and Friend, 1983; Hohl and Stossel, 1976; Hachler and Hohl, 1984). In addition, the response of tuber tissue to infection depends on the temperature and duration of storage (Allen and Friend, 1983). These aspects make it difficult to obtain information regarding the underlying recognition processes. The leaf studies illustrate the variability of host reactions to a particular race of *P. infestans*, probably reflecting the genetic diversity of the cultivars examined.

In cultivars possessing major gene resistance, the cellular responses of resistant and susceptible leaves are similar, but there is increased intercellular growth and sporulation in susceptible tissues compared with confinement of the pathogen in the resistant host (Hohl and Suter, 1976). Polygenic resistance is expressed somewhat differently at the cellular level and involves rapid necrosis of epidermal cells, with the fungus being contained in the epidermis or underlying palisade mesophyll cells (Coffey and Wilson, 1983).

The interaction of *P. megasperma* f. sp. *glycinea* with soybean has also been examined in detail (Stossel et al., 1980). The importance of the experimental tissue is again illustrated; in this case, etiolated hypocotyls show different reactions to the pathogen depending on the age of the tissue. The lower part of the hypocotyl is resistant to both compatible and incompatible races of the pathogen, while the younger tissue shows the expected resistance to the incompatible races only (Lazarovits et al., 1980). In this young tissue, fungal penetration proceeds via germ tubes which form appressoria, and then infection hyphae grow intercellularly through the epidermis to the underlying cell layers. Subsequent growth is mainly intercellular in the middle lamellae, although the incompatible race often grows within the plant cell wall. Both races form haustoria, but those of the incompatible race are frequently surrounded by necrotic host cytoplasm, suggesting that host cell death occurred before successful invasion (Stossel et al., 1981). In an incompatible interaction, fungal growth is mainly confined to the outer cell layers, and the cells in contact with incompatible hyphae are characteristically necrotic (Stossel et al., 1981). This study also illustrates that, while the expression of resistance and susceptibility may be consistent on a whole-tissue level, individual cells respond differently to contact with the pathogen. Not all cells in the compatible interaction are penetrated by the fungus, and some hyphae penetrate further than others in the incompatible interaction.

A final example of the diversity of this group of pathogens is a study on *Phytophthora cinnamomi*. This fungus has a wide host range and causes disease in many dicotyledonous plants. Although some host species show a greater degree of field tolerance than others, there are no genetically defined systems for study of the disease. Grasses normally survive in infested areas, and a detailed study on the interaction of this fungus with *Zea mays* was undertaken to determine how nonhost plants respond to infection to allow survival of the plant (Hinch et al., 1985; Wetherbee et al., 1985). Surprisingly, the fungus penetrated all the tissues of the root, even growing into the stele.

The fungus adopted different preferred patterns of growth through different tissues. In the epidermis, hyphae grow intercellularly during the first contact with any cell. On second or subsequent contacts with a cell, hyphae grow preferentially between the cell wall and plasma membrane of the affected cell rather than through the middle lamella. Within the outer cortex, both intercellular and intracellular growth occurs. In the inner cortex, growth is preferentially intercellular through the air spaces. In the stele the pattern changes again, and intracellular growth is most common. The response varies with the cell type. In the epidermis, inner cortex, and stele, larger papillae may be formed. In the outer cortex, papillae are less common, and much smaller, and thin, diffuse layers of electron-lucent material are characteristically formed. Thus no single recognition event can be identified which leads to survival of the nonhost. Indeed, as the fungus grows into the vascular tissue, no mechanism for survival is observed. It has been suggested that the plant probably survives by slowing down the invasion by cellular mechanisms and growing new lateral roots rapidly enough to overcome loss of function in other roots caused by infection.

These studies illustrate:

1. The stage of development and the condition of the experimental tissue at the time of inoculation may influence the outcome of the interaction.
2. Individual cells may respond differently to contact with the pathogen.
3. Plants may express different strategies for surviving an attack by *Phytophthora* spp. Host resistance may be a slowing down of the growth pattern seen is susceptible hosts; nonhost resistance may involve different mechanisms at the cellular level, such as more rapid necrosis of cells and containment of the infection. It may also involve the slowing down of an infection, coupled with rapid growth of new functional tissue to overcome loss of function in diseased tissue.

RELATING RECOGNITION EVENTS DEFINED CYTOLOGICALLY TO THE GENETICS OF HOST-PATHOGEN INTERACTIONS

Based on a consideration of the cytological studies, it is difficult to draw many generalizations regarding recognition events at the tissue level. This

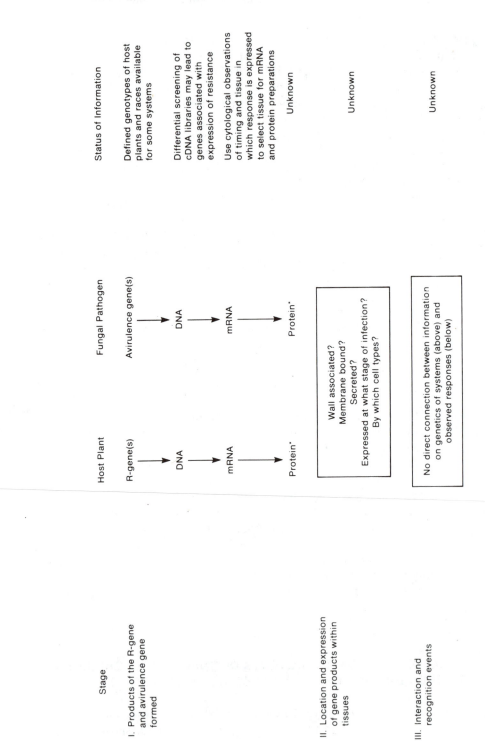

Stage	Host Plant	Fungal Pathogen	Status of Information
I. Products of the R-gene and avirulence gene formed	R-gene(s) → DNA → mRNA → Protein*	Avirulence gene(s) → DNA → mRNA → Protein*	Defined genotypes of host plants and races available for some systems Differential screening of cDNA libraries may lead to genes associated with expression of resistance Use cytological observations of timing and tissue in which response is expressed to select tissue for mRNA and protein preparations
II. Location and expression of gene products within tissues	Wall associated? Membrane bound? Secreted? Expressed at what stage of infection? By which cell types?		Unknown
III. Interaction and recognition events	No direct connection between information on genetics of systems (above) and observed responses (below)		Unknown

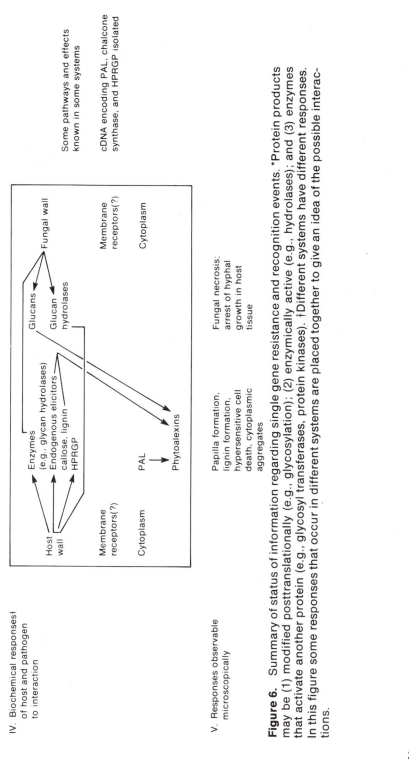

IV. Biochemical responses† of host and pathogen to interaction

Host wall → Enzymes (e.g., glycan hydrolases) / Endogenous elicitors / callose, lignin / HPRGP

Membrane receptors(?)
Cytoplasm → PAL → Phytoalexins

Glucans → Fungal wall
Glucan hydrolases

Membrane receptors(?)
Cytoplasm

Some pathways and effects known in some systems

cDNA encoding PAL, chalcone synthase, and HPRGP isolated

V. Responses observable microscopically

Papilla formation, lignin formation, hypersensitive cell death, cytoplasmic aggregates

Fungal necrosis; arrest of hyphal growth in host tissue

Figure 6. Summary of status of information regarding single gene resistance and recognition events. *Protein products may be (1) modified posttranslationally (e.g., glycosylation); (2) enzymically active (e.g., hydrolases); and (3) enzymes that activate another protein (e.g., glycosyl transferases, protein kinases). †Different systems have different responses. In this figure some responses that occur in different systems are placed together to give an idea of the possible interactions.

243

is because:

1. Different types of pathogenic fungi establish different types of nutritive and cellular relationships with their hosts.
2. The growth pathway of a particular fungus within host tissue may vary with such factors as the stage in the life cycle of the fungus which provides the inoculum, the particular host organ or tissue invaded, and the stage of maturity of the host tissue.
3. Resistance to pathogens may involve a number of different strategies which may or may not be common to the expression of host and nonhost resistance to a particular pathogen.
4. Resistance controlled by different gene loci may be expressed at different stages of infection in different tissues.
5. We have incomplete knowledge of the molecular components of the interacting surfaces and of the role these surface components play in recognition leading to restriction of fungal growth in resistant or nonhost plants.

Implicit in a discussion of recognition events in host-pathogen interactions is the idea that the primary recognition event is an interaction of host resistance gene products with fungal avirulence gene products. As no product of a host resistance gene (or a fungal avirulence gene) has yet been identified, we are not able to connect the genetic information with the cytological and biochemical evidence indicating that wall components play active roles in host-pathogen interactions.

Ultimately, descriptions of resistance and avirulence genes at the molecular level will allow studies on the nature and control of the expression of resistance. This will lead to research on how the gene products behave with respect to the surface components at host-pathogen interface(s). There is a sequential series of interactions in any particular host-pathogen system involving different stages of fungal growth and different host cell types. Each of these interactions may constitute a recognition event and have a response which may condition to some extent the outcome of the next interaction.

The technology needed to make progress in this field of research is now available. By careful choice of tissues in genetically well-defined systems, it will be possible to clone genes associated with the expression of resistance, although ascribing functions to the products of these genes may be a difficult task. With the availability of such cloned genes, the gulf between work on the classical genetics of host-pathogen interactions and observations of biochemical and cytological responses associated with resistance will gradually narrow. A scheme summarizing the state of information regarding single-gene resistance and the recognition mechanisms involved is presented in Fig. 6.

Work is currently progressing from two directions—one involving classical genetics using near-isogenic lines differing in a single resistance gene and the other involving the cloning of genes for biochemically observed effects such as phytoalexin and cell wall glycoprotein production. Various cDNA clones encoding phenylalanine ammonia lyase, chalcone synthase (Kuhn et al., 1983; Reimold et al., 1983), and cell wall glycoprotein (J. Varner, personal communication) are now available. The techniques of *in situ* hybridization of cDNA probes with mRNA in infected tissue now provide a link between classical cytology and molecular genetics. In the immediate future we can expect rapid progress in our understanding of the genetic basis of disease resistance and of the recognition events involved as the technology of molecular biology is applied to systems carefully defined genetically and cytologically.

REFERENCES

AGRIOS, G. N. 1978. *Plant Pathology*, Academic Press, New York.

AIST, J. R. 1983. Structural responses as resistance mechanisms, in: *The Dynamics of Host Defence* (J. A. Bailey and B. J. Deverall, eds.), pp. 33–70, Academic Press, New York.

AIST, J. R., AND ISRAEL, H. W. 1977a. Effects of heat shock inhibition of papilla formation on compatible host penetration by two obligate parasites, *Physiol. Plant Pathol.* 10:13–20.

AIST, J. R., AND ISRAEL, H. W. 1977b. Papilla formation: Timing and significance during penetration of barley coleoptiles by *Erysiphe graminis hordei*, *Phytopathology* 67(4):455–461.

ALBERTS, B., BRAY, D., LEWIS, J., RAFF, M. ROBERS, K., AND WATSON, J. D. 1983. *Molecular Biology of the Cell*, Garland, New York.

ALLEN, F. H. E., AND FRIEND, J. 1983. Resistance of potatoe tubers to infection by *Phytophthora infestans*: A structural study of haustorial encasement, *Physiol. Plant Pathol.* 22:285–292.

BACIC, A., AND CLARKE, A. E. 1983. The cell surface in plant recognition phenomena, in: *Vegetative Compatibility Responses in Plants* (R. Moore, ed.), pp. 139–160, Baylor University Press, Waco, Texas.

BAILEY, J. A. 1983. Biological perspectives of host-pathogen interactions, in: *The Dynamics of Host Defence* (J. A. Bailey and B. J. Deverall, eds.), pp. 1–32, Academic Press, New York.

BELL, A. A. 1981. Biochemical mechanisms of disease resistance. *Annu. Rev. Plant Physiol.* 32:21–81.

BOWLES, D. J. 1982. Membrane glycoproteins, in: *Encyclopaedia of Plant Physiology*, Vol. 13A (A. Pirson and M. H. Zimmerman, eds.), pp. 584–597, Springer-Verlag, Berlin.

BROWN, M. S., ANDERSON, R. G. W., AND GOLDSTEIN, J. L. 1983. Recycling receptors: The round trip itinerary of migrant membrane proteins, *Cell* **32**:663–667.

BRUCE, R. J., AND WEST, C. A. 1982. Elicitation of casbene synthetase activity in castor bean: The role of pectic fragments of the plant cell wall in elicitation by a fungal endopolygalacturonase, *Plant Physiol.* **69**:1181–1188.

BUSHNELL, W. R. 1972. Physiology of fungal haustoria, *Annu. Rev. Phytopathol.* **10**:151–176.

BUSHNELL, W. R. 1981. Incompatibility conditioned by the MLa gene in powdery mildew of barley: The halt in cytoplasmic streaming, *Phytopathology* **71**:1062–1066.

BUSHNELL, W. R. 1982. Hypersensitivity in rusts and powdery mildews, in: *Plant Infection* (Y. Asada, W. R. Bushnell, S. Ouchi, and C. P. Vance, eds.), p. 362, Japan Scientific Societies Press, Tokyo, and Springer-Verlag, Berlin.

BUSHNELL, W. R., AND BERQUIST, S. E. 1975. Aggregation of host cytoplasm and the formation of papillae and haustoria in powdery mildew of barley, *Phytopathology* **65**:310–318.

CARPITA, N., DUBULARSE, D., MONTEZINOS, D., AND DELMER, D. P. 1979. Determination of the pore size of cell walls of living plant cells, *Science* **205**:1144–1147.

CLARKE, A. E., AND GLEESON, P. A. 1981. Molecular aspects of recognition and response in the pollen-stigma interaction, in: *Recent Advances in Phytochemistry,* Vol. 15, *The Phytochemistry of Cell Recognition and Cell Surface Interactions* (F. A. Loewus and C. A. Ryan, eds.), pp. 161–211, Plenum Press, New York.

CLARKE, A. E., AND KNOX, R. B. 1978. Cell recognition in plants, *Q. Rev. Biol.* **53**:3–28.

CLARKE, A. E., ANDERSON, R. L., AND STONE, B. A. 1979. Form and function of arabinogalactans and arabinogalactan-proteins, *Phytochemistry* **17**:521–540.

CLARKE, A. E., ABBOTT, A., MANDEL, T., AND PETTITT, J. 1980. Organization of the wall layers of the stigmatic papillae of *Gladiolus gandavensis:* A freeze fracture study, *J. Ultrastruct. Res.* **73**:269–281.

CLARKE, A. E., ANDERSON, M. A., BACIC, T., AND HARRIS, P. J. 1986. Molecular basis of cell recognition during fertilization in higher plants. *J. Cell. Sci.* in press.

COFFEY, M. D., AND WILSON, U. E. 1983. An ultrastructural study of the late-blight fungus *Phytophthora infestans* and its interaction with the foliage of two potato cultivars possessing different levels of general (field) resistance, *Can. J. Bot.* **61**:2669–2685.

DALY, J. M. 1984. The role of recognition in plant disease, *Annu. Rev. Phytopathol.* **22**:273–307.

DARVILL, A. G., AND ALBERSHEIM, P. 1984. Phytoalexins and their elicitors—A defense against microbial infection in plants. *Annu. Rev. Plant Physiol.* **35**:243–275.

DEVERALL, B. J. 1977. *Defence Mechanisms of Plants,* Cambridge Monographs in Experimental Biology, No. 19, Cambridge University Press.

DOKE, N., AND TOMIYAMA, K. 1980. Effect of hyphal wall components from *Phytophthora infestans* on protoplasts of potato tuber tissues, *Physiol. Plant Pathol.* **16**:169–176.

EDWARDS, H. H. 1975. The ultrastructure of ML-a-mediated resistance to powdery mildew infection in barley, *Can. J. Bot.* **53**:2589–2597.

ELLINGBOE, A. H. 1981. Challenging concepts in host-pathogen genetics, *Annu. Rev. Phytopathol.* **19**:125–143.

ELLINGBOE, A. H., AND SLESINSKI, R. S. 1971. Genetic control of mildew development, in: *Barley Genetics II* (R. A. Nilan, ed.), pp. 472–474, Washington State University Press, Pullman.

ERWIN, D. C., BARTNICKI-GARCIA, S., AND TSAO, P. H. (EDS.). 1983. *Phytophthora: Its Biology, Taxonomy, Ecology, and Pathology,* American Phytopathological Society, St. Paul, Minnesota.

ESQUERRE-TUGAYE, M. T., LAFITTE, C., MAZAU, D., TOPPAN, A., AND TOUZE, A. 1979. Cell surfaces in plant-microorganism interactions. II Evidence for the accumulation of hydroxyproline-rich glycoproteins in the cell wall of diseased plants as a defence mechanism, *Plant Physiol.* **64**:320–326.

FINCHER, G. B., STONE, B. A., AND CLARKE, A. E. 1983. Arabinogalactan-proteins—Structure, biosynthesis and function, *Annu. Rev. Plant Physiol.* **34**:47–70.

GAY, J. L., AND MANNERS, J. M. 1981. Transport of host assimilates to the pathogen, in: *Effects of Disease on the Physiology of the Growing Plant* (P. G. Ayres, ed.), pp. 85–100, Cambridge University Press.

GILCHRIST, D. G., 1983. Molecular modes of action, in: *Toxins and Plant Pathogenesis* (J. M. Daly and B. J. Deverall, eds.), pp. 81–130, Academic Press, New York.

GILCHRIST, D. G., AND YODER, O. C. 1984. Genetics of host-parasite systems: A prospectus for molecular biology, in: *Plant-Microbe Interactions: Molecular and Genetic Perspectives,* Vol. 1 (T. Kosuge and E. W. Nester, eds.), pp. 69–92, Macmillan, New York.

GOLD, R. E., AND MENDGEN, K. 1984a. Cytology of basidiospore germination, penetration and early colonization of *Phaseolus vulgaris* by *Uromyces appendiculatus* var. *appendiculatus, Can. J. Bot.* **62**:1989–2002.

GOLD, R. E., AND MENDGEN, K. 1984b. Vegetative development of *Uromyces appendiculatus* var. *appendiculatus* in *Phaseolus vulgaris, Can. J. Bot.* **62**:2003–2010.

HACHLER, H., AND HOHL, H. R. 1984. Temporal and spatial distribution patterns of collar and papillae wall appositons in resistant and susceptible tuber tissue of *Solanum tuberosum* infected by *Phytophthora infestans, Physiol Plant Pathol.* **24**:107–118.

HAMMERSCHMIDT, R., LAMPORT, D. T. A., AND MULDOON, E. P. 1984. Cell wall hydroxyproline enhancement and lignin deposition as an early event in the resistance of cucumber to *Cladosporium cucumerium, Physiol. Plant Pathol.* **24**:43–47.

HARRIS, P. J., ANDERSON, M. A., BACIC, A., AND CLARKE, A. E. 1984. Cell-cell recognition in plants with special reference to the pollen-stigma interaction, in: *Surveys of Plant Molecular and Cellular Biology,* Vol. 1 (B. J. Miflin, ed.), pp. 161–203, Oxford University Press.

HAZEN, B. E., AND BUSHNELL, W. R. 1983. Inhibition of the hypersensitive reaction in barley to powdery mildew by heat shock and cytocalasin B, *Physiol. Plant Pathol.* **23:**421–438.

HEATH, M. C. 1971. Haustorial sheath formation in cowpea leaves immune to rust infection, *Phytopathology* **61:**383–388.

HEATH, M. C. 1974. Light and electron microscope studies of the interactions of host and non-host plants with cowpea rust—*Uromyces phaseoli* var. *vignae, Physiol. Plant Pathol.* **4:**403–414.

HEATH, M. 1977. A comparative study of non-host interactions with rust fungi, *Physiol. Plant Pathol.* **10:**73–88.

HEATH, M. C. 1979. Partial characterization of the electron-opaque deposits formed in the non-host plant, French bean, after cowpea rust infection, *Physiol. Plant Pathol.* **15:**141–148.

HEATH, M., 1980. Reactions of nonsuscepts to fungal pathogens, *Annu. Rev. Phytopathol.* **18:**211–236.

HEATH, M. C. 1981. A generalized concept of host-parasite specificity, *Phytopathology* **71:**1121–1123.

HENRY, R. J., SCHIBECI, A., AND STONE, B. A. 1983. Localization of β-glucan synthases on the membranes of cultured *Lolium multiflorum* (ryegrass) endosperm cells, *Biochem. J.* **209:**371–395.

HESLOP-HARRISON, J. 1983. Self-incompatibility: Phenomenology and physiology, *Proc. Roy. Soc. London Ser. B.* **218:**371–395.

HINCH, J. M., AND CLARKE, A. E. 1982. Callose formation in *Zea mays* as a response to infection with *P. cinnamoni, Physiol. Plant Pathol.* **21:**113–124.

HINCH, J. M., WETHERBEE, R., MALLETT, J. E., AND CLARKE, A. E. 1985. Response of *Zea mays* roots to infection with *Phytophthora cinnamomi.* I. The epidermal layer, *Protoplasma* **126:**178–187.

HOGENBOOM, N. G. 1983. Bridging a gap between related fields of research: Pistil-pollen relationships and the distinction between incompatibility and incongruity in non-functioning host-parasite relationships, *Phytopathology* **73:**381–383.

HOHL, H. R., AND STOSSEL, P. 1976. Host-parasite interfaces in a resistant and a susceptible cultivar of *Solanum tuberosum* inoculated with *Phytophthora infestans:* Tuber tissue, *Can. J. Bot.* **54:**900–912.

HOHL, H. R., AND STOSSEL, P. 1976. Host-parasite interfaces in a resistant and a susceptible cultivar of *Solanum tuberosum* inoculated with *Phytophthora infestans:* Leaf tissue, *Can. J. Bot.* **54:**1956–1970.

HORNBERG, C., AND WEILER, E. W. 1984. High affinity sites for abscisic acid on the plasmalemma of *Vicia faba* guard cells, *Nature* **310:**321–324.

HUNTER, T. 1984. Growth factors: The epidermal growth factor receptor gene and its product, *Nature* **311:**414–416.

JACOBS, M., AND GILBERTS, S. F. 1983. Basal localisation of the presumptive auxin carrier in pea stem cells, *Science* **220:**1297–1300.

Jin, D. F., and West, C. A. 1984. Characteristics of galacturonic acid oligomers as elicitors of casbene synthetase activity in castor bean seedlings, *Plant Physiol.* **74**:989–992.

Johnson, L. E. B., Bushnell, W. R., and Zeyen, R. J. 1979. Binary pathways for analysis of primary infection and host response in populations of powdery mildew fungi, *Can. J. Bot.* **57**:497–511.

Johnson, L. E. B., Bushnell, W. R., and Zeyen, R. J. 1982. Defense patterns in non-host higher plant species against two powdery mildew fungi. 1. Monocotyledonous species, *Can. J. Bot.* **60**:1068–1083.

Jørgensen, J. H. 1977. Spectrum of resistance conferred by ml-o, *Euphytica* **26**:55–62.

Jørgensen, J. H., and Mortensen, K. 1977. Primary infection by *Erysiphe graminis* f.sp. *hordei* of barley mutants with resistance in the ml-o locus, *Phytopathology* **67**:678–685.

Kauss, H. 1985. Role of Ca^{2+} in callose formation, *Proceedings of the John Innes Symposium,* in press.

Keen, N. T. 1982. Specific recognition in gene-for-gene host-parasite systems, *Adv. Plant Pathol.* **1**:35–82.

Kendra, D. F., and Hadwiger, L. E. 1984. Characterization of the smallest chitosan oligomer that is maximally antifungal to *Fusarium solani* and elicits pisatin formation in *Pisum sativum, Exp. Mycol.* **8**:276–281.

Kohle, H., Young, D. H., and Kauss, H. 1984. Physiological changes in suspension-cultured soybean cells elicited by treatment with chitosan, *Plant Sci. Lett.* **33**:221–230.

Kojima, M., Kawakita, K., and Uritani, I. 1982. Studies on a factor in sweet potato root which agglutinates spores of *Ceratocystis fimbriata,* Black rot fungus, *Plant Physiol.* **69**:474–478.

Kosuge, T. 1981. Carbohydrates in plant-pathogen interactions, in: *Encyclopedia of Plant Physiology,* New Series, Vo. 13B, *Plant Carbohydrates II: Extracellular Carbohydrates* (W. Tanner and F. A. Loewus, eds.), pp. 632–652, Springer-Verlag, Berlin.

Kuhn, D. N., Chappell, J., and Hahlbrock, K. 1983. Identification and use of cDNAs of phenylalanine ammonia-lyase and 4-coumarate: CoA ligase mRNAs in studies of the induction of phytoalexin biosynthetic enzymes in cultured parsley cells, in: *Structure and Function of Plant Genomes* (O. Ciferri and L. Dure, eds.), pp. 329–336, Plenum Press, New York.

Kunoh, H. 1981. Early stages of infection process of *Erysiphe graminis* on barley and wheat, in: *Microbial Ecology of the Phylloplane* (J. P. Blakeman, ed.), pp. 85–101, Academic Press, London.

Kunoh, H. 1982. Primary germ tubes of *Erysiphe graminis* conidia, in: *Plant Infection: The Physiological and Biochemical Basis* (Y. Asada, W. R. Bushnell, S. Ouchi, and C. P. Vance, eds.), pp. 45–59, Japan Scientific Societies Press, Tokyo, and Springer-Verlag, Berlin.

Larkin, P. J. 1981. Plant protoplast agglutination and immobilization, in: *Recent Advances in Phytochemistry,* Vol. 15, *The Phytochemistry of Cell Recogni-*

tion and Cell Surface Interactions (F. A. Loewus, and C. A. Ryan, eds.), pp. 135–160, Plenum Press, New York.

LAWRENCE, M. J., MARSHALL, D. F., CURTIS, V. E., AND FEARON, C. H. 1985. Gametophytic self-incompatibility re-examined: A reply, *Heredity* **54**:131–138, in press.

LAZAROVITS, G., STOSSEL, P. AND WARD, E. W. B. 1980. Age-related changes in specificity and glyceollin production in the hypocotyl reaction of soybeans to *Phytophthora megasperma* var. *sojae*, *Phytopathology* **70**:94–97.

LEACH, J. E., CANTRELL, M. A., AND SEQUIERA, L. 1982. Hydroxyproline-rich bacterial agglutinin from potato, *Plant Physiol.* **70**:1353–1358.

LEE, S. C., AND WEST, C. A. 1981. Polygalacturonase from *Rhizopus stolonifer*, an elicitor of casbene synthetase activity in castor bean (*Ricinus communis* L.) seedlings, *Plant Physiol.* **67**:640–645.

LIPPINCOTT, J. A., AND LIPPINCOTT, B. B. 1984. Concepts and experimental approaches in host-microbe recognition, in: *Plant-Microbe Interactions: Molecular and Genetic Perspectives*, Vol. 1 (T. Kosuge and E. W. Nester, eds.), pp. 195–214, MacMillan, New York.

MCNEILL, M. A., AND SELVENDRAN, R. R. 1980. Methylation analysis of cell wall material from parenchymatous tissues of *Phaseolus vulgaris* and *Phaseolus coccineus*, *Carbohydr. Res.* **79**:115–124.

MCNEILL, M., DARVILL, A. G., FRY, S. C., AND ALBERSHEIM, P. 1984. Structure and function of the primary cell walls of plants, *Annu. Rev. Biochem.* **53**:624–663.

MELLON, J. E., AND HELGESON, J. P. 1982. Interactions of a hydroxyproline-rich glycoprotein from tobacco callus with potential pathogens, *Plant Physiol.* **70**:401–405.

MENDGEN, K., LANGE, M., AND BRETSCHNEIDER, K. 1985. Quantitative estimation of the surface carbohydrates on the infection structures of rust fungi with enzymes and lectins, *Arch. Microbiol.* **140**:307–311.

MILLER, S. A., AND MAXWELL, D. P. 1984a. Light microscope observations of susceptible, host resistant, and nonhost resistant interactions of alfalfa with *Phytophthora megasperma*, *Can. J. Bot.* **62**:100–116.

MILLER, S. A., AND MAXWELL, D. P. 1984b. Ultrastructure of susceptible, host resistant, and nonhost resistant interactions of alfalfa with *Phytophthora megasperma*, *Can. J. Bot.* **62**:117–128.

MULCAHY, D. L., AND MULCAHY, G. B. 1983. Gametophytic self-incompatibility reexamined, *Science* **220**:1247–1251.

NESTLER, E. J., AND GREENGARD, P. 1983. Protein phosphorylation in the brain, *Nature* **305**:583–588.

NOZUE, M., TOMIYAMA, K., AND DOKE, K. 1979. Evidence for adherence of host plasmalemma to infecting hyphae of both compatible and incompatible races of *Phytophthora infestans*, *Physiol. Plant Pathol.* **15**:111–115.

O'NEILL, M. A., AND SELVENDRAN, R. R. 1980. Methylation analysis of cell-wall material from parenchymatous tissues of *Phaseolus vulgaris* and *Phaseolus coccineus*, *Carbohydr. Res.* **79**:115–124.

PETERS, B. M., CRIBBS, D. H., AND STELZIG, D. A. 1978. Agglutination of plant protoplasts by fungal cell-wall glucans, *Science* **201**:364–365.

PFEFFER, S., AND ULLRICH, A. 1985. Is the precursor a receptor? *Nature* **313**:184.

POLYA, G. M., MICUCCI, V., RAE, A. L., HARRIS, P. J., AND CLARKE, A. E. 1986. Calcium-dependent protein phosphorylation in germinated pollen of *Nicotiana alata,* an ornamental tobacco, *Plant Physiol.* in press.

RALL, L. B., SCOTT, J., BELL, G. I., CRAWFORD, R. J., PENSCHOW, J. D., NIALL, H. D., AND COGHLAN, J. P. 1985. Mouse prepro-epidermal and growth factor synthesis by the kidney and other tissues, *Nature* **313**:228–231.

REES, D. A. 1977. *Polysaccharide Shapes,* Outline Studies in Biology, Chapman and Hall, London.

REIMOLD, U., KROGER, M., KREUZALER, F., AND HAHLBROCK, K. 1983. Coding and 3′ non-coding nucleotide sequence of chalcone synthase mRNA and assignment of amino acid sequence of the enzyme, *EMBO J.* **2**(10):1801–1805.

ROLAND, J. -C., AND VIAN, B. 1981. Use of purified endopolygalacturonase for a topochemical study of elongating cell walls at the untrastructural level, *J. Cell Sci.* **40**:333–343.

RYAN, C. A. 1984. Systemic responses to wounding, in: *Plant-Microbe Interactions: Molecular and Genetics Perspectives,* Vol. 1 (T. Kosuge and E. W. Nester, eds.), pp. 307–320, Macmillan, New York.

SEQUIERA, L. 1978. Lectins and their role in host-pathogen specificity, *Annu. Rev. Phytopathol.* **16**:453–481.

SEQUIERA, L. 1984. Plant bacterial interactions, in: *Encyclopaedia of Plant Physiology,* Vol. 17 (H. F. Linskens and J. Heslop-Harrison, eds.), pp. 187–207.

SHERWOOD, R. T., AND VANCE, C. P. 1982. Initial events in the epidermal layer during penetration, in: *Plant Infection: The Physiological and Biochemical Basis* (Y. Asada, W. R. Bushnell, S. Ouchi, and C. P. Vance, eds.), pp. 362, Japan Scientific Societies Press, Tokyo, and Springer-Verlag, Berlin.

SKOU, J. P., JØRGENSEN, J. H., AND LILHOLT, U. 1984. Comparative studies on callose formation in powdery mildew compatible and incompatible barley, *Phytopathol. Z.* **109**:147–168.

SOLIDAY, C. L., FLURKEY, W. H., OKITA, T. W., AND KOLATTUKUDY, P. E. 1984. Cloning and structure determination of cDNA for cutinase, an enzyme involved in fungal penetration in plants, *Proc. Nat. Acad. Sci. USA* **81**:3939–3943.

SPENCER, D. M. 1978. *The Powdery Mildews,* Academic Press, London.

STOLZENBURG, M. C. 1983. The role of localized host cytoplasmic responses in barley powdery mildew resistance conditioned by the ml-o gene, M. S. Thesis, Department of Plant Pathology, Cornell University, Ithaca, New York.

STOLZENBURG, M. C., AIST, J. R., AND ISRAEL, H. W. 1984a. The role of papillae in resistance to powdery mildew conditioned by the ML-O gene in barley. I. Correlative evidence. *Physiol. Plant Pathol.* **25**:337–346.

STOLZENBURG, M. C., AIST, J. R., AND ISRAEL, H. W. 1984b. The role of papillae in resistance to powdery mildew conditioned by the ML-O gene in barley. II. Experimental evidence, *Physiol. Plant Pathol.* **25**:347–361.

STOSSEL, P., LAZAROVITS, G., AND WARD, E. W. B. 1980. Penetration and growth of compatible and incompatible races of *Phytophthora megasperma* var. *sojae* in soybean hypocotyl tissues differing in age, *Can. J. Bot.* **58**:2594–2601.

STOSSEL, P., LAZAROVITS, G., AND WARD, E. W. B. 1981. Electron microscope study of race-specific and age-related resistant and susceptible reactions of soybeans to *Phytophthora megasperma* var. *sojae, Phytopathology* **71**:617–623.

TEPFER, M., AND TAYLOR, I. E. P. 1981. The permeability of plant cell walls as measured by gel filtration chromatography, *Science* **213**:761–763.

TOPPAN, A., ROBY, D., AND ESQUERRE-TUGAYE, M. T. 1982. Cell surfaces in plant-microorganism interactions. III. *In vivo* effect of ethylene on hydroxy-proline-rich glycoprotein accumulation in the cell wall of diseased plants, *Plant Physiol.* **70**:82–86.

VAN HOLST, G. J., AND VARNER, J. E. 1984. Reinforced polyproline. II. Conformation in hydroxyproline-rich cell wall glycoprotein from carrot root, *Plant Physiol.* **74**:247–251.

VAN DER PLANK, J. E. 1982. *Host-Pathogen Interactions in Plant Disease,* Academic Press, London.

WALKER-SIMMONS, M., JIN, D., WEST, C. A., HADWIGER, L., AND RYAN, C. A. 1984. Comparison of proteinase inhibitor-inducing activities and phytoalexin elicitor activities of a pure fungal endopolygalacturanase, pectic fragments, and chitosans, *Plant Physiol.* **76**:833.

WESSELS, J. G. H., AND SIETSMA, J. H. 1981. Fungal cell walls, in: *Encyclopedia of Plant Physiology,* New Series, Vol. 13B, *Plant Carbohydrates II: Extracellular Carbohydrates* (W. Tanner and F. A. Loewus, eds.), Springer-Verlag, Berlin.

WEST, C. A. 1981. Fungal elicitors of the phytoalexin response in higher plants, *Naturwissenshaft* **68**:447–457.

WETHERBEE, R., HINCH, J. M., BONIG, I., AND CLARKE, A. E. 1985. Response of *Zea mays* roots to infection with *Phytophthora cinnamoni*. II. The cortex and stele, *Protoplasma* **126**:188–197.

WHATLEY, M. H., AND SEQUIERA, L. 1981. Bacterial attachment to plant cell walls, *Recent Adv. Phytochem.* **15**:213–240.

WOOLACOTT, B., AND ARCHER, S. A. 1984. The influence of the primary germ tube on infection of barley by *Erysiphe graminis* f.sp. *hordei, Plant Pathology* **33**:225–231.

YODER, O. C. 1980. Toxins in pathogenesis, *Annu. Rev. Phytopathol.* **18**:103–129.

YOSHIKAWA, M., KEEN, N. T., AND WANG, M. C. 1983. A receptor on soybean membranes for a fungal elicitor of phytoalexin accumulation, *Plant Physiol.* **73**:497–506.

ZEYEN, R. J., AND BUSHNELL, W. R. 1979. Papilla response of barley epidermal cells caused by *Erysiphe graminis*: Rate and method of deposition determined by microcinematography and transmission electron microscopy, *Can. J. Bot.* **57**:898–913.

Chapter 8

Pectic Enzymes and Bacterial Invasion of Plants

Alan Collmer

HISTORICALLY, THE SEARCH for the molecular attributes that enable certain microorganisms to invade, colonize, and cause disease in plants has begun with molecules that can be readily isolated from pathogen cultures and shown to be biologically active when applied to plant tissues. Demonstration that a particular molecule is capable of contributing to a disease process typically is followed by attempts to prove its necessity. As evidence for the importance of the disease molecule accumulates, additional questions begin to emerge concerning its role in the overall biology of the pathogen, specifically, the mechanism of production of the molecule, the activity of the molecule in natural ecological niches of the pathogen, and the operation of the molecule in concert with other disease factors produced by the pathogen. Indeed, the characterized disease molecule then becomes a loose thread which aids in unraveling the entire complex of factors that distinguishes the pathogen from related saprophytes.

This progression is exemplified by research on the pectolytic *Erwinia* spp. that cause soft-rot diseases in many plants. These diseases, which typically affect fleshy plant organs, are similar to the storage rots caused by many fungi; both groups of pathogens cause maceration and death of infected tissues, and both produce abundant quantities of pectic enzymes. Jones (1909) initially reported the production of pectic enzymes by a soft-rot *Erwinia* and observed microscopically the damage to middle lamellae and cell walls that occurred in advance of the bacteria as they invaded the intercellular spaces of carrot roots and during treatment with crude pectic enzyme preparations. Brown (1915) studied the effects of a pectolytic extract of *Botrytis cinerea* on potato tuber slices and demonstrated that treatments affecting enzyme activity similarly affected both maceration and host cell killing. This suggested that a single agent accounted for both disease processes. It was not until the 1970s that advanced protein purification techniques demonstrated that the pectic enzymes produced by soft-rot erwinias could indeed account for the major symptoms of the disease (Basham and Bateman, 1975a; Garibaldi and Bateman, 1971; Mount et al., 1970; Stephens and Wood, 1975). These findings were completely in accord with the growing knowledge of the structural importance of pectic polymers in the plant primary cell wall (which retains the osmotically fragile protoplast) and the middle lamella (which cements the cells into a coherent tissue) (Bateman and Basham, 1976; Keegstra et al., 1973).

These physiological studies suggested the broad importance of pectic enzymes in diseases caused by necrotrophic pathogens and the particular importance of the enzymes in diseases caused by soft-rot erwinias. But fundamental questions remained unanswered. Would an *Erwinia* mutant unable to produce one of these plant-disintegrating enzymes still be pathogenic? Do *Erwinia* pectic enzymes differ in their enzymic properties or mode of production from pectic enzymes produced by many saprophytes (or by pathogens that cause diseases different from soft rot)? Do pectic enzymes have other, perhaps more important, roles in the life cycle of soft-rot erwinias? Finally, is a soft-rot *Erwinia*, in its pathogenic association with plants, little different from a highly pectolytic version of its well-studied relative *Escherichia coli*?

Questions of this sort became accessible with the development of a conjugational transfer system in *Erwinia chrysanthemi* by Chatterjee and Starr (1977). Since then there has been an exponential increase in genetic research on soft-rot erwinias. Ten years ago the primary research tool in *Erwinia* laboratories was isolated pectic enzymes; now it is isolated pectic enzyme genes. The purpose of this chapter is to prepare the reader for the coming tidal wave of molecular biological research on soft-rot erwinias and their pectic enzymes. Thus, the chapter will emphasize (1) soft-rot erwinias as model pathogens, which by virtue of their relatively simple host interactions, genetic manipulability, and relatedness to the intensely studied *E. coli* (Chatterjee and Starr, 1980), hold the promise of having their phytopathogencity completely elucidated at the molecular level; and (2) pectic enzymes as model disease molecules with multiple effects on plant tissues, multiple roles in the life cycles of pathogens, and complex production mechanisms which also contribute to phytopathogenicity.

SOFT-ROT ERWINIAS: TAXONOMY AND ECOLOGY

The genus *Erwinia* is comprised of a diverse group of enterobacteria assembled in part because of their ecological association with plants (Lelliott and Dickey, 1984; Starr, 1981). The highly pectolytic species in the "carotovora" (soft-rot) group (Dye, 1969) are the most closely related to coliform enterobacteria like *E. coli* (Chatterjee and Starr, 1980; Starr, 1981). The three major erwinias within this group, *Erwinia chrysanthemi, E. carotovora* subsp. *carotovora,* and *E. carotovora* subsp. *atroseptica* have somewhat different host ranges and geographical distributions, partly as a result of ecologically important factors, particularly optimum temperature for growth (Perombelon and Kelman, 1980). *Erwinia chrysanthemi* differs from the two subspecies of *E. carotovora* (which will not be differentiated in this chapter) in several taxonomic features (Dickey, 1979; Lelliott, and Dickey, 1984; Mergaert et al., 1984) and, as will be discussed below, also differs in its production of pectic enzymes in a surprising number of ways.

Although soft-rot erwinias act as opportunistic pathogens when they cause rots in damaged plant storage organs, they can also be true pathogens causing systemic infections, vascular disorders, parenchymal necroses, and latent infections in many growing plants (Perombelon, 1982). *Erwinia chrysanthemi*, in particular, reveals a degree of specificity among different strains for certain hosts (Dickey, 1979, 1981; Lacy et al., 1979; Samson and Nassan-Agha, 1979). Because soft-rot erwinias are also competitive saprophytes in the rhizosphere (Stanghellini, 1982), it is apparent that they can associate with plants in a variety of relationships. Pectic enzyme production may be important in all these ecological relationships. Thus it is noteworthy that the ability of soft-rot erwinias to compete with soil microorganisms for the utilization of pectic polymers permits enrichment of their population in soil samples amended with D-galacturonan (Burr and Schroth, 1977; Meneley and Stanghellini, 1976).

SOFT-ROT ERWINIAS: GENETIC TOOLS

A powerful array of tools now permit *Erwinia* chromosomal genes to be marked by transposons, mobilized to other *Erwinia* strains, cloned into *E. coli*, or returned from *E. coli* to the *Erwinia* chromosome. Indeed, chromosomal genetics has been better developed and explored in soft-rot erwinias, particularly *E. chrysanthemi*, than in any other plant-pathogenic bacteria (Chatterjee and Starr, 1980; Chatterjee et al., 1981a; Leary and Fulbright, 1982). Transposon mutagenesis in soft-rot erwinias has been accomplished with Tn5 (Chatterjee et al., 1983; Zink et al., 1984), Tn9 (Van Gijsegem et al., 1985a), and phage Mu derivatives (Andro et al., 1984; Jayaswal et al., 1984). Chromosomal genes have been mobilized by Hfr and R plasmids (Chatterjee, 1980; Chatterjee and Starr, 1977; Chatterjee et al., 1981a; Guimares et al., 1979; Kotoujansky et al., 1982; Lacy, 1978; Pérombelon and Boucher, 1978), by generalized transducing phage (Chatterjee and Brown, 1980; Resibois et al., 1984), and by RP4::mini-Mu (Chatterjee et al., 1985a; Schoonejans and Toussaint, 1983). These techniques, involving the manipulation and exchange of genes between *Erwinia* strains, permit classical genetic analyses and have revealed fundamental similarities in the *E. coli* and *E. chrysanthemi* chromosomal maps.

A complementary approach involving molecular cloning of *Erwinia* genes into *E. coli* promises to be particularly useful in dissecting the complex genetics underlying the encoding, regulation, and export of pectic enzymes. As shown diagrammatically in Fig. 1, production of the full complement of pectic enzymes during pathogenesis depends on a series of cis-acting factors, e.g., the *pel* genes encoding various pectate lyase (PL) isozymes (discussed below), and trans-acting factors, e.g., genes whose products mediate export (*out*), regulation (*pec*), and possibly general pathogenicity (*vir*). This system presents problems difficult to overcome using classical genetic techniques. For example, because several *pel* genes contribute to the pectolytic phenotype of soft-rot erwinias, screening populations of random mutants for loss of extracellular pectolytic capacity yields mutants deficient in trans-acting factors rather than in *pel* genes. This problem is readily overcome by an established cycle of gene manipulations (Fig. 1) involving (1) molecular cloning of the desired *Erwinia* gene into *E. coli*, (2) analysis and mutagenesis of the cloned gene in *E. coli*, (3) mobilization of the mutated gene back to *Erwinia,* and (4) exchange recombination of the mutated gene into the *Erwinia* chromosome. Two examples will illustrate the particular ease and potential of experiments involving genetic exchanges between *E. chrysanthemi* and *E. coli.*

First, *Erwinia* genes can be mobilized into *E. coli* by the elegant technique of *in vivo* cloning with the RP4::mini-Mu plasmid pULB113. This derivative of the broad-host-range conjugative plasmid RP4 carries a deleted Mu prophage whose transpositional functions cause the formation of R primes carrying large segments [up to at least 100 kilobases (kb)] of the host chromosome (Van Gijsegem and Toussaint, 1982). These R prime plasmids can be mobilized to *E. coli* by conjugation, and those carrying the desired *Erwinia* genes selected by their ability to complement appropriate *E. coli* mutations. For example, by using *E. coli* recipients

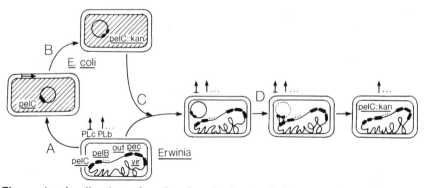

Figure 1. Application of molecular cloning techniques to the genetics of pectic enzyme production and pathogencity in soft-rot erwinias. Wild-type *Erwinia* is shown producing PL isozymes encoded by a series of *pel* genes. Other genes (*out, pec,* and *vir*) act in *trans* to permit export and regulation of PL and possible production of other pathogenicity determinants. Analysis of a given gene (e.g., *pelC*) involves (A) in vivo or in vitro cloning of the gene into *E. coli* (hatch-marked) where its expression and products can be studied in a genetic background lacking the trans-acting factors, e.g., the *out* genes permitting PL export through the outer membrane; (B) insertional inactivation of the cloned gene; (C) mobilization of the mutated gene in plasmid vector pBR322 back into *Erwinia*, where it can be maintained on the plasmid; or (D) exchanged into the chromosome by simultaneously selecting for marker resistance and vector instability. Alternative approaches and the particular opportunities that this well-established cycle of manipulations holds for *Erwinia* genetics are discussed in the text.

deficient in various genes involved in the utilization of hexuronates, Van Gijsegem and Toussaint (1983) cloned all the corresponding *E. chrysanthemi* genes by simple heterospecific matings.

Second, heterospecific matings between *E. chrysanthemi* and *E. coli* strains carrying the helper plasmids R64*drd*11 and pLVC9 [a derivative of pGJ28 (Van Haute et al., 1983) constructed by G. Warren] can be used to easily return *Erwinia* genes cloned in the universal plasmid vector pBR322 to *E. chrysanthemi* (Roeder and Collmer, 1985). Furthermore, pBR322 replicates in *E. chrysanthemi* and other erwinias (Hinton et al., 1985; Lacy and Sparks, 1979; Reverchon and Robert-Baudouy, 1985) but, as with *E. coli* (Jones et al., 1980), is rapidly lost from cultures maintained in phosphate-deficient media. This permits high frequencies of exchange recombination for marker exchange mutagenesis if the pBR322 derivative carries a selectable marker inserted in an *E. chrysanthemi* DNA fragment. As discussed below, this technique has been used to mutate a single PL isozyme gene in *E. chrysanthemi* (Roeder and Collmer, 1985).

Aside from the ease with which *Erwinia* genes can be cloned, manipulated, and returned to the *Erwinia* genome, there are additional advantages to using *E. coli* as a background against which to study *Erwinia* genes. As depicted in Fig. 1, pectic enzyme production and pathogenicity in soft-rot erwinias can be expected to involve complex interactions of cis- and trans-acting genetic elements. *Escherichia coli* provides an ideal background against which to reconstitute the relevant *Erwinia* genes and analyze their interactions. *Erwinia* genes are expressed well in *E. coli* (see section on molecular cloning of PL isozymes), and genes involved in phytopathogenicity and the utilization of pectate are apparently missing from *E. coli* (although *E. coli* is able to utilize galacturonate). Thus, for example, it may be possible to clone *out* and *pec* genes on the basis of their action *in trans* on the export or regulation of previously cloned *pel* genes in *E. coli*, as well as by more universal techniques involving the complementation of appropriate *Erwinia* mutations with cosmid libraries or the marking of target genes with transposons.

ERWINIA PECTIC ENZYMES

The Extracellular Pectic Enzyme Complex

The extracellular pectic enzymes produced by soft-rot erwinias have at least three distinguishable physiological activities. (1) They are catabolic enzymes enabling the bacteria to utilize D-galacturonan as a sole carbon source; (2) they participate in their own induction by generating assimilable products that signal the presence of an extracellular polymeric substrate; and (3) they have destructive effects on host tissues. Before exploring their role in these activities, the surprisingly complex battery of pectic enzymes excreted by soft-rot erwinias will be described.

The conversion of pectic polymers to intermediates of the glycolytic pathway is a multienzyme process. The extracellular phase begins with an attack on 1,4-glycosidic linkages within the polymer and is complete with the generation of oligogalacturonates that can be transported into the bacterial cell. As depicted in Fig. 2 (which also contains a table of enzyme ab-

Figure 2. Enzymes and intermediates involved in the degradation of pectic polymers and the induction of pectic enzymes (abbreviations tabulated below). Extracellular pectic polymers are endo-attacked by PL, PG, and PNL (if highly methoxylated) and exo-attacked by exoPG. The lyases generate an unsaturated product (see UDG). In *E. chrysanthemi* exoPG works in concert with PL to release both DG and UDG from pectate; in *E. carotovora* pectate is degraded by PL and PG. Smaller oligogalacturonates, particularly dimers, are then taken up by the cell. As shown in the intracellular pathway for DG catabolism, OGL activity is necessary for the formation of two ketodeoxy-uronates (DTH and DGH) which are implicated in PL induction. (DTH is depicted with C1 at the bottom to illustrate the isomerization step more clearly.) *(Shown on opposite page.)*

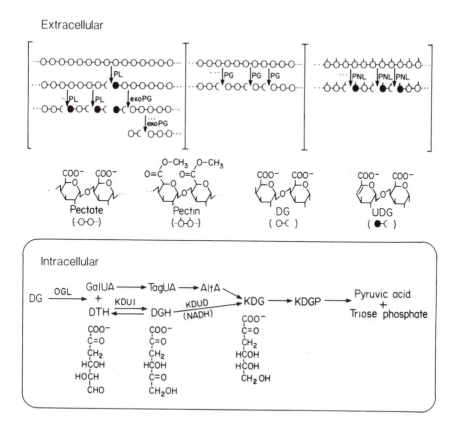

ABR.	INTERMEDIATES	ABBR.	ENZYMES
Pectate	D-Galacturonan	PL	Pectate lyase
Pectin	D-Galacturonan-methylesterified	PNL	Pectin lyase
		PG	Polygalacturonase
DG	Digalacturonate	exoPG	Exo-poly-α-D-galacturonosidase
UDG	Unsaturated digalacturonate		
GalUA	D-Galacturonate	OGL	Oligogalacturonide lyase
TagUA	D-Tagaturonate		
AltA	D-Altronate	KDUI	4-Deoxy-L-threo-5-hexosulose-uronate-ketol-isomerase ("ketodeoxyuronate iso-merase
DTH	4-Deoxy-L-threo-5 hexoseulose uronate		
DGH	3-deoxy-D-glycero-2,5-hexodiulosonate		
KDG	2-Keto-3-deoxy-D-gluconate	KDUD	2-Keto-3-deoxy-D-gluconate oxidoreductase (3-deoxy-D-glycero-2,5-hexodiulosonate dehydrogenase) ("ketodeoxyuronate dehydrogenase")
KDGP	2-Keto-3-deoxy-6-phospho-D-gluconate		

breviations), soft-rot erwinias excrete enzymes that cleave α-1,4-galacturonosyl linkages in D-galacturonans by hydrolysis (PG and exoPG) or by β-elimination (PL and PNL). The lyases have higher pH optima than the hydrolases (typically 8.5 versus 5.0) and can be specifically assayed in the presence of contaminating hydrolases because their 4,5-unsaturated reaction products absorb at 230 nm and react in the periodate–thiobarbituric acid assay (Bateman and Basham, 1976; Rombouts and Pilnik, 1980).

Extracellular pectic emzymes also differ in their action patterns. The exo-attacking enzyme, exoPG, cleaves dimers from the nonreducing end of the polymer. Endo-attacking enzymes (PL, PG, and PNL) generate a series of homologous oligomers after repeated cleavage of the polymer. Figure 2 depicts a generalized action pattern for these enzymes and shows the dimer as the smallest product. While this pattern appears accurate for *E. chrysanthemi* total extracellular PL, the pattern for individual isozymes varies (Collmer et al., 1982b); *E. carotovora* PG generates monomer galacturonate after exhaustive digestion (Nasuno and Starr, 1966); and the reaction products of *Erwinia* PNL remain to be characterized. One strain of *E. carotovora* has recently been reported to produce an extracellular exopectate lyase which degrades the polymer to unsaturated di- and trigalacturonates (Zink and Chatterjee, 1985).

PL has an absolute requirement for divalent cations and appears to be the major pectic enzyme produced by soft-rot erwinias, particularly *E. chrysanthemi*. The ability to produce at least low levels of PNL is also apparently common to all soft-rot erwinias (discussed below). On the other hand, *E. carotovora* and *E. chrysanthemi* differ in their production of extracellular pectic hydrolases: *E. carotovora* produces PG, and *E. chrysanthemi* produces exoPG (Chatterjee et al., 1981b; Collmer et al., 1982b; Nasuno and Starr, 1966; Stack et al., 1980).

The combined action on D-galacturonan of the extracellular enzymes produced by a single strain has been most extensively explored with *E. chrysanthemi* strain 630 (Fig. 2). The exoPG produced by this strain aggressively attacks D-galacturonan fragments bearing a 4,5-unsaturated terminus generated by prior PL cleavage (Collmer et al., 1982b). The two enzymes thus complement each other in generating saturated and unsaturated digalacturonates: PL randomly attacks internal linkages in the polymer and generates free ends for attack by exoPG which then efficiently releases an assimilable dimer with every cleavage. The combined action of endo- and exo-acting enzymes in the conversion of polymers to dimers (e.g., cellulose to cellobiose or starch to maltose) is common among microbial extracellular depolymerase systems. What is unique in *E. chrysanthemi* is that differing ratios of saturated and unsaturated digalacturonate are generated depending on the relative rate of PL activity (Collmer et al., 1982b). The possible significance of this in the regulation of PL synthesis is discussed below.

Pectate Lyase Isozymes: Multiplicity

Multiple forms of PL are resolved when *E. chrysanthemi* culture supernatants are subjected to isoelectric focusing. Studies employing sucrose gradient isoelectric focusing of a variety of *E. chrysanthemi* strains have revealed the following features. At least two isozymes of PL are produced by all strains of *E. chrysanthemi* (Garibaldi and Bateman, 1971; Pupillo et al., 1976). Based on isoelectric points and other properties, the isozymes have been classified into three groups: alkaline (pI 9.7–9.0), neutral (pI 8.7–7.8), and acidic (pI 4.7–5.0) (Pupillo et al., 1976). The production of acidic isozymes is more variable than the production of neutral or alkaline isozymes (Pupillo et al., 1976). Immunological reactions indicate that at least two of the isozymes produced by a given strain are antigenically distinct, while each may be similar to the isozyme of similar isoelectric point produced by other strains (Mazzucchi et al., 1974). The isozymes differ in their relative ability to reduce the viscosity of pectate solutions, alkaline isozymes generally being more effective. (The interactions of these isozymes with plant tissues will be discussed below.)

The application of high-resolution isoelectric focusing techniques has revealed additional complexities in PL isozyme profiles. When *E. chrysanthemi* PL isozymes are resolved by ultrathin-layer polyacrylamide gel isoelectric focusing and then activity-stained with a pectate-agarose overlay, additional forms not previously separated by sucrose gradient isoelectric focusing are found (Bertheau et al., 1984; Collmer et al., 1985). For example, strain CUCPB 1237, previously thought to produce one basic and one neutral isozyme (Collmer et al., 1982b; Garibaldi and Bateman, 1971), was shown to produce a basic doublet, a neutral doublet, an acidic isozyme, and several minor forms of PL (Collmer et al., 1985).

The PL isozyme profile of *E. carotovora* differs substantially from that of *E. chrysanthemi*. There are fewer isozymes, and they all have rather basic isoelectric points. Biochemical studies have suggested that some strains produce only a single extracellular form of PL (Mount et al., 1970; Stack et al., 1980), while others produce two (Davis et al., 1984) or three (Quantick et al., 1983). This discrepancy can be partially attributed to the use of different fractionation techniques and the similar biochemical properties of the isozymes. A survey of six strains of *E. carotovora* (including the type strains for subsp. *atroseptica* and subsp. *carotorora*), using activity-stained ultrathin-layer isoelectric focusing techniques (Ried and Collmer, 1985), revealed a PL isozyme profile similar to that observed by Quantick et al. (1983), suggesting that this may be the prevalent pattern (J. L. Ried and A. Collmer, unpublished observations). As will be discussed in the next section, cloned PL structural genes provide a powerful tool for studying the encoded isozymes.

Pectate Lyase Isozymes: Molecular Cloning

The pattern of PL isozyme production by *E. coli* clones containing *E. chrysanthemi* DNA indicates that multiple *pel* genes (PL structural genes) underlie at least a portion of the complexity in PL isozyme profiles. *Erwinia chrysanthemi* DNA libraries constructed in *E. coli* with cosmids (Keen et al., 1984; Reverchon et al., 1985), plasmids (Collmer et al., 1985), RP4::mini-Mu (Van Gijsegem et al., 1985b), and lambda vectors (Kotoujansky et al., 1985) reveal several features of the *pel* gene system: (1) The *pel* genes are expressed well in *E. coli* from native promoters, thus facilitating the screening of pectolytic clones with pectate indicator media and analysis of the encoded PL isozymes. (2) The PL isozymes produced by *E. coli* clones have the same molecular weights and isoelectric points as the corresponding isozymes produced by *E. chrysanthemi*. (3) There are at least five *pel* genes in some strains of *E. chrysanthemi*. By using high-resolution isoelectric focusing and activity-staining techniques to analyze the products of *pel* genes carried on large fragments of cloned *E. chrysanthemi* DNA, Van Gijsegem et al. (1985b) and Kotoujansky et al. (1985) showed that separate *pel* genes (*pelA* through *pelE*) encoded an acidic isozyme (PLa), a pair of neutral isozymes (PLb and PLc) and a pair of alkaline isozymes (PLd and PLe). The molecular cloning of pectic enzyme genes from *E. carotovora* has yielded somewhat different results. Lei et al. (1985) cloned three PL isozyme genes from one *E. carotovora* strain, but only single forms of PL, PG, and exopectate lyase were detected in a cosmid library of another strain (Zink and Chatterjee, 1985). Two PL isozymes were preliminarily reported in the plasmid library of a third *E. carotovora* strain (Roberts et al., 1984).

Molecular cloning allows chromosomal mapping and marker exchange mutagenesis of *pel* genes. Van Gijsegem et al. (1985b) used the frequency of RP4::mini-Mu-mediated cotransposition of *E. chrysanthemi pel* genes and markers required by a polyauxotrophic *E. coli* recipient to determine the following gene order: *met-ile-cya-(pelB, pelC) (pelA, pelD, pelE)-purE*. Although the chromosomal location of *pel* genes may not be the same in other *E. chrysanthemi* strains (Van Gijsegem et al., 1985b), arrangement of the genes in the two clusters described above as consistent with the results from all strains so far investigated (Collmer et al., 1985; Keen et al., 1984; Kotoujansky et al., 1985; Reverchon et al., 1985). The *pelC* gene of one *E. chrysanthemi* strain was inactivated by *in vitro* insertion of the kanamycin resistance determinant from pUC4K (Vieira and Messing, 1982). As discussed above (Fig. 1), the inactivated gene (carried on pBR322) was mobilized into *E. chrysanthemi* and exchanged into the chromosome. With another strain, the cloned *pelC* gene was inactivated by Mu-*lac* insertion, transformed into *E. chrysanthemi*, and exchanged into the chromosome by homologous recombination (Diolez and Coleno, 1985). Because the Mu-*lac* insertion produced a PLc–β-galactosidase hybrid protein, *pelC* expression in the mutant could be followed by monitoring β-galactosidase activity. Both of these *E. chrysanthemi* mutants produced all of the PL isozymes except PLc (Diolez and Coleno, 1985; Roeder and Collmer, 1985).

Although the occurrence of multiple chromosomal genes encoding the same enzyme is not unprecedented in bacteria (e.g., the ornithine carbamoyltransferase isozymes of *E. coli* K-12: Kikuchi and Gorini, 1975; Legrain et al., 1976), the duplication of enzyme structural genes is rare (Bachmann and Low, 1980; Riley and Anilionis, 1978). Two divergent models can be invoked to explain a functional basis for the proliferation of PL isozymes: (1) The isozymes are redundant, and multiple *pel* genes are needed only to give the bacterium sufficient pectolytic capacity. (2) Alternatively, the individual PL isozymes may have specialized activities in pathogenicity (e.g., in determination of host range) or in particular ecological niches of the bacterium. It may be significant that *E. chrysanthemi*, which has a more complex PL isozyme profile than *E. carotovora*, also has a higher pectolytic capacity and more evidence of host specificity among its strains.

INTERACTIONS OF *ERWINIA* PECTIC ENZYMES WITH PLANT TISSUES

The destructive effects of *Erwinia* pectic enzymes on parenchymatous plant tissues can be attributed (at least in part) to the structural importance and enzymatic accessibility of pectic polymers in the primary cell wall. While cellulose fibers and their ensheathing hemicellulose polymers also contribute to cell wall structure, the interconnecting pectic polymers in dicots are uniquely vulnerable to enzymic attack. Thus pectic enzymes are "wall-modifying" enzymes that enable other cell wall–degrading enzymes to gain access to their substrates (Bauer et al., 1973), and some *E. chrysanthemi* PL isozymes, purified free of other cell wall–degrading enzymes, have been shown able to cause massive destruction of isolated primary cell walls (Basham and Bateman, 1975b; Baker et al., 1980).

While there is general agreement on the activity of pectic enzymes in causing tissue maceration, the mechanism by which they kill plant cells has proven more controversial. Osmotic fragility resulting from enzymic

damage to the cell wall is one mechanism for which there is substantial evidence (Bateman and Basham, 1976). Plasmolyzed tissues are little affected by PL (as measured by the retention of electrolytes or vital stains).
But upon deplasmolysis, enzyme-treated tissues die, even after the
removal of PL and soluble reaction products. Furthermore, electrolyte
leakage is closely correlated with PL activity and the release of unsaturated galacturonates from treated tissue rather than with some other property of the pectolytic proteins (Basham and Bateman, 1975a,b; Stephens
and Wood, 1975).

More recent results, revealing that cell wall fragments released by pectic enzymes have a variety of biological activities on plants, raise the possibility that the mechanism of cell killing may be complex (McNeil et al.,
1984). That is, even though the evidence that pectic enzymes are capable
of rendering parenchymatous plant tissues osmotically fragile is unam-
biguous, it is still possible that, in the early stages of exposure to PL, the
reaction products are more damaging to the cells than the weakened cell
walls. An observation that potentially complicates this question is that
acid-released plant cell wall fragments have a more damaging effect on
turgid than on deplasmolyzed suspension-cultured sycamore cells, as determined by the inhibition of $[^{14}C]$ leucine incorporation into proteins
(Yamazaki et al., 1983).

Elicitation of defense reactions (phytoalexin and proteinase inhibitor
production) is one of the responses of plants to pectic enzymes and their
reaction products (Bishop et al., 1981; Hahn et al., 1981; Lee and West,
1981; McNeil et al., 1984). Although much of the research on this phenomenon has been done with PG isolated from fungi, two PL isozymes
from E. carotovora have been shown to release elicitors of soybean phytoalexin synthesis from D-galacturonan and soybean cell walls (Davis et
al., 1984). Erwinia carotovora is sensitive to the potato phytoalexin rishitin (Lyon and Bayliss, 1975), and rishitin accumulates in aerobically incubated tubers with restricted E. carotovora infections (Lyon et al.,
1975). The importance of this in soft-rot pathogenesis remains unclear,
especially since there appears to be a defense reaction (discussed below)
in aerated tubers that affects pectic enzyme activity directly.

Although PL isozymes are by definition able to attack internal
glycosidic linkages in D-galacturonan model substrates (typically derived
from citrus fruit), substantial differences exist in their ability to damage
host tissues. Among the E. chrysanthemi PL isozymes, the alkaline pI
forms are more efficient at reducing the viscosity of pectate solutions and
are also more destructive, particularly in maceration assays (Garibaldi
and Bateman, 1971). One acidic isozyme caused neither maceration nor
cell death in potato, carrot, or cucumber tissues, although it was able to
attack pectate derived from these tissues and was relatively efficient at
reducing the viscosity of pectate solutions (Garibaldi and Bateman,

1971). It is not known whether this is typical of the acidic pI PL isozymes produced by other strains of *E. chrysanthemi*, but it is interesting to note that with some strains of *E. chrysanthemi* this isozyme was produced only in host tissues (Pupillo et al., 1976). The cell-bound PL enzymes produced by the nonpathogenic enterobacteria *Klebsiella oxytoca* and *Yersinia enterocolitica* are also unable to macerate plant tissues (Bagley and Starr, 1979). What accounts for the differential ability of PL isozymes to damage host tissues? One possibility is that the more destructive isozymes are adapted to attack particular galacturonosyl linkages that are critical to cell wall and middle lamella structures, and that citrus D-galacturonan assays inadequately represent the action of the enzymes on the complex galacturonans in the cell wall. Alternatively, the more destructive isozymes may remain active longer in host tissues.

Plant tissues susceptible to *Erwinia* soft rots are not defenseless against *Erwinia* pectic enzymes. Maher and Kelman (1983) have reported an oxygen-dependent resistance of potato tubers to maceration by pectolytic culture filtrates of *E. carotovora*. Since active PL can be recovered from aerobically incubated tissues, this resistance mechanism does not involve irreversible enzyme inactivation. Conditions that reduce oxygen tension in potato tubers, e.g., a film of surface water (Burton and Wigginton, 1970), also render potatoes more susceptible to *Erwinia* soft rot (De Boer and Kelman, 1978; Lund and Wyatt, 1972), thus suggesting that this resistance mechanism may be an important determinant in soft-rot pathogenesis.

Quantick et al. (1983) have reported that only one of the three PL isozymes produced by a strain of *E. carotovora* in culture can be isolated from infected potato tissues. This is apparently due to selective inactivation of the isozymes by substances within host cells. When *E. carotovora* culture filtrates containing the three isozymes are incubated in a buffered medium with potato tuber slices, only the most alkaline isozyme can be recovered after a 24-hr incubation. However, all three isozymes can be recovered when the tissue is plasmolyzed. These observations suggest the operation of a resistance mechanism different from the one reported by Maher and Kelman (1983).

REGULATION OF PECTIC ENZYME PRODUCTION

The ability to produce pectic enzymes per se is clearly insufficient for a bacterium to act as a soft-rot pathogen, since only a few of the many pectolytic bacterial species are consistently associated with disease. While pathogens may produce pectic enzymes with unique properties (as discussed in the previous section), it is likely that the *regulation* of enzyme production is also an important factor in pathogenesis. The general importance of microbial enzyme regulation in pathogenicity is discussed in the excellent review by VanEtten and Kistler (1984).

Exploration of pectic enzyme regulatory mechanisms in erwinias is guided by the paradigm of catabolic enzyme regulation in *E. coli* (e.g., Freifelder, 1983). It is important to note, however, that efficient regulation of pectic enzymes in soft-rot erwinias presents two problems not encountered by *E. coli* in its regulation of intracellular enzymes like β-galactosidase. First, the enzymes function on an extracellular substrate. Consequently, the regulatory system must be sensitive to the environmental presence of a polymer that the bacteria cannot assimilate (for induction to occur), as well as to the subsequent rate of enzyme activity in that particular environment (to prevent wasteful overproduction). Second, pectic enzymes may have two independent roles (in saprophytic catabolism and in pathogenesis) in the life cycle of the bacterium, and each role may require a different mode of regulation. The next two sections will be concerned largely with the induction and self–catabolite repression of PL in its role as a catabolic enzyme permitting soft-rot erwinias to utilize D-galacturonan in culture. The third section will deal with the usual regulation of PNL.

Induction of Pectate Lyase

The capacity to produce large amounts of PL is a characteristic feature of soft-rot erwinias. Comparison of the rates of PL production on different carbon sources by *E. chrysanthemi, E. carotovora,* and various pectolytic saprophytes reveals two important features: (1) The basal level of PL synthesis in erwinias is substantial, particularly in *E. chrysanthemi;* and (2) the rate of PL synthesis rapidly increases by severalfold when erwinias are grown in media containing D-galacturonan (Chatterjee et al., 1979, Collmer et al., 1982a; Zucker and Hankin, 1970; Zucker et al., 1972). A similar stimulation of PL synthesis is observed when *E. chrysanthemi* is incubated with isolated plant cell walls (Collmer and Bateman, 1982). How are the bacteria able to recognize the presence of large or insoluble pectic polymers in their external milieu? Several observations suggest that the PL induction process has extracellular and intracellular phases, involves the basal activity of several enzymes in the pectate catabolic pathway, and necessitates intracellular formation of ketodeoxyuronate intermediates (Fig. 2).

The extracellular phase involves release from the polymer of oligogalacturonates (particularly dimers) that can be assimilated by the cell and converted to inducers. The need for this extracellular digestive phase in the induction process is indicated by several observations. When cultures are shifted to media containing D-galacturonan, a lag period precedes the stimulation of PL synthesis. In *E. carotovora* this lag is prolonged by the addition of EDTA, an inhibitor of PL activity, and abolished by the addition of unsaturated digalacturonate, a PL product (Tsuyumu, 1977). The lag period in *E. chrysanthemi* can also be abolished by digalacturonates but, apparently as a result of exoPG activity, inhibition of PL induction of EDTA is not observed unless the cultures are incubated with isolated plant cell walls rather than soluble D-galacturonan (Collmer and Bateman, 1982). Figure 3, showing the induction potential of various pectic compounds at different concentrations, provides further evidence for the importance of the dimers in PL induction in *E. chrysanthemi*. The dimers are effective inducers at concentrations 1/1000 of the weight-equivalent concentration required by the polymer to induce activity; the polymer becomes effective as an inducer only at concentrations approaching the K_m of the extracellular enzymes (Collmer and Bateman, 1981).

The higher induction potential of the dimers relative to the galacturonate monomer (Fig. 3) suggests that the inducer is a metabolite in the digalacturonate catabolic pathway. As shown in Fig. 2, the intracellular catabolism of either saturated or unsaturated digalacturonate generates two ketodeoxyuronates (DTH and DGH) that do not occur in the monomer galacturonate pathway (Kilgore and Starr, 1959; Preiss and

Figure 3. Relative ability of pectic compounds differing in degree of polymerization and concentration to induce PL in *E. chrysanthemi*. A culture grown in glycerol-minimal medium was centrifuged and resuspended in minimal medium containing digalacturonate (DG), unsaturated digalacturonate (UDG), galacturonate (GalUA), or D-galacturonan at various concentrations. After incubation for 3 hr, the cultures were assayed for PL activity. (From Collmer and Bateman, 1981.)

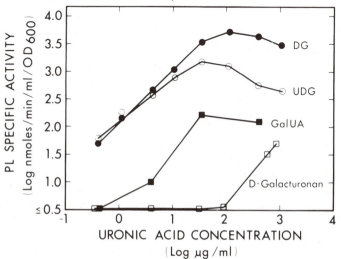

Ashwell, 1963a,b). If DTH and/or DGH are the actual inducers of PL, then mutants deficient in oligogalacturonide lyase (OGL) would be expected to be uninducible by digalacturonates, D-galacturonan, or isolated plant cell walls. This, indeed, is the case. OGL-deficient *E. chrysanthemi* mutants obtained from either chemical (Collmer and Bateman, 1981) or Tn5 (Chatterjee et al., 1985b) mutagenesis are uninducible by these pectic substrates but, like the wild type, can be induced by exogenous ketodeoxyuronates. Furthermore, a mutant deficient in ketodeoxyuronate dehydrogenase (and consequently expected to accumulate ketodeoxyuronates) is hyperinduced by digalacturonates (Chatterjee et al., 1985b).

Self–Catabolite Repression of Pectate Lyase

There are two conditions under which it is advantageous for a cell to repress synthesis of a catabolic enzyme even in the presence of an inducing substrate: (1) when a readily metabolized alternative carbon source, e.g., glucose, is available, and (2) when the enzyme is generating products at a rate exceeding the utilization ability of the cell. The latter phenomenon (self–catabolite repression) is particularly important in the regulation of extracellular enzymes which act on complex substrates in variable environments.

Repression of catabolic enzymes by glucose in *E. coli* is correlated with reduced levels of cyclic AMP (cAMP) and impaired transcription of a broad class of operons encoding catabolic enzymes (Robison et al., 1971). There is evidence that this universal bacterial regulatory system also controls PL synthesis in *E. carotovora*. When glucose is added to *E. carotovora* cultures, both PL production and intracellular cAMP levels decline. This decline in PL production can be reversed by adding cAMP to the medium (Hubbard et al., 1978). Furthermore, a mutant deficient in cAMP produces lower levels of PL and other catabolic enzymes unless exogenous cAMP is supplied (Mount et al., 1979).

Several lines of evidence suggest that in both *E. carotovora* and *E. chrysanthemi* PL is subject to cAMP-mediated self–catabolite repression when its product, unsaturated digalacturonate, is present in the medium at high concentrations (a condition expected to accompany excess PL production). Tsuyumu (1979) reported that the rate of PL production on 2 mM unsaturated digalacturonate was low and not further repressed by the addition of glucose during the first hour of incubation. Subsequently, as the unsaturated digalacturonate in the medium becomes depleted, the rate of PL production increases sharply and becomes the subject to demonstrable glucose repression. This lag in the stimulation of PL production by 2 mM unsaturated digalacturonate is abolished by adding cAMP.

In *E. chrysanthemi*, self–catabolite repression is produced by high concentrations of unsaturated but not saturated digalacturonate (Chatterjee et al., 1981b; Collmer and Bateman, 1981). The differential effect on PL production of the two dimers is evident in Fig. 3. Several observations suggest that this results from a more rapid cleavage of the unsaturated dimer by OGL (Collmer and Bateman, 1981; Moran et al., 1968), which then allows repressive rates of ketodeoxyuronate catabolism to occur. The differential effect is abolished by supplying the dimers at concentrations too low to support bacterial growth, by supplying both dimers at once (which results in levels of PL production equivalent to those obtained with the unsaturated dimer alone) or by adding cAMP (which results in levels of PL production equivalent ot those obtained with the saturated dimer alone) (Chatterjee et al., 1981b; Collmer and Bateman, 1981). Furthermore, a mutant that is partially deficient in OGL and able to grow slowly on unsaturated digalacturonate is hyperinduced, in comparison to the wild-type strain, by the unsaturated dimer (Collmer and Bateman, 1981). These results should be interpreted cautiously, since repressive effects of high concentrations of unsaturated digalacturonate have also been observed with a ketodeoxyuronate dehydrogenase–deficient *E. chrysanthemi* mutant which cannot utilize digalacturonates (Chatterjee et al., 1985b).

The differential production of PL in response to the two digalacturonates, which is less pronounced in *E. carotovora* (Chatterjee et al., 1981b), may have physiological significance in *E. chrysanthemi*. In environments unfavorable to PL activity the *E. chrysanthemi* extracellular pectic enzyme complex releases mostly saturated dimer from galacturonans (see Fig. 2). Thus it is possible that the cells are informed of changing needs for PL synthesis by the ratio of the two digalacturonates.

Pectin Lyase Production in Soft-Rot Erwinias

While the synthesis of PL is regulated much like that of other bacterial catabolic enzymes, PNL is produced more like a temperate phage is produced; i.e., it is induced by UVi light irradiation and other agents that damage or inhibit DNA synthesis. This puzzling phenomenon was discovered when Tomizawa and Takahashi (1971) observed that pectic enzyme production in *E. carotovora* was stimulated by bleomycin, mitomycin C, and naladixic acid. The increase in pectolytic activity was shown to be due to the production of PNL rather than to an increased production of PL or PG (Kamimiya et al., 1972; 1974). Several features of PNL production by soft-rot erwinias have been established:

1. All the strains of *E. chrysanthemi* and *E. carotovora* (including representatives from subsp. *carotovora* and *atroseptica*) so far in-

vestigated have been found to produce the enzyme (Itoh et al., 1980; Tsuyumu and Chatterjee, 1984).

2. PNL is not produced in response to pectic substrates (Tsuyumu and Chatterjee, 1984).
3. In many, but not all, strains of soft-rot erwinias, PNL is inducible by mitomycin C and related agents (Tsuyumu and Chatterjee, 1984).
4. PNL is coordinately induced with bacteriocins in *E. carotovora* (Itoh et al., 1980) and with a temperate phage in *E. chrysanthemi* (Tsuyumu and Chatterjee, 1984).

The contrasting responses of PNL and PL production to mitomycin C treatment in *E. chrysanthemi* and *E. carotovora* are shown in Table 1.

PNL induction thus appears to be part of the SOS response in soft-rot erwinias. The SOS regulatory system, which has been extensively explored in *E. coli*, controls a global set of cellular responses to agents that disrupt DNA metabolism. SOS functions include an enhanced DNA repair capacity, increased mutagenesis, filamentation, and prophage induction (Little and Mount, 1982). The *recA* and *lexA* proteins are key components of the SOS regulatory system. Inducing conditions activate protease activity in the *recA* protein (which also controls homologous recombination). The *recA* protease cleaves the *lexA* protein, a repressor of the many genes involved in SOS functions (Little et al., 1980). The *recA* protease activity is also known to cleave the phage lambda repressor, thus triggering the lytic cycle and allowing the phage to escape from a

Table 1 Production of PNL and PL in *E. chrysanthemi* and *E. carotovora* Following Mytomycin Treatment.[a]

| | ENZYME ACTIVITY[b] | | | |
| | PNL | | PL | |
Strain	Control	Mitomycin C	Control	Mytomycin C
Erwinia chrysanthemi	0.9	18.1	2.6	1.1
Erwinia carotovora	2.5	68.6	1.6	0.6

[a]Modified from Table 1 of Tsuyumu and Chatterjee (1984) and published with permission of the copyright owner, Academic Press, Inc. (London).

[b]Exponential-phase broth cultures in medium lacking pectic compounds were treated with mitomycin C. After cell lysis, the cultures were assayed for pectic enzyme activity. Activity is expressed as units per milliliter of culture, with 1 unit of lyase activiy defined as the amount ob ddunt of enzyme producing a change in absorbance at 235 nm of 1.0 at 30°C.

threatened cell (Roberts et al., 1978). There is also evidence that the *recA* function is required for PNL induction by DNA-damaging agents in *E. carotovora* (A. K. Chatterjee, personal communication).

One explanation for this phenomenon is that the PNL gene in soft-rot erwinias is located near or within a temperate phage or a defective phage (bacteriocin) and, as a consequence, induction results from a prophage or bacteriocin induction (Itoh et al., 1982). In light of the numerous differences in pectic enzyme production between *E. chrysanthemi* and *E. carotovora*, it is noteworthy that an SOS inducible PNL is a feature shared by certain strains in both groups of bacteria. It is also puzzling that PNL is poorly induced by mitomycin C in many strains of *E. chrysanthemi* and *E. carotovora* subsp. *atroseptica* (Tsuyumu and Chatterjee, 1984). Since PNL from other microorganisms has been shown to be very effective at macerating and killing plant tissues (Byrde et al., 1968; Hislop et al., 1979), it is possible that *Erwinia* PNL is an important disease factor. The induction and activity of PNL during soft-rot pathogenesis remains to be established.

EXPORT OF PECTATE LYASE

The export (secretion) of certain proteins to the periplasmic space or the inner or outer membrane is a general phenomenon in gram-negative bacteria that has attracted substantial genetic and biochemical research (Benson et al., 1985; Silhavy et al., 1983). Less common and less well understood is the ability of some gram-negative bacteria to export specific polypeptides beyond the outer membrane to the external milieu (Pugsley and Schwartz, 1985). The ability of soft-rot erwinias to export pectic enzymes seems essential to their pathogenicity. The pattern of PL localization (culture supernatant, periplasmic space, or cytoplasm) in a variety of pectolytic bacteria provides correlative support for the importance of PL export: Most of the PL remains cell-bound in nonpathogens, whereas much of it is released into the medium of soft-rot erwinias (although there is a substantial lag in PL export by *E. carotovora* that is not observed with *E. chrysanthemi* (Chatterjee et al., 1979). Furthermore, *E. chrysanthemi* mutants that are unable to export PL are nonpathogenic (Andro et al., 1984; Chatterjee and Starr, 1977, 1978).

Erwinia chrysanthemi selectively exports PL during logarithmic growth, while retaining cytoplasmic and periplasmic marker proteins (Andro et al., 1984; Chatterjee et al., 1979). This suggests the existence of an export machinery that interacts specifically with PL and other *E. chrysanthemi* extracellular proteins (e.g., cellulase). The pattern of PL localization in export-deficient (Out−) *E. chrysanthemi* mutants and pectolytic *E. coli* clones (Table 2) provides some insight into the export

Table 2 Distribution of PL in Subcellular Fractions of *E. chrysanthemi* Wild-type and Out⁻ Strains and of a pectolytic *E. coli* Clone.[a]

Strain[c]	PERCENTAGE OF TOTAL ACTIVITY[b]		
	Extracellular	Periplasmic	Cytoplasmic
Erwinia chrysanthemi WT	91	9	0
Erwinia chrysanthemi Out⁻	0	82	18
Eschericholi coli (pBR322::pelC)	3	90	7

[a]Modified from Table 2 of Andro et al. (1984) and Table 3 of Collmer et al. (1985) and published with permission of the copyright owner, the American Society for Microbiology.

[b]Subcellular fractions were prepared from *E. chrysanthemi* using lysozyme-EDTA treatment and from *E. coli* using osmotic shock. In all cases, at least 75% of the activity of the plasmid-encoded periplasmic marker protein, β-lactamase, was found in the periplasmic fraction.

[c]*Erwinia chrysanthemi* strains carried pULB113 (Amp^r). The Out— mutant was obtained by mutagenesis with the phage Mu derivative, mini-D108::Tn9 (Andro et al., 1984). *Escherichia coli* HB101 carried pBR322 with a 4.7-kb *E. chrysanthemi* DNA insert encoding PLc (Collmer et al., 1985).

process. PL accumulates in the periplasmic space of these clones and mutants (Andro et al., 1984; Keen et al., 1984; Collmer et al., 1985), suggesting that export occurs in two steps. In step I, PL is exported through the inner membrane to the periplasmic space; in step II, PL is exported through the outer membrane to the medium. A two-step process was originally proposed for the export of plasmid-encoded hemolysin in *E. coli* (Springer and Goebel, 1980).

According to this model, step I export is mediated by an *E. chrysanthemi* cellular machinery homologous to the Sec machinery that exports proteins out of the cytoplasm of *E. coli*. Several features of the process in *E. coli* are noteworthy: (1) A universal characteristic of secreted proteins is an amino-terminal signal peptide which is removed from the mature protein during the export process; (2) export information for many secreted proteins is confined to the amino-terminal portion; truncated polypeptides or hybrid proteins lacking much of the carboxy terminus are still localized correctly; and (3) the use of hybrid proteins in which an active β-galactosidase is fused with the carboxy terminus of a secreted protein has permitted selection for mutations in trans-acting genes that encode elements of the Sec machinery (Silhavy et al., 1983).

The ability of *E. coli* clones to efficiently secrete PL into the periplasmic space suggests that PL has a signal peptide that is recognized by the

E. coli Sec machinery (Collmer et al., 1985; Keen et al., 1984). Furthermore, for at least one of the *E. chrysanthemi* CUCPB 1237 PL isozymes (PLc), active proteins of apparently identical molecular weight can be isolated from the periplasmic space of *E. coli* or the medium of *E. chrysanthemi* cells containing the parental gene (Collmer et al., 1985). This suggests that processing of the putative PL signal peptide also occurs in *E. coli* Step II in this model involves the interaction of products of the *out* genes with specific regions or features of the PL polypeptide, resulting in the export of PL through the outer membrane. Since *E. coli* apparently lacks *out* genes, it has the same export phenotype as Out— *E. chrysanthemi* strains: Both organisms are unable to export PL beyond the periplasmic space.

GENETICS OF PECTIC ENZYME PRODUCTION AND PATHOGENICITY

Of the mutants altered in pectic enzyme production that have been characterized, only those that are deficient in the export of PL have altered pathogenic phenotypes. Chatterjee and Starr (1977, 1978) reported an ethyl methanesulfonate-induced *E. chrysanthemi* mutant that produced a reduced level of PL (but was apparently unaffected in exoPG production), released virtually no PL into the medium, and was unable to cause maceration in susceptible plant tissues. Andro et al. (1984) reported that an *out* mutation in which PL export but not synthesis was affected was similarly nonpathogenic (in the test plant *Saintpaulia ionantha*). Since Out— mutants are also unable to export cellulase and possibly other proteins involved in pathogenesis, their use in evaluating the role of PL in pathogenesis must be qualified.

Surprisingly, *E. chrysanthemi* mutants deficient in OGL (Chatterjee et al., 1985; Collmer and Bateman, 1981; Collmer et al., 1982a) and *E. carotovora* mutant deficient in cAMP (Mount et al., 1979: M. S. Mount, personal communication) have been found to retain the parental virulence in potato tuber maceration assays. It is noteworthy that OGL-deficient strains are unable to utilize D-galacturonan or digalacturonates. This suggests that utilization of the products of pectic digestion does not contribute to the ability of the bacteria to macerate susceptible tissues. PL production in OGL-deficient mutants is also uninducible by pectic compounds or isolated plant cell walls, which suggests either that the uninduced level of PL synthesis is sufficient for tissue maceration or that induction *in planta* is mediated by preformed inducers or occurs by a mechanism different from that studied in culture (Chatterjee et al., 1985; Collmer et al., 1982a). Interpretation of these results and similar findings with the cAMP-deficient *E. carotovora* mutant (which produces only reduced levels of PL in culture) requires a better understanding of the regulation of PL synthesis *in planta*.

Molecular cloning of *pel* genes now allows evaluation of the contribution of individual pectic enzymes to pathogenesis. The results so far are ambiguous. On the one hand, *E. coli* clones producing *Erwinia* PL isozymes are able to cause maceration in potato tubers (but only when artificially high levels of inoculum are applied to compromised host tissues) (Allen et al., 1984; Collmer et al., 1985; Keen et al., 1984), an observation that supports the importance of PL in soft-rot pathogenesis. On the other hand, marker exchange mutagenesis in *E. chrysanthemi* has revealed that a mutation in one of these cloned genes (the *E. chrysanthemi* CUCPB 1237 *pelC* gene) has no apparent effect on virulence as determined by potato tuber injection assays (Roeder and Collmer, 1985). Finally, there is a paucity of information on disease determinants other than pectic enzymes produced by soft-rot erwinias. These determinants may be revealed by the effect on plants of pectic enzyme-deficient *Erwinia* mutants or by the isolation of pectolytic nonpathogenic mutants.

CONCLUDING REMARKS

Soft-rot erwinias are not unique in their ability to produce pectic enzymes. Unlike many pathogen-produced disease molecules (e.g., toxins and phytohormones), pectic enzymes are produced by a wide array of microorganisms including other enterobacteria, many biotrophic *Pseudomonas* phytopathogens, and possibly *Rhizobium* spp. (Chatterjee et al., 1979; Hildebrand, 1971; Hubbell et al., 1978). What are the unique attributes that enable erwinias to cause soft rots and a variety of other plant diseases?

One possibility is that the types of pectic enzymes and the mechanisms of their production are important determinants in soft-rot pathogenesis. Certainly, export of the enzymes is essential to pathogenicity, but the importance of regulation is less clear. While *E. chrysanthemi* and *E. carotovora* can be distinguished from other pectolytic bacteria by the high basal level of synthesis and the rapid inducibility of PL in culture, the importance of this is questionable since various regulatory mutants producing only reduced levels of PL in culture retain their ability to cause soft rot. More research on pectic enzyme regulation *in planta* is demanded by these results and by the discovery that many soft-rot erwinias produce PNL in response to DNA-damaging agents. The intriguing discovery that some *E. chrysanthemi* strains produce at least five PL isozymes is hard to evaluate without examining additional mutants deficient in each of the isozymes and characterization of the enzymic properties of the isozymes. We also need to know more about the activity of the individual PL isozymes and other *Erwinia* pectic enzymes on native substrates, as well as the secondary effects of this activity on plant tissues.

A factor complicating interpretation of the pathogenic importance of individual pectin enzymes and production mechanisms is the substantial difference between the extracellular pectic enzyme systems of *E. chrysanthemi* and *E. carotovora*. Relative to *E. carotovora*, *E. crysthanemi* has a more complex PL isozyme profile, produces severalfold more PL (induced or uninduced), exports PL more rapidly, and produces exoPG instead of PG. Despite these differences, both species are capable of causing indistinguishable symptoms in certain host tissues such as the tubers and stems of potatoes.

Ultimately, evaluation of the genetics of pectic enzyme production and pathogenicity in soft-rot erwinias will require the development of phenotypic tests that better represent the range of selection pressures underlying the evolution of erwinias. Three classes of assays may be useful. Assays in the first class, typified by the injection of bacteria into anaerobically incubated whole-potato tubers, permit standardized evaluation of bacteria as opportunistic pathogens under conditions that accompany major storage losses. A limitation of these assays is that they bypass a gamut of obstacles (e.g., entry via the lenticels) commonly faced by bacteria in nature and consequently do not address the important question of how these opportunistic pathogens earn their opportunities. To correct for this, a second class of assays could specifically explore events that precede the development of storage rots. The third class would employ assays of selected strains causing systemic infections and vascular wilts, as well as other symptoms, in growing host tissues.

As pointed out by Kelman (1979), although plants have evolved in the presence of a wide array of free-living bacteria, there are surprisingly few phytopathogenic species. It is thus intriguing that pathogens like *E. chrysanthemi* are capable of a variety of relationships with plants. This pathogenic diversity against a highly manipulable genetic background makes soft-rot erwinias particularly alluring experimentally. Pectic enzymes, which historically have been biochemically approachable, are now being intensely explored genetically. We should soon know whether the activity of these enzymes has obscured other critical disease determinants or whether the enzymes are indeed a key to the invasion of plants by soft-rot erwinias.

ACKNOWLEDGMENTS

I thank Arun Chatterjee for fruitful discussions of many aspects of this chapter, John Payne for suggesting several improvements in the manuscript, and Karen Teramura for preparing the illustrations. Work in the author's laboratory was supported by grant 84-CRCR-1-1366 from the Competitive Research Grants Office of the U. S. Department of Agriculture.

REFERENCES

ALLEN, C., ROBERTS, D. P., FORD, M., STROMBERG, V. K., BERMAN, P. M., LACY, G. H., AND MOUNT, M. S. 1984. Lack of pathogenicity by *Escherichia coli* containing plasmids with genes mediating pectolytic enzyme production cloned from *Erwinia cartovora* subsp. *carotovora* strain EC14, *Phytopathology* **74:**816. Abstract.

ANDRO, T., CHAMBOST, J. -P., KOTOUJANSKY, A., CATTANEO, J., BERTHEAU, Y., BARRAS, F., VAN GIJSEGEM, F., AND COLENO, A. 1984. Mutants of *Erwinia chrysanthemi* defective in secretion of pectinase and cellulase, *J. Bacteriol.* **160:**1199–1203.

BACHMANN, B. J., AND LOW, K. B. 1980. Linkage map of *Escherichia coli* K-12, edition 6, *Microbiol. Rev.* **44:**1–56.

BAGLEY, S. T., AND STARR, M. P. 1979. Characterization of intracellular polygalacturonic acid *trans*-eliminase from *Klebsiella oxytoca, Yersinia enterocolitica,* and *Erwinia chrysanthemi, Curr. Microbiol.* **2:**381–386.

BAKER, C. J., AIST, J. R., AND BATEMAN, D. F. 1980. Ultrastructural and biochemical effects of endopectate lyase on cell walls from cell suspension cultures of bean and rice, *Can. J. Bot.* **58:**867–880.

BASHAM, H. G., AND BATEMAN, D. F. 1975a. Killing of plant cells by pectic enzymes: The lack of direct injurious interaction between pectic enzymes or their soluble reaction products and plant cells, *Phytopathology* **65:**141–153.

BASHAM, H. G., AND BATEMAN, D. F. 1975b. Relationship of cell death in plant tissue treated with a homogeneous endopectate lyase to cell wall degradation, *Physiol. Plant Pathol.* **5:**249–261.

BATEMAN, D. F., AND BASHAM, H. G. 1976. Degradation of plant cell walls and membranes by microbial enzymes, in: *Encyclopedia of Plant Physiology,* New Series, Vol. 4, *Physiological Plant Pathology* (R. Heitefuss and P. H. Williams, eds.), pp. 316–355, Springer-Verlag, New York.

BAUER, W. D., K. W. TALMADGE, K. KEEGSTRA, AND P. ALBERSHEIM. 1973. The structure of plant cell walls. II. The hemicellulose of the walls of suspension-cultured sycamore cells, *Plant Physiol.* **51:**174–187.

BENSON, S. A., HALL, M. N., AND SILHAVY, T. J. 1985. Genetic analysis of protein export in *Escherichia coli* K12, *Annu. Rev. Biochem.* **54:**101–134.

BERTHEAU, Y., MADGIDI-HERVAN, E., KOTOUJANSKY, A., NGUYEN-THE, C., ANDRO, T., AND COLENO, A. 1984. Detection of depolymerase isoenzymes after electrophoresis or electrofocusing, or in titration curves, *Anal. Biochem.* **139:**383–389.

BISHOP, P. D., MAKUS, D. K., PEARCE, G., AND RYAN, C. A. 1981. Proteinase inhibitor-inducing factor activity in tomato leaves resides in oligosaccharides enzymically released from cell walls, *Proc. Natl. Acad. Sci. USA* **78:**3536–3540.

BROWN, W. 1915. Studies in the physiology of parasitism. I. The action of *Botrytis cinera, Ann. Bot.* **29**:313–348.

BURR, T. J., AND SCHROTH, M. N. 1977. Occurrence of soft-rot *Erwinia* spp. in soil and plant material, *Phytopathology* **67**:1382–1387.

BURTON, W., AND WIGGINTON, M. J. 1970. The effect of a film of water upon the oxygen status of a potato tuber, *Potato Res.* **13**:180–186.

BYRDE, R. J. W., AND FIELDING, A. H. 1968. Pectin methyl *trans*-eliminase as the maceration factor of *Sclerotinia fructigena* and its significance in brown rot of apple, *J. Gen. Microbiol.* **52**:287–297.

CHATTERJEE, A. K. 1980. Acceptance by *Erwinia* spp. of the R plasmid R68.45 and its ability to mobilize the chromosome of *Erwinia chrysanthemi, J. Bacteriol.* **142**:111–119.

CHATTERJEE, A. K., AND BROWN, M. A. 1980. Generalized transduction in the pectolytic bacterium *Erwinia chrysanthemi, J. Bacteriol.* **143**:1444–1449.

CHATTERJEE, A. K., AND STARR, M. P. 1977. Donor strains of the soft-rot bacterium *Erwinia chrysanthemi* and conjugational transfer of the pectolytic capacity, *J. Bacteriol.* **132**:862–869.

CHATTERJEE, A. K., AND STARR, M. P. 1978. Genetics of pectolytic enzyme production in *Erwinia chrysanthemi*, in: *Proceedings of the Fourth International Conference on Plant Pathogenic Bacteria, Angers, France, 1978* (Station de Pathologie Végétale et Phytobactériologie, ed.), pp. 89–94, Institut National de la Recherche Agronomique, Angers, France.

CHATTERJEE, A. K., AND STARR, M. P. 1980. Genetics of *Erwinia* species, *Annu. Rev. Microbiol.* **34**:645–676.

CHATTERJEE, A. K., BUCHANAN, G. E., BEHRENS, M. K., AND STARR, M. P. 1979. Synthesis and excretion of polygalacturonic acid *trans*-eliminase in *Erwinia, Yersinia* and *Klebsiella* species, *Can. J. Microbiol.* **25**:94–102.

CHATTERJEE, A. K., BROWN, M. A., ZIEGLE, J. S., AND K. K. THURN. 1981a. Progress in chromosomal genetics of *Erwinia chrysanthemi*, in: *Proceedings of the Fifth International Conference on Plant Pathogenic Bacteria, Cali, Colombia, 1981* (J. C. Lozano, ed.), pp. 389–402, Centro Internacional de Agricultura Tropical, Cali, Colombia.

CHATTERJEE, A. K., THURN, K. K., AND TYRELL, D. J. 1981b. Regulation of pectolytic enzymes in soft rot *Erwinia*, in: *Proceedings of the Fifth International Conference on Plant Pathogenic Bacteria, Cali, Colombia, 1981* (J. C. Lozano, ed.), pp. 252–262, Centro Internacional de Agricultura Tropical, Cali, Colombia.

CHATTERJEE, A. K., THURN, K. K., AND FEESE, D. A. 1983. Tn5-induced mutations in the enterobacterial phytopathogen *Erwinia chrysanthemi, Appl. Environ. Microbiol.* **45**:644–650.

CHATTERJEE, A. K., ROSS, L. M., McEVOY, J. L., AND THURN, K. K. 1985a. pULB113, an RP4::mini-Mu plasmid, mediates chromosomal mobilization and R-prime formation in *Erwinia amylovora, E. chrysanthemi*, and subspecies of *E. carotovora, Appl. Environ. Microbiol.* **50**:1–9.

CHATTERJEE, A. K., THURN, K. K., AND TYRELL, D. J. 1985b. Isolation and characterization of Tn5 insertion mutants of *Erwinia chrysanthemi* that are deficient in polygalacturonate catabolic enzymes oligogalacturonate lyase and 3-deoxy-D-glycero-2,5-hexodiulosonate dehydrogenase, *J. Bacteriol.* **162**:708–714.

COLLMER, A., AND BATEMAN, D. F.. 1981. Impaired induction and self-catabolite repression of extracellular pectate lyase in *Erwinia chrysanthemi* mutants deficient in oligogalacturonide lyase, *Proc. Natl. Acad. Sci. USA* **78**:3920–3924.

COLLMER, A., AND BATEMAN, D. F. 1982. Regulation of extracellular pectate lyase in *Erwinia chrysanthemi:* Evidence that reaction products of pectate lyase and exo-poly-α-D-galacturonosidase mediate induction on D-galacturonan, *Physiol. Plant Pathol.* **21**:127–139.

COLLMER, A., BERMAN, P., AND MOUNT, M. S. 1982a. Pectate lyase regulation and bacterial soft-rot pathogenesis, in: *Phytopathogenic Prokaryotes,* Vol. 1 (M. S. Mount and G. H. Lacy, eds.), pp. 395–422, Academic Press, New York.

COLLMER, A., WHALEN, C. H., BEER, S. V., AND BATEMAN, D. F. 1982b. An exo-poly-α-D-galacturonosidase implicated in the regulation of extracellular pectate lyase production in *Erwinia chrysanthemi, J. Bacteriol.* **149**:626–634.

COLLMER, A., SCHOEDEL, C., ROEDER, D. L., RIED, J. L., AND RISSLER, J. F. 1985. Molecular cloning in *Escherichia coli* of *Erwinia chrysanthemi* genes encoding multiple forms of pectate lyase, *J. Bacteriol.* **16**:913–920.

DAVIS, K. R., LYON, G. D., DARVILL, A. G., AND ALBERSHEIM, P. 1984. Host-pathogen interactions. XXV. Endopolygalacturonic acid lyase from *Erwinia carotovora* elicits phytoalexin accumulation by releasing plant cell wall fragments, *Plant Physiol.* **74**:52–60.

DE BOER, S. H., AND KELMAN, A. 1978. Influence of oxygen concentration and storage factors on susceptibility of potato tubers to bacterial soft rot (*Erwinia carotovora*), *Potato Res.* **21**:65–80.

DICKEY, R. S. 1979. *Erwinia chrysanthemi*: A comparative study of phenotypic properties of strains from several hosts and other *Erwinia* species, *Phytopathology* **69**:324–329.

DICKEY, R. S. 1981. *Erwinia chrysanthemi*: Reaction of eight plant species to strains from several hosts and to strains of other *Erwinia* species, *Phytopathology* **71**:23–29.

DIOLEZ, A., AND COLENO, A. 1985. Mu-*lac* insertion-directed mutagenesis in a pectate lyase gene of *Erwinia chrysanthemi, J. Bacteriol.* **163**:913–917.

DYE, D. W. 1969. A taxonomic study of the genus *Erwinia*. II. The "carotovora" group, *New Zealand J. Sci.* **12**:81–97.

FREIFELDER, D. 1983. *Molecular Biology*, Jones and Bartlett, Boston.

GARIBALDI, A., AND BATEMAN, D. F. 1971. Pectic enzymes produced by *Erwinia chrysanthemi* and their effects on plant tissue, *Physiol. Plant Pathol.* **1**:25–40.

GUIMARES, W. V., PANOPOULOS, N. J., AND SCHROTH, M. N. 1979. Conjugative properties of *incP*, *incT*, *incF* plasmids *Erwinia chrysanthemi* and *Pseudomonas phaseolicola*, in: *Proceedings of the Fourth International Conference on Plant Pathogenic Bacteria, Angers, France, 1978* (Station de

Pathologie Végétale et Phytobactériologie, ed.), pp. 53–65, Institut National de la Recherche Agronomique, Angers, France.

HAHN, M. G., DARVILL, A. G., AND ALBERSHEIM, P. 1981. Host-pathogen interactions. XIX. The endogenous elicitor, a fragment of a plant cell wall polysaccharide that elicits phytoalexin accumulation in soybeans, *Plant Physiol.* **68:**1161–1169.

HILDEBRAND, D. C. 1971. Pectate and pectin gels for differentiation of *Pseudomonas* sp. and other bacterial plant pathogens, *Phytopathology,* **61:**1430–1436.

HINTON, J. C. D., Pérombelon, M. C. M., and Salmond, G. P. C. 1985. Efficient transformation of *Erwinia carotovora* subsp. *carotovora* and *E. carotovora* subsp. *atroseptica, J. Bacteriol.* **161:**786–788.

HISLOP, E. C., KEON, J.P.R., and FIELDING, A. H. 1979. Effects of pectin lyase from *Monilinia fructigena* on viability, ultrastructure and localization of acid phosphatase of cultured apple cells, *Physiol. Plant Pathol.* **14:**371–381.

HUBBARD, J. P., WILLIAMS, J. D., NILES, R. M., AND MOUNT, M. S. 1978. The relation between glucose repression of endopolygalacturonate *trans*-eliminase and adenosine 3′5′-cyclic monophosphate levels in *Erwinia carotovora, Phytopathology* **68:**95–99.

HUBBELL, D. H., MORALES, V. M., AND UMALI-GARCIA, M. 1978. Pectolytic enzymes in *Rhizobium, Appl. Environ. Microbiol.* **35:**210–213.

ITOH, Y., IZAKI, K., AND TAKAHASHI, T. 1980. Simultaneous synthesis of pectin lyase and carotovoricin induced by mitomycin C, nalidixic acid or ultraviolet light irradiation in *Erwinia carotovora, Agric. Biol. Chem.* **44:**1135–1140.

ITOH, Y., SUGIURA, J., IZAKI, K., AND TAKAHASHI, H. 1982. Enzymological and immunological properties of pectin lyases from bacteriocinogenic strains of *Erwinia carotovora, Agric. Biol. Chem.* **46:**199–205.

JAYASWAL, R. K., BRESSAN, R. A., AND HANDA, A. K. 1984. Mutagenesis of *Erwinia carotovora* subsp. *carotovora* with bacteriophage Mu dl (Ap^r *lac* c*ts*62): Construction of *his-lac* gene fusions, *J. Bacteriol.* **158:**764–766.

JONES, I. M., PRIMROSE, S. B., ROBINSON, A., AND ELLWOOD, D. C. 1980. Maintenance of some ColE1-type plasmids in chemostat culture, *Mol. Gen. Genet.* **180:**579–584.

JONES, L. R., 1909. The bacterial soft rots of certain vegetables. II. Pectinase, the cytolytic enzyme produced by *Bacillus carotovorus* and certain other soft-rot organisms, *Vt. Agric. Exp. Sta. Bull.* **147:**283–360.

KAMIMYA, S., IZAKI, K., AND TAKAHASHI, H. 1972. A new pectolytic enzyme in *Erwinia aroideae* formed in the presence of nalidixic acid, *Agric. Biol. Chem.* **36:**2367–2372.

KAMIMYA, S., NISHIYA, T., IZAKI, K., AND TAKAHASHI, H. 1974. Purification and properties of a pectin *trans*-eliminase in *Erwinia aroideae* formed in the presence of nalidixic acid, *Agric. Biol. Chem.* **38:**1071–1078.

KEEGSTRA, K., TALMADGE, K. W., BAUER, W. D., AND ALBERSHEIM, P. 1973. The structure of plant cell walls. III. A model of the walls of suspension-cultured sycamore cells based on interconnections of the macromolecular components, *Plant Physiol.* **51:**188–197.

KEEN, N. T., DAHLBECK, D., STASKAWICZ, B., AND BELSER, W. 1984. Molecular cloning of pectate lyase genes from *Erwinia chrysanthemi* and their expression in *Escherichia coli, J. Bacteriol.* **159**:825–831.

KELMAN, A. 1979. How bacteria induce disease, in: *Plant Disease, An Advanced Treatise,* Vol. 4, *How Pathogens Induce Disease* (J. G. Horsfall and E. B. Cowling, eds.), pp. 181–202, Academic Press, New York.

KIKUCHI, A., AND GORINI, L. 1975. Similarity of genes *argF* and *argI, Nature* **256**:621–624.

KILGORE, W. W., AND STARR, M. P. 1959. Catabolism of galacturonic and glucuronic acids by *Erwinia carotovora, J. Biol. Chem.* **234**:2227–2235.

KOTOUJANSKY, A., LEMATTRE, M., AND BOISTARD, P. 1982. Utilization of a thermosensitive episome bearing transposon Tn*10* to isolate Hfr donor strains of *Erwinia carotovora* subsp. *chrysanthemi, J. Bacteriol.* **150**:122–131.

KOTOUJANSKY, A., DIOLEZ, A., BOCCARA, M., BERTHEAU, Y., ANDRO, T., AND COLENO, A. 1985. Molecular cloning of *Erwinia chrysanthemi* pectinase and cellulase structural genes, *EMBO J.* **4**:781–785.

LACY, G. H. 1978. Genetic studies with plasmid RP1 in *Erwinia chrysanthemi* strains pathogenic on maize, *Phytopathology* **68**:1323–1330.

LACY, G. H., AND SPARKS, R. B. 1979. Transformation of *Erwinia herbicola* with plasmid pBR322 deoxyribonucleic acid, *Phytopathology* **69**:1293–1297.

LACY, G. H., HIRANO, S. S., VICTORIA, J. I., KELMAN, A., AND UPPER, C. D. 1979. Inhibition of soft-rotting *Erwinia* spp. strains by 2,4-dihydroxy-7-methoxy-2*H*-1,4-benzoxazin-3(4*H*)-one in relation to their pathogenicity on *Zea mays, Phytopathology* **69**:757–763.

LEARY, J. V., AND FULBRIGHT, D. W. 1982. Chromosomal genetics of *Pseudomonas* spp. and *Erwinia* spp., in: *Phytopathogenic Prokaryotes,* Vol. 2 (M. S. Mount and G. H. Lacy, eds.), pp. 229–253, Academic Press, New York.

LEE, S. C., AND WEST, C. A. 1981. Polygalacturonase from *Rhizopus stolonifer,* an elicitor of casbene synthetase activity in castor bean (*Ricinus communis* L.) seedlings, *Plant Physiol.* **67**:633–639.

LEGRAIN, C., STALON, V., AND GLANSDORFF, N. 1976. *Escherichia coli* ornithine carbamoyltransferase isoenzymes: Evolutionary significance and the isolation of λ *argF* and λ *argI* transducing bacteriophages, *J. Bacteriol.* **128**:35–38.

LEI, S. P., LIN, H. C., HEFFERNAN, L., AND WILCOX, G. 1985. Cloning of the pectate lyase genes from *Erwinia carotovora* and their expression in *Escherichia coli, Gene* **35**:63–70.

LELLIOTT, R. A., AND DICKEY, R. S. 1984. *Erwinia,* in: *Bergey's Manual of Systematic Bacteriology,* 9th ed. (N. R. Kreig, ed.), pp. 469–476, Williams and Wilkins, Baltimore.

LITTLE, J. W., AND MOUNT, D. W. 1982. The SOS regulatory system of *Escherichia coli, Cell* **29**:11–22.

LITTLE, J. W., EDMISTON, S. H., PACELLI, L. Z., AND MOUNT, D. W. 1980. Cleavage of the *Escherichia coli lexA* protein by the *recA* protease, *Proc. Natl. Acad. Sci. USA* **77**:3225–3229.

LUND, B. M., AND WYATT, G. M. 1972. The effect of oxygen and carbon dioxide concentrations on bacterial soft rot of potatoes. I. King Edward potatoes inoculated with *Erwinia carotovora* var. *atroseptica, Potato Res.* **15**:174–179.

LYON, G. D., AND BAYLISS, C. E. 1975. Effect of rishitin on *Erwinia carotovora* var. *atroseptica* and other bacteria. *Physiol. Plant Pathol.* **6**:177–186.

LYON, G. D., LUND, B. M., BAYLISS, G. E., AND WYATT, G. M. 1975. Resistance of potato tubers to *Erwinia carotovora* and formation of rishitin and phytuberin in infected tissue, *Physiol. Plant Pathol.* **6**:43–50.

MAHER, E. A., AND KELMAN, A. 1983. Oxygen status of potato tuber tissue in relation to maceration by pectic enzymes of *Erwinia carotovora, Phytopathology* **73**:536–539.

MAZZUCCHI, U., ALBERGHINA, A., AND GARIBALDI, A. 1974. Comparative immunological study of pectic lyases produced by soft rot coliform bacteria, *Phytopathol. Mediterr.* **13**:27–35.

MCNEIL, M., DARVILL, A. G., FRY, S. C., AND ALBERSHEIM, P. 1984. Structure and function of the primary cell walls of plants, *Annu. Rev. Biochem.* **53**:625–663.

MENELEY, J. C., AND STANGHELLINI, M. E. 1976. Isolation of soft-rot *Erwinia* spp. from agricultural soils using an enrichment technique, *Phytopathology* **66**:367–370.

MERGAERT, J., VERDONCK, L., KERSTERS, K., SWINGS, J., BOEUFGRAS, J. -M., AND DE LEY, J. 1984. Numerical taxonomy of *Erwinia* species using API systems, *J. Gen. Microbiol.* **130**:1893–1910.

MORAN, F., NASUNO, S., AND STARR, M. P. 1968. Oligogalacturonide *trans*-eliminase of *Erwinia carotovora, Arch. Biochem. Biophys.* **125**:734–741.

MOUNT, M. S., BATEMAN, D. F., AND BASHAM, H. G. 1970. Induction of electrolyte loss, tissue maceration, and cellular death of potato tissue by an endopolygalacturonate *trans*-eliminase, *Phytopathology* **60**:924–931.

MOUNT, M. S., BERMAN, P. M., MORTLOCK, R. P., AND HUBBARD, J. P. 1979. Regulation of endopolygalacturonate *trans*-eliminase in a adenosine 3′,5′-cyclic monophosphate deficient mutant of *Erwinia carotovora, Phytopathology* **69**:117–120.

NASUNO, S., AND STARR, M. P. 1966. Polygalacturonase of *Erwinia carotovora, J. Biol. Chem.* **241**:5298–5306.

Pérombelon, M. C. M. 1982. The impaired host and soft rot bacteria, in: *Phytopathogenic Prokaryotes*, Vol. 2 (M. S. Mount and G. H. Lacy, eds.), pp. 55–69, Academic Press, New York.

Pérombelon, M. C. M., and Boucher, C. 1978. Developing a mating system in *Erwinia carotovora*, in: *Proceedings of the IVth International Conference on Plant Pathogenic Bacteria, Angers, France*, pp. 47–52.

Pérombelon, M. C. M., and Kelman, A. 1980. Ecology of the soft rot erwinias, *Annu. Rev. Phytopathol.* **18**:361–387.

PREISS, J., AND ASHWELL, G. 1963a. Polygalacturonic acid metabolism in bacteria. I. Enzymatic formation of 4-deoxy-L-*threo*-5-hexoseulose uronic acid, *J. Biol. Chem.* **238**:1571–1576.

PREISS, J., AND ASHWELL, G. 1963b. Polygalacturonic acid metabolism in bacteria. II. Formation and metabolism of 3-deoxy-D-glycero-2,5-hexodiulosonic acid, *J. Biol. Chem.* **238**:1577–1583.

PUGSLEY, A. P., AND SCHWARTZ, M. 1985. Export and secretion of proteins by bacteria, FE*MS Microbiol. Rev.* **32**:3–38.

PUPILLO, P., MAZZUCCHI, U., AND PIERINI, G. 1976. Pectic lyase isozymes produced by *Erwinia chrysanthemi* Burkh. *et al.* in polypectate broth or in *Dieffenbachia* leaves, *Physiol. Plant Pathol.* **9**:113–120.

QUANTICK, P., CERVONE, F., AND WOOD, R. K. S. 1983. Isoenzymes of a polygalacturonate *trans*-eliminase produced by *Erwinia atroseptica* in potato tissue and in liquid culture, *Physiol. Plant Pathol.* **22**:77–86.

RESIBOIS, A., COLET, M., FAELEN, M., SCHOONEJANS, E., AND TOUSSAINT, A. 1984. EC2, a new generalized transducing phage of *Erwinia chrysanthemi*, *Virology* **137**:102–112.

REVERCHON, S. AND ROBERT-BAUDOUY, J. 1985. Genetic transformation of the phytopathogenic bacteria, *Erwinia chrysanthemi*, *Biochimie* **67**:253–257.

REVERCHON, S., HUGOUVIEUX-COTTE-PATTAT, N., AND ROBERT-BAUDOUY, J. 1985. Cloning of genes encoding pectolytic enzymes from a genomic library of the phytopathogenic bacterium, *Erwinia chrysanthemi*, *Gene* **35**:121–130.

RIED, J. L., AND COLLMER, A. 1985. An activity stain for the rapid characterization of pectic enzymes in isoelectric focusing and sodium dodecyl sulfate polyacrylamide gels, *Appl. Environ. Microbiol.* **50**:615–622.

RILEY, M., AND ANILIONIS, A. 1978. Evolution of the bacterial genome, *Annu. Rev. Microbiol.* **32**:519–560.

ROBERTS, D. P., BERMAN, P. M., LACY, G. H., MOUNT, M. S., AND ALLEN, C. 1984. *Erwinia carotovora* subsp. *carotovora* DNA encoding pectate lyases cloned into plasmid pBR322, *Phytopathology* **74**:797. Abstract.

ROBERTS, J. W., ROBERTS, C. W., AND CRAIG, N. L. 1978. *Escherichia coli recA* gene product inactivates phage λ repressor, *Proc. Nat. Acad. Sci. USA* **75**:4714–4718.

ROBISON, A. G., BUTCHER, R. W., AND SUTHERLAND, E. W. 1971. *Cyclic AMP,* Academic Press, New York.

ROEDER, D. L., AND COLLMER, A. 1985. Marker-exchange mutagenesis of a pectate lyase isozyme gene in *Erwinia chrysanthemi, J. Bacteriol.* **164**:51–56.

ROMBOUTS, F. M., AND PILNIK, W. 1980. Pectic enzymes, in: *Microbial Enzymes and Bioconversions: Economic Microbiology,* Vol. 5 (A. H. Rose, ed.), pp. 227–282, Academic Press, New York.

SAMSON, R., AND NASSAN-AGHA, N. 1979. Biovars and serovars among 129 strains of *Erwinia chrysanthemi*, in: *Proceedings of the Fourth International Conference on Plant Pathogenic Bacteria, Angers, France, 1978* (Station de Pathologie Végétale et Phytobactériologie, ed.), pp. 547–553, Institute National de la Recherche Agronomique, Angers, France.

SCHOONEJANS, E., AND TOUSSAINT, A. 1983. Utilization of plasmid pULB113 (RP4::mini-Mu) to construct a linkage map of *E. carotovora* subsp. *chrysanthemi, J. Bacteriol.* **154**:1489–1492.

SILHAVY, T., BENSON, S. A., AND EMR, S. D. 1983. Mechanisms of protein localization, *Microbiol. Rev.* **47**:313–344.

SPRINGER, W., AND GOEBEL, W. 1980. Synthesis and secretion of hemolysin by *Escherichia coli, J. Bacteriol.* **144**:53–59.

STACK, J. P., MOUNT, M. S., BERMAN, P. M., AND HUBBARD, J. P. 1980. Pectic enzyme complex from *Erwinia carotovora*: A model for degradation and assimilation of host pectic fractions, *Phytopathology* **70**:267–272.

STANGHELLINI, M. E. 1982. Soft-rotting bacteria in the rhizosphere, in: *Phytopathogenic Prokaryotes,* Vol. I (M. S. Mount and G. H. Lacy, eds), pp. 249–261, Academic Press, New York.

STARR, M. P. 1981. The genus *Erwinia,* in: *The Prokaryotes: A Handbook on Habitats, Isolation, and Identification of Bacteria* (M. P. Starr, H. Stolp, J. H. Trüper, A. Balows, and H. G. Schlegel, eds.), pp. 1260–1271, Springer-Verlag, New York.

STEPHENS, G. J., AND WOOD, R. K. S., 1975. Killing of protoplasts by soft-rot bacteria, *Physiol. Plant Pathol.* **5**:165–181.

TOMIZAWA, H., AND TAKAHASHI, H. 1971. Stimulation of pectolytic enzyme formation of *Erwinia aroideae* by nalidixic acid, mitomycin C and bleomycin, *Agric. Biol. Chem.* **35**:191–200.

TSUYUMU, S. 1977. Inducer of pectic acid lyase in *Erwinia carotovora, Nature* **269**:237–238.

TSUYUMU, S. 1979. "Self-catabolite repression" of pectate lyase in *Erwinia carotovora, J. Bacteriol.* **137**:1035–1036.

TSUYUMU, S., AND CHATTERJEE, A. K. 1984. Pectin lyase production in *Erwinia chrysanthemi* and othe soft-rot *Erwinia* species, *Physiol. Plant Pathol.* **24**:291–302.

VANETTEN, H. D., AND KISTLER, H. C. 1984. Microbial enzyme regulation and its importance for pathogenicity, in: *Plant-Microbe Interactions: Molecular and Genetic Perspectives,* Vol. 1 (T. Kosuge and E. W. Nester, eds.), pp. 42–68, Macmillan, New York.

VAN GIJSEGEM, F., AND TOUSSAINT, A. 1982. Chromosome transfer and R-prime formation by an RP4::mini-Mu derivative in *Escherichia coli, Salmonella typhimurium, Klebsiella pneumoniae,* and *Proteus mirabilis, Plasmid* **7**:30–44.

VAN GIJSEGEM, F., AND TOUSSAINT, A. 1983. *In vivo* cloning of *Erwinia carotovora* genes involved in the catabolism of hexuronates, *J. Bacteriol.* **154**:1277–1235.

VAN GIJSEGEM, F., HUGOUVIEUX-COTTE-PATTAT, N., AND ROBERT-BAUDOUY, J. 1985a. Isolation and characterization of *Erwinia chrysanthemi* mutants defective in degradation of hexuronates, *J. Bacteriol.* **161**:702–708.

VAN GIJSEGEM, F., TOUSSAINT, A., AND SCHOONEJANS, E. 1985b. *In vivo* cloning of the pectate lyase and cellulase genes of *Erwinia chrysanthemi, EMBO J.* **4**:787–792.

VAN HAUTE, E., JOOS, H., MAES, M., WARREN, G., VAN MONTAGU, M., AND SCHELL, J. 1983. Intergeneric transfer and exchange recombination of restriction fragments cloned in pBR322: A novel strategy for the reversed genetics of the Ti plasmids of *Agrobacterium tumefaciens*, *EMBO J.* **2:**411–417.

VIEIRA, J., AND MESSING, J. 1982. The pUC plasmids, an M13mp7-derived system for insertion mutagenesis and sequencing with synthetic universal primers, *Gene,* **19:**259–268.

YAMAZAKI, N., FRY, S. C., DARVILL, A. G., AND ALBERSHEIM, P. 1983. Host-pathogen interactions. XXIV. Fragments isolated from suspension-cultured sycamore cell walls inhibit the ability of the cells to incorporate [^{14}C] leucine into proteins, *Plant Physiol.* **72:**864–869.

ZINK, R. T., AND CHATTERJEE, A. K. 1985. Cloning and expression in *Escherichia coli* of pectinase genes of *Erwinia carotovora* subsp. *carotovora,* *Appl. Environ. Microbiol.* **49:**714–747.

ZINK, R. T., KEMBLE, R. J., AND CHATTERJEE, A. K. 1984. Transposon Tn5 mutagenesis in *Erwinia carotovora* subsp. *carotovora* and *E. carotovora* subsp. *atroseptica, J. Bacteriol.* **157:**809–814.

ZUCKER, M., AND HANKIN, L. 1970. Regulation of pectate lyase synthesis in *Pseudomonas fluorescens* and *Erwinia carotovora, J. Bacteriol.* **104:**13–18.

ZUCKER, M., HANKIN, L., AND SANDS, D. 1972. Factors governing pectate lyase synthesis in soft rot and non-soft rot bacteria, *Physiol. Plant Pathol.* **2:**59–67.

Chapter 9

Systemic Movement of Viruses

Tetsuo Meshi and Yoshimi Okada

IN PLANT VIRUS COMBINATIONS that lead to systemic diseases in plants, viruses replicate in the cells infected first and the move to other noninfected cells.There are two types of virus movement in plants: (1) slow cell-to-cell movement from an infected cell to adjacent cells, presumably through plasmodesmata, and (2) long-distance movement which occurs fairly rapidly from infected leaves to other tissues throughout the vascular system.

On the other hand, it has been shown that there are many virus-host combinations in which viruses cannot spread systemically. Practically or phenomenologically they can be classified into three major groups. In the first group accumulation is confined to the cells first infected and adjacent cells do not contain any virus. In this case, virus movement is blocked at the level of cell-to-cell movement and the plant appears to be immune to the virus. In the second group, the virus can move from cell to cell but is localized in a small area around the infection center, often causing a local necrotic lesion. This situation is referred to as hypersensitivity. In the third group, the virus can move from cell to cell within the inoculated leaves, but not to other uninoculated leaves.

Virus movement in plants is evidently affected by many factors, such as virus genes, the genetic background and developmental stage of the host plant, and various environmental conditions such as temperature, light, and so on. If the conditions change, virus movement and consequently the final distribution of a virus in a plant may also change. Thus the spread of a virus is a complicated process. At present, our understanding of virus movement at the molecular level is far from complete. However, it is evident that virus movement is not necessarily correlated with the extent of virus replication or accumulation in infected cells or with the appearance of disease symptoms.

GENERAL ASPECTS OF VIRUS MOVEMENT

Movement of Viruses in Whole Plants

In the classical experiments of Samuel (1934), tobacco mosaic virus (TMV), upon inoculation into a terminal leaflet of a middle-aged tomato plant, moved first to the root, then to the young leaves, and then to the leaflets close to the infected one. Some time later the middle-aged and older leaves became infected.

Dawson et al. (1975) developed a differential temperature treatment which allows synchronous virus replication in whole leaves. Lower tobacco leaves were inoculated mechanically with TMV and maintained at 25°C, upper uninoculated leaves were maintained at 5°C. At the the latter temperature TMV cannot multiply, but infectious materials can move into the upper young leaves. After the upper uninoculated leaves were shifted to a permissive temperature, replication began simultaneously in all infected cells of the upper leaves that TMV had reached. The characteristic pattern of necrosis appearing in the upper leaves after this treatment reflected the spread of infectious materials from lower inoculated leaves to upper uninoculated leaves (Dorokhov et al., 1981). The infectious materials were localized mainly in the vascular system and not in the mesophyll cells after 7–10 days of differential temperature treatment. About 90% of the mesophyll cells of the upper leaves were infected 3 days after being transferred to the permissive temperature. The infection spread first to the basal areas of the uppermost leaves and later to the basal parts of the lower leaves.

Ohashi and Shimomura (1971) reported the spread of infection in the tobacco leaf after point inoculation with TMV. A necrotic lesion reflecting virus spread and replication, can be induced by heat treatment (50°C, 2 min) of TMV-inoculated leaves of a systemic host or by transferring inoculated leaves of a loal lesion host from 30 to 23°C. The lesion enlarged first to some extent around the center of the infection point, then spread along the lateral and the main veins, and then around the main vein.

Long-distance movement of viruses occurs through the phloem; because virus movement is influenced by the flow of metabolites (Bennett, 1940), some viruses are found in the sieve elements (Esau and Cronshaw, 1967; Esau et al., 1967), and symptoms appear in tissues along the phloem and not in the xylem during the course of infection (Hatta and Matthews, 1974). Furthermore, when detached leaves or stems were placed in a virus-containing solution, the virus moved into the leaves, probably through the xylem but no infection occurred. In contrast, a few viruses have been found in the xylem (Chambers and Francki, 1966).

In cell-to-cell movement, viruses have been thought to move through the plasmodesmata, which are protoplastic strands extending from one cell to another through the cell wall (Fig.1) (Allison and Shalla, 1974; Gunning and Roberts, 1976). The most direct evidence supporting this idea is electron microscopic observation of viral particles in plasmodesmata (Figs. 2 and 3) (Esau et al., 1967; de Zoeten and Gaard, 1969; Kitajima and Lauritis, 1969; Alison and Shalla, 1974; Weintraub et al., 1976). However, the viral form moving from cell to cell may not be the viral particle (to be discussed in detail later).

Factors Affecting Virus Movement

The spread of a virus is evidently affected by many environmental and experimental conditions such as temperature, light, water potential, materials supplying energy, hormones, and the developmental stage and senescence of the host plant (Cheo, 1971; Colhoun, 1973; Matthews, 1981; Weststeijn, 1981; Wu and Dimitman, 1984). Each factor affects others, and all influence the metabolism of host plants. The resultant change in virus spread depends on alteration of the host defense mechanisms in one case and on the secondary effects of the stream of metabolites in the host in another case.

Genes of host plants and viruses also affect the extent of virus movement, described in detail below.

Form in which a Virus Moves

The form in which a virus moves has not been demonstrated conclusively. As mentioned above, it has long been thought that virus particles are the transported form. Virus particles have been observed within the plas-

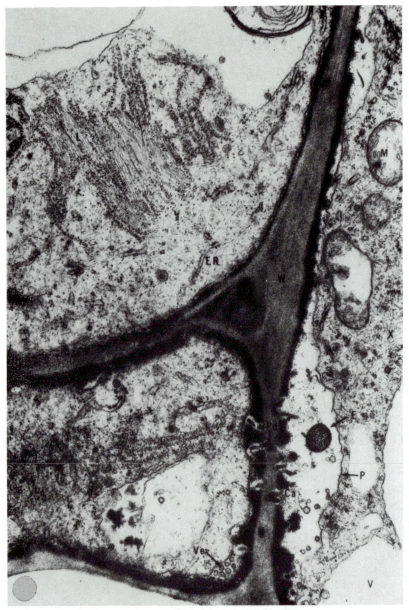

Figure 1. Plasmodesmata in the wall of virus-infected leaf cells from
Gomphrena globosa. This figure shows portions of a bundle sheath
parenchyma cell (right) and two spongy parenchyma cells from the zone immediately adjacent to necrotic cells of a lesion caused by potato virus X on a
leaf of *G. globosa.* Several plasmodesmata (Pd) can be seen as channels extending through the cell wall (W), deposits of callose (Ca), a β-1, 3 glucan

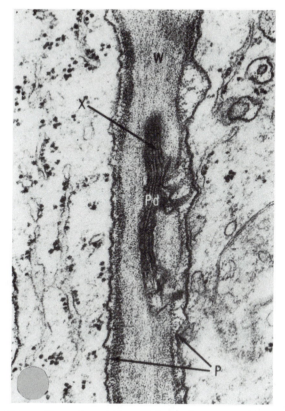

Figure 2. A plasmodesmal complex (Pd) can be seen with particles (X) of potato virus X. The plasmodesmal complex is imbedded in the cell wall (W) and appears in longitudinal section. Its central cavity is filled with virus particles, which appear as elongated strands. The sample was taken from a leaf from *G. globosa* that had been inoculated with potato virus X. Also shown is plasmalemma (P). Magnification: × 42,600. (From Allison and Shalla, 1974; reproduced with the permission of A. Allison and The American Phytopathological Society.)

occur between the plasmalemma (P) and the primary cell wall (W). An extracytoplasmic sac of virus (E) is embedded in the callose. Virus particles (X) are found scattered in the cytoplasm which also contains laminate inclusion components (L). Also shown are endoplasmic reticulum, (ER); mitochondrion, (M); vacuole, (V); vesicles, (Ve). Magnification: × 35,000. (From Allison and Shalla, 1974; reproduced with the permission of A. Allison and The American Phytopathological Society.)

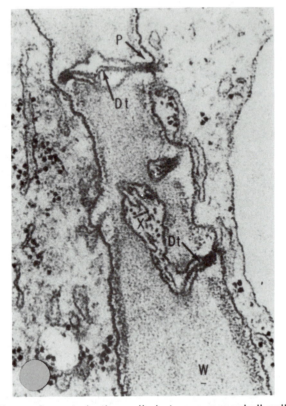

Figure 3. Plasmodesmata in the walls between mesophyll cells in the peripheral zone of a virus-caused lesion on a *G. globosa* leaf. In the upper plasmodesma, the neck is distended slightly beyond the cell wall (W), but the plasmalemma (P) is continuous with the lining of the plasmodesmal canal and cavity. The desmotubule (Dt) appears to traverse the entire canal of the plasmodesma. In the lower plasmodesma, virus particles (X) appear in longitudinal and cross-section view; the desmoluble (Dt) is still evident in the right-hand terminus of the plasmodesma. Magnification: × 55,200. (From Allison and Shalla, 1974; reproduced with the permission of A. Allison and The American Phytopathological Society.)

modesmata and phloem of infected plants (see above). However, coat protein-defective mutants (Siegel et al., 1962) and the larger RNA, RNA1, of tobacco rattle virus (Lister, 1969) can infect cells and spread from cell to cell. The former cannot produce complete virus particles, and the latter does not code for coat protein. These observations indicate that viral coat protein is not essential for virus cell-to-cell movement in all systems.

However, it cannot be concluded that naked viral single-stranded RNA or a replicative intermediate is the form involved in cell-to-cell movement.

Dorokhov et al. (1983) have recently isolated a new type of informosome-like ribonucleoprotein (vRNP) particles from TMV-infected tobacco leaves. vRNP particles have a filamentous structure and a buoyant density of about 1.40 g/cm³ differ from mature TMV particles which have a rigid rod-shaped structure and a buoyant density of 1.33 g/cm³. The appearance of vRNP is specific to the TMV infection and not influenced by the presence of actinomycin D. vRNP particles are not precursors of mature virions and are thought to be involved in translation (Dorokhov et al., 1984a). The major components of vRNP are polypeptides with molecular weights of 17,500 (TMV coat protein), 31,000, 37,000, 39,000, and 76,000, and genomic and intermediate-class subgenomic RNAs. Dorokhov et al. (1984b) suggested the involvement of vRNP both in cell-to-cell movement and in long-distance movement based on the following observations. With the use of a temperature-sensitive (ts) mutant defective in cell-to-cell movement (details given later), they observed a correlation between the blockage of virus movement and the absence of vRNP at a restrictive temperature. vRNP but not virus particles were detected in the upper uninoculated leaves under different temperature conditions, where potentially infectious materials exist. The systemic spread and the detection of vRNP in the upper uninoculated leaves were also correlated. From these results, vRNP is taken to be a candidate for the form in which virus genetic materials move through the vascular system and plasmodesmata.

VIRUS MOVEMENT AND THE HOST RANGE

There are two types of resistance in plants, even in species that show phenotypically the same immune response to a given virus infection: (1) true immunity, in which the virus cannot replicate in any of the cells of the plant; and (2) blocking of the cell-to-cell movement of the virus, in which the virus can replicate in an infected cell but cannot move from the cell first infected to other uninfected cells.

Beier et al. (1977) surveyed 1031 lines of cowpea (*Vigna sinensis*) and observed the response to mechanical inoculation of the SB isolate of cowpea mosaic virus (CPMV) on the primary leaves of seedlings. They classified 65 lines as immune because they showed no symptoms and yielded no virus, but protoplasts prepared from 54 out of 55 immune lines were found to be susceptible to CPMV SB.

Similar results have been reported. The cowpea plant is not a natural host of the common strain of TMV; it is characterized as being a subliminally infected host, and only a miniscule amount TMV is recovered from inoculated leaves. Koike et al. (1976) and Huber et al. (1981) showed that mesophyll protoplasts isolated from cowpea leaves were susceptible to infection with the common strain of TMV.

Sulzinski and Zaitlin (1982) investigated the replication of TMV in subliminally infected hosts—cowpea and cotton. They isolated mesophyll protoplasts after mechanical inoculation with TMV at intervals and counted the number of protoplasts in which TMV coat protein was detected by fluorescent antibodies. Only 1 in $5-15 \times 10^4$ protoplasts was infected with TMV, and this number remained unchanged from immediately after inoculation up to at least 11 days after inoculation. These observations indicated that subliminally infected and (possibly many) immune hosts can support virus multiplication in individual cells, but that the virus cannot move from the infected cells to other uninfected cells.

Although it is not known what constitutes a barrier to virus movement in such phenotypically immune hosts, this barrier can be overcome by the presence of a helper virus. Hamilton and Dodds (1970) and Dodds and Hamilton (1972) showed that TMV could replicate and spread systemically in barley (*Hordeum vulgaris L.,* reported to be immune to TMV) when TMV was coinoculated into the primary leaves with barley stripe mosaic virus (BSMV). The systemic spread of TMV in barley was also stimulated by brome mosaic virus (BMV), and the two viruses replicated independently (Hamilton and Nichols, 1977).

As described earlier, viruses generally cannot infect upper uninoculated leaves via the conducting system when plants are placed in a virus suspension. Taliansky et al. (1982b,c) showed that when upper leaves were preinfected with a helper virus, a virus in solution could spread into the upper leaves and infect them, even if the plant was a nonhost for the virus. For example, DMV spread in tomato (*Lycopersicon esculentum L.*) and bean (*Phaseolus vulgaris L.*) after preinfection with TMV tomato strain L and cowpea strain Cc, respectively. TMV common strain *P. vulgare* infected wheat (*Triticum vulgare L.*) with BSMV as a helper. In Tm-2 tomato, a line of tomato cultivars resistant to TMV infection (details given below), TMV L spread after preinfection with potato virus X (PVX).

These data showing that nonspreading viruses can multiply and spread in nonhost plants in the presence of a helper virus, indicate that a virus-specific transport function determines the host range of a given virus. The observation that distinctly different viruses can function as helpers or can complement a defect in a nonspreading virus in a nonhost suggests that the mechanim of cell-to-cell movement is common enough to many, if not all, groups of plant viruses (Taliansky et al., 1982b). The viral genes responsible for the helper activity will be described later.

HOST GENES AFFECTING VIRUS MOVEMENT

The genotypes of hosts influence the response to viral infection. Studies on host genotypes have focused mainly on resistance to virus infection, which is important in plant breeding. However, little is known about the products of the genes and their functions. The molecular mechanisms involved in the host response are extremely complicated. Phenotypically, the same response could be caused by other combinations of distinctly different factors. In this section, from the viewpoint of virus spread in host plants, we describe three different cases of genes affecting the final distribution of viruses in plants.

The *N* Gene in Tobacco

Several tobacco varieties carrying the *N* gene, e.g., *Nicotiana glutinosa* and *N. tabacum* L. cv. Xanthi-nc and cv. Samsun NN, respond to infection by TMV with the formation of a necrotic local lesion (for a review see Sela, 1981). The local lesion can be seen about 30 hr after inoculation and increases in size until 8–10 days after inoculation. Finally, viruses are confined to a small area around the center of infection of the inoculated leaves, and no systemic spread occurs. This hypersensitive reaction is affected by temperature and the developmental stage of the plant. At high temperatures, above 28°C, and in young seedlings a hypersensitive response is not observed.

Otsuki et al. (1972) showed that in TMV-infected Samsun NN tobacco virus multiplication started readily after inoculation but decreased after the lesions formed. Protoplasts isolated from Samsun NN tobacco leaves were susceptible to TMV, and no necrotic response was detected (Otsuki et al., 1972; Loebenstein et al., 1980). An auxinlike regulator, 2,4-dichlorophenoxyacetic acid (2,4-D), reduces the lesion size at low concentrations but promotes it at higher concentrations (van Loon, 1979). In the presence of 2,4-D the multiplication of TMV in protoplasts of Samsun NN was the same as that in Samsun (deficient in the *N* gene) but in the absence of 2,4-D the multiplication rate was reduced (Loebenstein et al., 1980). These observations suggest that virus multiplication is not directly correlated with necrotization but that the reduced rate of multiplication is one of the factors controlling the final localization of the virus. Cell walls or cell-to-cell contacts may be necessary for localization and necrotization.

Several days after TMV inoculation into hypersensitive tobacco leaves, uninfected parts of inoculated leaves and upper uninoculated leaves show resistance to subsequent infection with TMV, as well as to other viruses causing a reduction in the size or number of lesions (Ross, 1961). This phenomenon is referred to as systemic acquired resistance. Actinomycin D and ultraviolet (UV) irradiation are known to cause lesion enlargement and to reduce this resistance (Loebenstein et al 1968, 1970). Novel transcription and protein synthesis seem to be necessary for the establishment of systemic acquired resistance. Several proteins, referred to as pathogenesis-related (PR or b) proteins, have been found in resistant leaves (van Loon and van Kammen, 1970). These PR proteins can be induced by other stimuli, e.g., polyacrylic acid, an interferon inducer in animals (Gianinazzi and Kassanis, 1974; Cassells et al., 1978), and acetylsalicylic acid (White, 1979). PR proteins have also been found in Samsun tobacco (not carrying the N gene) after infection with tobacco necrosis virus (TNV). Therefore PR proteins are not direct products of the N gene. While hypersensitive tobaccos form necrotic local lesions with all strains of TMV, certain tobacco varieties not containing the N gene also respond by lesion formation to infection with certain TMV strains. TMV strain flavum is restricted to necrotic local lesions in White Burley tobacco leaves and induces a response similar to systemic acquired resistance in hypersensitive tobaccos (Fraser, 1979). This indicates that systemic acquired resistance does not depend on whether a given tobacco carries the N gene.

Schuster and Flemming (1976) demonstrated the existence of a barrier in nonnecrotized cells around local lesions of TMV, which reduced the transport of substrates. Deposition of callose (β-1,3-glucan) in cells around local lesions of TMV-infected NN tobacco leaves may cause the final viral localization by physical means by leading to thickening of cell walls and probably blockage of the plasmodesmata (Shimomura and Dijkstra, 1975). Favali et al. (1978) showed that sieve elements in the area around the lesions in $N.$ $glutinosa$ became occluded by callose. Confinement of a virus in a small area around the infected center results after many molecular processes have occurred. Callose deposition has been observed around the necrotic lesions formed in other virus-host combinations (Allison and Shalla, 1974; Wu and Dimitman, 1970). The deposition of callose may be involved in viral localization, but this response is not N gene-specific.

Many studies have reported metabolic alterations in or near local necrotic lesions, including protein synthesis (van Loon and van Kammen, 1970) and increases in the level of phenolic compounds (Duchesne et al., 1977; Weststeijn, 1978), permeability of the cell membrane (Weststeijn, 1978), and hormone concentrations (Balázs et al., 1969). The N gene product is thought to be a kind of inducer responsible for a series of factors governing the final virus localization, necrotic reactions, and systemic acquired resistance. It is known that these phenotypically visible responses can be induced by other stimuli. The genetics of the N gene and molecular characterization of PR proteins have already been surveyed in this series by Gianinazzi (1984) and Ryan (1984).

Tm-2 and Tm-2² in Tomato

Another class of resistance genes is known in tomato. Two genes, *Tm-2* and *Tm-2²*, confer a hypersensitive response to infection with TMV (Cirulli and Alexander, 1969; Pelham, 1972). *Tm-2²* is an allele of *Tm-2* (Alexander, 1971) but induces more extreme hypersensitivity (Pelham, 1972), and no virus isolates have been reported that can overcome the *Tm-2²* effect. A tomato breeding line homozygous for *Tm-2* is immune to infection with TMV L (Motoyoshi and Ohshima, 1975). No virus accumulation or symptoms are observed. However, TMV L can replicate in leaf mesophyll protoplasts prepared from these resistant tomatoes (Motoyoshi and Ohshima, 1975, 1977). Although this situation is similar to the case of a tobacco carrying the *N* gene, the *Tm-2* and *Tm-2²* effects on TMV L are more directly involved in the blockage of cell-to-cell movement of the virus. As described earlier, PVX can complement the *Tm-2* effect on TMV L by working as a helper virus (Taliansky et al., 1982b). It is possible to obtain a virus that can overcome the resistance of Tm-2 tomato spontaneously, and therefore one or a few viral protein(s) are involved in the interaction of host component(s) determining the resistance of the Tm-2 tomato. The system using TMV L, and L-derived mutant that overcomes Tm-2 resistance, and Tm-2 tomato will be a good system in which to investigate the virus-host interaction with respect to cell-to-cell movement.

Cowpea Chlorotic Mottle Virus in Cowpea

Many strains of cowpea chlorotic mottle virus (CCMV) cause no symptoms in cowpea PI 186465. When CCMV type strain T was inoculated into the primary leaves of this plant, the virus could be isolated from the primary leaves but not from the trifoliate leaves (Wyatt and Kuhn, 1979). The virus accumulation rate was low, 12–20 times less than the rate measured during the rapid phase of replication in susceptible cowpeas. Viruses isolated from CCMV T-inoculated PI 186465 leaves showed about a 90% deficiency with respect to RNA 3 compared with the viruses purified from susceptible cowpeas. The resistance of PI 186465 appears to be expressed against virus replication and systemic spread (Wyatt and Khun, 1979, 1980). Kuhn et al. (1981) analyzed the genetic control of symptom expression, movement, and virus accumulation in cowpea plants inoculated with CCMV. Although virus accumulation and replication appeared to be controlled by the action of several genes, virus movement was controlled by a single dominant gene (*Mv*). Cowpea plants having homozygous recessive (*mv,mv*) genes like PI 186465, can restrict the systemic spread of the virus.

VIRUS GENES INVOLVED IN MOVEMENT

Evidently, several products of a virus affect virus movement. In the case of TMV at least the coat protein and 30K protein are responsible for its systemic movement.

Coat Protein

All TMV mutants defective in coat protein are known to spread within inoculated leaves but usually do not move into upper and lower uninoculated leaves (Siegel et al., 1962). This suggests that coat protein is involved in the long-distance movement of TMV but not in cell-to-cell movement. Coat protein is one of the constituents of the informosome-like vRNP found by Dorokhov et al. (1983), which is suggested to be the form in which the virus moves. Dorokhov et al. (1984b) and Taliansky et al. (1982) revealed that a defect in the long-distance movement of coat protein mutant Ni118 (Wittman et al. 1965) is complemented by Ni2519 or Ls1 which are defective in cell-to-cell movement but can produce normal coat protein (see below). Viral RNA in vRNP of Ni118 was more sensitive to ribonuclease attack than that of wild-type *T. vulgare* (Dorokhov et al., 1983). Dorokhov et al. (1984b) suggested that the function of coat protein in long-distance movement is to protect the genomic RNA from nucleolytic attack during movement through the system.

30K Protein

TMV Ls1 is a spontaneous *ts* mutant derived from wild-type tomato strain L isolated by Nishiguchi et al. (1978). In inoculated tomato leaves, Ls1 does not spread at 32°C but does so at 22°C. However, it can replicate in isolated protoplasts at both temperatures at the same rate as in the parent strain L. When the lower epidermis was peeled off at 24-hr intervals after inoculation with Ls1 and infected cells were visualized using fluorescent antibodies, no increase in the number of fluorescent spots consisting of more than two fluorescent cells was observed when tomato plants were maintained at 32°C (Nishiguchi et al., 1980). On the other hand, in tomatoes inoculated with Ls1 and maintained at 22°C and those inoculated with L, the spread of infection was observed. These observations lead to the conclusion that the defect in Ls1 is in the virus cell-to-cell movement.

Taliansky et al. (1982c) showed that the defect in the transport function in Ls1 is complemented by a helper virus. When Ls1 was coinoculated with common strain *T. vulgare* or PVX, which can move from cell to cell at a high temperature, Ls1 also spread. When upper leaves were preinoculated with a helper virus (e.g., L or Ni118) and the stems placed in a solution containing Ls1, Ls1 spread into the upper leaves at a nonpermissive temperature. On the contrary, Ls1 acted as a helper at 24°C but not at 32°C (Taliansky et al., 1982c). The observation that viruses defective in cell-to-cell movement can be complemented by a helper virus suggests that the virus codes for a trans-acting diffusible material responsible for the transport function, tentatively called a transport protein. The available data indicate that the 30K protein is the transport protein.

Leonard and Zaitlin (1982) compared tryptic peptide maps of TMV L and Ls1 polypeptides. They found a difference only in the maps of the 30K proteins. This was further supported by Ohno et al. (1983) who determined nucleotide sequences including that of the 30K and coat protein cistrons of Ls1 using cloned cDNA copies and compared the results with those for the parent strain L. Only one amino acid substitution was found between the two within the sequenced region and Pro was substituted for Ser in the 30K protein.

Ni2519 is a nitrous acid-induced mutant derived from a parent strain, A14 (a common strain) selected to be temperature-sensitive for lesion formation on hosts carrying the *N* gene (Wittman et al., 1965). Lesions of Ni2519 expand at 25°C (although they are smaller than those of the wild type) but not at 33°C (Jockusch, 1968). At a nonpermissive temperature, functional coat protein and actinomycin D-resistant RNA synthesis was observed in isolated Ni2519 infected cells, but no coat protein was detected in the infected leaves (Bosch and Jockusch, 1972). This observation led to the suggestion that like Ls1, Ni2519 was defective in cell-to-cell movement. Actually, the defect in cell-to-cell movement in Ni2519 was complemented by helper viruses such as wild-type common strain *T. vulgare* when coinoculated. However, the defect in long-distance movement of Ni2519 could not be complemented, although Ni2519 complemented Ni118 (Taliansky et al., 1982a). Ni2519 has an additional cis-dominant mutation in virus assembly which cannot be complemented by any TMV strain (Taliansky et al., 1982a). Zimmern and Hunter (1983) showed that both defects could be explained as an A-to-G transition in the 30K protein cistron, which caused an amino acid substitution of Arg for Gly in the 30K protein and probably at the same time a structural alteration in RNA close to the assembly origin.

The 30K protein is translated from a subgenomic mRNA called I_2-RNA (Bruening et al., 1976) and is easily detected as a product using either an *in vitro* system of a wheat germ extract or a rabbit reticulocyte lysate. However *in vivo* detection of the 30K protein was not accomplished easily, mainly because of the small amount in the infected tissues. In TMV-infected tobacco leaves, Joshi et al. (1983) first detected a polypeptide thought to be the 30K protein specific to TMV infection, but only for short periods, 16.5–24 hr after inoculation. Ooshika et al. (1984) and Kiberstis et al. (1983) independently prepared antibodies against a synthetic peptide corresponding to the C-terminal 16- and 11-amino-acid sequence of the 30K protein predicted from nucleotide sequence data on the 30K protein cistron. The antisynthetic peptide antibodies reacted specifically with the 30K protein synthesized *in vitro*. With the use of these antibodies, the 30K protein was revealed to be definitely synthesized in TMV-infected tobacco protoplasts.

Watanabe et al. (1984) studied the expression of TMV coding proteins and their mRNAs in protoplasts in which TMV multiplied snychronously. The 30K protein and its subgenomic mRNA were synthesized only between 2 and 9 hr after inoculation and not detected later than 12 hr after inoculation, while other proteins and their mRNAs were synthesized continuously. It was suggested that this transient expression of the 30K protein was regulated at the level of mRNA synthesis. In this protoplast system, synthesized 30K protein was found in the nuclear fraction (Watanabe et al., unpublished observations). Pulse-chase experiments showed that the 30K protein was rather stable. It was detected about 2 days after inoculation in the nuclear fraction. Therefore the amount of 30K protein synthesized at the early stage of infection may be enough for it to function.

The action of the 30K protein is still obscure. Shalla et al. (1982) showed that the number of plasmodesmata connecting an Ls1-infected cell to an adjacent cell was reduced at a nonpermissive temperature. The reduction in the number of plasmodesmata may be partly involved in restriction of the movement of Ls1 at high temperatures. However, this phenomenon cannot explain the compete blocking of virus movement, because some plasmodesmata persist at high temperatures. Taliansky et al. (1982) suggested that the transport protein (probably the 30K protein) opens the gates (the barrier to cell-to-cell movement determined by the host). It will be informative to identify the component(s) with which the 30K protein interacts. The behaviors of Ls1 at a nonpermissive temperature and of L in Tm-2 tomato are similar. The product of Tm-2 may interact with the 30K protein.

Thus far, the amino acid sequences of the 30K proteins of TMV common strains *T. vulgare* and OM, tomato strain L, and cowpea strain Cc have been determined (Meshi et al., 1982a,b; Goelet et al., 1982;

Takamatsu et al., 1983). Commonly, the 30K protein is basic, being composed of about 270 amino acid residues. Interestingly, the homology among the 30K proteins is very low, especially in the C-terminal one-third, compared with the homology of other proteins (130K and 180K and coat proteins). Nevertheless the defect in the 30K protein apparently is complemented not only by the 30K protein of other TMV strains but also by the proteins of corresponding size of other plant viruses.

In an infected plant, especially one growing outdoors, many mutants coexist with the wild type. However, most of them probably cannot compete effectively with the wild type as long as they have a common host. Such mutants will be expected to appear and disappear in due course. If a mutant could acquire a new host by means of a mutation, it would evolve beyond the constraints of the wild type, e.g., the constraints of cross-protection. The mutant would then become a new strain. In view of the observed relationship of the 30K protein to the host range and the variations in this protein, which do not cause a defect in virus replication, a mutant coding for a changed 30K protein could possibly change the host range. The divergence found among 30K proteins might effect the establishment of new strains during the course of evolution.

Other Virus Genes

Wyatt and Kuhn (1980) isolated strain R of CCMV from resistant cowpea PI 186465. CCMV R replicates rapidly in inoculated leaves and can move systemically, inducing disease symptoms. Pseudorecombinant studies with RNAs of strains T and R showed that RNA 1 influenced both the symptoms and systemic movement. Replication in the resistant cowpeas was influenced by either RNA 1 or 3, but the increased level of replication in inoculated leaves was not sufficient for systemic movement (Wyatt and Wilkinson, 1984).

REFERENCES

ALEXANDER, L. J. 1971. Host-pathogen dynamics of tobacco mosaic virus on tomato, *Phytopathology* **61**:611–617.

ALLISON, A. V. AND SHALLA, T. A. 1974. The ultrastructure of local lesions induced by potato virus X: A sequence of cytological events in the course of infection, *Phytopathology* **64**:784–793.

BALÁZS, E., GÁBORJÁNYI, R., TÓTH, A., AND KIRÁLY, Z. 1969. Ethylene production in Xanthi tobacco after systemic and local virus infections, *Acta Phytopathol. Acad. Sci. Hung.* **4**:355–358.

BEIER, H., SILER, D. J., RUSSELL, M. L., AND BRUENING, G. 1977. Survey of susceptibility to cowpea mosaic virus among protoplasts and intact plants from *Vigna sinensis* lines, *Phytopathology* **67**:917–921.

BENNETT, C. W. 1940. Relation of food translocation to movement of virus of tobacco mosaic, *J. Agric. Res.* **60**:361–390.

BOSCH, F. X., AND JOCKUSCH H. 1972. Temperature-sensitive mutants of TMV: Behaviour of a non-coat protein mutant in isolated tobacco cells, *Mol. Gen. Genet.* **116**:95–98.

BRUENING, G., BEACHY, R. N., SCALLA, R., AND ZAITLIN, M. 1976. *In vitro* and *in vivo* translation of the ribonucleic acids of a cowpea strain of tobacco mosaic virus, *Virology* **71**:498–517.

CASSELS, A. C., BARNETT, A., AND BARLASS, M. 1978. The effect of polyacrylic acid treatment on the susceptibility of *Nicotiana tabacum* cv. Xanthi-nc to tobacco mosaic virus, *Physiol. Plant Pathol.* **13**:13–21.

CHAMBERS, T. C., AND FRANCKI, R. I. B. 1966. Localization and recovery of lettuce necrotic yellows virus from xylem tissues of *Nicotiana glutinosa*, *Virology* **29**:673–676.

CHEO, P. C. 1971. Effect in different plant species of continuous light and dark treatment on tobacco mosaic virus replicating capacity, *Virology* **46**:256–265.

CIRULLI, M., AND ALEXANDER, L. J. 1969. Influence of temperature and strain of tobacco mosaic virus on resistance in a tomato breeding line derived from *Lycopercicon peruvianum*, *Phytopathology* **59**:1287–1297.

COLHOUN, J. 1973. Effects of environmental factors on plant disease, *Annu. Rev. Phytopathol.* **11**:343–364.

DAWSON, W. O., SCHLEGEL, D. E., AND LUNG, M. C. Y. 1975. Synthesis of tobacco mosaic virus in intact tobacco leaves systemically inoculated by differential temperature treatment, *Virology* **65**:565–573.

DE ZOETEN, G. A., AND GAARD, G. 1969. Possibilities of inter- and intracellular translocation of some icosahedral plant viruses, *J. Cell Biol.* **40**:814–823.

DODDS, J. A., and HAMILTON, R. I. 1972. The influence of barley stripe mosaic virus on the replication of tobacco mosaic virus in *Hordeum vulgare L*, *Virology* **50**:404–411.

DOROKHOV, Y. L., MIROSHNICHENKO, N. A., ALEXANDROVA, N. M., AND ATABEKOV, J. G. 1981. Development of systemic TMV infection in upper noninoculated tobacco leaves after differential temperature treatment, *Virology* **108**:507–509.

DOROKHOV, Y. L., ALEXANDROVA, N. M. MIROSHNICHENKO, N. A., AND ATABEKOV, J. G. 1983. Isolation and analysis of virus-specific ribonucleoprotein of tobacco mosaic virus-infected tobacco, *Virology* **127**:237–252.

DOROKHOV, Y. L., ALEXANDROVA, N. M., MIROSHNICHENKO, N. A., AND ATABEKOV, J. G. 1984a. Stimulation by aurintricarboxylic acid of tobacco mosaic virus-specific RNA synthesis and production of informosome-like infection-specific ribonucleoprotein, *Virology* **135**:395–405.

DOROKHOV, Y. L., ALEXANDROVA, N. M., MIROSHNICHENKO, N. A., AND ATABEKOV, J. G. 1984b. The informosome-like virus-specific ribonucleoprotein (vRNP) may be involved in the transport of tobacco mosaic virus infection, *Virology* **137**:127–134.

DUCHESNE, M., FRITIG, B., AND HIRTH, L. 1977. Phenylalanine ammonia-lyase in tobacco mosaic virus-infected hypersensitive tobacco: Density-labeling evidence of *de novo* synthesis, *Biochem. Biophys. Acta* **485**:465–481.

ESAU, K., AND CRONSHAW, J. 1967. Relation of tobacco mosaic virus to the host cells, *J. Cell Biol.* **33**:665–678.

ESAU, K., CRONSHAW, J., AND HOEFERT, L. L. 1967. Relation of beet yellows virus to the phloem and to movement in the sieve tube, *J. Cell Biol.* **32**:71–87.

FAVALI, M. A., CONTI, G. G., AND BASSI, M. 1978. Modification of the vascular bundle ultrastructure in the "resistant zone" around necrotic lesions induced by tobacco mosaic virus, *Physiol. Plant Pathol.* **13**:247–251.

FRASER, R. S. S. 1979. Systemic consequences of the local lesion reaction to tobacco mosaic virus in a tobacco variety lacking the *N* gene for hypersensitivity, *Physiol. Plant Pathol.* **14**:383–394.

GIANINAZZI, S. 1984. Genetics and molecular aspects of resistance induced by infection or chemicals, in: *Plant-Microbe Interactions: Molecular and Genetic Perspectives* (T. Kosuge and E. W. Nester, eds.) Vol. 1, pp. 321–342, Macmillan, New York.

GIANINAZZI, S., AND KASSANIS, B. 1974. Virus resistance induced in plants by polyacrylic acid, *J. Gen. Virol.* **23**:1–9.

GOELET, P., LOMONOSSOFF, G. P., BUTLER, P. J. G., AKAM, M. E., GAIT, M. J., AND KARN, J. 1982. Nucleotide sequence of tobacco mosaic virus RNA, *Proc. Natl. Acad. Sci. USA* **79**:5818–5822.

GUNNING, B. E. S., AND ROBERTS, A. W. 1976. *Intercellular Communication in Plants: Studies on Plasmodesmata*, Springer-Verlag, Berlin.

HAMILTON, R. I., AND DODDS, J. A. 1970. Infection of barley by tobacco mosaic virus in single and mixed infection, *Virology* **42**:266–268.

HAMILTON R. I., AND NICHOLS C. 1977. The influence of bromegrass mosaic virus on the replication of tobacco mosaic virus in *Hordeum vulgare*, *Phytopathology* **67**:484–489.

HATTA, T., AND MATTHEWS, R. E. F. 1974. The sequence of early cytological changes in Chinese cabbage leaf cells following systemic infection with turnip yellow mosaic virus, *Virology* **59**:383–396.

HUBER, R., HONTELEZ, J., AND VAN KAMMEN, A. 1981. Infection of cowpea protoplasts with both the common strain and the cowpea strain of TMV, *J. Gen. Virol.* **55**:241–245.

JOCKUSCH, H. 1968. Two mutants of tobacco mosaic virus temperature-sensitive in two different functions, *Virology* **35**:94–101.

JOSHI, S., PLEIJ, C. W. A., HAENNI, A. L., CHAPEVILLE, F., AND BOSCH, L. 1983. Properties of the tobacco mosaic virus intermediate length RNA-2 and its translation, *Virology* **127**:100–111.

KIBERSTIS, P. A., PESSI, A., ATHERTON E., JACKSON, R., HUNTER, T., AND ZIMMERN, D. 1983. Analysis of *in vitro* and *in vivo* products of the TMV 30K open reading frame using antisera raised against a synthetic peptide, *FEBS Lett.* **164:**355–360.

KITAJIMA, E. W., AND LAURITIS, J. A. 1969. Plant virions in plasmodesmata, *Virology* **37:**681–685.

KOIKE, M., HIBI, T., AND YORA, K. 1976. Infection of cowpea protoplasts by tobacco mosaic virus, *Ann. Phytopathol. Soc. Japan* **42:**105.

KUHN, C. W., WYATT, S. D., AND BRANTLEY, B. B. 1981 Genetic control of symptoms, movement, and virus accumulation in cowpea plants infected with cowpea chlorotic mottle virus, *Phytopathology* **71:**1310–1315.

LEONARD, D. A., AND ZAITLIN, M. 1982. A temperature-sensitive strain of tobacco mosaic virus defective in cell-to-cell movement generates an altered viral-coded protein, *Virology* **117:**416–424.

LISTER, R. M. 1969. Tobacco rattle, NETU, viruses in relation to functional heterogeneity in plant viruses, *Fed. Proc.* **28:**1875–1889

LOEBENSTEIN, G., RABINA, S., AND VAN PRAAGH, T. 1968. Sensitivity of induced localized acquired resistance to actinomycin D, *Virology* **34:**264–268.

LOEBENSTEIN, G., CHAZAN, R., AND EISENBERG, M. 1970. Partial suppression of the localizing mechanism to tobacco mosaic virus by UV irradiation. *Virology* **41:**373–376.

LOEBENSTEIN, G., GERA, A., BARNETT, A., SHABTAI, S., AND COHEN, J. 1980. Effect of 2,4-dichlorophenoxyacetic acid on multiplication of tobacco mosaic virus in protoplasts from local-lesion and systemic-responding tobaccos, *Virology* **100:**110–115.

MATTHEWS, R. E. F. 1981. *Plant Virology,* 2nd ed., Academic Press, New York.

MESHI, T., OHNO, T., AND OKADA, Y. 1982a. Nucleotide sequence and its character of cistron coding for the 30K protein of tobacco mosaic virus (OM strain), *J. Biochem.* **91:**1441–1444.

MESHI, T., OHNO, T., AND OKADA, Y. 1982b. Nucleotide sequence of the 30K protein cistron of cowpea strain of tobacco mosaic virus. *Nucleic Acids Res.* **10:**6111–6117.

MOTOYOSHI, F., AND OSHIMA, N. 1975. Infection with tobacco mosaic virus of leaf mesophyll protoplasts from susceptible and resistant lines of tomato, *J. Gen. Virol.* **29:**81–91.

MOTOYOSHI, F., AND OSHIMA, N. 1977. Expression of genetically controlled resistance to tobacco mosaic virus infection in isolated tomato leaf mesophyll protoplasts, *J. Gen. Virol.* **34:**499–506.

NISHIGUCHI, M., MOTOYOSHI, F., AND OSHIMA, N. 1978. Behaviour of a temperature sensitive strain of tobacco mosaic virus in tomato leaves and protoplasts, *J. Gen. Virol.* **39:**53–61.

OHASHI, Y., AND SHIMOMURA, T. 1971. Necrotic lesion induced by heat treatment on leaves of systemic host infected with tobacco mosaic virus, *Ann. Phytopathol. Soc. Japan* **37:**22–28.

OHNO, T., TAKAMATSU, N., MESHI, T., OKADA, Y., NISHIGUCHI, M., AND KIHO, Y. 1983. Single amino acid substitution in 30K protein of TMV defective in virus transport function, *Virology* **131:**255–258.

OOSHIKA, I., WATANABE, Y., MESHI, T., OKADA, Y., IGANO, K, INOUYE, K., AND YOSHIDA, N. 1984. Identification of the 30K protein of TMV by immunoprecipitation with antibodies directed against a synthetic peptide, *Virology* **132**:71–78.

OTSUKI, Y., SHIMOMURA, T., AND TAKEBE, I. 1972. Tobacco mosaic virus multiplication and expression of the *N* gene in necrotic responding tobacco varieties, *Virology* **50**:45–50.

PELHAM, J. 1972. Strain-genotype interaction of tobacco mosaic virus in tomato, *Ann. Appl. Biol.* **71**:219–228.

ROSS, A. F. 1961. Systemic acquired resistance induced by localized virus infections in plants, *Virology* **14**:340–358.

RYAN, C. A. 1984. Systemic response to wounding, in: *Plant-Microbe Interactions: Molecular and Genetic Perspectives* (T. Kosuge and E. W. Nester, eds.), Vol. 1, pp. 307–320, Macmillan, New York.

SAMUEL, G. 1934. The movement of tobacco mosaic virus within the plant, *Ann. Appl. Biol.* **21**:90–111.

SCHUSTER, G., AND FLEMMING, M. 1976. Studies on the formation of diffusion barriers in hypersensitive hosts of tobacco mosaic virus and the role of necrotization in the formation of diffusion barriers as well as in the localization of virus infections, *Phytopathol. Z.* **87**:345–352.

SELA, I. 1981. Plant-virus interactions related to resistance and localization of viral infections, *Adv. Virus Res.* **26**:201–237.

SHALLA, T. A., PETERSEN, L. J., AND ZAITLIN, M. 1982. Restricted movement of a temperature-sensitive virus in tobacco leaves in association with a reduction in numbers of plasmodesmata, *J. Gen. Virol.* **60**:355–358.

SHIMOMURA, T., AND DIJKSTRA, J. 1975. The occurrence of callose during the process of local lesion formation, *Neth. J. Plant Pathol.* **81**:107–121.

SIEGEL, A., ZAITLIN, M., AND SEHGAL, O. P. 1962. The isolation of defective tobacco mosiac virus strains, *Proc. Natl. Acid. Sci. USA* **48**:1845–1851.

SULZINSKI, M. A., AND ZAITLIN, M. 1982. Tobacco mosaic virus replication in resistant and susceptible plants: In some resistant species virus is confined to a small number of initially infected cells, *Virology* **121**:12–19.

TAKAMATSU, N., OHNO, T., MESHI, T., AND OKADA, Y. 1983. Molecular cloning and nucleotide sequence of the 30K and the coat protein cistron of TMV (tomato strain) genome, *Nucleic Acids Res.* **11**:3767–3778.

TALIANSKY, M. E., ATABEKOVA, T. I., KAPLAN, I. B., MOROZOV, S. YU., MALYSHENKO, S. I., AND ATABEKOV, J. G. 1982a. A study of TMV *ts* mutant Ni2519. 1. Complementation experiments, *Virology* **118**:301–308.

TALIANSKY, M. E., MALYSHENKO, S. I., PSHENNIKOVA, E. S., AND ATABEKOV, J. G. 1982b. Plant virus-specific transport function. 2. A factor controlling virus host range, *Virology* **122**:327–331.

TALIANSKY, M. E., MALYSHENKO, S. I., PSHENNIKOVA, E. S., KAPLAN, I. B., ULANOVA, E. F., AND ATABEKOV, J. G. 1982c. Plant virus-specific transport function. 1. Virus genetic control required for systemic spread, *Virology* **122**:318–326.

VAN LOON, L. C. 1979. Effects of auxin on the localization of tobacco mosaic virus in hypersensitively reacting tobacco, *Physiol. Plant Pathol.* **14**:213–226.

VAN LOON, L. C., AND VAN KAMMEN, A. 1970. Polyacrylamide disc electrophoresis of the soluble leaf proteins from *Nicotiana tabacum* var. 'Samsun' and 'Samsun NN.' 2. Changes in protein constitution after infection with tobacco mosaic virus, *Virology* **40**:199–211.

WATANABE, Y., EMORI, Y., OOSHIKA, I., MESHI, T., OHNO, T., AND OKADA, Y. 1984. Synthesis of TMV-specific RNAs and proteins at the early stage of infection in tobacco protoplasts: Transient expression of the 30K protein and its mRNA, *Virology* **133**:18–24.

WEINTRAUB, M., RAGETLI, H. W. J., AND LEUNG, E. 1976. Elongated virus particles in plasmodesmata, *J. Ultrastruct. Res.* **56**:351–364.

WESTSTEIJN, E. A. 1978. Permeability changes in the hypersensitive reaction of *Nicotiana tabacum* cv. Xanthi nc. after infection with tobacco mosaic virus, Physiol. Plant Pathol. **13**:253–258.

WESTSTEIJN, E. A. 1981. Lesion growth and virus localizations in leaves of *Nicotiana tabacum* cv. Xanthi nc. after inoculation with tobacco mosaic virus and incubation alternately at 22°C, and 32°C, *Physiol. Plant Pathol.* **18**:357–368.

WHITE, R. F. 1979. Acetylsalicylic acid (aspirin) induces resistance to tobacco mosaic virus in tobacco, *Virology* **99**:410–412.

WITTMANN, H. G., WITTMANN-LIEBOLD, B., AND JAUREGUI-ADELL, J. 1965. Die primare Proteinstruktur temperatur-sensitiver Mutanten des Tabakmosaikvirus 1. Spontanmutanten, *Z. Naturforsch* **20b**:1221–1234.

WU, J. H., AND DIMITMAN, J. E. 1970. Leaf structure and callose formation as determinants of TMV movement in bean leaves as revealed by UV irradiation studies, *Virology* **40**:820–827.

WU, J. H., AND DIMITMAN, J. E. 1984. Increased resistance to the spread of tobacco mosaic virus in Pinto bean leaves caused by sugar and light, *Phytopathol. Z.* **110**:37–48.

WYATT, S. D., AND KUHN, C. W. 1979. Replication and properties of cowpea chlorotic mottle virus in resistant cowpeas. *Phytopathology* **69**:125–129.

WYATT, S. D., AND KUHN, C. W. 1980. Derivation of a new strain of cowpea chlorotic mottle virus from resistant cowpeas, *J. Gen. Virol.* **49**:289–296.

WYATT, S. D., AND WILKINSON, T. C. 1984. Increase and spread of cowpea chlorotic mottle virus in resistant and fully susceptible cowpeas, *Physiol. Plant Pathol.* **24**:339–345.

ZIMMERN, D., AND HUNTER, T. 1983. Point mutation in the 30K open reading frame of TMV implicated in temperature-sensitive assembly and local lesion spreading of mutant Ni 2519, *EMBO J.* **2**:1893–1900.

SECTION IV

Viruses

Chapter 10

Viruses of a *Chlorella*-like Green Alga

James L. Van Etten, Yuannan Xia, and Russel H. Meints

VIRUSES THAT INFECT CYANOBACTERIA (blue-green algae) are well known and resemble typical bacteriophages (for a review see Sherman and Brown, 1978). In contrast, viruses or viruslike particles (VLPs) infecting eukaryotic algae are poorly characterized. Since 1971 about 30 reports have appeared suggesting that many eukaryotic algae harbor polyhedral particles which resemble viruses. These reports mention at least 27 genera of eukaryotic algae, and the reader is referred to the excellent reviews of Lemke (1976), Sherman and Brown (1978), and Dodds (1979, 1983) for a complete listing. With the few exceptions described in the next section, this research has consisted solely of ultrastructural studies. Characterization of these VLPs was not pursued primarily because they could not be obtained in sufficient quantities for biochemical studies. Several factors usually contributed to this problem: (1) Only a few algal cells contained particles, (2) the cells contained particles only at one stage of the algal life cycle, (3) the cells that had particles did not lyse, and (4) in most cases the particles were not infectious. These observations plus the fact that many of the particles were present in multicellular filamentous algae, hindered the development of a sensitive biological assay for them.

However, these difficulties are no longer a problem for a family of viruses which infect and replicate in certain strains of unicellular, eukaryotic, exsymbiont, *Chlorella*-like green algae. Viruses which infect *Chlorella*, strain NC64A, can be produced in large quantities and assayed by plaque formation. The first of these viruses was discovered less than 5 years ago in *Chlorella*-like algae symbiotic with *Hydra viridis*. The ecology and evolutionary relationships of these viruses are intriguing, and they can also serve as useful models in the study of plant gene regulation and expression.

This chapter will briefly describe the viruses or VLPs of other eukaryotic algae which have been partially characterized, but will focus primarily on the *Chlorella* viruses.

OTHER EUKARYOTIC ALGAL VIRUSES OR VLP

Gibbs and his colleagues (Gibbs et al., 1975; Skotnicki et al., 1976) isolated and characterized the first infectious virus in a eukaryotic alga. This single-stranded RNA (ssRNA)-containing virus replicates in the large filamentous green alga *Chara corallina*. The particles are rigid helical rods, 532 nm long and 18 nm wide, and resemble tobacco mosaic virus (TMV). This virus is about twice the length of TMV (300 nm) and contains a correspondingly larger ssRNA genome (MW 3.6×10^6 versus 2.1×10^6 for TMV RNA). The virus capsid protein is slightly smaller than TMV capsid protein [16.5 versus 17.5 kilodaltons (kd) for TMV]. The virus reacts with antisera to 2 strains of TMV but not with antisera to 10 other strains of TMV. It can be transmitted to virus-free *C. corallina* by injection, and the inoculated algae become chlorotic 7–8 days after infection.

Dodds and his colleagues (Dodds and Cole, 1980; Cole et al., 1980) have characterized a noninfectious VLP from the filamentous green alga *Uronema gigas*. These VLPs accumulate at a low concentration in liquid growth medium at a rate which parallels algal growth. The particles are seen in about 1% of the cells in the germling or in the young filament stage; they are not detected in cells of mature filaments. Purified particles are polyhedra 390 nm in diameter; they have a 15-nm-thick multilaminate shell and a core made up of two distinct materials. An interesting feature is the presence of a 1-μm-long tail, with a central swelling, that is attached to about 10% of the particles. The particles contain double-stranded DNA (dsDNA) and at least 10 structural proteins; the major protein weighs about 45 kd.

Hoffman and Stanker (Hoffman and Stanker, 1976; Stanker and Hoffman, 1978; Stanker et al., 1981) have characterized a noninfectious VLP from the multicellular filamentous green alga *Cylindrocapsa geminella*. Like the *U. gigas* VLPs, *C. geminella* VLPs were detected in only 5–10% of the cells at the single-celled germling stage. The particles are polyhedra, 200–239 nm in diameter, having a 14- to 16-nm-thick multilaminate shell and a dense fibrillar core. Purified particles contain dsDNA (MW $175–190 \times 10^6$) and at least 10 structural proteins.

A common feature of the *U. gigas* and *C. geminella* systems is that VLP-containing cells typically lack nuclei. Cells containing the particles are believed to lyse. The absence of a nucleus in VLP-containing cells has also been noted in many of the other ultrastructural studies mentioned above. Considerable effort was made to increase the titer of the VLPs in both *U. gigas* and *C. geminella* by varying factors such as light and temperature. While a short heat treatment increased the titer severalfold in both algae, yields remained poor.

A virus which infects four isolates of the marine unicellular nanoflagellate *Micromonas pusilla* has been described by Mayer and Taylor (1979). Unlike typical algae, *M. pusilla* lacks a discernible cell wall. The virus is a polyhedron 130–135 nm in diameter. Waters and Chan (1982) developed an endpoint titration assay for the virus and demonstrated that it had a latent period of about 7 hr. Viral infection reduces CO_2 fixation by the host and ultimately kills it. The host cannot be grown axenically; however, *M. pusilla* can be grown monoxenically in the presence of certain bacterial species.

Finally, Gromov and Mamkaeva (1981) described a virus which infects zoospores of the green alga *Chlorococcum minutum*. This virus has a hexagonal head measuring 220×180 nm and, most interestingly, a tail which is believed to invaginate into the head of the viral particle. The tail is presumably ejected from the head during infection of the host. This intriguing virus merits further investigation.

VIRUSES OF *CHLORELLA*-LIKE GREEN ALGAE

Discovery of the Viruses

The freshwater hydroid *H. viridis* harbors a eukaryotic *Chlorella*-like green alga in a mutually beneficial symbiotic relationship (Trench, 1979). Typically, each gastrodermal cell contains 10–20 algae, and a hydra contains $0.75–2.5 \times 10^5$ algae under optimal laboratory conditions. The algae fix CO_2 in light, but symbiosis can also be maintained in dark-grown hydras (Pardy, 1974). Although hydras can be grown separately from the algae if appropriately fed, attempts to grow the algae free of the host have generally been unsuccessful. The inability to culture these symbiotic algae *in vitro* was generally assumed to be a nutritional problem. This concept changed, however, when large (185 nm in diameter), polyhedral-shaped VLPs were found in the algae within 3–6 hr after isolation from the hydra. Replication of these particles lyses the entire population of isolated algae within 24 hr (Meints et al., 1981). These VLPs were named HVCV-1, for *H. viridis Chlorella* virus.

HVCV-1 sediments at ca. 2600S in sucrose density gradients and has a density of 1.295 g/ml in CsCl. HVCV-1 contains a large dsDNA genome of ca. 250 kilo base pairs (kbp) (estimated by summing DNA restriction fragments) and at least 19 structural proteins which vary in apparent molecular weight from 10.3 to 82 kd. The major viral protein of 46 kd stains with Schiffs reagent, suggesting that it is a glycoprotein (Van Etten et al., 1981).

Six other hydra isolates and one isolate of the protozoan *Paramecium bursaria*, which also contains symbiotic *Chlorella*-like algae, were examined for VLP. In all cases, isolation of the symbiotic algae led to the appearance of large dsDNA-containing VLPs within 24 hr (Van Etten et al., 1982). Restriction endonuclease analysis of their dsDNAs revealed four distinct VLPs. The distributional relationships of the VLPs were especially interesting. A VLP identical to HVCV-1 was found in *Chlorella* isolated from a hydra obtained from a lake in Massachusetts in 1981. A second *Chlorella* VLP, designated HVCV-2, was found in hydras isolated in 1981 from lakes in North Carolina, Nebraska, and Massachusetts. A third *Chlorella* VLP, designated HVCV-3, was present in two hydra isolates from England. The presence of identical genomes in VLPs isolated from algae from hydras from diverse geographic locations suggests that VLPs have an ancient relationship with their algal hosts and that they are genetically stable.

The DNA restriction pattern of the VLPs from the *P. bursaria Chlorella*-like algae was different from that of the HVCV viruses, and these VLPs were designated PBCV-1, for *P. bursaria Chlorella* virus. Kawakami and Kawakami (1978) also observed a virus infecting a *Chlorella*-like alga in a *P. bursaria* strain isolated in Japan. In this case, high numbers of viral particles were observed both inside and outside the paramecia. These viral particles, which have been described only ultrastructurally, were similar in size and morphology to PBCV-1.

The origin of the VLPs in *Hydra-Chlorella* systems is unknown, but particles in intracellular algae in the hydra have yet to be observed. One possibility is that the VLPs are lysogenic in the algae as long as the algae remain in a symbiotic relationship with their host. Removal of the algae from the hydra could induce the lytic phase. Perhaps this lysogenic VLP codes for a gene product that is expressed at the algal surface and involved in maintainance of the symbiotic state between the algae and the hydra. Alternatively, the hydra could produce a gene product which maintains this presumed lysogenic state. Another possibility is that the VLPs replicate in the hydra and infect the *Chlorella* during isolation. However, VLPs have not been seen in hydra cells. Finally VLPs may exist in small numbers within the hydra gut where algae are occasionally released after gastrodermal cell death. This still remains a possibility, even though VLPs have never been observed in the hydra gut and have not been isolated from spent culture media. Lysogeny or viral replication in the hydra can be tested experimentally by isolating total algal DNA or DNA from aposymbiotic hydras and probing for viral DNA sequences with ^{32}P-labeled viral DNA. These experiments have not been done.

In summary, *Chlorella* VLPs are common in these symbiotic systems. This is the simplest explanation for our general inability to culture symbiotic *Chlorella* from hydras. Whether the viral association is absolutely required for symbiosis is unknown. However, the relationship in *P. bursaria* is not as strong, since several investigators have cultured *Chlorella*-like algae free of paramecia (Loefer, 1936; Siegel, 1960; Weis, 1978).

Development of a Culturable System for PBCV-1

It is necessary to grow large numbers of hydras or paramecia to obtain large quantities of HVCV and PBCV VLPs. However, the maintainance of large colonies of hydras is time-consuming since they must be fed and bathed regularly. Therefore, a search was made for a culturable *Chlorella* that would support viral replication.

A number of cultured *Chlorella* isolates reputedly derived from invertebrate hosts, as well as two free-living *Chlorella* strains, were inoculated with the four viruses. Two of the *Chlorella* isolates (NC64A and ATCC-30562), both originally isolated from *P. bursaria*, lysed after inoculation with PBCV-1 and released large quantities of VLPs (Van Etten et al., 1983a). The progeny VLPs were infectious and identical to the input particles. Thus Koch's postulates were fulfilled, and PBCV-1 is an authenic virus. More importantly, PBCV-1 forms plaques on lawns of these *Chlorella* isolates, which provides a sensitive biological assay for PBCV-1. None of the other viruses replicated in the *Chlorella* isolates.

The plaque assay for PBCV-1, together with the ability to synchronously infect the algae allows one to study PBCV-1 with techniques used with bacteriophage. For example, one-step growth studies, life cycle studies, and genetic studies are now feasible for PBCV-1. One disadvantage of the system, however, is the lack of a sexual stage in the host *Chlorella*. The genus *Chlorella* encompasses a biochemically heterogeneous collection of unicellular green algae all of which replicate only asexually (Shihra and Krauss, 1965; DaSilva and Gyllenberg, 1972; Hellmann and Kessler, 1974; Yamada and Sakaguchi, 1982).

Properties of PBCV-1

General Characteristics of PBCV-1

PBCV-1 is easy to purify, and 80–105 A_{260} units (7–10 mg of PBCV-1) and $1.4–2 \times 10^{12}$ plaque-forming units (PFUs) of purified PBCV-1 are obtained from 1 liter of lysate. Purified virus can be stored at 4°C for at least 1 year without detectable loss of infectivity. However, freezing rapidly destroys viral infectivity.

PBCV-1 contains 64% protein, 21–25% dsDNA, and 5–10% lipid (Van Etten et al., 1983b; Skrdla et al., 1984). The virus contains at least 50 structural proteins which range in apparent molecular weight from 10 to 135 kd. One 54-kd protein comprises about 40% of the total viral protein. This protein and a second protein of 135 kd stain with Schiff's reagent, suggesting they are glycoproteins. These two glycoproteins and two others (14.5 and 71 kd) are labeled when the intact virus is treated with ^{125}I and are probably surface proteins (Skrdla et al., 1984).

The major lipids in PBCV-1 are phosphatidylcholine, phosphatidylethanolamine, and one unidentified component (Skrdla et al., 1984). Viral infectivity is rapidly destroyed by treatment with chloroform and more slowly with ethyl ether or toluene; thus the lipid component is required for infectivity. In contrast, detergents such as 2% Triton X-100, 2% Nonidet P-40, and 2% sodium deoxycholate have no effect on viral infectivity. However, PBCV-1 slowly loses infectivity in 2% sodium N-lauryl sarcosine and rapidly in 2% sodium dodecyl sulfate. Ultrastructural studies on PBCV-1 indicate that it is surrounded by a multilaminate shell, and so we suspect that the lipid component is located internal to a protein shell.

Not much is known about the structure of the large PBCV-1 dsDNA genome (ca. 300 kbp as estimated from summing restriction fragments). The DNA probably consists of a single molecule, but attempts to isolate intact DNA molecules from PBCV-1 have not been successful, presumably because of its large size. Attempts to label distinct ends of the DNA by 5'- or 3'-end labeling with ^{32}P have also been unsuccessful. Thus it is possible that the DNA is either linear with blocked ends, circular, or circularly permuted. A restriction map of the viral DNA is being prepared, and its completion will provide information on genome structure.

The molecular weight of PBCV-1 has been estimated by two methods. First, since PBCV-1 contains 21–25% DNA, which has a minimum molecular weight of about 1.9×10^8, the molecular weight of PBCV-1 is calculated to be $7.6-9 \times 10^8$ daltons (Van Etten et al., 1983b). Second, the viral molecular weight was estimated by field-flow fractionation, and a value of 1.0×10^9 was obtained (Yonker et al., 1985). Even though the results obtained with both methods are similar, they require certain assumptions and must be considered tentative.

If a viral molecular weight of 1×10^9 and an extinction coefficient of $A_{0.1\%}^{260} = 10.7$ are used (uncorrect for light scattering), one A_{260} unit of PBCV-1 contains 5.6×10^{10} virus particles. Experimentally $1.5 - 3 \times 10^{10}$ PFUs per A_{260} unit of PBCV-1 are obtained. Thus 27–50% of the viral particles form plaques.

Growth Cycle of PBCV-1

One-step growth experiments reveal that progeny virus particles are first released about 3–4 hr after infection, and virus release is completed by 6–8 hr (Van Etten et al., 1983b). Mechanical disruption of the cells releases infectious virus particles 30–50 min before spontaneous lysis. PBCV-1 also replicates in dark-grown *Chlorella* or in light-grown cells treated with the photosynthetic inhibitor 3- (3′,4′-dichlorophenyl)- 1,1-dimethylurea just before virus infection. In both cases the kinetics of viral release are similar to those obtained with untreated light-grown cells, and so host photosynthesis is not essential for PBCV-1 replication. However, the burst size, which is 200–350 PFUs per cell in light-grown cells, is reduced by about 50% in dark-grown cells. PBCV-1 infection immediately inhibits host CO_2 fixation. Like bacteriophages, the virus replicates most efficiently in actively growing *Chlorella* and poorly in stationary-phase cells.

Attachment of PBCV-1 to Chlorella NC64A

PBCV-1 attaches rapidly to *Chlorella* NC64A with an adsorption rate of 5×10^{-9}/ml/min. PBCV-1 attachment is specific, since the virus does not attach to any of 10 other *Chlorella* strains or species (Meints et al., 1984). The nature of the viral receptor is not known; in fact, nothing is known about the chemical composition of *Chlorella* NC64A cell walls. The receptor is stable to heat and organic solvents, since PBCV-1 attaches equally well to isolated walls of *Chlorella* even if the walls are first heated to 100°C or extracted with methanol. Treatment of the walls with a number of proteases has no effect on viral attachment. However, exposure of the walls to HC1, pH 1, 0.5 M H_2SO_4, or 0.2 M NaOH for 1 hr at 100°C inactivates the receptor (J. L. Van Etten, unpublished data). Treatment of wall preparations with the lectins wheat germ agglutinin, ricin, *Lens culinaris* lectin, and concanavalin A in the presence or absence of their specific hapten inhibitors also inactivates the receptor. These results suggest that the lectins are nonspecific in their binding to the algal walls and that lectin binding merely occludes viral attachment to the walls. With the exception of high concentrations (400 mM) of glucosamine and mannosamine, simple sugars such as glucose, N-acetylglucosamine, galactose, maltose, sucrose, fucose, rhamnose, arabinose, xylose, and N-acetylgalactosamine have no effect on viral attachment (J. L. Van Etten, unpublished data).

The attachment and initial infection of *Chlorella* NC64A by PBCV-1 are shown in Fig. 1A. The virus always attaches to the wall via one of its hexagonal vertices. Attachment is followed by dissolution of the host wall at the point of viral attachment and entry of the viral DNA into the cell. An empty capsid is left on the surface of the host (Meints et al., 1984). Since the virus attaches to and digests *Chlorella* NC64A wall fragments previously heated to 100°C, the virus probably contains the hydrolytic enzyme(s). However, empty viral capsids are not observed after PBCV-1 attaches to wall fragments; thus the release of PBCV-1 DNA may require a host function(s).

Site of PBCV-1 Replication

The intracellular site of PBCV-1 replication is unknown. Electron micrographs of cells taken 3 hr after viral infection reveal that virus assembly occurs in viral organizing centers located in the cytoplasm (arrow in Fig. 1B). Typically, intact nuclei are present, but not always at this stage of infection. In contrast, mitochondria and chloroplasts appear structurally intact.

PBCV-1 does not require a functional nucleus for replication, since it replicates in cells which have previously been inactivated by ultraviolet (UV) light. However, viral replication is slow (having a latent period of 16–20 hr) and the burst size is low (2–10 PFUs per cell) in UV-treated cells (J. L. Van Etten, unpublished data). Viral proteins are synthesized on cytoplasmic ribosomes and not on organelle ribosomes, since cycloheximide but not chloramphenicol inhibits viral replication (Skrdla et al., 1984).

Viral DNA Synthesis

Studies on DNA synthesis following PBCV-1 infection have led to the following observations and conclusions (Van Etten et al., 1984):

1. PBCV-1 infection immediately inhibits host DNA synthesis.
2. Viral DNA synthesis begins about 45 min after infection and requires *de novo* synthesis of a protein or proteins. This implies that PBCV-1 directs the synthesis of its own DNA polymerase or at least an essential component required for viral DNA synthesis.

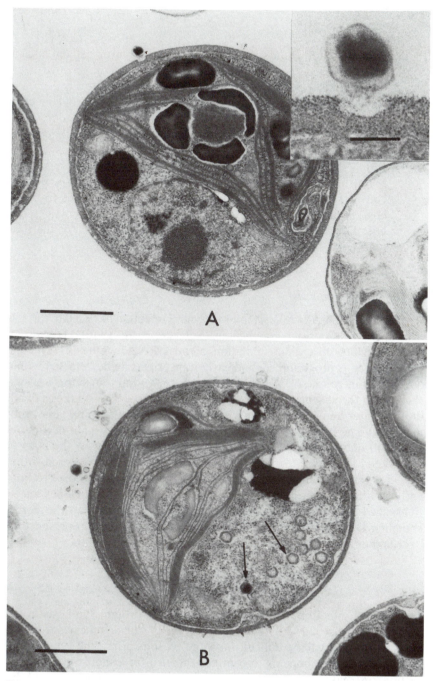

Figure 1. (A) Infection of *Chlorella* NC64A by PBCV-1. *Note:* The virus always attaches to the host wall by one of its hexagonal verticies and digests the host wall at the point of attachment. The size of the marker in the insert represents 100 nm. (B) Assembly of progeny viruses about 4 hr after PBCV-1 infection (arrows). The size of the marker represents 1 μm.

3. Viral infection results in a decrease in host chloroplast and nuclear DNA within 1–2 hr after infection (Fig. 2). Presumably, a virus-encoded enzyme(s) is, at least partially, responsible for this decrease, since the addition of cycloheximide at the time of infection prevents host DNA degradation. Also, the host DNA decreases, albeit more slowly, if UV-inactivated algae are inoculated with PBCV-1 (J. L. Van Etten, unpublished data).

4. PBCV-1 infection increases the total DNA in the *Chlorella* three to fourfold by 4 hr after infection. Thus PBCV-1 synthesis requires considerable energy and deoxynucleotide intermediates from the host. Some of these intermediates presumably come from degraded host DNA, but recycling them can supply only 25% of the total required.

Viral RNA Synthesis

Studies on RNA synthesis following viral infection have led to the following observations and conclusions (Schuster et al., 1986). (1) Viral infection results in a rapid inhibition of host RNA synthesis. (2) Viral transcription is programmed, and early transcripts can be detected within 5 min after infection. (3) A few, but not all, early viral transcripts are synthesized in the absence of de novo protein synthesis. The synthesis of additional, later transcripts depends on the translation of an early gene product(s). (4) The transition from early to late viral transcription occurs between 40 and 60 min after infection, which correlates with the start of viral DNA synthesis. (5) The early or late genes are not clustered within the viral genome, since several PBCV-1 DNA clones hybridize to both early and late transcripts. (6) Three of four PBCV-1 DNA clones tested hybridized to transcripts which additively were larger than the corresponding DNA probe. This could reflect RNA processing, presence of overlapping genes, or transcription from both DNA strands.

Viral Protein Synthesis

PBCV-1 infection immediately inhibits total protein synthesis in the host by at least 75%. However, analysis of the synthesis of virus-specific proteins will be difficult for three reasons. First, the synthesis of some host proteins appears to continue (at least for a short time) after infection. Second, the specific labeling of viral proteins is low, even with long labeling periods (greater than 1 hr) and high-specific-activity amino acids. Thus viral infection may result in the degradation of some host proteins and the recycling of amino acids. Third, PBCV-1 obviously encodes for many proteins (its genome is large enough to code for 200–300 proteins), which makes it difficult to resolve them on gels.

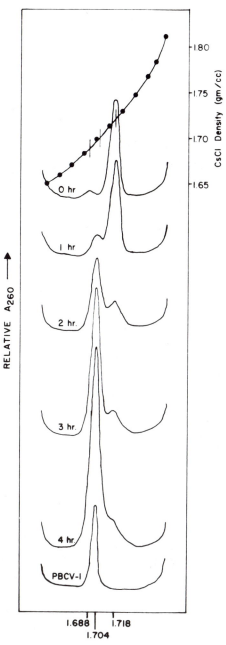

Figure 2. Cesium chloride equilibrium centrifugation of DNA isolated from an equal number of *Chlorella* NC64A cells at various times after infection with PBCV-1. The top drawing is from uninfected cells and in descending order from cells at 1, 2, 3, and 4 hr after infection. The bottom drawing is DNA isolated from purified virus. Nuclear DNA bands at 1.718 g/ml, chloroplast DNA at 1.688 g/ml, and viral DNA at 1.704 g/ml. *Note:* viral DNA clearly begins to accumulate by 1 hr post infection and host nuclear DNA decreases by 1 to 2 hr post infection. [Reproduced with permission from Academic Press: Van Etten, Burbank et al., *Virology,* **134**:443–449 (1984).]

317

ADDITIONAL VIRUSES THAT INFECT *CHLORELLA* NC64A

Discovery and Properties

In the fall of 1983 we collected 11 freshwater samples from ponds and
rivers in North and South Carolina, Illinois, and Nebraska and titered
them against *Chlorella* NC64A. To our surprise 4 of the 11 samples
produced plaques which varied in number from ca. 60 to 4×10^4 PFUs/ml
of freshwater (Van Etten et al., 1985a). Six plaques were picked from
each of the four samples, and the viruses were characterized. Like
PBCV-1, all the viruses contained large dsDNA genomes, were polyhedra, and were chloroform-sensitive. Furthermore, the viral DNAs were
extensively but not completely homologous with PBCV-1 DNA. However, many of the viruses, even those isolated from the same water sample,
could be distinguished from each other and from PBCV-1 by at least one
of the following characteristics: plaque size (the isolates from South
Carolina formed smaller plaques), DNA restriction patterns, and resistance to certain DNA restriction endonucleases. In fact, the viral DNAs
were grouped into five classes based on the sensitivity or resistance of
their DNAs to 15 restriction endonucleases (Van Etten et al., 1985a).

The resistance of the viral DNAs to certain restriction endonucleases
suggests that they contain modified bases. This was verified by analyses
of 11 of these viral DNAs which revealed that all of them contained 5-
methylcytosine (5mC) and some of them also contained N^6-me-
thyladenine (6mA) (Van Etten et al., 1985c). The concentrations of
methylated bases varied from 0.3 to 13% 5mC and up to 8% 6mA
depending on the virus. No other unusual bases were detected.

Nuclear DNA from the host *Chlorella* contains 21% 5mC and traces
of 6mA (Van Etten et al., 1985c). However, at least some, if not all, of
the 5mCs in the host DNA reside in base sequences different from those
present in the viruses. The two restriction endonucleases *Msp*I and
*Hpa*II recognize the base sequence CCGG, but *Msp*I will not cleave it
if the internal carbon is methylated. Both enzymes hydrolyze all of the
viral DNAs; owever, only *Msp*I completely hydrolyzes the host nuclear
DNA indicating that, like many eukaryotic DNAs, host DNA contains
5mC in the C^mCGG sequence. In contrast, viral DNAs do not contain
5mC in the CCGG sequence. This result suggests that the viruses encode for unique methyltransferase enzymes.

Do Viruses Code for Modification and Restriction Systems?

As mentioned in the section on viral DNA synthesis, PBCV-1 infection
degrades host nuclear and chloroplast DNA beginning about 1 hr after
viral infection (Fig. 2). Furthermore, the enzymes(s) responsible for this

decrease is probably virus-encoded. To identify this enzyme(s) the appearance of a nonspecific DNase in PBCV-1-infected cells was examined. However, no new nonspecific DNases were found (J. L. Van Etten, unpublished data).

Since PBCV-1 might encode for unique methyltransferase enzymes, we anticipated that PBCV-1 might also encode for an enzyme(s) with restriction endonuclease-like activity. Enzyme extracts were prepared from uninfected *Chlorella* and from cells 3 hr after PBCV-1 infection by procedures similar to those used to isolate bacterial restriction endonucleases. An enzyme fraction from PBCV-1-infected cells, but not from uninfected cells, degraded host DNA but not PBCV-1 DNA. This enzyme cleaved unmethylated pBR322 DNA to discrete DNA fragments identical in size to those produced by the bacterial restriction endonuclease *Mbo*I. *Mbo*I recognizes the sequence GATC and cleaves DNA on the 5' of guanylate. The PBCV-1-induced enzyme cleaved at the same position and produced the same sticky ends as *Mbo*I (J. L. Van Etten, manuscript in preparation).

Enzyme extracts prepared from cells infected with two other *Chlorella* viruses, NC-1A and I1-3A, also degraded the host DNA but not its parent viral DNA; furthermore, both enzymes digested unmethylated pBR322 DNA into discrete fragments. Interestingly, the sizes of the pBR322 DNA fragments obtained with both enzymes were different from each other and from those obtained with the PBCV-1-induced enzyme.

PBCV-1-infected cells also contain methyltransferase activity. With unmethylated lambda DNA as a substrate enzyme extracts prepared from PBCV-1-infected cells contain much higher methyltransferase activity than similar extracts from uninfected cells. Furthermore, the resultant methylated lambda DNA resists cleavage by the PBCV-1-induced restriction endonuclease (J. L. Van Etten, unpublished data). Thus the PBCV-1-induced methyltransferase adds methyl groups at specific sites which confer resistance to DNA cleavage by the PBCV-1-induced restriction endonuclease.

These results suggest that in response to viral infection, *Chlorella* NC64A produces unique methyltransferase and restriction endonuclease enzymes. Furthermore, the enzymes resemble bacterial methyltransferases and restriction endonuclease. From an evolutionary standpoint it will be interesting to compare the *Chlorella* viral genes to corresponding bacterial genes.

Restriction digestion and modification can also explain why host DNA is degraded during PBCV-1 infection and PBCV-1 DNA is not. PBCV-1 DNA is resistant to the restriction endonuclease *Mbo*I but sensitive to *DPN*I. In contrast, host nuclear DNA is sensitive to *Mbo*I and resistant to *Dpn*I. Both enzymes recognize the GATC base sequence, but *Mbo*I cleaves only DNA lacking 6mA in this sequence, whereas *Dpn*I cleaves only DNA containing 6mA in this sequence. Thus PBCV-1 DNA contains G^mATC, and the host DNA has the GATC sequences. Presumably, to protect against restriction digestion, the PBCV-1-induced methyltransferase adds methyl groups to A in the GATC sequence of the viral DNA, but not the host DNA. *In vivo* the PBCV-1-induced restriction endonuclease probably cleaves host DNA to specific fragments which are then digested by nonspecific host DNases; these nucleotides may be recycled into viral DNA.

Ecological Considerations

Freshwater samples have not previously been assayed for eukaryotic algal viruses because sensitive biological assays were not available. Now they can be easily screened for these viruses with the *Chlorella* NC64A plaque assay. An additional 35 freshwater samples from 10 states were examined for plaque-forming viruses. Thirteen of these samples from five states (Nebraska, Alabama, Massachusetts, New York, and California) contained at least 1 PFU/10 ml of water (Van Etten et al., 1985b). Like PBCV-1, all are large dsDNA-containing polyhedra. Presumably, these viral DNAs also contain various levels of 5mC and/or 6mA, because all the DNAs are resistant to one or more DNA restriction endonucleases.

Virus titer fluctuated in water samples collected at monthly intervals from seven locations in central Illinois (Van Etten et al., 1985b). Qualitative differences also were found since plaque size differed in some samples.

Several points deserve discussion. First, viruses that infect *Chlorella* NC64A are common in nature and occur in concentrations as high as 4×10^4 PFUs/ml (Van Etten et al., 1985a). The ecological role of these viruses in the aquatic environment is unknown. The natural host for these viruses could either be exsymbiont *Chlorella*, a natural, free-living *Chlorella* sp., some other eukaryotic alga, or an entirely different organism.

Second, the viral genomes are highly diverse as judged by restriction patterns. This may reflect variability due to genetic recombination, or strain incompatibility due to different restriction-modification systems. These possibilities can be tested by simultaneously infecting *Chlorella* NC64A with two different viruses and examining the progeny. The variability of these viral genomes contrasts dramatically with the extreme stability of the viral genomes of VLPs isolated from *H. viridis* symbiotic *Chlorella*-like algae. (See section on discovery of the viruses.)

Third, despite the widespread occurrence of viruses infecting *Chlorella* NC64A, viruses which form plaques on other *Chlorella* sp. (10 were tested) or *Chlamydomonas* (two species) have not been found (Van

Etten et al., 1985a; Van Etten et al., 1985b). Such viruses are perhaps not common in nature, exist as lysogens in nature and only infrequently enter a lytic phase, or appropriate hosts have not been tested.

Lysogeny could explain the infrequent appearance of VLPs in algal cells and the restriction of viruses to certain stages of algal development which have been observed by other investigators. The apparent lack of infectivity of these previously observed VLPs is also consistent with lysogeny. The VLPs might either infect the host and resume a lysogenic relationship or be excluded by other preexisting lysogenic viruses, as occurs in lysogenic bacteria (Luria et al., 1978).

Finally, the algae might harbor viruses in a carrier-state type of relationship (also called pseudolysogeny) where, at any one time, a small population of algae are continually infected by the virus. This possibility is reasonable, since this type of relationship between PBCV-1 and *Chlorella* NC64A has been found. In this one instance, *Chlorella* NC64A continually produced low concentrations of PBCV-1 (ca. 1×10^5 PFUs/ml) even after repeated subculturing of the alga over a 6-month period. Single-colony isolates of these *Chlorella* did not produce PBCV-1; however, inoculation of these isolates with PBCV-1 reinitiated the whole process (J. Van Etten, unpublished data).

Classification of the *Chlorella* Viruses

The *Chlorella* viruses have several properties in common with the Iridoviridae, also called icosahedral cytoplasmic deoxyriboviruses (Matthews, 1982; Goorha and Granoff, 1979). Iridoviruses are widely distributed in nature and commonly infect insects and some animals. Common properties between the Iridoviridae and the *Chlorella* viruses include icosahedral morphology, a large dsDNA genome, and a lipid component comprising 5–10% of the viral weight. The most studied member of this family is frog virus 3 (FV 3). FV 3, like PBCV-1, is one of the few known viruses infecting a eukaryotic organism which contains methylated bases (5mC) in its genomic DNA (Willis and Granoff, 1980); FV 3 also encodes a unique methyltransferase enzyme (Willis et al., 1984). However, FV 3 and PBCV-1 differ in at least two ways. First, FV 3 is typically uncoated inside the host cells (Houts et al., 1974; Kelly, 1975), whereas PBCV-1 is uncoated at the cell surface (Meints et al., 1984). Second, FV 3 cannot initiate replication in UV-inactivated host cells (Goorha et al., 1977); in contrast, PBCV-1 initiates and completes viral replication (albeit slowly and with a small burst size) in UV-inactivated *Chlorella* (J. L. Van Etten, unpublished data). FV 3 DNA did not hybridize with PBCV-1 DNA, even under low stringency conditions (J. L. Van Etten, unpublished data).

Most likely the *Chlorella* viruses will eventually be classified in the Iridoviridae as a separate genus.

CONCLUDING REMARKS

There are a number of reasons why these *Chlorella* viruses are interesting and deserve investigation.

1. They allow examination of a new virus-host relationship.
2. Since the host cells can be synchronously infected with virus and a definable product (i.e., progeny virus) appears several hours later, they serve as an excellent model system for studying gene regulation and expression in a photosynthetic eukaryote.
3. These viruses are unlike most viruses infecting higher plants, since they contain a dsDNA genome. Thus it is possible that portions of the viral genomes such as origin(s) of replication or promoters may be useful in constructing vectors for transferring genes into other algae or even into higher plants.
4. The viruses contain an enzyme(s) which digests the host cell wall and may be a new source of plant cell wall-degrading enzymes.
5. Preliminary evidence indicates that the viruses encode for unique methyltransferase and restriction endonuclease enzymes.
6. The common occurrence of VLPs in *Chlorella*-like algae in hydras suggests that these particles may be essential for this symbiosis.
7. Since viruses that infect *Chlorella* NC64A are common in freshwater, an understanding of their ecology is of interest. We encourage more investigators to work on these fascinating viruses.

ACKNOWLEDGMENTS

We thank Dwight Burbank, Kit Lee, Anne Schuster, Lois Girton, and Merri Skrdla for help with much of the experimental work on these viruses, and Les Lane, Myron Brakke, and Ted Pardy for many helpful and stimulating discussions.

Research from our laboratories has been supported by grants from the Department of Energy and from the National Institutes of Health.

REFERENCES

COLE, A., DODDS, A. J., AND HAMILTON, R. I. 1980. Purification and some properties of a double-stranded DNA containing virus-like particle from *Uronema gigas*, a filamentous eucaryotic green alga, *Virology* **100**:166–174.

DaSilva, E. J., and Gyllenberg, H. G. 1972. A taxonomic treatment of the genus *Chlorella* by the technique of continuous classification, *Arch. Mikrobiol.* **87:**99–117.

Dodds, J. A. 1979. Viruses of marine algae, *Experimentia* **35:**440–442.

Dodds, J. A. 1983. New viruses of eukaryotic algae and protozoa, in: *A Critical Appraisal of Viral Taxonomy* R. E. F. Matthews, ed., pp. 177–188, CRC Press, Boca Raton, Florida.

Dodds, J. A., and Cole, A. 1980. Microscopy and biology of *Uronema gigas*, a filamentous eukaryotic green alga, and its associated tail virus-like particle, *Virology* **100:**156–165.

Gibbs, A., Skotnicki, A. H., Gardiner, J. E., Walker, E.S., and Hollings, M. 1975. A tobamovirus of a green alga, *Virology* **64:**571–574.

Goohra, R., and Granoff, A. 1979. Icosahedral cytoplasmic deoxyriboviruses, in: *Comprehensive Virology*, Vol. 14 (H. Fraenkel-Conrat and R. R. Wagner, eds. pp. 347–399, Plenum Press, New York.

Goorha, R., Willis, D. B., and Granoff, A. 1977. Macromolecular synthesis in cells infected by frog virus 3. VI. Frog virus 3 replication is dependent on the cell nucleus, *J. Virol.* **21:**802–805.

Gromov, B. V., and Mamkaeva, K. A. 1981. A virus infection in the synchronized population of *Chlorococcum minutum* zoospores, *Arch. Hydrobiol. Suppl.* **60**(3):252–259.

Hellmann, V. and Kessler, E. 1974. Physiologische and biochemische Beitrage zur Taxonomie der Gattung *Chlorella* VIII. Die Basenzusammensetzung der DNS, *Arch. Microbiol.* **95:**311–318.

Hoffman, L. R., and Stanker, L. H. 1976. Virus-like particles in the green alga *Cylindrocapsa, Can. J. Bot.* **54:**2827–2841.

Houts, G. E., Gravell, M., and Granoff, A. 1974. Electron microscopic observation of early events of frog virus 3 replication. *Virology* **58:**589–594.

Kawakami, H., and Kawakami, N. 1978. Behavior of a virus in a symbiotic system, *Paramecium bursaria*-zoochlorella. *J. Protozool.* **25:**217–225.

Kelly, D. C. 1975. Frog virus 3 replication: Electron microscope observations on the sequence of infection in chick embryo fibroblasts. *J. Gen. Virol.* **26:**71–86.

Lemke, P. A. 1976. Viruses of eukaryotic microorganisms. *Annu. Rev. Microbiol.* **30:**105–145.

Loefer, J. 1936. Isolation and growth characteristics in the "zoochlorella" of *Paramecium bursaria. Am. Midl. Nat.* **70:**184–188.

Luria, S. E., Darnell, J. E., Baltimore, D., and Campbell, A. 1978. General Virology, John Wiley, New York.

Matthews, R. E. F. 1982. Classification and nomenclature of viruses, *Intervirology* **17:**1–198.

Mayer, J. A., and Taylor, F. J. R. 1979. A virus which lyses the marine nanoflagellate *Micromonas pusilla, Nature* **281:**299–301.

Meints, R. H., Van Etten, J. L., Kuczmarski, D., Lee, K., and Ang, B. 1981. Viral infection of the symbiotic *Chlorella*-like alga present in *Hydra viridis, Virology* **113:**698–703.

MEINTS, R. H., LEE, K., BURBANK, D. E., AND VAN ETTEN, J. L. 1984. Infection of a *Chlorella*-like alga with the virus, PBCV-1: Ultrastructural studies, *Virology* **138**:341–346.

PARDY, R. L. 1974. Some factors affecting the growth and distribution of the algal endosymbionts of *Hydra viridis*, *Biol. Bull. (Woods Hole, Mass.)* **147**:105–118.

SHERMAN, L. A., AND BROWN, R. M. 1978. Cyanophages and viruses of eukaryotic algae, in: *Comprehensive Virology*, Vol. 12 (H. Fraenkel-Conrat and R. R. Wagner, eds.), pp. 145–233, Plenum Press, New York.

SHIHRA, I. , AND KRAUSS, R. W. 1965. *Chlorella*: Physiology and taxonomy of forty one isolates, University of Maryland, College Park.

SIEGEL, R. 1960. Hereditary endosymbiosis in *Paramecium bursaria*, *J. Protozool.* **25**:366–370.

SKOTNICKI, A., GIBBS, A. J., AND WRIGLEY, W. G. 1976. Further studies on *Chara corrallina* virus, *Virology* **75**:457–468.

SKRDLA, M. P., BURBANK, D. E., XIA, Y., MEINTS, R. H., AND VAN ETTEN, J. L. 1984. Structural proteins and lipids in a virus, PBCV-1, which replicates in a *Chlorella*-like alga, *Virology* **135**:308–315.

STANKER, L. H., AND HOFFMAN, L. R. 1978. A simple histological assay to detect virus-like particles in the green alga *Cylindrocapsa* (Chlorophyta), *Can. J. Bot.* **57**:838–842.

STANKER, L. H., HOFFMAN, L. R., AND MACLEOD, R. 1981. Isolation and partial chemical characterization of a virus-like particle from a eukaryotic alga, *Virology* **114**:357–369.

TRENCH, R. K. 1979. The cell biology of plant-animal symbiosis. *Annu. Rev. Plant Physiol.* **30**:485–531.

VAN ETTEN, J. L., MEINTS, R. H., BURBANK, D. E., KUCZMARSKI, D., CUPPELS, D. A., AND LANE, L. C. 1981. Isolation and characterization of a virus from the intracellular green alga symbiotic with *Hydra viridis*, *Virology* **113**:704–711.

VAN ETTEN, J. L., MEINTS, R. H., KUCZMARSKI, D., BURBANK, D. E., AND LEE, K. 1982. Viruses of symbiotic *Chlorella*-like algae isolated from *Paramecium bursaria* and *Hydra viridis*, *Proc. Natl. Acad. Sci. USA* **79**:3867–3871.

VAN ETTEN, J. L., BURBANK, D. E., KUCZMARSKI, D., AND MEINTS. R. H. 1983a. Virus infection of culturable *Chlorella*-like algae and development of a plaque assay, *Science* **219**:994–996.

VAN ETTEN, J. L., BURBANK, D. E., XIA, Y., AND MEINTS, R. H. 1983b. Growth cycle of a virus, PBCV-1, that infects *Chlorella*-like algae, *Virology* **126**:177–125.

VAN ETTEN, J. L., BURBANK, D. E., JOSHI, J., AND MEINTS, R. H. 1984. DNA synthesis in a *Chlorella*-like alga following infection with the virus PBCV-1, *Virology* **134**:443–449.

VAN ETTEN, J. L., BURBANK, D. E., SCHUSTER, A. M., AND MEINTS, R. H. 1985a. Lytic viruses infecting a *Chlorella*-like alga, *Virology* **140**:135–143.

VAN ETTEN, J. L., VAN ETTEN, C. H., JOHNSON, J. K., AND BURBANK, D. E.

1985b. A survey for viruses from freshwater that infect a eukaryotic *Chlorella*-like alga, *Appl. Environ. Microbiol.,* **49:**1326–1328.

VAN ETTEN, J. L., SCHUSTER, A. M., GIRTON, L., BURBANK, D. E., SWINTON, D., HATTMAN, S. 1985c. DNA methylation of viruses infecting a eukaryotic *Chlorella*-like green alga, *Nucleic Acids Res.* **13:**3471–3498.

WATERS, R. E., CHAN, A. T. 1982. *Micromonas pusilla* virus: The virus growth cycle and associated physiological events within the host cells—Host range mutation, *J. Gen. Virol.* **63:**199–206.

WEIS, D. S. 1978. Correlation of infectivity and concanavalin A agglutinability of algae exsymbiotic from *Paramecium bursaria, J. Protozool.* **25:**366–370.

WILLIS, D. B., AND GRANOFF, A. 1980. Frog virus 3 is heavily methylated at CpG sequences, *Virology* **107:**250–257.

WILLIS, D. B., GOORHA, R., AND GRANOFF, A. 1984. DNA methyltransferase induced by frog virus 3, *J. Virol.* **49:**86–91.

YAMADA, T. AND SAKAGUCHI, K. 1982. Comparative studies on *Chlorella* cell walls: Induction of protoplast formation, *Arch. Microbiol.* **132:**10–13.

YONKER, C. R., CALDWELL, K. D., GIDDINGS, J. C. AND VAN ETTEN, J. L. 1985. Physical characterization of PBCV-1 virus by sedimentation field flow fractionation *J. Virol. Methods,* **11:**145–160.

Chapter 11

Retrolike Viruses in Plants

Tom J. Guilfoyle

RETROVIRUSES ARE A GROUP of animal viruses (Retroviridae) that contain single-stranded (ss) RNA genomes which are replicated through double-stranded (ds) DNA intermediates (for reviews see Varmus, 1983; Varmus and Swanstrom, 1982). Because genetic information is encoded on an RNA genome that is transcribed into a DNA genomic intermediate, these viruses encode an RNA-dependent DNA polymerase (reverse transcriptase) to carry out this information transfer. The retrovirus replication cycle can be summarized as:

$$\text{ss(+)RNA} \underset{4}{\overset{1}{\circlearrowright}} \text{dsDNA} \xrightarrow{2} \text{ss(+)RNA} \xrightarrow{3} \text{protein}$$

where (1) virus-encapsidated single-stranded plus-strand RNA is reverse-transcribed to double-stranded DNA. The double-stranded DNA is transcribed by host cell DNA-dependent RNA polymerase II to single-stranded plus-strand RNA that may be (4) encapsidated or (2) act as mRNA and (3) be translated by host cell machinery into viral proteins.

Varmus (1983) has categorized several features that are characteristic of retroviruses, and some of them are summarized below. Virions 80–120 nm in diameter are made up of a nucleoprotein core (40–70 nm in diameter) covered by an envelope derived from the plasma membrane of the host cell. The virion is composed of about 60% protein, 30% lipid, 5% carbohydrate, and 1% RNA. The encapsidated genomic RNA consists of two single-stranded subunits of 8–10 kilobases (kb). Low-molecular-weight cellular RNAs may also be encapsidated. The polypeptides required for virus replication are encoded on three genes organized on the RNA genome in a 5'-to-3' direction as *gag, pol,* and *env.* These genes encode nucleocapsid core protein, reverse transcriptase, and envelope-associated glycoproteins, respectively.

The initial events in the replication of retroviruses appear to occur in the cytoplasm of infected cells where the single-stranded /plus-strand RNA genome is reverse-transcribed into double-stranded linear DNA (Varmus et al., 1974). Since reverse transcriptase is encapsidated in the virus particle, this step in the life cycle of the virus need not require host cell proteins; however, it is possible that host cell components facilitate this process. The double-stranded DNA may then be transported to the nucleus where it can be converted to closed circular double-stranded DNA (Guntaka et al., 1976; Fritsch and Temin, 1977). It is proposed that either the closed circular DNA or the linear DNA is then integrated into host chromosomal DNA (proviral DNA) where it can be expressed from viral transcriptional promoters utilizing host cell RNA polymerase II (for a review see Varmus, 1983). There is no evidence that viral DNA is expressed or replicated in the nonintegrated state. Additional details on retrovirus replication strategy and life cycle are addressed below, where informative comparisons are made with retrolike viruses (caulimoviruses) from plants.

Two additional groups or families of viruses which are not true retroviruses nevertheless share some striking similarities with retroviruses. These viruses are animal hepatitis B viruses and plant caulimoviruses. For the purpose of comparative virology, I refer to these viruses as retrolike viruses. The most striking feature common to retroviruses, hepatitis B viruses, and caulimoviruses is reverse transcription of genomic RNAs and RNA replication intermediates. The major focus of this chapter is limited to caulimoviruses and, more specifically, cauliflower mosaic virus (CaMV), but when appropriate, I will point out the similarities and differences between retrolike caulimoviruses and hepatitis B viruses as well as true retroviruses.

PROPERTIES OF CAULIMOVIRUSES

Discovery and Classification

Although host range (Tompkins, 1937; Broadbent, 1957), diagnostic (Robb; 1963), and serological studies (Brunt, 1966; Pirone et al., 1961) provided preliminary evidence for a distinct group of related viruses that infected crucifers and certain other plants, the discovery that CaMV contained a double-stranded DNA genome (Shepherd et al., 1968, 1970; Shepherd and Wakeman, 1971; Russell et al., 1971) proved that these viruses represented a novel group of plant viruses. At least 13 viruses have been identified that appear to contain double-stranded DNA genomes (Hull and Davies, 1983), and these viruses make up the caulimovirus group. Unlike the situation with animal viruses, double-stranded DNA viruses are relatively rare in plants and represent only 3% of the total number of plant viruses that have been classified (Hull and Davies, 1983). Details on the general properties of the caulimovirus type member, CaMV, as well as the other caulimoviruses, can be found in a number of reviews (Shepherd, 1976, 1979; Hull, 1979; Hohn et al., 1982; Howell, 1982; Hull and Davies, 1983; Howell, 1985; Hull and Covey, 1985).

Infectivity Characteristics of Caulimoviruses

In nature, CaMV is transmitted by aphids (*Myzus persicae*), and the host range includes a variety of species in the family Cruciferae. In the laboratory, the virus is generally transmitted mechanically, and the most common hosts are turnip (*Brassica rapa* L. cv. Just Right) and mustard (*B. pervirdis* Bailey cv. Tendergreen). Certain isolates of CaMV have been reported to infect *Nicotiana clevelandii* (Hills and Cambell, 1968) and *Datura stramonium* (Lung and Pirone, 1972) in the family Solanaceae. Other caulimoviruses infect species in a wide variety of plant families including Caryophyllaceae, Compositae, Amaranthaceae, Chenopodiaceae, Nyctaginaceae, Rosaceae, Scrophulariaceae, Ericaceae, Euphorbiaceae, and Plantaginaceae (Shepherd, 1976; Hull and Davies, 1983); however, the host range for any one member of the caulimoviruses is generally restricted to a single family of plants. Recently, a caulimovirus, soybean chlorotic mottle virus, which is not serologically related to CaMV, was reported to infect soybean (*Glycine max*), green bean (*Phaseolus vulgaris*), cowpea (*Vigna unguiculata*), and lablab bean (*Dolichos lablab*) in the family Leguminoseae (Iwaki et al., 1984).

Symptoms in plants chronically infected with caulimoviruses include changes in leaf pigmentation (e.g., vein clearing), wrinkling of leaves, and stunting of growth. At the cytological level, a diagnostic feature of caulimovirus infection is the presence of inclusion bodies or viroplasms in infected cells (Rubio-Huertos, 1972; Robb, 1963, 1964; Shepherd, 1976; Shalla et al., 1980). Inclusion bodies are electron-dense structures that take a variety of ovoid, elliptical, or lobed shapes, are not surrounded by a membrane, may be vacuolated or nonvacuolated, and may or may not be packed with virus particles. The size and constitution of inclusion bodies is determined in part by the viral isolate, host plant species, and stage of infection (Shepherd, 1976; Shalla et al., 1980). At least two virus-encoded polypeptides, virion capsid or coat protein and viroplasm matrix or inclusion body protein, are associated with purified inclusion bodies (Shepherd et al., 1980; Shockey et al., 1980; Xiong et al., 1982), and a third virus-encoded polypeptide appears to affect inclusion body morphology and may or may not be firmly associated with inclusion bodies (Givord et al., 1984). Genetic evidence that virus-encoded inclusion body protein plays a role in the expression of disease symptoms in CaMV-infected plants has recently been reported (Daubert et al., 1984).

Composition of CaMV Virions

CaMV is isometric, has a diameter of about 50 nm, and is composed of a 42-kilodalton (kd) coat protein (Al Ani et al., 1979) and an 8-kilobase-pair (kbp) double-stranded DNA (Franck et al., 1980; Gardner et al., 1981). No host cell histones or high mobility group (HMG) proteins are associated with CaMV DNA in virus particles (T. Guilfoyle, unpublished results). DNA-dependent DNA polymerase and/or RNA-dependent DNA polymerase activities have been reported to be associated with purified virions and virion-like particles (Guilfoyle et al., 1983b; Menissier et al., 1984; Marsh et al., 1985). The encapsidated viral genomic DNA is circular but not covalently closed (Hull and Howell, 1978; Volovitch et al., 1978). The viral DNA of most CaMV isolates contains three site-specific discontinuities (Fig. 1). One DNA strand of the CaMV genome, referred to as α-strand DNA, possesses one discontinuity, and the site of the break in this DNA strand is designated as map position 0/1.0 on the circular genome (Hohn et al., 1982). The complementary strand of β-strand DNA usually contains discontinuities at map units 0.20 and 0.53 (Volovitch et al., 1978; Franck et al., 1980). One isolate of CaMV (CM4-184) contains a deletion of 421 bp in and around map position 0.20 (Howarth et al., 1981), and in this case the β-strand DNA contains only a single site-specific discontinuity. The discontinuities or breaks in the circular duplex DNA are not true nicks or gaps, but rather the 3' and 5' termini of the broken DNA strand overlap by about 8–20

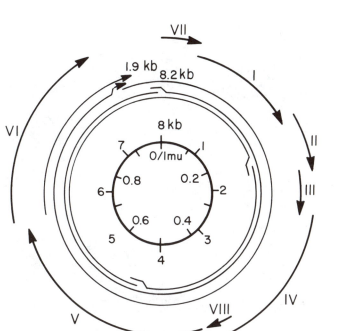

Figure 1. Map of the CaMV genome, RNA transcripts, and open reading frames (ORFs). Inner circle shows map units (mu) and size of the CaMV genome in kilobases (kb). The next circle outward illustrates the site-specific discontinuities (⎓⎓⎓) in the α- and β-DNA strands of the double-stranded DNA genome of virion CaMV DNA. The major RNA transcripts of 8.2 and 1.9 kb are depicted with light arrows (5′⟶3′). ORFs I through VIII are represented with heavy arrows (amino-terminus⟶carboxyl-terminus).

nucleotides (Richards et al., 1981). While the position of the first deoxyribonucleotide at the 5′ terminus is fixed at a specific nucleotide position, the nucleotide position of the 3′ terminus is somewhat variable. The 5′ termini are reported to contain one or more covalently attached ribonucleotides (Guilley et al., 1983).

A peculiar characteristic of encapsidated genomic DNA is the prevalence of topological by knotted or tangled molecules (Volovitch et al., 1978; Hohn et al., 1982; Menissier et al., 1983; Civerolo and Lawson, 1978). These molecules appear as knotted or rosettelike molecules in the electron microscope (Menissier et al., 1983; Guilfoyle et al., 1983a). Knotted molecules sediment more rapidly than unknotted circular or linear CaMV DNAs on sucrose gradients, and DNA molecules with different amounts of topological knotting can be resolved from one another on agarose gels (Fig. 2) (Volovitch et al., 1978; Menissier et al., 1983). The resolution of knotted circular CaMV DNAs on agarose gels appears

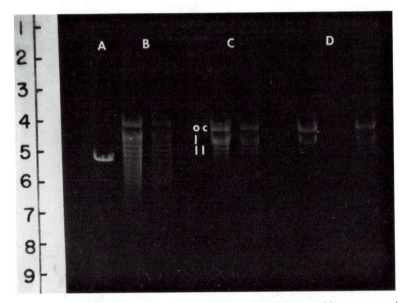

Figure 2. Agarose slab gel of virion CaMV DNA fractionated by sucrose density gradient centrifugation. After purification of CaMV DNA on cesium chloride gradients, the DNA was fractionated on 5–20% exponential sucrose density gradients into fast (B), intermediate (C), and slow (D) sedimenting fractions. Fractions from the gradients were electrophoresed on 0.8% agarose gels using a *tris*-acetate buffer, and DNA bands were visualized by staining with ethidium bromide. Col EI linear DNA was electrophoresed as a standard (A). Open circular (oc), linear (I), and "knotted" circular (II) CaMV DNAs are labeled. "Knotted" circular CaMV DNAs are recovered in the more rapidly sedimenting fractions of the sucrose density gradient.

similar to the separation of covalently closed circular DNAs possessing different amounts of supercoiling (Keller, 1975), but in the case of CaMV DNA, all DNA molecules contain site-specific discontinuities. Hohn et al. (1983) have proposed that this knotting occurs during the process of viral genome replication.

At least some virus particles or viruslike particles contain subgenomic DNAs with covalently attached ribonucleotides at their 5'-termini (Turner and Covey, 1984; Guilfoyle et al., 1983b; Marsh et al., 1985). These subgenomic DNAs are about 700 nucleotides in length, and after ribonuclease digestion the molecules are reduced in size by about 100 nucleotides (Turner and Covey, 1984; Marsh et al., 1985). The small encapsidated DNAs are sequence-specific and map to an approximate position between nucleotides 7435 and 8031 (Nucleotide positions reported here may vary slightly with different isolates of CaMV) on the circular genomic map (Covey et al., 1983; Turner and Covey, 1984; Guilfoyle et al., 1983b; Marsh et al., 1985). Other subgenomic single-stranded minus-strand DNAs (i.e., DNA complementary to CaMV polyadenylated RNA; see section on CaMV replication), partially double-stranded CaMV DNAs, and DNA-RNA hybrids have also been detected in virions and virionlike particles (Marsh et al., 1985). The encapsidated subgenomic CaMV DNAs may result from the packaging of defective genomes or nonspecific packaging of incomplete or degraded CaMV genomes, or these DNAs and DNA-RNA hybrids may represent replication intermediates (Marsh et al., 1985).

VIRION-ENCODED POLYPEPTIDES

Direct sequencing of the entire CaMV DNA genome (isolate Cabb B-S) allowed Franck et al. (1980) to determine open reading frames (ORFs) that could potentially code for polypeptides (Fig. 1). All large ORFs are encoded on α-strand DNA, and six ORFs were originally identified (Frank et al., 1980). Two additional small ORFs encoded on α-strand DNA were pointed out by Hohn et al. (1982). Additional sequencing results from two other CaMV isolates (CM1841 and D/H) revealed the same ORFs as those mapped on isolate Cabb B-S (Gardner et al., 1981; Balazs et al., 1982).

ORF I starts at nucleotide 331, ends at nucleotide 1347, and could code for a protein with a molecular weight of about 38 kd (Franck et al., 1980). At this time, a polypeptide corresponding to this ORF has been identified neither in CaMV-infected plants nor in *in vitro* translations using polyadenylated RNA isolated from CaMV-infected tissues.

ORF II begins at nucleotide 1328 and ends at nucleotide 1828 (Franck et al., 1980) This ORF could code for an 18-kd polypeptide. The protein encoded by ORF II is thought to be an aphid transmission factor, since DNA deletions within this ORF result in a loss of aphid transmissibility (Woolston et al., 1983; Armour et al., 1983). A naturally occurring deletion mutant, CM4-184, which contains a 421-bp deletion within ORF II (Howarth et al., 1981), is not transmitted by aphids (Hull, 1980). Wooston et al. (1983) identified an 18-kd polypeptide associated with viroplasms of plants infected with CaMV isolates carrying an ORF II and showed that this polypeptide was absent when infection was carried out with CaMV DNA where ORF II was completely deleted. Construction of hybrid genomes between a Campbell isolate, which is not aphid-transmissible and not deleted with respect to ORF II, and a Cabb B-JI isolate, which is aphid-transmissible, demonstrated that aphid transmissibility and production of an

18-kd polypeptide was dependent on a DNA restriction fragment containing ORF I and II from the aphid-transmissible isolate (Woolston et al., 1983). Givord et al. (1984) constructed a CaMV mutant DNA from an aphid-transmissible isolate by deleting nucleotides 1534–1642 in the center of ORF II. This deletion mutant was not aphid-transmissible, but a polypeptide of about 16 kd was detected in some viroplasm preparations in comparison to an 18-kd polypeptide detected in viroplasms induced by undeleted viral isolates. In other experiments, a 12-kd polypeptide was detected with this deletion mutant, and it was suggested that the 16-kd deletion product may be unstable in comparison to the 18-kd polypeptide. Additional studies with this deletion mutant indicated that ORF II may also function in viroplasm stability and constitution (Givord et al., 1984).

A 15-kd polypeptide could be produced from ORF III which starts at nucleotide 1812 and ends at nucleotide 2219 (Franck et al., 1980). This 15-kd polypeptide has been detected in CaMV-infected plants by utilizing an antibody raised against a synthetic peptide of 19 amino acids which corresponded to the amino acid sequence (i.e., deduced from the DNA sequence) of the amino-terminal region of ORF III (Xiong et al., 1984). Although the function of this polypeptide has not been determined, a functional ORF III appears to be required for viral infectivity (Lebeurier et al., 1982; Daubert et al., 1983; Dixon et al., 1983).

Nucleotides from 2168 to 3670 could encode ORF IV a polypeptide of about 57 kd (Franck et al., 1980). Based on the amino acid sequence of ORF IV, which contains a high lysine content, Franck et al. (1980) suggested that this ORF might correspond to the viral capsid or coat protein. CaMV coat protein is reported to contain 18% lysine on a molar basis (Brunt et al., 1975). Since the major caspid protein of CaMV virions appears to be 42 kd (Al Ani et al., 1979), the 57-kd polypeptide encoded on ORF IV must be processed to a 42-kd polypeptide. Hahn and Shepherd (1982) have presented evidence that a primary translation product of about 58 kd is used in viral assembly and that this polypeptide is proteolytically processed *in vivo* to lower-molecular-weight capsid polypeptides. Additional evidence that ORF IV is the viral coat protein comes from studies reported by Daubert et al. (1982). Recombinant plasmids containing ORF IV produced polypeptides in *Escherichia coli* that reacted with antibodies raised against CaMV.

ORF V spans nucleotides 3591-5672 and could produce a polypeptide of about 79 kd (Franck et al., 1980). The amino acid sequence derived from the DNA sequence of ORF V has some homology with reverse transcriptases encoded by retroviruses and hepatitis B viruses of animals (Toh et al., 1983; Patarca and Haseltine, 1984) and the *copia* element of *Drosophila melanogaster* (Saigo et al., 1984). In addition, the molecular weight of the putative polypeptide corresponding to ORF V is similar in size to the *pol* gene product encoded by retroviruses (Varmus, 1983). The proposed role of reverse transcriptase in CaMV replication is discussed below.

Nucleotide positions 5713–7338 delimit ORF VI which produces a polypeptide of about 61 kd (Franck et al., 1980). This polypeptide has been identified in CaMV-infected plants and is thought to be a major component of inclusion bodies (Al Ani et al., 1980; Odell and Howell, 1980; Covey and Hull, 1981; Xiong et al., 1982). Because of its association with purified inclusion bodies, this polypeptide is generally referred to as the inclusion body or viroplasm protein. The inclusion body protein is thought to be a matrix or structural protein as opposed to an enzyme, but a possible enzymic function for this polypeptide cannot be ruled out at this time.

Little information is available on ORFs VII and VIII, with the exception of their nucleotide positions on the CaMV genome which span nucleotides 13–301 and 3264–3589, respectively (Hohn et al., 1982). ORF VII could code for a polypeptide of about 11 kd, while ORF VIII could produce a polypeptide of about 12 kd. Neither of these polypeptides has been detected in CaMV-infected plants. Evidence that the expression of ORF VII may not be required for viral infectivity was provided by Dixon et al. (1983) who demonstrated that insertions of 10 or 22 bp within this ORF did not interfere with infectivity. These authors also showed that a 105-bp deletion in ORF VII did not block virus viability.

Figure 1 shows that six of the eight ORFs overlap or adjoin one another. ORF VIII is completely contained in ORF IV, although the reading frames of these two ORFs differ. All three reading frames are used to produce the clustered ORFs I–V; therefore these five putative polypeptides could not arise from a single polyprotein.

TRANSCRIPTION OF CaMV

The most abundant CaMV-specific polyadenylated RNAs found in infected plants are 8.2 and 1.9 kb (Odell and Howell, 1980; Odell et al., 1981; Covey and Hull, 1981; Guilley et al., 1982). These two transcripts are both encoded by α-strand DNA (Howell and Hull, 1978; Hull et al., 1979). Stable transcripts from the β-strand DNA have not been detected. Since *in vitro* translation of the 1.9-kb polyadenylated RNA is inhibited by 7-methyl-GTP, it is likely that this RNA is 5′-capped (Odell and Howell, 1980). Whether the 8.2-kb RNA is capped has not been determined because this RNA is not efficiently translated *in vitro*. Minor

CaMV-specific RNA transcripts have also been reported and mapped on the CaMV genome (Guilley et al., 1982; Condit et al., 1983; Condit and Meagher, 1983), but how these minor RNAs arise and what function they might play in the virus life cycle or propagation is unclear. To date, only one polyadenylated RNA, the 1.9-kb transcript, has been shown to be a functional mRNA, and this RNA species is translated into the inclusion body protein encoded on ORF VI (Odell and Howell, 1980; Covey and Hull, 1981; Xiong et al., 1982). There is no direct evidence to suggest that any viral transcript is spliced; however, the possibility of RNA splicing has been suggested by recent experiments with specific CaMV isolates (Hirochika et al., 1985).

Both the 8.2- and 1.9-kb polyadenylated RNAs have been mapped on the viral genome (Fig. 1) (Covey et al., 1981; Dudley et al., 1982; Guilley et al., 1982). The length of the 8.2-kb transcript is greater than that of the genome by 180 nucleotides, and the 5' terminus occurs at nucleotide position 7435 while the 3' terminus [minus the poly (A) tail] maps to nucleotide position 7615 (Guilley et al., 1982). The 1.9-kb transcript has a 5' terminus at nucleotide position 5764 and a 3' terminus identical to that of the 8.2-kb transcript (Guilley et al., 1982). The promoters for the 8.2- and 1.9-kb RNAs are located in the intergenic regions (i.e., regions outside of ORFs) of the CaMV genome (Fig. 1). At 31 or 33 nucleotides upstream of the two major polyadenylated transcripts' 5' ends (i.e., cap sites) "TATA box" sequences have been identified (Guilley et al., 1982), and these promoters have been shown to function *in vitro* (Guilley et al., 1982; Suzich and T. Guilfoyle, unpublished results) as well as *in vivo* (Koziel et al., 1984; Brisson et al., 1984; Odell et al., 1985). Odell et al. (1985) have pointed out additional sequences which may play some role in promoter strength upstream of the 8.2-kb start or cap site (at positions −46 to −105 from the 5' terminus of the 8.2-kb transcript), and these include a "CCAAT box," an inverted repeat, and a possible enhancer sequence (GTGGATTG in CaMV corresponding to the consensus core of animal enhancer sequences, GTGGAAAG or GTGGTTTG). A polyadenylation signal can be found upstream of the 3' end of the 8.2- and 1.9-kb transcripts (Guilley et al., 1982). Individual mRNAs for ORFs I–V, VII, and VIII have not been identified, and it has been suggested that the 8.2-kb transcript may function as a polycistronic mRNA (Hull, 1984; Dixon and Hohn, 1984).

Transcription of the CaMV genome occurs in nuclei of infected cells (Guilfoyle, 1980, 1985). Nuclei purified from CaMV-infected tissues and incubated in an *in vitro* transcription cocktail synthesize RNA transcripts that hybridize with restriction fragments encompassing the entire viral genome. *In vitro* transcription is restricted to the α strand of the CaMV genome. CaMV-specific *in vitro* transcription is inhibited by α-amanitin at concentrations that parallel the inhibition of plant nuclear RNA polymerase II. The monovalent cation concentration required for optimal *in vitro* transcription of CaMV DNA in isolated nuclei is identical to that observed with host RNA polymerase II engaged on the nuclear chromatin template, suggesting that CaMV DNA may be associated with chromatin proteins in the nucleus of infected cells. That multiple copies of CaMV DNA are present in nuclei is suggested by the observation that 3–5% of the total *in vitro* transcripts are CaMV-specific (Guilfoyle et al., 1983b).

The nature of the CaMV DNA template transcribed in host cell nuclei was identified by Olszewski et al. (1982). Unencapsidated covalently closed circular CaMV DNA in infected plants was originally reported by two laboratories (Olszewski and Guilfoyle., 1983; Menissier et al., 1982), and Olszewski et al. (1982) further demonstrated that the closed circular DNA was compartmentalized in infected cell nuclei, that it was an 8-kb monomer-length CaMV DNA, that it was associated with chromatin or chromatin-like proteins, and that some of the molecules were associated with host RNA polymerase II engaged in the process of transcription on α-strand DNA. This closed circular DNA has been referred to as a minichromosome (Olszewski et al., 1982; Guilfoyle et al., 1983a,b) and probably represents the major, if not only, functional CaMV DNA template for transcription in infected cells. The CaMV minichromosome has a nucleosome repeat similar to that of host cell chromatin (Olszewski et al., 1982), and molecules, which are released in cleared nuclear lysates and are of appropriate size, appear to have a nucleosome-like structure when observed in the electron microscope (Menissier et al., 1983). The covalently closed nature of the CaMV DNA template used for transcription helps explain how an 8.2-kb transcript can be generated. In the absence of a closed circular template, a transcript initiated at nucleotide position 7435 (i.e., the transcriptional start site of the 8.2-kb transcript) would presumably run off the viral DNA template at the site where the discontinuity in the α-strand DNA is located. Since the template utilized for transcription lacks this discontinuity, RNA polymerase II can traverse the entire CaMV genome.

Attempts to detect CaMV DNA integrated into host nuclear DNA in infected turnip leaves have not been successful (G. Hagen and T. Guilfoyle, unpublished results). However, these findings do not rule out the possibility that covalently closed circular or other forms of CaMV DNA might integrate into the nuclear genome. If CaMV DNA integrated randomly at few sites per nucleus or in a small proportion of nuclei in the total cell population, integration events would go undetected. Since CaMV can propagate through nondividing cells where integration events might be limited if not prohibited (for a discussion on retrovirus integration see Varmus, 1983) and since dividing clonal cell lines (e.g., plant tumors) do not arise during CaMV infections, the chances of detecting CaMV DNA integration events in intact plants might be remote.

REPLICATION OF CaMV

Subcellular Compartmentation

The subcellar compartment where CaMV replication occurs in infected cells has not been clearly identified. There is preliminary evidence for CaMV replication in inclusion bodies (Kamei et al., 1969 Favali et al., 1973; Modjtahedi et al., 1984; Mazzodini et al., 1985) and nuclei (Ansa et al., 1982; Guilfoyle et al., 1983b). The failure to resolve which subcellular compartment is the site for viral replication may result from analysis of impure organellar fractions, failure to discriminate between authentic replication complexes and those artificially resulting from cell disruption, failure to differentiate between DNA repair and DNA replication, or utilization of alternative modes for replication by the virus (e.g., replication by reverse transcriptase in inclusion bodies to produce virion DNA for encapsidation and replication by DNA polymerase α in the nucleus to produce minichromosomes).

Replication Complexes

Putative replication complexes have been isolated from cleared lysates of subcellular fractions containing a combination of nuclei and inclusion bodies (Pfeiffer and Hohn, 1983; Guilfoyle et al., 1983b; Pfeiffer et al., 1984; Marsh et al., 1985). Pfeiffer and Hohn (1983) first demonstrated that cleared lysates fractionated on sucrose gradients contained CaMV-specific replication activity that sedimented more slowly than CaMV minichromosomes. They concluded that CaMV minichromosomes were not templates for CaMV replication and presented preliminary evidence that replication might occur by reverse transcription of the 8.2-kb CaMV RNA transcript. The CaMV replication activity reported by Pfeiffer and Hohn (1983) was inhibited 40% by actinomycin D and 50% by ribonuclease A. Subsequent to these initial studies, Guilfoyle et al. (1983b) showed that two distinct CaMV-specific replication activities could be detected on sucrose gradients of cleared nuclear lysates. The slower sedimenting activity had properties somewhat analogous to those described by Pfeiffer and Hohn (1983) while the faster sedimenting activity was located in fractions with a higher density than CaMV minichromosomes and at a density similar to that of CaMV virions.

Marsh et al. (1985) demonstrated that the replication activity in the virion portion of the sucrose gradient was resistant to both ribonuclease A and deoxyribonuclease I (suggesting that the replicating complexes were encapsidated) but was partially inhibited by actinomycin D. With the use of strand-specific probes, Marsh et al. (1985) showed that minus-strand DNA synthesis (i.e., synthesis of α-strand DNA complementary to 8.2-kb polyadenylated RNA) was highly resistant to actinomycin D inhibition, while plus-strand DNA synthesis was highly sensitive to inhibition by this drug. This pattern of inhibition by actinomycin D is consistent with a reverse-transcription model for CaMV replication where minus-strand DNA is synthesized on an RNA template and thus resistant to actinomycin D, while plus-strand DNA is synthesized on a minus-strand DNA template and is sensitive to this drug. In contrast to the faster sedimenting activity, the slower sedimenting CaMV replication activity was largely resistant to ribonuclease but strongly inhibited by deoxyribonuclease, and minus-strand DNA synthesis and plus-strand DNA synthesis were equally sensitive to actinomycin D.

Recent results reported by Pfeiffer et al. (1984) indicate that the slower sedimenting replication complexes synthesize minus- and plus-strand DNAs that are equally sensitive to inhibition by actinomycin D, but minus-strand DNA synthesis is more strongly inhibited (80% inhibition) by ribonuclease treatment than plus-strand DNA synthesis (30% inhibition). Neither of these observations is totally consistent with a reverse-transcription model for CaMV replication; however, the authors speculate that CaMV reverse transcriptase may be directly inhibited by actinomycin D. The possible reasons for inhibition of plus-strand DNA synthesis by ribonuclease were not addressed by the authors.

The results on putative replication complexes reported by Pfeiffer and Hohn (1983) and Pfeiffer et al. (1984) are not completely consistent with those reported by Guilfoyle et al. (1983b) and Marsh et al. (1985). More information is required to determine whether viral replication occurs on unencapsidated, partially encapsidated, or fully encapsidated templates. The inconsistencies on replication complexes observed by Pfeiffer et al. (1984) and Marsh et al. (1985) might be related to such factors as artifactual association of reverse transcriptase or other cellular DNA polymerases with nucleic acids and other cellular components during tissue disruption or dissociation of authentic replication complexes during purification. It is also possible that the replication complexes observed represent different stages in the replication process or defective replication events occurring in chronically infected cells.

Accumulation of Replication Intermediates

Hull and Covey (1983b), Guilley et al. (1983), and Pfeiffer and Hohn (1983) proposed that CaMV replication occurs via reverse transcription of the 8.2-kb polyadenylated RNA. A model for this strategy of replication is presented in Figs. 3 and 4 and is described in detail below. Some of the results supporting this model come from the identification of subgenomic CaMV DNAs in infected cells. Covey et al. (1983) described a number of small, unencapsidated DNAs that map between the discontinuity in the α-strand DNA and the transcription start site for the 8.2-kb transcript. One of these small DNAs (referred to as sa DNA) is 725 nucleotides in length and contains covalently attached RNA. sa DNA has the same polarity as α-strand DNA and is thus complementary to the 8.2-kb RNA transcript. A double-stranded DNA composed of three DNA subunits, which is related to sa DNA but which lacks attached RNA, was also observed by Covey et al. (1983). Guilley et al. (1983) reported that a single-stranded DNA similar to sa DNA was also found in virions. Turner and Covey (1984) suggested that the sa DNA found in virions represented a "strong-stop" DNA similar to that observed with retroviruses (Varmus, 1983) (Fig. 4). In searching for replication intermediates, Marsh et al. (1985) identified a heterogeneous population of single-stranded minus-strand CaMV DNAs (or minus-strand DNA–plus-strand RNA hybrids) ranging in size from less than the size of sa DNA to full genome length 8-kb single-stranded DNA in virions or virion-like particles. In addition, partially double-stranded DNAs with minus-strand DNA tails or regions were detected in these particles (Marsh et al., 1985). Asymmetric production of minus strand CaMV DNA has also been detected in infected protoplasts (Maule, 1985; Thomas et al., 1985).

Hull and Covey (1983a) and Marco and Howell (1984) have reported on putative replication intermediates that appear to be unencapsidated. Molecules observed included double-stranded DNAs spanning (in a counterclockwise direction on the circular CaMV genome) the discontinuity in α-strand DNA to the discontinuity at map position 0.53 or 0.20 on the β-strand DNA and partially double-stranded DNAs as described above, but with single-stranded extensions of the α-strand DNA beyond map positions 0.53 and 0.20. Such molecules are consistent with replication being primed at the site-specific discontinuities where α-strand DNA is the first strand to be synthesized using the 8.2-kb polyadenylated RNA as template, followed by β-strand synthesis using the α-strand DNA as template. Some of these molecules could arise artificially, however, by double-stranded breaks occurring at the sites of single-stranded discontinuities.

Whether any of the putative replication intermediates found to be encapsidated or unencapsidated are authentic replication intermediates remains to be substantiated. It is just likely that the putative intermediates described are simply defective genomes or CaMV degradation products. It will be necessary to follow the kinetics of replication and virion assembly *in situ* (e.g., possibly by using synchronous infection of protoplasts;

Maule, 1985) and/or to follow replication *in vitro* using reconstituted components to unequivocally demonstrate the exact pathway of a CaMV replication.

Reverse Transcriptase

Several laboratories have presented preliminary results suggesting that a DNA polymerase or reverse transcriptase is found in CaMV-infected tissues and protoplasts but is absent or modified in uninfected tissues (Guilfoyle et al., 1983b; Volovitch et al., 1984; Pfeiffer et al., 1984). Chromatography of turnip DNA polymerases on heparin-Sepharose (Guilfoyle et al., 1983b; Mazzolini et al., 1985; Thomas et al., 1985) or phosphocellulose (Volovitch et al., 1984) showed similar alterations in the elution profile following the infection of turnip leaves with CaMV. The DNA polymerase activity restricted to CaMV-infected plants was shown by Volovitch et al. (1984) to copy both oligo (dG)-primed poly(rC) templates and mRNAs. These results do not prove, however, that the novel DNA polymerase observed in CaMV-infected plants is reverse transcriptase. It is possible that the new DNA polymerase activity observed following CaMV infection is due to the cellular pathology induced by viral infection. Furthermore, Pfeiffer et al. (1984) claim that the unique DNA polymerase activity described by Volovitch et al. (1984) in CaMV-infected tissues is in fact found in both uninfected and infected plants.

Pfeiffer et al. (1984) used activity gels (DNA polymerase activity renatured in gels following polyacrylamide gel electrophoresis of cell-free extracts) to show that nuclear extracts from CaMV-infected plants, when compared to uninfected plants, contained different proportions of a 75- and a 110-kd DNA polymerase. It is important to note, however, that although the ratios of these two forms of DNA polymerase varied in infected and uninfected plants, both DNA polymerase activities were detected in healthy and CaMV-infected plants. Activities corresponding to 75- and 110-kd polypeptides were also detected in sucrose density gradient fractions containing putative CaMV replication complexes, and the activity associated with the 75-kd polypeptide appeared to be enriched in the most active CaMV-specific replication fractions. By using activity gels, Menissier et al. (1984) identified a 74-kd DNA polymerase associated with purified CaMV virions. Whether the 74- to 75-kd polypeptide observed on activity gels from putative replication complexes or purified virions corresponds to CaMV-encoded reverse transcriptase cannot be confirmed by the analyses described above. The nature of this polypeptide and the identity of the virus-encoded reverse transcriptase will probably become clear only after antibody probes for ORF V become available.

Replication Strategy and Propagation of CaMV

Based on the above results, the CaMV replication cycle can be summarized as:

$$\text{dsDNA} \underset{2}{\overset{1}{\circlearrowleft}} \text{ss(+)RNA} \xrightarrow{3} \text{protein}$$

where virus-encapsidated open circular double-stranded DNA after conversion to covalently closed circular DNA is (1) transcribed by host nuclear RNA polymerase II to a greater than genome length single-stranded plus-strand RNA. The single-stranded RNA may then be (2) reverse-transcribed to open circular double-stranded DNA to complete the replication cycle. Single-stranded plus-strand RNAs may also serve as templates for translation into virus-encoded proteins.

Although the normal route in nature for inoculation of host plants with CaMV is via aphid transmission, mechanical inoculation with cell sap, purified virions, purified virion DNA, and cloned CaMV DNA are effective (reviewed by Shepherd, 1976; Hohn et al., 1982; Howell, 1982; Guilfoyle, 1985). With whole plants, mechanical inoculation can be carried out with open circular viral DNA (possessing site-specific discontinuities), with linearized and closed circular monomeric CaMV DNAs excised from recombinant plasmids (Howell et al., 1980; Lebeurier et al., 1980; Walden and Howell, 1982; Lebeurier et al., 1982; Hohn et al., 1982), and with intact recombinant plasmids containing partial or complete tandem dimers (i.e., nested CaMV genomes) of CaMV DNA (Lebeurier et al., 1982; Walden and Howell, 1983). These studies indicate that neither site-specific discontinuities observed with encapsidated DNA nor viral associated components, including RNAs, subgenomic DNAs, and proteins, are required for infectivity. It is worth noting, however, that progeny virus produced from plants inoculated with cloned DNAs contain monomeric CaMV DNA with the normal site-specific discontinuities (Howell et al. 1980, 1981; Walden and Howell, 1982). In addition to infectivity results obtained with whole plants, some success has been reported with protoplasts prepared from host plants inoculated with purified virions (Howell and Hull, 1978; Furusawa et al., 1980; Maule, 1983), purified virion DNA (Yamaoka et al., 1982), and *E. coli* spheroplasts harboring recombinant CaMV DNA in the form of a tandem dimer (Tanaka et al., 1984). Infection with *E. coli* spheroplasts is probably possible only with recombinant plasmids containing partial or tandem dimers (or higher polymers), since it appears from the results of Walden and Howell (1983) that an intragenomic recombination which results in the excision of monomeric CaMV DNA from the recombinant plasmid is the infective molecule. Although protoplasts can be infected with CaMV or CaMV DNA, it is questionable whether CaMV DNA can be maintained in cultured plant cells for extended periods of time. It is possible that the replication strategy of CaMV is not compatible with host DNA replication, and cells containing CaMV may be selected against in continuous culture.

The life cycle of CaMV is schematically represented in Fig. 3, and the details of the life cycle and replication strategy (Fig. 4) are summarized below. Once the CaMV DNA enters host cells (whether directly or after the encapsidated DNA is released from the virion), it is likely that some of the molecules become compartmentalized in nuclei. Depending on the structure of the infective DNA, 3'-exonuclease digestion of overhanging ends at site-specific discontinuities and ligation at nicks produce covalently closed circular monomeric CaMV DNA. During this process, the DNA probably becomes associated with histones and other nuclear proteins to form CaMV minichromosome. At least a portion of the minichromosomes acquires RNA polymerase II and the necessary transcription factors to facilitate transcription from α-strand DNA and produces at least two major RNA transcripts of 8.2 and 1.9 kb. Whether the minichromosome is capable of replicating (e.g., using host DNA polymerase α and host replication factors) to produce additional minichromosomes is unlikely (especially in mature nondividing leaf cells), but no definitive information is currently available.

After transport from the nucleus to the cytoplasm, the 1.9-kb transcript can be translated on 80S ribosomes to produce an inclusion body or viroplasm protein. Whether other virus-encoded polypeptides (e.g., coat protein, aphid transmission factor, and reverse transcriptase) are translated in the cytoplasm from nonpolyadenylated or polyadenylated templates (which have gone undetected to date) or from a polycistronic 8.2-kb template has not been determined. Regardless of the mechanism for translation of virus-encoded polypeptides, these proteins are probably required for replication, packaging, and cell-to-cell movement of the virus.

At least one mode of replication of CaMV appears to involve reverse transcription of the major 8.2-kb RNA transcript to produce double-stranded monomeric CaMV DNA with the site-specific discontinuities (Guilley et al., 1983; Hull and Covey, 1983b, c; Pfeiffer and Hohn, 1983; Hohn et al., 1983; Howell et al., 1983, Hull and Covey, 1985; Hohn et al., 1985). This replication strategy is schematically diagrammed in Fig. 4. In this model for CaMV replication, the primer used to initiate the replication process is tRNA[met] which hybridizes through its 3' terminus with a stretch of 15 complementary ribonucleotides on the 8.2-kb RNA transcript. The priming site is located about 600 nucleotides from the 5' terminus of the 8.2-kb RNA. The virus-encoded reverse transcriptase

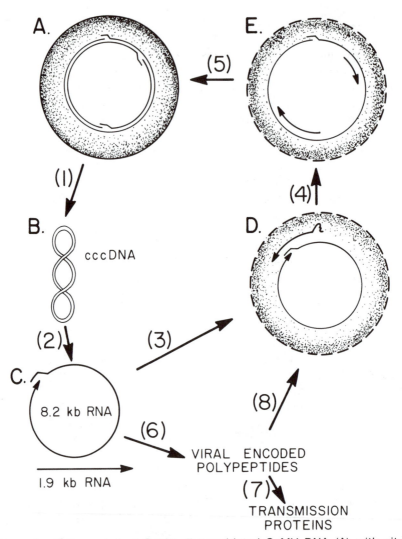

Figure 3. "Life cycle" of CaMV. Encapsidated CaMV DNA (A) with site-specific discontinuities represents the normal infectious form of CaMV DNA which is transmitted to host plants by aphids. After entering susceptible plant cells and removal of the coat protein, the open circular CaMV DNA is transported (1) to the nucleus and converted to covalently closed circular double stranded DNA (cccDNA) (B). The cccDNA is in the form of a minichromosome and serves as a template (2) for RNA polymerase II which produces major RNA transcripts of 8.2 and 1.9 kb (C). These polyadenylated RNAs and possibly other RNAs may be translated (6) into polypeptides that function in virus transmission (7) (e.g., cell-to-cell movement, aphid transmission) or virus replication (8) (e.g., reverse transcriptase). The greater than genome length 8.2 kb RNA transcript serves as the template (3, 4, and 5) for reverse transcriptase which generates the open circular virion CaMV DNA (D, E, and A; see Fig. 4) and completes the virus replication cycle. Mature virus (A) is depicted as a closed circle with stippling. The possibility that viral replication may take place in virion or virionlike particles is illustrated by the broken circles with stippling (D and E).

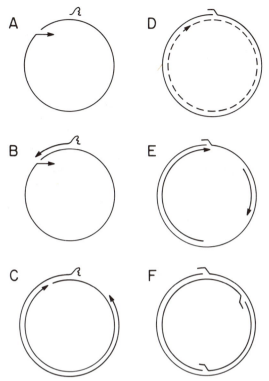

Figure 4.   Reverse transcriptase model for CaMV replication. (A) The 8.2-kb RNA (light arrow, 5' $\longrightarrow$ 3') with a terminal redundancy of 180 nucleotides contains a sequence complementary to 15 nucleotides in the 3'-terminus of tRNA[met]. This sequence in the 8.2-kb RNA is located about 600 nucleotides from its 5'-end and serves as a priming site for reverse transcriptase. (B) With the tRNA primer ( ), reverse transcriptase begins minus strand DNA synthesis (heavy arrow 5 $\longrightarrow$ 3') and continues minus strand DNA synthesis until the template is exhausted at the 5'-end of the RNA. This 700-nucleotide minus strand DNA with covalently attached RNA is called "strong stop" DNA and is found within and outside virions. (C) Ribonuclease H, which is presumably a component of the reverse transcriptase, digests the RNA portion of the DNA/RNA hybrid generated in (B). This allows the 3'-end of the 8.2-kb RNA to hybridize with the 3'-end of "strong stop" minus strand DNA, since the 5'- and 3'-ends of the RNA contain the same 180 nucleotide sequence. After this strand switch, minus strand DNA synthesis can continue until the complete minus strand of 8 kb, the α strand, is produced (D). This completed minus strand DNA may be free, (D) associated with parts of 8.2 kb RNA (broken circle, $--\rightarrow$), or (E) associated with growing plus strand DNAs. Plus strand DNAs are initiated at the sites of the two discontinuities in β-strand DNA and are presumably primed by pieces of 8.2-kb RNA complementary to minus strand DNA within the regions of the site-specific discontinuities. Minus strand DNA, which may be completed or in the process of synthesis itself, serves as the template for plus strand DNA synthesis. Plus strand DNA synthesis continues and must also make a strand switch at the discontinuity in the completed minus strand. The completion of the plus strand DNA with slight extensions beyond the 5'-termini (this also occurs with minus strand DNA) at the site-specific discontinuities results in the production of the open circular virion DNA (F).

uses this primer to initiate reverse transcription of the 8.2-kb RNA, and the minus-strand DNA is propagated until the RNA template is exhausted at its 5' end. The product of this reaction is a minus-strand DNA of about 600 deoxyribonucleotides with covalently attached tRNA[met] (about 75 ribonucleotides) at its 5' terminus and represents a structure comparable to strong-stop DNA of retroviruses (for reviews see Varmus and Swanstrom, 1982; Varmus, 1983).

A considerable amount of encapsidated and unencapsidated strong-stop CaMV DNA is detected in infected cells (Covey et al., 1983; Guilfoyle et al., 1983b; Turner and Covey, 1984; Marsh et al., 1985), however, this product represents a defective intermediate of replication unless it base-pairs with the 3' terminus of the same or possibly another 8.2-kb transcript. This base-pairing is possible because of the 180-bp repeat at the 5' and 3' ends of the 8.2-kb RNA and is probably facilitated by ribonuclease H activity associated with the reverse transcriptase enzyme (see Varmus and Swanstrom, 1982; Varmus, 1983) which degrades the RNA moiety of the strong-stop DNA-RNA hybrid. The jump from the 5' to the 3' end of the 8.2-kb RNA allows reverse transcription to continue around the greater than genome length RNA transcript. This reverse transcription might continue until the synthesis of full-genome-length 8-kb minus-strand DNA is completed. A heterogeneous population of minus-strand DNAs, including 8-kb minus-strand DNA, has been detected in virus and viruslike particles, and these molecules appear to be in the form of DNA-RNA hybrids (Marsh et al., 1985; Marsh and Guilfoyle, unpublished results). Alternatively, plus-strand DNA synthesis might be initiated before completion of minus-strand DNA synthesis. Some unencapsidated intermediates have been detected which support this replication strategy (Hull and Covey, 1983a; Marco and Howell, 1984). Regardless of whether plus-strand DNA synthesis occurs during or after completion of minus-strand DNA synthesis or whether both alternatives are possible, plus-strand DNA synthesis is initiated at nucleotides 4218 and 1632 which correspond to the two site-specific discontinuities in β-strand DNA. The primers for plus-strand DNA synthesis have not been identified, but may be short stretches of 8.2-kb RNA which are produced by ribonuclease H digestion and are complementary to minus-strand DNA within regions around nucleotides 4218 and 1632. These regions are rich in guanine and adenine bases (Hohn et al., 1983). Once initiated, plus-strand synthesis can continue, using the newly synthesized minus-stand DNA as template.

The template for second or plus-strand synthesis is exhausted at the priming site for minus-strand synthesis, and a second jump from the 5' end to the 3' end of the minus-strand DNA must occur at this point. The 3' end of the minus-strand DNA could base-pair with the 3 ' end of the propagating plus-strand DNA in the region of the minus-strand priming site and facilitate this second jump. Plus-strand DNA synthesis could then continue until it reaches the second priming site for plus-strand DNA synthesis at nucleotide position 1632. In the meantime, the remainder of the plus-strand DNA could be synthesized from the priming site at nucleotide 1632 to the discontinuity at nucleotide 4218. Slight overlaps (i.e., the heterogeneous 3'-ends which overhang the 5' ends) at the site-specific discontinuities are most likely produced by strand displacement and the slight extension of 3'ends of the elongating DNA chains. Most of the ribonucleotides making up the primers appear to be removed from the mature encapsidated CaMV DNAs, but the interrupted DNA strand at each discontinuity may retain a ribonucleotide at its 5' terminus (Guilley et al., 1983).

COMPARISON OF CaMV WITH AUTHENTIC RETROVIRUSES AND HEPATITIS B VIRUSES

CaMV has many characteristics in common with animal retroviruses and/or hepatitis B viruses. The most striking similarity is that CaMV as well as retroviruses (for reviews see Varmus and Swanstrom, 1982; Varmus, 1983) and hepatitis B viruses (Summers et al, 1975; Summers and Mason, 1982a) replicate their genomes through the process of reverse transcription. In each case, a reverse transcriptase is encoded on the viral genome, and some homology exists among amino acid sequences of the enzymes in the three families of viruses (Toh et al., 1983; Patarca and Haseltine, 1984). A greater than genome length RNA transcript with terminal redundancies has been directly demonstrated as the template in the case of retroviruses and has been postulated as the template for reverse transcription in the case of CaMV (Guilley et al., 1983; Hull and Covey, 1983b, c; Pfeiffer and Hohn, 1983) and hepatitis B viruses (Summers and Mason, 1982a; Cattaneo et al., 1984).

Like retroviruses, CaMV appears to use a tRNA primer for initiating minus-strand DNA synthesis (Guilley et al., 1983; Hull and Covey, 1983b, c; Pfeiffer and Hohn, 1983). This contrasts with hepatitis B viruses which probably use a protein to prime minus-strand DNA synthesis (Molnar-Kimber et al., 1983). Primers for plus-strand DNA synthesis also appear to be RNA molecules in retroviruses and CaMV, and DNA sequence homology among putative plus-strand DNA priming sites in CaMV, retroviruses, and hepatitis B viruses have been pointed out by Hohn et al. (1983). Another similarity shared by retroviruses and CaMV is the production and accumulation of strong-stop minus-strand DNA (Covey et al., 1983; Turner and Covey 1984; Marsh et al., 1985). For each virus, this DNA species is thought to arise as a "runoff" DNA transcript from the 5' end of the RNA template. Strong-stop DNA accumulates if the reverse transcriptase fails to make the jump or switch from the 5' end to the 3' end of the RNA template during minus-strand DNA synthesis.

Unlike retroviruses, CaMV and hepatitis B viruses encapsidate DNA genomes rather than RNA genomes. Both CaMV and hepatitis B virus package open circular double-stranded or partially double-stranded DNA. A 3.2-kb circular DNA which is only partially double-stranded is packaged into hepatitis B virions (for reviews see Summers and Mason, 1982b; Marion and Robinson, 1983; Tiollais et al., 1985). Minus-strand DNA of hepatitis B virus in virions is complete except for a site-specific discontinuity, but plus-strand DNA is heterogeneous in length and may range from 50 to 100% of 3.2 kb (Summers et al., 1975). In hepatitis B viruses, the 5' end of the heterogeneous plus-strand DNA occurs at a unique site about 300 bp from the 5' end of the complete minus-strand DNA, and this structure provides for a cohesive overlap which allows a circular duplex to be formed with a heterogeneous single-stranded region (for reviews see Summers and Mason, 1982b; Marion and Robinson, 1983). This contrasts with CaMV which encapsidates completely double-stranded 8-kb DNA with site-specific discontinuities having a triple-stranded character.

Additional similarities between hepatitis B viruses and CaMV are apparent. Both viruses replicate their genomes asymmetrically and produce free minus-strand DNA or minus-strand DNA hybridized with the viral RNA which acts as the template for reverse transcription (Mason et al., 1982; Blum et al., 1984; Miller et al., 1984a; Marsh et al., 1985; Tiollais et al., 1985). With each virus, this free minus-strand DNA can be found in virus core or viruslike particles. Hepatitis B virions contain a DNA polymerase (reverse transcriptase) capable of synthesizing minus-strand DNA from an RNA template, converting DNA-RNA hybrids to double-stranded DNA, and filling in plus-strand DNA within single-stranded regions of encapsidated DNA (Summers and Mason, 1982a; Miller et al, 1984b). Marsh et al. (1985) have described a somewhat analogous situation with CaMV where at least some viruses or viruslike particles contain a reverse transcriptase-like activity which appears to use an RNA template for minus-strand DNA synthesis and a DNA template for plus-strand DNA synthesis.

Some other analogies between retroviruses and CaMV are worth noting. Both retroviruses (for reviews see Varmus and Swanstrom, 1982; Varmus, 1983) and CaMV (Howell et al., 1981, 1983; Walden and Howell, 1982, 1983) show high rates of recombination, and in each case this could occur during the process of reverse transcription where the growing minus-strand DNA switches from one RNA template to another. In these cases, the recombination is intermolecular, and infectious monomeric genomes can be recovered from pairs of defective genomes used for inoculation. Whether some type of intramolecular recombination can occur to produce less than genome length viral DNAs is not clear; however, nuclei of cells infected with retroviruses (Guntaka et al., 1976) or CaMV (Olszewski et al., 1982) contain a population of subgenomic covalently closed circular viral DNAs, and these might arise from intramolecular recombinaton events from defective replication events and/or from reverse transcription of spliced RNAs. Subgenomic supercoiled DNAs have also been observed in CaMV-infected protoplasts (Rollo and Covey, 1985).

A major difference between CaMV and the animal viruses discussed here is the apparent lack of integration by CaMV into host nuclear genomes compared to retroviruses and hepatitis B viruses. The lack of integration observed with CaMV may, however, simply result from insufficient methodology to detect integration events (see discussion on CaMV integration above) or may be a distinguishing characteristic of this virus.

CaMV AS A VECTOR

Although it is not the purpose of this chapter to review the studies on CaMV as a vector for delivering foreign DNA into plant cells, it is worth noting that these studies indicate that the CaMV genome tolerates little manipulation (see Gronenborn et al., 1981; Hohn et al., 1982; Howell, 1982; Hull and Davies, 1983; Daubert et al. 1983; Dixon et al., 1983). Much of the intransigence of the CaMV genome to genetic manipulation may be explained by its strategy of replication. Since CaMV replicates as an extrachromosomal element, it is necessary to conserve all functions required for this replication (virus-encoded reverse transcriptase, priming sites, end transcriptional start and stop sites) or associated with this replication (possibly the viral inclusion body protein, the coat protein, and other open reading frames of unknown function). Although the overall strategy for the generation and translation of virus-encoded mRNAs is not completely clear, it may also be necessary to conserve the position and spacing of open reading frames to facilitate the processing and/or translation of viral mRNAs (e.g., for reading frames I to V). Because of their propensity for intermolecular (and possibly intramolecular) recombination, modified CaMV genomes may have to be delivered in a fashion that prohibits this recombinogenic behavior (e.g., this becomes especially important if helper virus genomes are used in combination with the CaMV genome carrying the foreign DNA). Withstanding the considerable problems in manipulating the CaMV genome for vector studies, some success using CaMV as vector has recently been achieved (Brisson et al., 1984), and advancing knowledge of the life cycle of the virus may provide the necessary insight to make CaMV a practical and useful vector.

ACKNOWLEDGMENTS

The author thanks Dr. Gretchen Hagen for a critical reading of the manuscript. Research reported here from the author's laboratory was supported by grants from Agrigenetics, the U.S. Department of Agriculture Competitive Research Grants Office, and the National Institutes of Health.

REFERENCES

AL ANI, R., PFEIFFER, P., AND LEBEURIER, G. 1979. The structure of cauliflower mosaic virus. II. Identity and location of the viral polypeptides, *Virology* **93**:188–197.

AL ANI, R., PFEIFFER, P., WHITECHURCH, D., LESOT, A., LEBEURIER, G., AND HIRTH, L. 1980. A virus specified protein produced upon infection by cauliflower mosaic virus (CaMV), *Ann. Virol. (Inst. Pasteur)* **131E**:33–53.

ANSA, O. A., BOWYER, J. W., AND SHEPHERD, R. J. 1982. Evidence for replication of cauliflower mosaic virus DNA in plant nuclei, *Virology* **121**:147–156.

ARMOUR, S. L., MELCHER, U., PIRONE, T. P., LYTTLE, D. J., AND ESSENBERG, R. C. 1983. Helper component for aphid transmission encoded by region II of cauliflower mosaic virus DNA, *Virology* **129**:25–30.

BALAZS, E., GUILLEY, H., JONARD, C., AND RICHARDS, K. 1982. Nucleotide sequence of DNA from an altered-virulence isolate D/H of the cauliflower mosaic virus, *Gene* **19**:239–249.

BLUM, H. E., HASSE, A. T., HARRIS, J. D., WALKER, D., AND VYAS, G. N. 1984. Asymmetric replication of hepatitis B virus DNA in human liver: Demonstration of cytoplasmic minus-strand DNA by blot analysis and *in situ* hybridization, *Virology* **139**:87–96.

BRISSON, N., PASZKOWSKI, J. R., PENSWICK, B., GRONENBORN, B., POTRYKUS, L., AND HOHN, T. 1984. Expression of a bacterial gene in plants by using a viral vector, *Nature* **310**:511–514.

BROADBENT, L. 1957. *Investigations of Virus Diseases of Brassica Crops*, Cambridge University Press, London.

BRUNT, A. A. 1966. Partial purification, morphology, and serology of dahlia mosaic virus, *Virology* **28**:778–779.

BRUNT, A. A., BARTON, R. J., TREMAINE, J. H., AND STACE-SMITH, R. 1975. The composition of cauliflower mosaic virus protein, *J. Gen. Virol.* **27**:101–106.

CATTANEO, R., WILL, H., AND SCHALLER, H. 1984. Hepatitis B virus transcription in the infected liver, *EMBO J.* **3**:2191–2196.

CIVEROLO, E. L., AND LAWSON, R. H. 1978. Topological forms of cauliflower mosaic virus nucleic acid, *Phytopathology* **68:**101–109.

CONDIT, C., AND MEAGHER, R. B. 1983. Multiple discrete 35S transcripts of cauliflower mosaic virus, *J. Mol. Appl. Genet.* **2:**301–314.

CONDIT, C., HAGEN, T. J., MCKNIGHT, T. D., AND MEAGHER, R. B. 1983. Characterization and preliminary mapping of cauliflower mosaic virus transcripts, *Gene* **25:**101–108.

COVEY, S. N., AND HULL, R. 1981. Transcription of cauliflower mosaic virus DNA: Detection of transcripts, properties, and location of the gene encoding the virus inclusion body protein, *Virology* **111:**463–474.

COVEY, S. N., LOMONOSSOFF, G. P., AND HULL, R. 1981. Characterization of cauliflower mosaic virus DNA sequences which encode major polyadenylated transcripts, *Nucleic Acids Res.* **9:**6735–6747.

COVEY, S. N., TURNER, D., AND MULDER, G. 1983. A small DNA molecule containing covalently linked ribonucleotide originates from the large intergenic region of the cauliflower mosaic virus genome, *Nucleic Acids Res.* **11:**251–264.

DAUBERT, S., RICHINS, R., SHEPHERD, R., AND GARDNER, R. 1982. Mapping of the coat protein gene of CaMV by its expression in a prokaryotic system, *Virology* **122:**444–447.

DAUBERT, S., SHEPHERD, R. J., AND GARDNER, R. C. 1983. Insertional mutagenesis of the cauliflower mosaic virus genome, *Gene* **25:**201–208.

DAUBERT, S., SCHOELZ, J., DEBAO, L., AND SHEPHERD, R. 1984. Expression of disease symptoms in cauliflower mosaic virus genomic hybrids, *J. Mol. Appl. Genet.* **2:**537–547.

DIXON, L. K., AND HOHN, T. 1984. Initiation of translation of the cauliflower mosaic virus genome from a polycistronic mRNA: Evidence from deletion mutagenesis, *EMBO J.* **3:**2731–2736.

DIXON, L. K., KOENIG, K., AND HOHN, T. 1983. Mutagenesis of cauliflower mosaic virus, *Gene* **25:**189–199.

DUDLEY, R. K., ODELL, J. T., AND HOWELL, S. H. 1982. Structure and 5′-termini of the large and 19S RNA transcripts encoded by the cauliflower mosaic virus genome, *Virology* **117:**19–28.

FAVALI, M. A., BASSI, M., AND CONTI, G. G. 1973. A quantitative autoradiographic study of intracellular sites for replication of cauliflower mosaic virus, *Virology* **53:**115–119.

FRANCK, A., GUILLEY, H., JONARD, G., RICHARDS, K., AND HIRTH, L. 1980. Nucleotide sequence of cauliflower mosaic virus DNA, *Cell* **21:**285–294.

FRITSCH, E., AND TEMIN, H. M. 1977. Inhibition of viral DNA synthesis in stationary chicken embryo fibroblasts infected with avian retroviruses, *J. Virol.* **24:**461–469.

FURUSAWA, I., YAMAOKA, N., OKUNO, T., YAMAMOTO, M., KOHNO, M., AND KUNOH, H. 1980. Infection of turnip *Brassica rapa* cultivar *perviribis* protoplasts with cauliflower mosaic virus, *J. Gen. Virol.* **48:**431–435.

GARDNER, R. C., HOWARTH, A. J., HAHN, P., BROWN-LEUDI, M., SHEPHERD, R.

J., AND MESSING, J. 1981. The complete nucleotide sequence of an infectious clone of cauliflower mosaic virus by M13mp7 shot gun sequencing, *Nucleic Acids Res.* **9**:2871–2888.

GARFINKEL, D. J., BOEKE, J. D., AND FINK, G. R. 1985. Ty element transposition: Reverse transcriptase and Virus-like particles, *Cell* **42**:507–517.

GIVORD, L., XIONG, C., GIBAND, M., KOENIG, I., HOHN, T., LEBEURIER, G., AND HIRTH, L. 1984. A second cauliflower mosaic virus gene product influences the structure if the viral inclusion body. *EMBO J.* **3**:1423–1427.

GRONENBORN, B., GARDNER, R. C., SCHAFER, S., AND SHEPHERD, R. J. 1981. Propagation of foreign DNA in plants using cauliflower mosaic virus as a vector, *Nature* **294**:773–776.

GUILFOYLE, T. J. 1980. Transcription of the cauliflower mosaic virus genome in isolated nuclei from turnip leaves, *Virology* **107**:71–80.

GUILFOYLE, T. J. 1985. Propagation of DNA viruses, in: *Methods in Enzymol., Plant Molecular Biology,* (H. Weissbach and A. Weissbach, eds.), Academic Press, New York, in press.

GUILFOYLE, T. J., OLSZEWSKI, N., AND HAGEN, G. 1983a. The cauliflower mosaic virus minichromosome, in: *Genome Organization and Expression in Plants* (L. S. Dure and O. Ciferri, eds.), pp. 419–425, Plenum Press, New York.

GUILFOYLE, T., OLSZEWSKI, N., HAGEN, G., KUZJ, A., AND McCLURE, B. 1983b. Transcription and replication of the cauliflower mosaic virus genome: Studies with isolated nuclei, nuclear lysates, and extranuclear cell-free leaf extracts, in: *Plant Molecular Biology* (R. Goldberg, ed.), pp. 117–136, Alan R. Liss, New York.

GUILLEY, H., DUDLEY, R. K., JONARD, G., BALAZS, E., AND RICHARDS, K. 1982. Transcription of cauliflower mosaic virus DNA: Detection of promoter sequences, and characterization of transcripts, *Cell* **30**:763–773.

GUILLEY, H., RICHARDS, K. E., AND JONARD, G. 1983. Observations concerning the discontinuous DNAs of cauliflower mosaic virus, *EMBO J.* **2**:277–282.

GUNTAKA, R. V., RICHARDS, O. C., SHANK, P. R., KUNG, H. J., DAVIDSON, N., FRITSCH, E., BISHOP, J. M., AND VARMUS, H. E. 1976. Covalently closed circular DNA of avian sarcoma virus: Purification from nuclei of infected quail tumor cells and measurement by electron microscopy and gel electrophoresis, *J. Mol. Biol.* **106**:337–357.

HAHN, P., AND SHEPHERD, R. J. 1982. Evidence for a 58-kilodalton polypeptide as a precursor of the coat protein of cauliflower mosaic virus, *Virology* **116**:480–488.

HILLS, G. J., AND CAMBELL, R. N. 1968. Morphology of broccoli necrotic yellows virus, *J. Ultrastruct. Res.* **24**:134–144.

HIROCHIKA, H., TAKATSUJI, H., UBASAWA, A., AND IKEDA, J. E. 1985. Site specific selection in cauliflower Mosaic Virus DNA: Possible involvement of RNA splicing and reverse transcription, *EMBO J.* **4**:1673–1680.

HOHN, T., RICHARDS, K., AND LEBEURIER, G. 1982. Cauliflower mosaic virus on

its way to becoming a useful plant vector, *Curr. Top. Microbiol. Immunol.*
96:193–236.

HOHN, T., PIETRZAK, M., DIXON, L., KOENIG, I., PENSWICK, J., HOHN, B., AND
PFEIFFER, P. 1983. Involvement of reverse transcription in cauliflower mosaic virus replication, in: *Plant Molecular Biology* (R. Goldberg, ed.), pp.
147–155, Alan R. Liss, New York.

HOHN, T., HOHN, B., AND PFEIFFER, P. 1985. Reverse transcription in CaMV,
TIBS 205–209.

HOWARTH, A. J., GARDNER, R. C., MESSING, J., AND SHEPHERD, R. J. 1981.
Nucleotide sequences of naturally occurring deletion mutants of cauliflower
mosaic virus, *Virology* **112**:678–685.

HOWELL, S. H. 1982. Plant molecular vehicles: Potential vectors for introducing
foreign DNA into plants, *Annu. Rev. Plant Physiol.* **33**:609–650.

HOWELL, S. H. 1985. The molecular biology of plant DNA viruses, *CRC Crit.
Rev. Plant Sci.* **2**:287–316.

HOWELL, S. H., AND HULL, R. 1978. Replication of cauliflower mosaic virus and
transcription of its genome in turnip leaf protoplasts. *Virology* **86**:468–481.

HOWELL, S. H., WALKER, L. L., AND DUDLEY, R. K. 1980. Cloned cauliflower
mosaic virus DNA infects turnips, *Science* **208**:1265–1267.

HOWELL, S. H., WALKER, L. L., AND WALDEN, R. M. 1981. Rescue of *in vitro*
generated mutants of cloned cauliflower mosaic virus genome in infected
plants, *Nature* **293**:483–486.

HOWELL, S. H., WALDEN, R. M., AND MARCO, Y. 1983. Recombination and
replication of cauliflower mosaic virus DNA, in: *Plant Molecular Biology*
(R. Goldberg, ed.) pp. 137–146, Alan R. Liss, New York.

HULL, R. 1979. The DNA of plant DNA viruses, in: *Nucleic Acids in Plants* (T.
C. Hall and J. W. Davies, eds.), pp. 3–29, CRC Press, Boca Raton, Florida.

HULL, R. 1980. Structure of cauliflower mosaic virus genome III.Restriction endonuclease mapping of thirty-three isolates. *Virology* **100**:76–90.

HULL, R. 1984. A model for the expression of CaMV nucleic acid, *Plant Mol.
Biol.* **3**:121–125.

HULL, R., AND COVEY, S. N. 1983a. Characterization of cauliflower mosaic virus
DNA forms isolated from infected turnip leaves, *Nucleic Acids Res.*
11:1881–1895.

HULL, R., AND COVEY, S. N. 1983b. Does cauliflower mosaic virus replicate by
reverse transcription?, *Trends Biochem. Sci.* **3**:254–256.

HULL, R., AND COVEY, S. N. 1983c. Replication of cauliflower mosaic virus
DNA, *Sci. Prog. Oxford* **68**:403–422.

HULL, R., AND COVEY, S. N. 1985. Cauliflower Mosaic Virus: Pathways of infection, *Bioessays* **3**:160–163.

HULL, R., AND DAVIES, J. W. 1983. Genetic engineering with plant viruses and
their potential as vectors, *Adv. Virus Res.* **28**:1–33.

HULL, R., AND HOWELL, S. H. 1978. Structure of the cauliflower mosaic virus genome. II. Variation in DNA structure and sequence between isolates, *Virology* **86**:482–493.

HULL, R., COVEY, S. N., STANLEY, J., AND DAVIES, J. W. 1979. The polarity of the cauliflower mosaic virus genome, *Nucleic Acids Res.* **7**:669–677.

IWAKI, M., ISOGAWA, Y., TSUZUKI, H., AND HONDA, Y. 1984. Soybean chlorotic mottle, a new caulimovirus on soybean, *Plant Disease* **68**:1009–1011.

KAMEI, T., RUBIO-HUERTOS, M., AND MATSUI, C. 1969. Thymidine-³H uptake by X-bodies associated with cauliflower mosaic virus infection, *Virology* **37**:506–508.

KELLER, W. 1975. Determination of the number of superhelical turns in simian virus 40 DNA by gel electrophoresis, *Proc. Natl. Acad. Sci. USA* **72**:4876–4880.

KOZIEL, M. G., ADAMS, T. L., HAZLET, M. A., DAMM, D., MILLER, J., DAHLBECK, D., JAYNE, S., AND STASKAWICZ, B. J. 1984. A cauliflower mosaic virus promoter directs expression of kanamycin resistance in morphogenic transformed plant cells, *J. Mol. Appl. Genet.* **2**:549–562.

LEBEURIER, G., HIRTH, L., HOHN, T., AND HOHN, B. 1980. Infectivities of native and cloned DNA of cauliflower mosaic virus, *Gene* **12**:139–146.

LEBEURIER, G. L., HIRTH, L., HOHN, T., AND HOHN, B. 1982. *In vivo* recombination of cauliflower mosaic virus DNA, *Proc. Natl. Acad. Sci. USA* **79**:2932–2936.

LUNG, M. C. Y., AND PIRONE, T. P. 1972. *Datura stramonium*, a local lesion host for certain isolates of cauliflower mosaic virus, *Phytopathology* **62**:1473–1474.

MARCO, Y., AND HOWELL, S. H. 1984. Intracellular forms of viral DNA consistent with a model of reverse transcriptional replication of the cauliflower mosaic virus genome, *Nucleic Acids Res.* **12**:1517–1528.

MARION, P. L., AND ROBINSON, W. S. 1983. Hepadna viruses: Hepatitis B and related viruses, *Curr. Top. Microbiol. Immunol.* **105**:99–121.

MARSH, L., KUZJ, A., AND GUILFOYLE, T. 1985. Identification and characterization of cauliflower mosaic virus replication complexes—Analogy to hepatitis B viruses, *Virology* **143**:212–223.

MASON, W., ALDRICH, C., SUMMERS, J., AND TAYLOR, J. M. 1982. Asymmetric replication of duck hepatitis B virus DNA in liver cells: free minus strand DNA, *Proc. Natl. Acad. Sci. USA* **79**:3997–4001.

MAULE, A. J. 1983. Infection of protoplasts from several *Brassica* species with cauliflower mosaic virus following inoculation using polyethylene glycol, *J. Gen. Virol.* **64**:2655–2660.

MAULE, A. J. 1985. Partial characterization of different classes of viral DNA and kinetics of DNA synthesis in turnip protoplasts infected with cauliflower mosaic virus, *Plant Molec. Bio.* **5**:25–34.

MAZZOLINI, L., BONNEVILLE, J. M., VOLOVITCH, M., MAGAZIN, M., AND YOT, P. 1985. Strand specific viral DNA synthesis in purified viroplasms isolated from turnip leaves infected with cauliflower mosaic virus, *Virology* **145**:293–303.

MENISSIER, J., LEBEURIER, G., AND HIRTH, L. 1982. Free cauliflower mosaic virus supercoiled DNA in infected plants, *Virology* **117**:322–328.

MENISSIER, J., DE MURCIA, G., LEBEURIER, G., AND HIRTH, L. 1983. Electron microscopic studies of the different topological forms of the cauliflower mosaic virus DNA: Knotted encapsidated DNA and nuclear minichromosome, *EMBO J.* **2**:1067–1071.

MENISSIER, J., LAQUEL, P., LEBEURIER, G., AND HIRTH, L. 1984. A DNA polymerase activity is associated with cauliflower mosaic virus, *Nucleic Acids Res.* **12**:8769–8778.

MILLER, R. H., TRAN, C. -T., AND ROBINSON, W. S. 1984a. Hepatitis B virus particles of plasma and liver contain viral DNA-RNA hybrid molecules, *Virology* **139**:53–63.

MILLER, R. H., MARION, P. L., AND ROBINSON, W. S. 1984b. Hepatitis B viral DNA-RNA hybrid molecules in particles from infected liver are converted to viral DNA molecules during an endogenous DNA polymerase reaction, *Virology* **139**:64–72.

MODJTAHEDI, N., VOLOVITCH, M., SOSSOUNTOV, L., HABRICOT, Y, BONNEVILLE, J. M., AND YOT, P. 1984. Cauliflower mosaic virus-induced viroplasms support viral DNA synthesis in a cell-free system, *Virology* **133**:289–300.

MOLNAR-KIMBER, K. L., SUMMERS, J., TAYLOR, J. M., AND MASON, W. 1983. Protein covalently bound to minus-strand DNA intermediates of duck hepatitis B virus, *J. Virol.* **45**:165–172.

ODELL, J. T., AND HOWELL, S. H. 1980. The identification, mapping and characterization of mRNA for p66 cauliflower mosaic virus-coded protein, *Virology* **102**:349–359.

ODELL, J. T., DUDLEY, R. K., AND HOWELL, S. H. 1981. Structure of the 19S RNA transcript encoded by the cauliflower mosaic virus genome, *Virology* **111**:377–385.

ODELL, J. T., NAGY, F., AND CHUA, N. -H. 1985. Identification of DNA sequences required for activity of the cauliflower mosaic virus 35S promoter, *Nature* **313**:810–812.

OLSZEWSKI, N. E., AND GUILFOYLE, T. J. 1983. Nuclei purified from cauliflower mosaic virus-infected turnip leaves contain subgenomic covalently closed circular cauliflower mosaic virus DNAs, *Nucleic Acids Res.* **11**:8901–8914.

OLSZEWSKI, N., HAGAN, G., AND GUILFOYLE, T. J. 1982. A transcriptionally active covalently closed minichromosome of cauliflower mosaic virus DNA isolated from infected turnip leaves, *Cell* **29**:395–402.

PATARCA, R., AND HASELTINE, W. A. 1984. Matters arising, *Nature* **309**:288.

PFEIFFER, R., AND HOHN, T. 1983. Involvement of reverse transcription in the replication of cauliflower mosaic virus: A detailed model and test of some aspects, *Cell* **33**:781–789.

PFEIFFER, R., LAQUEL, P., AND HOHN, T. 1984. Cauliflower mosaic virus replication complexes: Characterization of the associated enzymes and of the polarity of the DNA synthesized *in vitro*, *Plant Mol. Biol.* **3**:261–270.

PIRONE, T. P., POUND, G. S., AND SHEPHERD, R. J. 1961. Properties and serology of purified cauliflower mosaic virus, *Phytopathology* **51**:541–546.

RICHARDS, K. E., GUILLEY, H., AND JONARD, G. 1981. Further characterization of the discontinuities of CaMV DNA, *FEBS Lett.* **134**:67–70.

ROBB, S. M. 1963. A method for the detection of dahlia mosaic virus in *Dahlia, Ann. Appl. Biol.* **52**:145–148.

ROBB, S. M. 1964. Location, structure, and cytochemical staining reactions of the inclusion bodies found in *Dahlia variabilis* infected with dahlia mosaic virus, *Virology* **23**:141–144.

ROLLO, F., AND COVEY, S. N. 1985. Cauliflower mosaic virus DNA persists as supercoiled forms in cultured turnip cells. *J. Gen. Virol.* **66**:603–608.

RUBIO-HUERTOS, M., CASTRO, S., FUJISAWA, I., AND MATSUI, C. 1972. Electron microscopy of the formation of carnation etched ring virus intracellular inclusion bodies, *J. Gen. Virol.* **15**:257–260.

RUSSELL, G. J., FOLLETT, E. A. C., SUBAK-SHARPE, J. H., AND HARRISON, B. D. 1971. The double-stranded DNA of cauliflower mosaic virus, *J. Gen. Virol.* **11**:129–138.

SAIGO, K., KUGIMIYA, W., MATSUO, Y., INOUNYE, S., YOSHIOKA, K., AND YUKI, S. 1984. Identification of the coding sequence for a reverse transcriptase-like enzyme in a transposable genetic element in *Drosophila melanogaster, Nature* **312**:659–661.

SHALLA, T. A., SHEPHERD, R. J., AND PETERSON, L. J. 1980. Comparative cytology of nine isolates of cauliflower mosaic virus, *Virology* **102**:381–388.

SHEPHERD, R. J. 1976. DNA viruses of higher plants, *Adv. Virus Res.* **20**:305–339.

SHEPHERD, R. J. 1979. DNA plant viruses, *Annu. Rev. Plant Physiol.* **30**:405–423.

SHEPHERD, R. J., AND WAKEMAN, R. J. 1971. Observation on the size and morphology of cauliflower mosaic virus deoxyribonucleic acid, *Phytopathology* **61**:188–193.

SHEPHERD, R. J., WAKEMAN, R. J., AND ROMANKO, R. R. 1968. DNA in cauliflower mosaic virus, *Virology* **36**:150–152.

SHEPHERD, R. J., BRUENING, G. E., AND WAKEMAN, R. J. 1970. Double-stranded DNA from cauliflower mosaic virus, *Virology* **41**:339–347.

SHEPHERD, R. J., RICHINS, R., AND SHALLA, T. A. 1980. Isolation and properties of the inclusion bodies of cauliflower mosaic virus, *Virology* **102**:389–400.

SHOCKEY, M. W., GARDNER, C. D., MELCHER, U., AND ESSENBERG, R. C. 1980. Polypeptides associated with inclusion bodies from leaves of turnip infected with cauliflower mosaic virus, *Virology* **105**:575–581.

SUMMERS, J., AND MASON, W. S. 1982a. Replication of the genome of a hepatitis B-like virus by a reverse transcription of an RNA intermediate, *Cell* **29**:199–201.

SUMMERS, J., AND MASON, W. S. 1982b. Properties of hepatitis B-like viruses related to their taxonomic classification, *Hepatology* **1**:179–183.

SUMMERS, J., O'CONNELL, A., AND MILLMAN, J. 1975. Genome of a hepatitis B virus: Restriction enzyme cleavage and structure of the DNA isolated from Dane particles, *Proc. Natl. Acad. Sci. USA* **72**:4597–4601.

TANAKA, N., IKEGAMI, M., HOHN, T., MATSUI, C., AND WATANABE, I. 1984. *E. coli* spheroplast-mediated transfer of cloned cauliflower mosaic virus DNA into plant protoplasts, *Mol. Gen. Genet.* **195**:378–380.

THOMAS, C. M., HULL, R., BRYANT, S. A., AND MAUL, A.J. 1985. Isolation of a fraction from cauliflower mosaic virus-infected protoplasts which is active in the synthesis of (+) and (−) strand viral DNA and reverse transcription of primed RNA templates, *Nucleic Acid Res.* **13**:4557–4576.

TIOLLIAS, P., POURCEL, C. AND DE JEAN, A. 1985. The hepatitis B virus, *Nature* **317**:489–495.

TOH, H., HAYASHIDA, H., AND MIYATA, T. 1983. Sequence homology between retroviral reverse transcriptase and putative polymerases of hepatitis B virus and cauliflower mosaic virus, *Nature* **305**:827–829.

TOMPKINS, C. M. 1937. A transmissible mosaic disease of cauliflower, *J. agric. Res.* **55**:33–46.

TURNER, D. S., AND COVEY, S. N. 1984. A putative primer for the replication of cauliflower mosaic virus by reverse transcription is virion-associated, *FEBS Lett.* **165**:285–289.

VARMUS, H. E. 1983. Retroviruses, in: *Mobile Genetic Elements* (J. A. Shapiro, ed.), pp. 411–502, Academic Press, New York.

VARMUS, H. E., AND SWANSTROM, R. 1982. Replication of retroviruses, in: *Molecular Biology of Tumor Viruses: RNA Tumor Viruses* (R.A. Weiss, N. Teich, H. E. Varmus, and J. M. Coffin, eds.), pp. 369–512, Cold Spring Harbor Laboratory, Cold Spring Harbor, New York.

VARMUS, H. E., GUNTAKA, R. V., FAN, W. J. W., HEASLEY, S., AND BISHOP, J. M. 1974. Synthesis of viral DNA in the cytoplasm of duck embryo fibroblasts and in enucleated cells after infection by avian sarcoma virus, *Proc. Natl. Acad. Sci. USA* **71**:3874–3878.

VOLOVITCH, M., DRUGEON, G., AND YOT, P. 1978. Studies on the single-stranded discontinuities of the cauliflower mosaic virus genome, *Nucleic Acids Res.* **5**:2913–2925.

VOLOVITCH, M., MODJTAHEDI, N., YOT, P., AND BRUN, G. 1984. RNA-dependent DNA polymerase activity in cauliflower mosaic virus-infected plant leaves, *EMBO J.* **3**:309–314.

WALDEN, R. M., AND HOWELL, S. H. 1982. Intergenomic recombination events among pairs of defective cauliflower mosaic virus genomes in plants, *Mol. Appl. Genet.* **1**:447–456.

WALDEN, R. M., AND HOWELL, S. H. 1983. Uncut recombinant plasmids bearing nested cauliflower mosaic virus genomes infect plants by intragenomic recombination, *Plant Mol. Biol.* **2**:27–31.

WOOLSTON, C. J., COVEY, S. N., PENSWICK, J. R., AND DAVIES, J. W. 1983. Aphid transmission and a polypeptide are specified by a defined region of the cauliflower mosaic virus genome, *Gene* **23**:15–23.

XIONG, C., MULLER, S., LEBEURIER, G., AND HIRTH, L. 1982. Identification by immunoprecipitation of cauliflower mosaic virus *in vitro* major translation product with a specific serum against viroplasm protein, *EMBO J.* **1:**971–976.

XIONG, C., LEBEURIER, G., AND HIRTH, L. 1984. Detection *in vivo* of a new gene product (gene III) of caulflower mosaic virus, *Proc. Natl. Acad. Sci. USA* **81:**6608–6612.

YAMAOKA, N., FURUSAWA, I. I., AND YAMAMOTO, M. 1982. Infection of turnip protoplasts with cauliflower mosaic virus DNA *Virology* **122:**503–505.

Chapter 12

Viruses of Plant Pathogens

Matthew J. Dickinson

OVER RECENT YEARS, evidence has accumulated that many plant-pathogenic organisms are infected by viruses, and that in several instances these infections cause a dramatic reduction in the virulence of the pathogen. Consequently, it has been proposed that some plant diseases could be controlled biologically using specific viruses which infect the disease agents. While such an approach to disease control is economically and ecologically attractive, it is first necessary to obtain a comprehensive understanding of virus-host interactions and the mechanisms of pathogenicity.

In many instances the evidence of virus infections has been limited to electron microscopic observations of viruslike particles (VLPs) in thin sections of infected tissues. However, some viruses infecting phytopathogens have been isolated and characterized. This chapter will attempt to review the molecular properties and genetics of these viruses and, where relevant, the properties of viruses infecting related but nonphytopathogenic hosts. The second part of the chapter is devoted to a discussion of the potential of viruses as agents for the biological control of plant pathogens.

VIRUSES OF PHYTOPATHOGENIC BACTERIA

Table 1 shows the range of plant diseases of economic importance which are caused by bacteria. The *Pseudomonas, Xanthomonas, Corynebacteria, Nocardia, Mycoplasma,* and *Spiroplasma* include many species which are not phytopathogens but are infected by viruses which share characteristics with those of phytopathogenic species. For a comprehensive review on the biological properties of bacteriophages, readers are referred to Adams (1959).

The isolation of bacterial viruses (bacteriophages) requires that the host be propagated to a higher titer. In the case of some organisms, such as a number of species of the wall-less *Mycoplasma* and *Spiroplasma,* this is not always possible because of their fastidious growth requirements. Bacteriophages are generally isolated by plating cell culture supernatants onto lawns of susceptible bacteria. Infectious virus forms individual plaques, consisting of small areas of lysed bacteria or cells with a retarded growth rate, and these plaques can be used as a source of inoculum to infect more cells and therefore to increase the virus titer. Following the lysis of these cells, the virus is separated from cellular debris by low-speed centrifugation or precipitation with polyethylene glycol and may then be further purified by ultracentrifugation. Since virus particles have specific densities, they form bands in density gradients. Self-generating gradients of cesium salts (chloride, sulfate, or formate) are generally used, although some viruses are more stable in a nonionic gradient such as metrizamide.

Purified virus may be disrupted by a number of procedures to permit analysis of the protein and nucleic acid components to be undertaken. Many phages can be disrupted by being shaken with phenol, which denatures the protein. The nucleic acid can then be isolated in the aqueous phase, while the denatured protein remains at the interface between the phases. Often a detergent such as sodium dodecyl sulfate (SDS), or proteases, are included in this disruption procedure to aid in protein denaturation.

Bacteriophages of various morphologies occur, which may contain single-stranded (ss) or double-stranded (ds) DNA or RNA. The convenient criterion of gross particle morphology has been used to classify these phages into a number of taxonomic groups (Bradley, 1967). Reanny and Ackermann (1982) have extended this classification to the groups shown in Table 2. Tailed phages, which are the most common type, nearly al-

Table 1 Common Diseases Caused by Phytopathogenic Bacteria.

Gram-negative aerobic rods and cocci

Pseudomonas aptata	Sugar beet bacterial blight
Pseudomonas lachrymans	Cucurbits angular leaf spot
Pseudomonas mors-prunorum	Cherry bacterial canker
Pseudomonas phaseolicola	French bean common blight
Pseudomonas savastanoi	Olive knot
Pseudomonas solanacearum	Tobacco bacterial wilt
Pseudomonas syringae	Tobacco wildfire
Xanthomonas campestris	Brassica black rot, rice bacterial leaf blight, rice bacterial leaf streak
Xanthomonas malvacearum	Cotton bacterial blight
Xanthomonas citri	Citrus canker
Xanthomonas manihotis	Cassava bacterial blight
Xanthomonas phaseoli	French bean common blight, French bean fuscous blight
Xanthomonas pruni	Bacterial spot on peaches and plums
Agrobacterium tumefaciens	Crown gall in sugar beet
Agrobacterium rhizogenes	Apple hairy root

Gram-negative facultatively anaerobic rods

Erwinia amylovora	Pear and apple fire blight
Erwinia atroseptica	Potato blackleg
Erwinia carotovora	Potato blackleg
Erwinia chrysanthemi	Vascular wilt of chrysanthemum
Erwinia nigrifluens	Oozing sap from acorns
Erwinia rhapontica	Crown rot of rhubarb
Erwinia stewartii	Maize bacterial wilt
Erwinia tracheiphila	Cucurbits bacterial wilt

Actinomycetes and related organisms

Corynebacterium flaccumfaciens	French bean wilt
Corynebacterium insidiosum	Alfalfa wilt
Corynebacterium michiganense	Tomato bacterial canker
Corynebacterium sepedonicum	Potato ring spot
Nocardia vaccinii	Galls on blueberry plants
Streptomyces scabies	Potato tuber scab
Streptomyces tumuli	Sugar beet raised scab

Mycoplasmas

Mycoplasma-like organisms	Clover phyllody, aster yellows
Spiroplasma citri	Citrus stubborn, corn stunt, horseradish brittle root

ways contain linear dsDNA genomes and are subdivided according to tail morphology. The remaining groups, D to G, are relatively rare and are distinct from each other.

The infection of a susceptible host by a bacteriophage generally follows a temporarilly ordered sequence of events. Attachment of the

Table 2 Morphological Groups of Bacteriophages.

GROUP	FAMILY	PROPERTY	GENOME[a]	EXAMPLE
A	Myoviridae	Long, contractile tails	ds linear DNA	T4
B	Styloviridae	Long, noncontractile tails	ds linear DNA	Lambda
C	Podoviridae	Short tails	ds linear DNA	T7
D1	Microviridae	Cubic	ss circular DNA	ϕX174
D3	Corticoviridae	Cubic	ds circular DNA	PM2
D4	Tectiviridae	Cubic	ds linear DNA	PR4
E1	Leviviridae	Cubic	ss linear RNA	MS2
E2	Cystoviridae	Cubic	ds segmented RNA	ϕ6
F1	Inoviridae	Filamentous	ss circular DNA	fd
F2	Plectrovirus	Filamentous	ss circular DNA	MVL-1
G	Plasmaviridae	Pleomorphic	ds circular DNA	MVL-2

[a]ds, double-stranded; ss, single-stranded.

phage occurs at the cell surface, with different phages utilizing different receptors. The phage nucleic acid then enters the cell and in the process becomes physically separated from the protein capsid. Depending on the nature of the phage and the physiological condition of the host, the phage may enter one of two developmental cycles. Virulent phages redirect the biosynthesis of the infected bacterium entirely to phage formation. Once the virions have matured intracellularly, they lyse the cell from within and release a burst of infectious phage. Temperate phages on the other hand are able to establish lysogeny, and one of two alternative sequences of events may follow infection. Either the phage replicates intracellularly and causes cell lysis, or it establishes a condition of lysogeny. In lysogeny the phage genome is carried and replicated in the bacterial cell from generation to generation without lysis. The bacteria become immune to the phage, and cells grow normally.

Filamentous phages such as M13 and fd, which infect *Escherichia coli*, have a different, nonlytic mode of infection. Following intracellular replication, these particles extrude through the cell membrane without damaging it. Consequently, infected cells remain viable, although they generally exhibit a retarded growth rate (Denhardt et al., 1978).

In this section, viruses of phytopathogenic bacteria are grouped according to their morphology, and each of the groups is discussed in terms of the genome structure and infectious cycle of its members.

Group A: Myoviridae

Myoviridae are phages with long contractile tails. The best characterized examples of this group are the T-even phages of *E. coli*, which possess linear dsDNA genomes and have a lytic mode of infection. Similar phages have been isolated from the phytopathogens *Xanthomonas campestris* (Liew and Alvarez, 1981) and *X. pruni* (Ghei et al., 1968). Like the T-even phages they contain linear dsDNA genomes.

Group A phages have also been isolated from phytopathogenic *Pseudomonas* and *Agrobacterium* species. They have been observed in *Mycoplasma* (Howard et al., 1980) but not as yet in any phytopathogenic species. However, no phytopathogenic mycoplasmas have been studied in detail because of the difficulties in culturing them, and it is possible that group A phages will be found in some of them.

Group B: Styloviridae

Styloviridae, such as coliphage lambda, are characterized by long noncontractile tails. Similar phages have been observed in phytopathogenic species of *Agrobacterium, Erwinia, Corynebacteria, Spiroplasma,* and *Xanthomonas.*

Two types of group B phages have been isolated from *Agrobacterium tumefaciens.* Phages $PB2_1$ (Stonier et al., 1967), PB_2A, PB6 (Ω), PS8, and PV-1 (Vervilet et al., 1975) have hexagonal heads (80 nm long and 71 nm wide) and long flexible tails (280 nm long and 18 nm wide). These phages, like coliphage lambda, contain linear dsDNA genomes of MW ~46.4 × 10^6 which possess 5' overlapping ssDNA at either end. These cohesive ends enable the linear molecules to circularize. In lambda, circularization of the genome following its penetration into the host cell is a prerequisite for its integration into the host genome (Campbell, 1971). The mechanism by which this occurs has been extensively discussed elsewhere (Hendrix et al., 1983), and it is part of the process by which the phage is maintained in the host cell in the lysogenic state. The phages of *Agrobacterium* are also temperate, and it is likely that circularization of their genomes by cohesive end joining is part of their integration process.

A second group of smaller B-type phages has also been isolated from *A. tumefaciens.* PB6-806 (Vervilet et al., 1975) and ψ (Expert and Tournier, 1982) are believed to be the same phage and are particles composed of hexagonal heads (68 × 68 nm) and tails 130 nm long and 13 nm wide. In the virions the phage genome is a linear dsDNA molecule of MW 25.2 × 10^6. Nucleic acid hybridization experiments have shown that there is no detectable DNA homology between the genome of PB6-806 and that of the larger type-B agrobacterium phage PB6 (Ω).

The genome of ψ has been physically mapped using restriction endonucleases which cleave DNA molecules into specific nucleotide sequences to form unique fragments. The size of the fragments produced by a specific restriction endonuclease can be determined by electrophoresis on agarose gels. If a number of enzymes are used both singly and in combination, it is possible to order their cleavage sites relative to each other. By ordering the sites a physical map of the genome may be constructed.

Restricted endonuclease digestion can also be used to detect terminal redundancy and circular permutation in phage genomes. These phenomena arise as a consequence of sequential packaging of the genome from a concatameric replicative form DNA, as happens for phage P22 of *Salmonella typhimurium* (Tye et al., 1974). Packaging starts at a fixed point on the concatamer, the *pac* site, and for a genome to have a 3% terminal redundancy, the concatamer is cut sequentially at sites separated by 103% of the contour length of the restriction map. This nucleic acid is then packaged into the protein shell in the process of headful packaging (Streisinger et al., 1967). If genomic DNA from a circularly permuted terminally redundant phage is cut with restriction endonucleases, some fragments appear in submolar amounts and some are heterogeneous in size reflecting the differences between individual phage particles (Jackson et al., 1978). Phage ψ was found to have such a genome structure by this technique and, like phage P22, is probably packaged from concatameric DNA by a headful mechanism.

Phage ψ is temperate and, like that of lambda, its genome is integrated into the genome of the host at a unique site on lysogenic bacteria. This was shown by combining the techniques of restriction digestion of DNA and nucleic acid hybridization. If a linear phage genome is cut by a restriction enzyme for which there is one site, two fragments will be produced. If the host genome containing this DNA is cut with the same enzyme, the two fragments will be longer because some host DNA will be attached to the ends. These fragments will be detectable above the background of other fragments produced on digestion of host DNA because they will hybridize with radioactively labeled phage DNA. By expanding this approach, and using a number of different enzymes, it is possible to map integrated phage genomes in fine detail and to establish their site or sites of integration into the host genome. This has been determined in some detail for phage ψ, and this phage has furthermore been shown to frequently induce mutations in its host upon lysogenization (Expert et al., 1982).

B-type phages have also been isolated from a number of *Xanthomonas* species. However, only SBX-1 from *Xanthomonas* sp1 has been observed to occur naturally with the bacterium while it is infecting plants. This phage contains a dsDNA genome (Dunleavy and Urs, 1973) but has not been extensively characterized. Likewise, the B-type phages of *Corynebacteria michiganense* (Echandi and Sun,1973;Wakimoto et al., 1969) and *Erwinia amylovora* (Ritchie and Klos, 1979) have not been studied in any detail. B-type phages of *Spiroplasma citri* have been observed only in cultures and in thin sections of salivary glands of leafhoppers injected with the organism (Townsend et al., 1977). B-type phages have been isolated from a number of species of *Nocardia,* but not as yet from the only phytopathogenic species, *N. vaccinii.* However, this species has not been extensively studied.

Group C: Podoviridae

Podoviridae have been found in phytopathogenic *Erwinia, Spiroplasma,* and *Xanthomonas.* Probably the best characterized of these are the short-tailed polyhedral viruses isolated from *S. citri.* These virus particles measure 37–44 × 35–37 nm and have short wedge-shaped tails 13–18 nm long.

Two forms of the virus have been isolated, although they are morphologically indistinguishable. The two types possess serologically related coat proteins but have different genome structures and modes of infection (Dickinson et al., 1984). The first type contains a linear dsDNA genome of about 20 kilobase pairs (kbp) (Cole et al., 1977). Evidence from denaturation and reannealing experiments has suggested that the viral genome is circularly permuted with a 5% terminal redundancy (Cole et al., 1978) These results are supported by restriction endonuclease digestion of genomic DNA from two similar viruses called *ag* and AV9/3, which results in fragments of heterogeneous size in nonstoichiometric amounts (M. J. Dickinson, unpublished observations). It is likely that the genomes of these viruses are packaged by a headful mechanism similar to that observed phage P22.

The properties of a second type of short-tailed polyhedral spiroplasma virus have recently been reported (Dickinson et al., 1984). This virus, called *ai*, has a linear dsDNA genome of 16 kbp which can circularize as a result of the presence of cohesive ends. Concatameric replicative form DNA has been isolated from infected cells, suggesting that replication may occur via a rolling circle mechanism. Coliphage lambda is packaged from concatamers produced this way, as molecules of the same length with identical cohesive end structures, by a site-specific cleavage mechanism. The phage-coded enzyme terminase cuts concatamers to introduce staggered nicks at the cohesive end site, and each unit length is packaged into proheads (Feiss and Becker, 1983). in order to produce *ai* virion DNA with cohesive end structures, the concatameric replicative form DNA may be packaged by a similar mechanism.

Virus *ai* has antigens in common with the viruses *ag* and AV9/3 which possess 20-kbp genomes, and purified *ai* DNA hybridizes with specific pieces of DNA from these related viruses (Dickinson et al., 1984). It has been postulated that the genes coding for the structurally related polypeptides are within the regions of the genome that show homology.

The significance of the cohesive end structure in *ai* appears to be in the capacity of the virus, like lambda, to lysogenize its hosts. Stable lysogens of *S. citri* have been isolated from surviving growth after viral infection (Dickinson and Townsend, 1984a). These lysogens release virus at a low level and are resistant to superinfection by it, but not to infection by the serologically related viruses *ag* and AV9/3. Immunity to superinfection appears to be associated with a conversion of the receptors for *ai* on the surface of the spiroplasma so that the virus is unable to attach.

With use of hybridization method described earlier for detecting integrated virion DNA in host genomes, it has been shown that all the strains of *S. citri* that cause citrus stubborn disease which have been examined contain *ai* sequences integrated into their genomes. but not all are lysogens. This integrated DNA appears to represent a deleted form of the viral genome. Deleted forms of temperate phages have been found in other bacteria, e.g., *E. coli* K12 which sometimes contains deleted forms of lambda (Fischer-Fantuzzi and Calef, 1964). These cryptic lambda prophages do not confer immunity to superinfection and cannot be induced by ultraviolet. It appears that they represent integrated prophages which, through deletion, have lost their ability to self-replicate and to escape the bacterial regulation mechanism. The results for *ai* are consistent with a cryptic prophage occurring in spiroplasmas. Two conceptually distinct interpretations of cryptic prophages have been proposed. They may be a form of genetic debris or confer some function on the cells which contain them. It has not been established whether they have any function in spiroplasmas.

Group D

This type of phage has not been detected in any phytopathogenic bacteria, although there are examples in other species of *Pseudomonas*. The significance of this observation, if any, is not known.

Group E: Cystoviridae

In contrast, Cystoviridae have been found only in phytopathogenic species of *Pseudomonas*. Phage $\phi6$ is a morphologically unique lipid-containing phage of *Pseudomonas phaseolicola,* which contains a segmented dsRNA genome. The structure of the phage has been studied by controlled disruption of the particles with a detergent and isolation of subviral particles by rate zonal centrifugation (Bamford and Palva, 1980). The particle itself is composed of 25% lipid, 13% RNA, and 62% protein. There are 10 structural proteins: P1, MW 93,000; P2, MW 88,000; P3, MW 84,000; P4, MW 36,800; P5, MW 24,000; P6, MW 21,000; P7, MW 19,900; P8, MW 10,500; P9, MW 8700; P10, MW 6000; one precursor polypeptide: P11, MW 25,200; and one nonstructural polypeptide: P12, MW 20,100. After disruption with a detergent, the particle splits into a nucleocapsid fraction comprising RNA with proteins P1, P2, P4, P7, P8, and half of P5, and a membrane fraction with P3, P6, P9, P10, and the other half of P5.

It has been shown that P1, P2, P4, and P7 are early proteins which are probably responsible for the polyhedral shape of the nucleocapsid, which they form by combining with P8, P8 itself being present on the nucleocapsid surface. The remaining structural polypeptides form part of the viral membrane, and P3 is found on the surface of the whole virion. The phage is known to attach specifically to the pili of sensitive host bacteria (Cuppels et al., 1979). After attachment the virion envelope fuses with the bacterial membrane, and the nucleocapsid penetrates into the cell. P3 and P6 are needed for this penetration, while P5 is responsible for the lytic activity of the phage.

The genome of $\phi6$ is a dsRNA molecule with a nucleotide composition of 27.3% C, 21.8% A, 28.9% G, and 22.0% U (Semancik et al., 1973). This G + C content of 56% is notably high for dsRNA viruses. The RNA is in the form of three distinct species of MW 4.5, 2.8, and 2.2 $\times 10^6$. Studies with nonsense and temperature-sensitive mutants have indicated that the largest RNA species codes for the nucleocapsid core proteins P1, P2, P4, and P7. Another RNA codes for P3 and P6, which are involved in phage attachment, and the third codes for P8 and other envelope proteins (Mindich et al., 1976).

The replication of $\phi6$ requires RNA polymerase, and it is believed that P1 and P2 are constituents of this enzyme since they have been found to be essential for both RNA transcription and replication. Complete ssRNA transcripts have been found for all three genome components associated with polysomes early in infection, indicating that transcription occurs by a semiconservative mechanism and that the mRNA species are polycistronic (Van Etten et al., 1980). The RNA polymerase is believed to function in a manner similar to that of viruses of the fungus *Penicillium stoleniferum* (Buck, 1979). This enzyme catalyzes replication semiconservatively by a two-step process. First, transcription occurs from dsRNA in which the newly synthesized strand displaces one strand of the dsRNA and becomes part of the duplex. Synthesis of dsRNA then uses the displaced strand as a template. The mechanism for control of replication versus transcription is not as yet known, but it has been postulated for *Penicillium* that host factors are involved.

The original host of $\phi6$ was *P. phaseolicola* HB10Y (Vidaver et al., 1973). A study on six resistant mutants by Cuppels et al. (1979) showed one particular mutant that produced large amounts of infectious particles during growth without any effect on the cell growth rate. Phages have been found to persistently infect bacteria in one of two general ways (Barksdale and Arden, 1974). In lysogeny, the potential for producing virus is transmitted to all the progeny. In the more unstable state of pseudolysogeny, the phage multiplies in only a fraction of the bacterial cells and uninfected cells can be isolated. Both of these states result in a resistance to superinfection and the potential for releasing the carried phage in culture. Lysogens can generally be distinguished by their retention of these properties following repeated cultivation in phage antiserum. Cultivation of pseudolysogens in antiserum results in a decrease in, or elimination of, the pseudolysogen and an accumulation of, or replacement by, host cells. In the case of $\phi6$-producing mutants, the phage appears to be carried in the pseudolysogenic state.

A second type of $\phi6$-producing mutant has also been isolated by

Romantschuk and Bamford (1981). In this case the phenotype results in a decrease in sensitivity to the phage lytic enzyme needed for both penetration and release of the virus. Since higher than normal lytic enzyme concentrations are needed for lysis, the mutants have a prolonged life cycle and a larger number of intracellular phage particles.

Group F: Inoviridae and Plectroviruses

Group F filamentous phages of two types have been found. The F1 type, or Inoviridae, have been found in *Xanthomonas oryzae* and *X. citri*. These phages, Xf and Cf, respectively, have similar morphologies, being about 1000 nm long (Kuo et al., 1969; Dai et al., 1980), and are serologically related. The coat protein of Xf has been sequenced and is unusually high in hydrophobic amino acids. It has a molecular weight of 4850 and contains no histidine, cystine, or phenylalanine residues, but contains 29% hydrophobic amino acids, as opposed to the corresponding figure of 20% for coliphage fd.

Like filamentous coliphages, both these *Xanthomonas* phages contain circular ssDNA genomes, of about 7.6 kbp, and replicate via double-stranded replicative forms. The genome of Cf has been physically mapped (Yang and Kuo, 1984).

The F2 type or plectroviruses, have been found only in *Spiroplasma* and *Mycoplasma*. Viruslike particles have been observed in mycoplasmas associated with clover phyllody (Gourett et al., 1973) and aster yellows (Allen, 1972). They are morphologically identical to the MVL-1-type viruses isolated from nonphytopathogenic mycoplasmas measuring 70–90 nm in length and 16 nm in diameter. These MVL-1 viruses possess circular ssDNA genomes of 4.5 kbp and replicate via double-stranded forms. It is likely that the viruses from the phytopathogenic strains will be found to have similar properties.

A rod-shaped virus has recently been isolated and purified from *S. citri* (Dickinson and Townsend, 1984b). This virus, called *aa*, measures $240–260 \times 15–19$ nm and contains a circular ssDNA genome of 8.5 kbp which has been physically mapped. Like other filamentous phages, this virus replicates via a covalently closed circular double-stranded molecule and a nicked form of this molecule.

All these filamentous phages of phytopathogenic bacteria appear to have properties similar to those of filamentous coliphages, although none has been studied in extensive detail. The properties of these coliphages have been extensively reviewed (Denhardt et al, 1978). They persist in bacteria by infecting only a proportion of the *E. coli* cells. They attach to the sex pili, the presence of which is dependent on the presence of specific plasmids. After adsorption, the ssDNA is converted to double-stranded covalently closed circular RF I. This RF I is nicked by a highly specific virus-coded nuclease and replicates via a rolling circle mechanism to produce RF II molecules. The role of the single-stranded nick is not known, but the RF II finally replicates to produce the single-stranded virion DNA.

The virus assembles by complexing of the nucleic acid with viral proteins (which have been inserted into the cell membrane), and then extrudes through the membrane. The viral coat protein appears to span the membrane with an acidic amino end on the outside, a carboxy end on the inside, and a central section consisting of hydrophobic amino acids. These phages are nonlytic, and about 200 particles may be produced per cell generation without destruction of the host cell.

Filamentous mycoplasma and spiroplasma viruses are similarly nonlytic and have a persistent mode of infection. Phages Cf and Xf are also nonlytic. It is possible that the mode of infection of these viruses has many similarities to that of filamentous coliphages.

Group G: Plasmaviridae

Plasmaviridae have been found in mycoplasmas, but not as yet in any phytopathogenic species.

VIRUSES OF PLANT PATHOGENIC FUNGI

Viruslike particles were first observed in fungi over 20 years ago (Gandy and Hollings, 1962) and have since been found in many species, including many phytopathogens. However, infection of a fungus with a virus is not normally associated with lysis, and there is generally no noticeable morphological change in the host. Consequently, many viruses are difficult to isolate and can often be detected only as VLPs in electron micrographs of ultrathin sections of the fungus. For a recent discussion of the general properties of fungal viruses, readers are referred to Lemke (1979).

Physicochemical Properties

The most commonly isolated mycoviruses are isometric nucleoprotein particles 25–50 nm in diameter. These particles typically contain a segmented genome of dsRNA. A number of morphologically distinct VLPs have been observed in some fungi, including filamentous particles (Dieleman-van Zaayen, 1967) and enveloped herpes-type particles

(Kazama and Schornstein, 1973). However, none of these VLPs have as yet been isolated and characterized, nor observed in phytopathogenic fungi.

Periconia circinata is the causal agent of root rot and crown rot of sorghum. Its phytopathogenicity results from the production of secondary metabolites which are toxic to plants. Non–toxin-producing isolates of the fungus are nonpathogenic (Scheffer and Pringle, 1961). dsRNA virus particles have been found in both pathogenic and nonpathogenic isolates of *P. circinata* and, conversely, isolates of both kinds have been found without viruses. The viruses are isometric particles about 32 nm in diameter and contain dsRNA (Dunkle, 1974), but none have been extensively characterized.

As with *P. circinata*, VLPs have been found in other toxin-producing fungi, such as *Pyricularia oryzae* (rice blast fungus) and *Helminthosporium maydis* (corn blight fungus), but there appears to be no correlation between pathogenicity and the presence of virus, and it is more likely that the different phenotypes are related to different host genotypes.

In *Rhizoctonia solani,* a common soil pathogen that causes seed decay, root rot, and other diseases in a variety of crop plants, a positive correlation has been observed between hypovirulence and the presence of VLPs. When 13 strains were tested for the presence of dsRNA, the 3 that were positive were all less virulent than the others (Castanho et al., 1978). The dsRNA of one of these strains was found to be in three components of MW 2.2, 1.5, and 1.1×10^6, and this strain was used successfully in laboratory tests to prevent sugar beet damping off by a virulent strain of the fungus in soil inoculated with both strains.

The association of VLPs with hypovirulence has also been reported in *Endothia parasitica*, the casual agent of chestnut blight. This fungus, which attacks chestnut trees through broken branches, breaks in the bark, and woodpecker holes, caused an epidemic which eliminated chestnut trees as a major forest tree in the eastern United States between 1904 and 1950. However, in Italy in 1951 a chestnut coppice was found which, although infected with the fungus, was remarkably healthy. This hypovirulent fungus spread so rapidly that it is now believed that chestnut blight is no longer of epidemic proportions in Italy.

When it was found that a mixed infection with hypovirulent and virulent isolates of the fungus resulted in low virulence, it was suggested that cytoplasmic factors were responsible for the hypovirulence and that these factors were transmitted following hyphal anastomosis (cell fusion). This was confirmed by Moffitt and Lister (1973), who used an antibody prepared against synthetic dsRNA to show that two hypovirulent strains contained dsRNA while two virulent strains did not. A more comprehensive study reported by Day and Dodds (1979), indicated that all hypovirulent strains possessed dsRNA which appears to be present in one to three major components per strain. The estimated size range is MW 3.0–3.3 × 10⁶. It has been further demonstrated that this dsRNA is transmitted coincidentally with hypovirulence.

Attempts have been made to associate this dsRNA with VLPs. No icosahedral particles have been observed, but an unusual clublike particle appears to be qualitatively associated with this dsRNA. This VLP is morphologically similar to a number of particles found in diseased mushrooms (Lemke, 1979). However, the mushroom particles have not as yet been associated with dsRNA.

Ustilago maydis is the causal agent of corn smut and exhibits a killer system similar to that of the yeast *Saccharomyces cerevisiae* (Tipper and Bostian, 1984). In *S. cerevisiae*, three types of cells exist: Killer cells secrete a toxin to which they are immune, neutral cells are unable to produce a toxin but are immune to it, and sensitive cells cannot produce toxin and are sensitive to it. This trait, like hypovirulence in *E. parasitica*, was found to be cytoplasmically inherited and was eventually shown to be determined by two dsRNA species called P1 and P2. These species are separately encapsidated in identical protein coats, resulting in the two particles V1 and V2. Killer and neutral cells have V1 and V2, while sensitive cells possess either V1 or neither. The difference between killer and neutral cells appears to be due to the presence of different V2 particles or arises because neutrals have a recessive allele at one of two nuclear genes (the killer expression loci).

The killer system of *U. maydis* is similar to this. Three types of killer cells exist, P1, P4, and P6, which produce specific toxins but are sensitive to the other killer cells. The corresponding neutral cells are called P3 (1), P(4), and P3(6). Wood and Bozarth (1973) showed that both P1 and P3(1) carried spherical VLPs 41 nm in diameter. VLPs were initially found to be absent in sensitive strains, but subsequent reports have shown that some sensitive strains contain unencapsidated dsRNA (Koltin and Day, 1976). It has been suggested that these strains contain cytoplasmically inherited mechanisms that suppress killer and immunity functions. dsRNAs have now been found in all the killer and neutral strains of *U. maydis*, and each phenotype has its own characteristic dsRNA sizes. Owing to the use of different size markers by different groups of researchers and inaccurate estimates of the sizes of these dsRNA species, results reported in the literature have sometimes been inconsistent. However, the neutral strains invariably possess fewer dsRNA species than the corresponding killer strains.

Gaeumannomyces graminis is the causal agent of the take-all disease of cereals. This disease often increases in severity in the field over the first three crops and then declines in intensity. When VLPs were discovered in weakly pathogenic isolates, it was suggested that viruses could be the cause of disease decline in the field. However, Rawlinson et al. (1973)

carried out a more extensive study and failed to confirm this pattern, so the significance of these VLPs remains obscure.

In a study on 22 isolates of *G. graminis* from Indiana wheat, Indiana soybean, and English barley, Frick and Lister (1978) showed the presence of VLPs banding at 1.33–1.39 g/cm^3 in CsCl gradients in 19 isolates, and these particles were in the size classes 35, 39, and 41–42 nm. In a more comprehensive study, Buck et al. (1981a) characterized 13 viruses and dsRNA virus variants from *G. graminis* var. *tritici*. These viruses were classified into three distinct serogroups with different physical properties, although all three types might be present in a single fungal cell. Group 1 consists of 35-nm-diameter particles composed of one polypeptide species in the capsid of MW 54–60 × 10^3 and two to four dsRNA components of MW 1.0–1.3 × 10^6. Group-2 particles are of similar size but have a larger single capsid polypeptide of MW 68–73 × 10^3 which often forms aggregates. They contain two to four dsRNA components of MW 1.4–1.6 × 10^6. Group-3 particles are larger (40 nm in diameter) and contain three structural polypeptides of MW 78, 83, and 87 × 10^3 and two dsRNA species of MW 3.2 and 4.3 × 10^6. Buck (1984) has recently isolated a fourth serogroup of virus particles with a diameter of 29 nm, a single capsid protein of MW 66 × 10^3, and a single dsRNA component of MW 1.2 × 10^6. It is noteworthy that the full coding capacity of this RNA is required to produce the capsid protein, and it has been suggested that this particle may be a satellite virus dependent on the presence of other viruses for its replication.

Based on this work, Buck has proposed that at least two basic types of isometric dsRNA mycoviruses occur. The first possess undivided genomes of polycistronic RNA and, although only one particle is needed for infection, other viral functions, such as the killer protein of yeast, may be carried on satellite RNA. Mycoviruses of the second type possess divided genomes of monocistronic dsRNA.

Replication of Mycoviruses

Nearly all known dsRNA mycoviruses have a latent mode of infection in that they do not affect host primary metabolism. Transmission of virus therefore occurs only intracellularly, either within a cell during growth and cell division or between hosts via hyphal anastomosis. Because a mature filamentous fungus consists of a differentiated mycelium, it is difficult to obtain a synchronous infection with virus, and the replication cycle is hard to follow. When viruses have been observed intracellularly in fungal hyphae, it has not been possible to establish the age of the particular hyphae. However, in some work on *Penicillium*, virus particles were found to occur as aggregates in the cytoplasm of the actively growing part of the hyphae, just behind the apical tip (Border et al., 1972).

Most information on the replication of mycoviruses has been derived from *in vitro* studies with virion-associated RNA polymerases the characterization of virus replicative intermediates, and *in vitro* protein synthesis using viral mRNAs. The replication of phytopathogenic fungi has not been studied in detail, but RNA polymerase activity has been detected in group-1 and -2 viruses of *G. graminis*, and full-length transcripts have been shown to occur in infected cells (Buck et al., 1981b). This RNA polymerase has properties similar to those of the polymerase of *P. stoleniferum* viruses, and it is believed that replication occurs by a similar semiconservative mechanism as described earlier.

The assembly of fungal viruses is as yet unstudied but is likely to be a well-coordinated process. It is hoped that techniques being developed for the study of plant viruses, using protoplasts, may be applicable to fungi, and that such techniques will prove useful in elucidating the replication cycle of mycoviruses.

BIOLOGICAL SIGNIFICANCE OF VIRUSES IN PLANT PATHOGENS

Viruses of plant pathogens have been shown to occur in many shapes and sizes and to possess varied genome structures. However, one property common to a number of these viruses is that they have been observed in connection with altered symptoms of the pathogen.

Vidaver (1976) has reviewed some of the evidence for viruses affecting pathogenicity and has noted a number of reservations about their possible use in the biological control of plant diseases. One of these concerns the other effects that viral infection may have on the host. Virus-resistant mutants could be present in a bacterial population that accumulates as the sensitive cells are lysed. These resistants may be as virulent as the sensitive bacteria, and the virus would consequently lose its effectiveness as a biological control agent. Alternatively, if the virus were to lysogenize the host at a low level, lysogens would accumulate as a consequence of being immune to superinfection and the biological control capacity of the virus would again become useless. An example of lysogenization affecting susceptibility to viruses has been reported by Goto and Starr (1972). *Xanthomonas phaseoli* and *X. begoniae* were lysogenized by temperate *X. citri* phages, and although in this case their phytopathogenicity was not affected the lysogens became less susceptible to a number of other phages.

A second drawback concerning biological control with viruses is the possibility that they may transduce various characters from one host to another. Generalized transduction has been reported in *Erwinia chrysan-*

themi, and ultraviolet light was found to stimulate the frequency of transduction (Chatterjee and Brown, 1980). The occurrence of generalized transduction by phages in the field on a large scale could radically alter the genetics of the bacterial population.

A third problem with biological control is the effect of environmental factors such as temperature on the infectivity of the virus. For example, phage Xp3A of *X. pruni* was found to lyse the host at 20 and 27°C, but not at 35°C (Civerolo, 1973).

Bacteriocins have been proposed as an alternative means of biological control. They have been isolated from a number of bacteria and have specific host ranges (Vidaver, 1976). However, before they can be used as bacteriocides, they must first be extensively purified, since they are often found associated with inhibitors. As with viruses, much more work is required to elucidate the mode of action of bacteriocins before they can be used as bacteriocides in the field.

In the case of fungi, host-specific toxins, the equivalent of bacteriocins, have also been reported. However, like that of bacteriocins, the potential use of these toxins is limited.

Some success has been reported in the use of viruses to control plant diseases. Attenuation of the symptoms of *X. oryzae* has been reported following the immersion of intact or injured portions of root seedlings in phage suspensions for 12 hr before bacterial inoculation (Ranga Reddy et al., 1977). In *E. amylovora*, a type-C bacteriophage, PEa1(h), with a polyhedral head and a spikelike tail has been isolated, which codes for a polysaccharide-depolymerizing enzyme (Ritchie and Klos, 1979). This enzyme strips capsules of virulent strains *in vitro*, and colonies which are resistant to the phage have been found to be avirulent.

Although the use of viruses to protect crops from bacterial and fungal diseases is sound in principle, results have not always been consistent. Furthermore, the problem of infecting the hosts with cell-free virus preparations has not been solved on the laboratory scale, let alone in the field. Some of the most promising and safest applications of biological control with viruses may lie in the individual treatment of trees or other perennial woody crops by direct means.

The control of chestnut blight fungus (*E. parasitica*) may be an example of this and has apparently been successful in Italy without human intervention. In other cases, such as *Ceratocystis ulmi*, the causal agent of Dutch elm disease, an as yet uncharacterized heritable agent, which does not appear to be a virus, causes hypovirulence and may be applicable as a control agent (Gibbs and Brasier, 1973).

The control of bacterial diseases of some fruit trees may also be feasible using phages. *Xanthomonas citri*, which causes cankers on citrus trees, is infected by a temperate phage PXC7 (Wu, 1972). When PXC7 lysogenizes the host XCJ19, the colony type of the bacterium alters from smooth to dwarf. Colonies do, however, frequently revert to smooth, and this may or may not result in a loss of resistance to the phage. When lysogenic dwarf convertants were introduced into the leaves of orange plants, canker development was found to be retarded.

More recently, a similar pattern of conversion of cell type following lysogenization has been reported in *S. citri* (Dickinson and Townsend, 1984a). Different isolates of *S. citri*, the causal agent of citrus stubborn, normally produce a severe infection of the test plant *Cathoranthus roseus* (Madagascan periwinkle), which invariably proves lethal in 8–10 weeks at 30°C. Virus particles are not normally associated with the spiroplasmas in these infected plants. However, Alivizatos et al. (1982) infected *C. roseus* with a strain of *S. citri*, SPV-3, which produced unusually mild symptoms and failed to kill the host. They observed large numbers of short-tailed polyhedral VLPs associated with the spiroplasmas in these plants. *Cathoranthus roseus* plants grafted first with the lethal strain of *S. citri*, SP-A, and subsequently with SPV-3 developed mild symptoms and showed a large number of VLPs in electron micrographs. Plants infected with SP-A alone showed severe symptoms and died within a few weeks, and no VLPs could be seen. Although spiroplasmas were not completely eliminated from plants with mild symptoms, most plants survived for several years, and some superficially showed a total recovery.

Subsequently, the virus *ai* was isolated from strain SPV-3, and this phage has been shown to lysogenize its host (Dickinson and Townsend, 1984a). If lysogenization in *S. citri* is responsible for the remission of symptoms in plants infected with SPV-3, it would account for the observations of Alivizatos et al. (1982) that spiroplasmas did not disappear completely from plants and that spiroplasmas isolated from these plants were initially resistant to the virus.

Control of spiroplasma and mycoplasma diseases with viruses is particularly appealing because of the limited success obtained with conventional means of control. The breeding of varieties of plants which are resistant to spiroplasmas and/or the leafhopper vector has proved largely ineffective. The burning of infected crops does not eliminate the reservoir of infection in the vector, and by the time new plants reach maturity they may well have been reinfected. Remission of the symptoms of several mycoplasmal plant diseases has been demonstrated following tetracycline treatment (Sinha, 1979) but if treatment is not continuous the symptoms often reappear. Because of the lack of a cell wall, spiroplasmas are resistant to many antibiotics which kills other bacteria. Furthermore, the potential application of antibiotic control is limited, since the cost of repeated spraying would be prohibitive and the treatment of soil has proved ineffective. It has also been suggested that antibiotic residues could induce resistance in the soil population of microbes around the treated trees and that the eating of fruit containing these residues might induce resistance in microbes of significance in human medicine.

To control spiroplasma disease with virus, the safest and most feasible method of introducing the virus would be to infect bud wood with the lysogen and then graft it onto the infected plant. The virus spontaneousely released from the lysogen would probably kill most sensitive cells and lysogenize the remainder. However, further research will be required before such an approach becomes a reality.

CONCLUDING REMARKS

At present, only a limited number of viruses of phytopathogens have been studied in detail. However, many of the techniques currently being used and developed in other spheres of virology could be readily applied to these viruses. In particular, it would be useful to determine some of the functions coded for by these virion genomes, as some of them may prove to be responsible for the attenuation of host pathogenicity and therefore of potential use in the development of viruses for biological control.

Among the techniques which could help to achieve this are the isolation of virus mutants lacking specific functions, followed by various complementation analyses of these mutants. Mutant studies have already proved useful in studies bacteriophage $\phi6$ (Mindich et al., 1976), and with present techniques it is possible to produce mutations at specific predetermined sites in a viral genome. Furthermore, with the advent of molecular cloning and nucleotide sequencing, many virus genomes have now been completely sequenced, and there is no reason why those of viruses of phytopathogens could not be sequences in the future. Such sequences could be used to establish which regions of the genome are likely to code for specific products, and if these regions can then be cloned into expression vectors, it may be possible to determine the nature of the products. Cloning into bacteriophage vectors has already proved useful in some work on *Erwinia stewartii* (Coplin, 1979). Bacteriophage Mu was used to introduce plasmid DNA into the bacterium, and a similar approach could be developed to introduce defined regions of phytopathogenic virus genomes into bacterial cells for study of their expression.

Advances in studying the molecular aspects of phytopathogenic viruses, coupled with progress in the general biology of the viruses and their hosts, may in the future result in the effective and safe control of a number of plant diseases. A more detailed knowledge of virus-host interactions may also be useful in studying the molecular genetics of host phytopathogenicity. There is certainly much of significance to be learned from further studies on the viruses of plant pathogens.

ACKNOWLEDGMENTS

The author would like to thank Dr. Rod Townsend for suggesting improvements to the manuscript, and Ms. Burnadette Youens for typing the manuscript.

REFERENCES

ADAMS, M. H. (ED) 1959. *Bacteriophages,* Interscience, New York.

ALIVIZATOS, A. S., TOWNSEND, R. AND MARKHAM, P. G. 1982. Effects of infection with a spiroplasma virus on the symptoms produced by *Spiroplasma citri. Ann. Appl. Biol.* **101**:85–91.

ALLEN, T. C. 1972. Bacilliform particles within asters infected with a western strain of aster yellows. *Virology* **47**:491–493.

BAMFORD, D. H., AND PALVA, E. T. 1980. Structure of the lipid-containing bacteriophage $\phi 6$: Disruption by Triton X-100 treatment. *Biochem. Biophys. Acta* **601**:245–259.

BARKSDALE, L., AND ARDEN, S. B. 1974. Persisting bacteriophage infections, lysogeny and phage conversions. *Annu. Rev. Microbiol.* **28**:265–299.

BORDER, D. J., BUCK, K. W., CHAIN, E. B., KEMPSON-JONES, G. F., LHOAS, P., AND RATTI, G. 1972. Viruses of *Penicillium* and *Aspergillus* species. *Biochem. J.* **127**:4P–6P.

BRADLEY, D. E. 1967. Ultrastructure of bacteriophages and bacteriocins. *Bacteriol. Rev.* **31**:230–314.

BUCK, K. W. 1979. Replication of dsRNA mycoviruses, in: *Viruses and plasmids in fungi* (P. Lemke, ed.), pp. 94–151, Marcel Dekker, New York.

BUCK, K. W. 1984. A new double-stranded RNA virus from *Gaeumannomyces graminis. J. Gen. Virol* **65**:987–990.

BUCK, K. W., ALMOND, M. R., McFADDEN, J. J. P., ROMANOS, M. A., AND RAWLINSON, C. J. 1981a. Properties of thirteen viruses and virus variants obtained from eight isolates of the wheat take-all fungus *Gaeumannomyces graminis. J. Gen. Virol.* **53**:235–245.

BUCK, K. W., ROMANOS, M. A., McFADDEN, J. J. P., AND RAWLINSON, C. J. 1981b. *In vitro* transcription of dsRNA by virion associated RNA polymerases of viruses from *Gaeumannomyces graminis. J. Gen. Virol.* **57**:157–168.

CAMPBELL, A. 1971. Genetic structure, in: *The Bacteriophage Lambda,* (A. D. Hershey, ed.), pp.13–44, Cold Spring Harbor Laboratory, Cold Spring Harbor, New York.

CASTANHO, B., BUTLER, E. E., AND SHEPHERD, R. J. 1978. The association of double-stranded RNA with *Rhizoctonia* decline. *Phytopathology* **68**: 1515–1519.

CHATTERJEE, A. K., AND BROWN, M. A. 1980. Generalised transduction in the enterobacterial phytopathogen *Erwinia chrysanthemi.* *J. Bacteriol.* **143:** 1444–1449-

CIVEROLO, E. L. 1973. Relationships of *Xanthomonas pruni* bacteriophages of bacterial spot disease in *Prunus. Phytopathology* **63:**1279–1284.

COLE, R. M., MITCHELL, W. O., AND GARON, C. F. 1977. *Spiroplasmavirus citri* 3: Propagation, purification, proteins and nucleic acid. *Science* **198:** 1262–1263.

COLE, R. M., GARON, C. F., MITCHELL, W. O., JABLONSKA, E., AND RANHAND, J. M. 1978. Spiroplasma viruses: Current status. Fourth international Congress for Virology, p. 175, 1978. Abstract.

COPLIN, D. L. 1979. Introduction of bacteriophage Mu into *Erwinia stewartii* by use of an RK2::Mu hybrid plasmid. *J. Gen. Microbiol.* **113:**181–184.

CUPPELS, D. A., VIDAVER, A. K., AND VAN ETTEN, J. L. 1979. Resistance to bacteriophage φ6 by *Pseudomonas phaseolicola. J. Gen. Virol.* **44:**493–504.

DAI, H., CHIANG, K. -S., AND KUO, T.-T. 1980. Characterisation of a new filamentous phage Cf from *Xanthomonas citri. J. Gen. Virol.* **46:**277–289.

DAY, P. R., AND DODDS, J. A. 1979. Viruses of plant pathogenic fungi, in: *Viruses and Plasmids in Fungi* (P. A. Lemke, ed.), pp.201–238, Marcel Dekker, New York.

DENHARDT, D., DRESSLER, D. H., AND RAY, D. S. (EDS.) 1978. *The Single-Stranded DNA Phages,* Cold Spring Harbor Laboratory, Cold Spring Harbor, New York.

DICKINSON, M. J., AND TOWNSEND, R. 1984a. Integration of a temperate phage infecting *Spiroplasma citri. Isr. J. Med. Sci.* **20:**785–787.

DICKINSON, M. J., AND TOWNSEND, R. 1984b. Characterisation of the genome of a rod-shaped virus infection *Spiroplasma citri. J. Gen. Virol.* **65:**1607–1610.

DICKINSON, M. J., TOWNSEND, R., AND CURSON, S. J. 1984. Characterisation of a virus infecting the wall-free prokaryote *Spiroplasma citri. Virology* **135:**524–535.

DIELEMAN-VAN ZAAYEN, A. 1967. Virus-like particles in a weed mould growing on mushroom trays. *Nature* **216:**595–596.

DUNKLE, L. D. 1974. Double-stranded RNA mycovirus in *Periconia circinata. Physiol. Plant Pathol.* **4:**107–116.

DUNLEAVY, J. M., AND URS, N. V. R. 1973. Isolation and characterisation of bacteriophage SBX-1 and its bacterial host, both endemic in soybeans. *J. Virol.* **12:** 188–193.

ECHANDI, E., AND SUN, M. 1973. Isolation and characterisation of a bacteriophage for the identification of *Corynebacterium michiganense. Phytopathology* **63:**1398–1401.

EXPERT, D., AND TOURNEUR, J. 1982. ψ, a temperate phage of *Agrobacterium tumefaciens,* is mutagenic. *J. Virol.* **42:**283–291.

EXPERT, D., RIVIERE, F., AND TOURNEUR, J. 1982. The state of phage ψ DNA in lysogenic cells of *Agrobacterium tumefaciens. Virology* **121:**82–94.

FEISS, M., AND BECKER, A. 1983. DNA packaging and cutting, in: *Lambda II* (R. W. Hendrix, J. W. Roberts, F. W. Stahl, and R. A. Weisberg, eds.), pp. 305–330, Cold Spring Harbor Laboratory, Cold Spring Harbor, New York.

FISCHER-FANTUZZI, L., AND CALEF, L. 1964. A type of lambda prophage unable to confer immunity. *Virology* **23**:209–216.

FRICK, L. J., AND LISTER, R. M. 1978. Serotype variability in virus-like particles from *Gaeumannomyces graminis*. *Virology* **85**:504–517.

GANDY, D. G., AND HOLLINGS, M. 1962. Die-back of mushrooms: A disease associated with a virus. *Rep. Glasshouse Crops Res. Inst.* **1961**:103–108.

GHEI, O. K., EISENSTARK, A., TO, C. M., AND CONSIGLI, R. A. 1968. Structure and composition of *Xanthomonas pruni* bacteriophage. *J. Gen. Virol.* **3**:133–136.

GIBBS, J. N., AND BRASIER, C. M. 1973. Correlation between cultural characters and pathogenicity in *Ceratocystis ulmi* from Britain, Europe, and America. *Nature* **241**:381–383.

GOTO, M., AND STARR, M. P. 1972. Lysogenisation of *Xanthomonas phaseoli* and *X. begoniae* by temperate *X. citri* bacteriophages. *Ann. Phytopathol. Soc. Japan* **38**:267–274.

GOURETT, J. P., MAILLET, P. L., AND GOURANTON, J. 1973. Virus-like particles associated with mycoplasmas of clover phyloddy in the plant and in the insect vector. *J. Gen. Microbiol.* **74**:241–249.

HENDRIX, R. W., ROBERTS, J. W., STAHL, F. W., AND WEISBERG, R. A. (EDS.) 1983. *Lambda II*, Cold Spring Harbor Laboratory, Cold Spring Harbor, New York.

HOWARD, C. J., GOURLAY, R. N., AND WYLD, S. G. 1980. Isolation of a virus MVBr1 from *Mycoplasma bovirhinis*. *FEMS Microbiol. Lett.* **7**:163–165.

JACKSON, E. N., JACKSON, D. A., AND DEANS, R. J. 1978. *Eco*R1 analysis of bacteriophage P22 DNA packaging. *J. Mol. Biol.* **118**:365–388.

KAZAMA, F. Y., AND SCHORNSTEIN, K. L. 1973. Ultrastructure of a fungus herpes-type virus. *Virology* **52**:478–487.

KOLTIN, Y., AND DAY, P. R. 1976. Suppression of the killer phenotype in *Usilago maydis*. *Genetics* **82**:629–637.

KUO, T. -T., HUANG, T. -C., AND CHOW, T. -Y. 1969. A filamentous bacteriophage from *Xanthomonas oryzae*. *Virology* **39**:548–555.

LEMKE, P. (ED.) 1979. *Viruses and Plasmids in Fungi*, Marcel Dekker, New York.

LIEW, K. W., AND ALVAREZ, A. M. 1981. Biological and morphological characterisation of *Xanthomonas campestris* bacteriophages. *Phytopathology* **71**:269–273.

MINDICH, L., SINCLAIR, J. F., AND COHEN, J. 1976. The morphogenesis of bacteriophage φ6: Particles formed by nonsense mutants. *Virology* **75**:224–231.

MOFFIT, E. M., AND LISTER, R. M. 1973. Detection of mycoviruses using antiserum specific for dsRNA. *Virology* **52**:301–304.

RANGA REDDY, P., RAYCHAUDHURI, S. P., AND RAO, Y. P. 1977. Effects of bacteriophages in the infectivity of *Xanthomonas oryzae,* the incitant of bacterial leaf blight in rice. *Z. Pflanzenkr. Pflanzenschutz* **34**:592–596.

RAWLINSON, C. J., HORNBY, D., PEARSON, V., AND CARPENTER, J. M. 1973. Virus-like particles in the take-all fungus *Gaeumannomyces graminis. Ann. Appl. Biol.* **74**:197–209.

REANNY, D. C., AND ACKERMANN, H. L. 1982. Comparative biology and evolution of bacteriophages. *Adv. Virus Res.* **27**:205–280.

RITCHIE, D. P., AND KLOS, E. J. 1979. Some properties of *Erwinia amylovora* bacteriophages. *Phytopathology* **69**:1078–1083.

ROMANTSCHUK, M., AND BAMFORD, D. H. 1981. φ6-resistant phage-producing mutants of *Pseudomonas phaseolicola. J. Gen. Virol.* **56**:287–295.

SCHEFFER, R. P., AND PRINGLE, R. B. 1961. A selective toxin produced by *Periconia circinata. Nature* **191**:912–913.

SEMANCIK, J. S., VIDAVER, A. K., AND VAN ETTEN, J. L. 1973. Characterisation of a segmented double-helical RNA from bacteriophage φ6. *J. Mol. Biol.* **78**:617–626.

SINHA, R. L. 1979. Chemotherapy of mycoplasmal plant diseases, in: *The Mycoplasmas,* Vol. 3 (R. F. Whitcomb and J. G. Tully, eds.), pp. 310–335, Academic Press, New York.

STONIER, T., MCSHARRY, J., AND SPEITEL, T. 1967. *Agrobacterium tumefaciens* Conn. IV. Bacteriophage PB2₁ and its inhibitory effect on tumor induction. *J. Virol.* **1**:268–273.

STREISINGER, G., EMRICH, J., AND STAHL, H. 1967. Chromosome structure in phage T4. III. Terminal redundancy and length determination. *Proc. Natl. Acad. Sci. USA* **57**:292–295.

TIPPER, D. J., AND BOSTIAN, K. A. 1984. Double-stranded ribonucleic acid killer systems in yeasts. *Microbiol. Rev.* **48**:125–156.

TOWNSEND, R., MARKHAM, P. G., PLASKITT, K. A., AND DANIELS, M. J. 1977. Isolation and characterisation of a non-helical strain of *Spiroplasma citri. J. Gen. Microbiol.* **100**:15–21.

TYE, B. K., HUBERMAN, J., AND BOTSTEIN, D. 1974. Non-random circular permutation of phage P22 DNA. *J. Mol. Biol.* **85**:501–532.

VAN ETTEN, J. L., BURBANK, D. E., CUPPELS, D. A., LANE, L. C., AND VIDAVER, A. K. 1980. Semiconservative synthesis of single-stranded RNA by bacteriophage φ6 RNA polymerase. *J. Virol.* **33**:769–773.

VERVILET, G., HOLSTERS, M., TEUCHY, H., VAN MONTAGU, M. AND SCHELL, J. 1975. Characterisation of different plaque-forming and defective temperate phages in *Agrobacterium tumefaciens. J. Gen. Virol.* **26**:33–48.

VIDAVER, A. K. 1976. Prospects for control of phytopathogenic bacteria by bacteriophages and bacteriocins. *Annu. Rev. Phytopath.* **14**:451–465.

VIDAVER, A. K., KOSKI, R. K., AND VAN ETTEN, J. L. 1973. Bacteriophage φ6: A lipid-containing virus of *Pseudomonas phaseolicola. J. Virol.* **11**:799–805.

WAKIMOTO, S., UMATSU, T., AND MIZUKAMI, T. 1969. Bacterial canker disease of tomato in Japan. 2. Properties of bacteriophages specific for *Corynebac-*

terium michiganense (Smith) Jensen. *Ann. Phytopathol. Soc. Japan.* **35**:168–173.

WOOD, H. A., AND BOZARTH, R. F. 1973. Heterokaryon transfer of virus-like particles associated with a cytoplasmically inherited determinant in *Ustilago maydis*. *Phytopathology* **63**:1019–1021.

WU, W. C. 1972. Phage-induced alterations of cell disposition, phage adsorption and sensitivity, and virulence in *Xanthomonas citri*. *Ann. Phytopathol. Soc. Japan* **38**:333–341.

YANG, M. -K., AND KUO, T. -T. 1984. A physical map of the filamentous bacteriophage Cf genome. *J. Gen. Virol.* **65**:1173–1181.

SECTION V

Plant Response to Stress

Chapter 13

Hydrolytic Enzymes in Plant Disease Resistance

Thomas Boller

THE ARSENAL OF ANIMAL DEFENSES against microbial pathogens consists mainly of preexisting or inducible proteins (e.g., immunoglobulins, lysozyme, lactoferrin, myeloperoxidase). For plants, in contrast, secondary metabolites are thought to be the most important in resistance to pathogens (Bell, 1981; Stoessl, 1983) both in the form of preinfectional preexisting compounds (Schlösser, 1980) and in the form of postinfectional compounds, i.e., phytoalexins induced in response to a pathogen attack (Dixon et al., 1983; Darvill and Albersheim, 1984). Furthermore, the secondary metabolites of plants have been recognized as important weapons against herbivores (Rosenthal and Janzen, 1979). The question arises whether plants possess, in addition to their overwhelming variety of secondary metabolites, a similar array of preexisting or inducible "defensive proteins," comparable to immunoglobulins, which have no function in the primary metabolism of the organism.

Several classes of plant proteins have been studied in this regard.

Plant lectins have been likened to antibodies in view of their highly specific binding sites for sugar moieties (Etzler, 1985); their possible role in plant-pathogen interactions has been reviewed (Sequeira, 1978; Pistole, 1981). Inhibitors of microbial pectinases have been found in various plants (Albersheim and Anderson, 1971; Abu-Goukh et al., 1983). Proteinase inhibitors can be induced by wounding in some plants; they have no affinity for the plant's own proteinases but block proteinases of herbivores and pathogens and might thereby contribute to resistance (Ryan, 1984). The hypersensitive reaction of plants to incompatible viral or microbial pathogens sometimes causes the induction of large amounts of new soluble proteins. These abundant new proteins (called PR or b proteins) have no known enzymic role and no known way of interacting with viruses or pathogens, but their presence in the plant is often correlated with resistance (Van Loon, 1982; Gianinazzi, 1984).

In this chapter, an additional class of plant proteins is discussed, namely, plant hydrolases with a potential role in defense. An earlier review of this subject has appeared (Pegg, 1977). Some plant hydrolases, e.g., chitinase, can be considered "secondary proteins" since they have no known role in the primary metabolism of the plant. However, with their established ability to attack important microbial structures (e.g., the chitin in fungal cell walls), these hydrolases can be considered defense weapons against microbial pathogens. This hypothesis is reinforced by the finding that such hydrolases are often induced in the plant in response to a pathogen attack. While these hydrolases are secondary with regard to the primary metabolism, they might truly be primary with regard to plant disease resistance.

PLANT HYDROLASES WITH A POTENTIAL FUNCTION IN DISEASE RESISTANCE

A role in disease resistance has been proposed or might be suspected for a number of plant hydrolases. Here the discussion is limited to enzymes with the ability to attack specific structures of pathogens directly. The interesting subject of plant hydrolases which liberate toxic metabolites from preformed secondary compounds in the plant (Schlösser, 1980; Stoessl, 1983) is outside the scope of this chapter. The list that follows is not intended to be complete nor has it been shown conclusively that any of these hydrolases play a decisive role in a particular plant in resistance to a particular pathogen. However, in analogy to an argument often found in discussions of secondary metabolites (Rosenthal and Janzen, 1979), it is difficult to imagine that the large amounts of "secondary hydrolases" which preexist in some plants or are induced by infections in others have no vital function.

Chitinase and Lysozyme

Chitinase has often been studied in plants as a secondary hydrolase involved in defense. Its substrate, chitin (Fig. 1A), does not occur in higher plants but is present in the cell walls of many fungi. Hence this enzyme is a prime candidate for consideration as an enzyme directed against pathogens. Plant chitinase was first discovered in bean seeds (Powning and Irzykiewicz, 1965) and has been purified from several plants (Molano et al., 1979; Pegg and Young, 1982; Boller et al., 1983; Tsukamoto et al., 1984; Wadsworth and Zikakis, 1984). The physical and biochemical properties of plant chitinases have been summarized (Boller, 1985). In no case has an endogenous plant substrate been found for purified chitinase (Molano et al., 1979; Boller et al., 1983). However, the purified enzyme has been shown to attack and partially digest the cell walls of potentially pathogenic fungi (Young and Pegg, 1982; Boller et al., 1983). Bean chitinase has been shown to possess, in addition, hydrolyzing the peptidoglycan lysozyme activity (Fig. 1B) in bacterial cell walls (Boller et al., 1983).

Ever since Fleming (1922) discovered lysozyme in animal tissues and secretions, it has been considered an important defense of animals against bacterial infection (Jollès and Jollès, 1084). Fleming (1922) also remarked that he found "a very definite though not very strong" bacteriolytic activity in turnip. Later, certain plant latices were found to be rich sources of lysozyme. The enzyme was purified from the latex of *Carica papaya* (Howard and Glazer, 1969), *Ficus* (Glazer et al., 1969), and *Hevea brasiliensis* (Tata et al., 1983); in each case it was shown that the purified lysozyme exhibited a very high chitinase activity. The "lysozymes" from turnip (Burnier et al., 1971) and, more recently, from a *Rubus* cell culture (Bernasconi et al., 1985) have also been purified and have been found to act as endochitinases. Obviously, lysozyme activity and chitinase activity are closely related. Vertebrate lysozymes possess some chitinase activity, however, their hydrolytic activity is much greater with peptidoglycan than with chitin (Jollès and Jollès, 1984. [In addition, a separate, unrelated chitinase has been found in vertebrates (Lundblad et al., 1979).] It is interesting that all plant "lysozymes" characterized thus far behave differently: they are primarily chitinases because they hydrolyze chitin much more quickly than peptidoglycan.

β-1,3-Glucanase

β-1,3-Glucanase is often abundant in higher plants (Clarke and Stone, 1962) and has been purified from several sources (Moore and Stone,

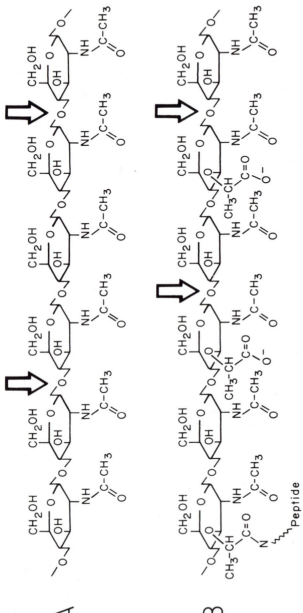

Figure 1. Chitin and bacterial peptidoglycan, two substrates for plant chitinase. (A) Chitin, a β-(1,4)-polymer of N-acetylglucosamine, is found in fungi and animals (e.g., insects) but not in higher plants. Plant chitinase acts as an endochitinase; arrows indicate a possible sequence of scissions (Molano et al., 1979; Boller et al., 1983). (B) Bacterial peptidoglycan, a β-(1,4)-linked polysaccharide with alternating N-acetylglucosamine and N-acetylmuramic acid units, is the backbone of bacterial cell walls. A part of the N-acetylmuramic acid units is peptidically linked to a short branched peptide. Plant chitinase, like other lysozymes, cleaves the bond between C-1 of N-acetylmuramic acid and C-4 of N-acetylglucosamine. In contrast to egg white lysozyme, it attacks only at unsubstituted N-acetylmuramic acid units, as indicated by the arrows (Howard and Glazer, 1979).

1972a; Wong and Maclachlan, 1979a,b; Young and Pegg, 1981; Keen and Yoshikawa, 1983; Shinshi and Kato, 1983; Felix and Meins, 1985). The physical and biochemical properties of these β-1,3-glucanases have been reviewed (Boller, 1985).

In contrast to chitinase, substrates of β-1,3-glucanase also occur endogenously in higher plants. Callose (β-1,3-glucan) is present in sieve tubes, in cell wall appositions formed in response to wounding, and in primary cell walls. Therefore β-1,3-glucanases have been much studied with regard to their function in the turnover and degradation of these structures (Clarke and stone, 1962; Ballance and Manners, 1978; Wong and Maclachlan, 1980).

However, the cooccurrence and coregulation of β-1,3-glucanase and chitinase in many situations indicates a common secondary function for the two enzymes (Abeles et al., 1971; Pegg and Young, 1981; Mauch et al., 1984; See Fig. 5 below). Purified β-1,3-glucanase from tomato has been shown to attack and partially digest the cell walls of a potentially pathogenic fungus, *Verticillium albo-atrum*; the degradation of fungal cell walls is synergistically stimulated by chitinase (Young and Pegg, 1982). Purified β-1,3-glucanases have also been shown to release elicitors of phytoalexin formation from isolated fungal cell walls (Keen and Yoshikawa, 1983).

Cellulase

Cellulase is important in cell wall degradation in plants and has been studied primarily with regard to abscission. As reviewed by Sexton and Roberts (1982), two types of cellulase have been found in plants. In addition to an abscission-specific cellulase induced in abscission zones, a different type of cellulase occurs constitutively in plant tissues; the function of the constitutive cellulase is unknown. In the context of this chapter, it is interesting that members of a much studied group of phytopathogenic fungi, the Oomycetes, contain cellulose rather than chitin in their cell walls (Wessels and Sietsma, 1981). It might be speculated that the cell walls of such fungi are also a target for plant cellulase.

Glycosidases Hydrolyzing Synthetic Aryl Glycosides

A great number of studies on plant glycosidases have been performed with simple artificial substrates, particularly with aryl glycosides (phenyl or *P*-nitrophenyl glycosides). The occurrence and biochemistry of these enzymes has recently been reviewed (Dey and Campillo, 1984). In view of the 550 references cited in this review, it is no exaggeration to say that dozens of plant enzymes have been studied for each of the commercially available glycosidase substrates. In most cases, such enzymes have been examined with regard to the primary metabolism, i.e., with regard to cell wall turnover or degradation of endogenously formed glycosides. For some of these enzymes, potential substrates also exist in the walls of fungal pathogens.

Unfortunately, little is known about the substrate specificity of many of these enzymes. Enzymes degrading aryl glycosides often fail to attack polysaccharides; conversely, most enzymes degrading polysaccharides do not hydrolyze the artificial substrates. Therefore, Dey and Campillo (1984) often note that the physiological role of the glycosidases studied with artificial substrates is unclear or unknown.

α-Mannosidase and β-N-Acetylglucosaminidase

Most plants contain high activities of α-mannosidase and β-*N*-acetylglucosaminidase (Dey and Campillo, 1984). Both enzymes occur in the central vacuoles of plant cells (Boller and Kende, 1979). In addition, α-mannosidase has been found in plant cell walls (Van der Wilden and Chrispeels, 1983). α-Mannosidase and β-*N*-acetylglucosaminidase probably have their primary function in the turnover of plant glycoproteins. In addition, they might have a secondary function; e.g., they might attack specific cell wall structures of plant pathogens, a possibility that has not been tested. Interestingly, in the symbiotic cells of legume root nodules, high activities of α-mannosidase are present in the peribacteroid space enclosing the bacteroids (Mellor et al., 1984). Purified α-mannosidase from beans agglutinates lymphocytes and is mitogenic, like the plant lectin concanavalin A (Paus and Steen, 1978).

α-Glucosidase and β-Glucosidase

Edreva and Georgieva (1980) found a 40- to 80-fold increase in the activity of α-glucosidase and β-glucosidase (measured with aryl glycosides as substrates) in tobacco infected with *Peronospora tabacina*. The activities did not increase in response to abiotic stress or during the course of senescence. Histochemically, the increased β-glucosidase activity in infected plants was localized in the mesophyll cells adjacent to the invading mycelium. The authors suggested that the increase in glycosidase activity was part of the defense mechanism of tobacco against the fungus.

Trehalase

Trehalase has been found in tissue cultures of several plants (Veluthambi et al., 1981). Its substrate, trehalose, has never been conclusively demonstrated in higher plants but is common in fungi and bacteria. Since trehalose was found to be toxic for *Cuscuta* and other plants with little trehalase activity, Veluthambi et al. (1981) proposed that trehalase functions in plants to provide protection from the potentially harmful effect of microbial trehalose. Streeter (1982) found trehalase in soybean root nodules and obtained evidence that a large part of the activity was of plant origin. Since symbiotic bacteroids accumulate large amounts of trehalose, plant trehalase might function, at some stage, in retrieving the assimilates stored by the symbiont in the form of trehalose.

Enzymes Degrading Bacterial Polysaccharides

Many bacteria possess species-specific cell surface polysaccharides that may be important in host-symbiont recognition in the symbioses of legumes and *Rhizobium* species (Bauer, 1981). Specific plant hydrolases may be important in this context. Dazzo et al. (1982) found that clover roots released enzymes that specifically altered the surface structure of the capsules of infective *Rhizobium* cells. Solheim and Fjellheim (1984) isolated various polysaccharides from two *Rhizobium* species, *R. trifolii* (the symbiont of clover) and *R. leguminosarum* (the symbiont of pea). They then incubated the isolated polysaccharides with root homogenates from clover and pea. In both homogenates, they found enzymes which degraded the polysaccharides. The activity of these apparently highly specific hydrolases was detected by a sensitive equilibrium dialysis technique; it could not be detected by viscometry or by the determination of reducing sugars. This indicated that degradation of the *Rhizobium* polysaccharides was limited and that only a few specific bonds were hydrolyzed. Interestingly, each plant homogenate degraded polysaccharides from its nonsymbiont *Rhizobium* species to a much lesser extent than those from its symbiont *Rhizobium* species

From this work, it is apparent that highly specific hydrolases might play an important role in host-symbiont and, by analogy, host-pathogen interactions. Assay techniques of the type developed by Solheim and

Fjellheim (1984) and by White et al. (1984) are promising for studies involving the discovery and analysis of such specific hydrolases.

Proteinases

Plant proteinases are primarily involved in cellular protein turnover. Plant
tissues often have comparatively low proteinase activity, sufficient to account for the observed protein turnover rates. As reviewed recently
(Boller, 1986), diseased plants often show increased proteinase activity;
the increased activity is most often due to the appearance of proteinases
secreted by the pathogens causing the disease. In these cases, the proteinases found in the interaction are produced by the pathogen, and the
plant might defend itself against them by producing proteinase inhibitors,
as discussed by Ryan (1984) in this series. However, some plants, e.g.,
many Bromeliaceae, contain high proteinase activity in their vegetative
tissues. In such cases, the proteinases might have secondary functions in
defense in plant-herbivore interactions (Boller, 1986). Another possibility, which has not been examined thus far, is that such proteinases attack
and degrade structural or enzymic proteins secreted by potential pathogens.

Ribonuclease

Although ribonuclease has a primary function in the cellular turnover of
nucleic acids, it is often induced to high levels in wounded or infected
plants (Wilson, 1975; Sacher et al., 1979, 1982). The amount of
ribonuclease present in wound-stimulated sweet potato has been reported to degrade all the cellular RNA in a few seconds (Sacher et al.,
1982). Does this excess ribonuclease possess secondary functions in
defense? There have been suggestions that ribonuclease might have
some importance in defense against viruses (Wilson, 1975), but evidence
for this is lacking.

Phosphatases

Acid phosphatase is mentioned here because it is characteristically active
in higher plants and because its release from intracellular compartments
in response to infections has been investigated (Pitt, 1973; Pitt and
Stewart, 1980). A large part of the activity of acid phosphatase is
localized in the cell wall; it is the most abundant wall-bound hydrolase in

some tissues (Lamport and Catt, 1981). Cell wall-bound acid phosphatase has been purified (Crasnier et al., 1980). Root acid phosphatase has been implicated in the mobilization of phosphate from organic phosphorus sources in the soil; however, there seems to be an enormous excess of acid phosphatase activity for this presumed function (Dracup et al., 1984). The significance of the presence of acid phosphatase in the walls of leaf cells is even less clear, since it is generally assumed that most cellular phosphate esters do not leave the cells. Does phosphatase have a function in plant-microbe interactions? Many fungi contain phosphorylated cell wall components (Wessels and Sietsma, 1981), and it is conceivable that plant acid phosphatase attacks them. However, it is not known whether this can cause an imbalance in the fungus; furthermore, many fungi secrete acid phosphatases themselves in order to gain access to sources of organic phosphorus.

Lipases and Phospholipases

Recently, the major soluble protein of potato tubers, patatin, which accounts for 20–30% of the total soluble protein, has been found to be a lipid acyl hydrolase (Racusen, 1984). While the primary function of such an enzyme might be in membrane turnover, much less enzyme protein would be required to fulfill such a function. Lipid acyl hydrolase degrades all available endogenous lipids in potato homogenates within a few seconds, even at 0°C (Galliard,1971). Lipid acyl hydrolase might have a secondary function as a defense against microbial invaders (Racusen, 1984). Interestingly, potato tuber disks are stimulated to form phytoalexins when treated with arachidonic acid, a free fatty acid liberated by lipid acyl hydrolase from the mycelium of *Phytophthora infestans* (Bostock et al., 1981).

Ribosome-Inactivating Proteins

Bacteria are known to produce proteinacous toxins as a defense against other bacteria. Some of these toxins are known to be enzymes with highly specific targets; e.g., colicin is a hydrolase which inactivates bacterial ribosomes by nucleolytic cleavage of a single site in the 16S RNA of the ribosome (Koninsky, 1982). Similarly, some plants produce toxic proteins that enzymically attack eukaryotic ribosomes (Jiménez and Vázquez, 1985). Best known among them are the seed proteins ricin and abrin (Olsnes et al., 1974), which are extremely toxic to animals, and may function in defense against herbivores. They consist of two subunits, a B chain which acts as a lectin, attaching the protein to animal cell membranes, and an A chain which enters the cells and enzymically destroys ribosomes (Fig. 2). Other plants, e.g., *Phytolacca* (Ready et al., 1983) and others (Stirpe et al., 1983), contain a ribosome-inactivating protein similar to the A chain of ricin and abrin but have no B component. The site of enzymic action of all these proteins is not known; however, it is likely that they act as hydrolytic enzymes which recognize and split a covalent bond in the 60S subunit of the ribosome (Olsnes et al., 1975; Ready et al., 1983; Jiménez and Vázquez, 1985).

These proteins have been implicated in the resistance of plants to virus infections (Stirpe et al., 1983). Here they may serve as examples of defensive hydrolases with specific targets; such highly specialized hydrolases might be of importance in plant-pathogen interactions also.

Localization of Plant Hydrolases with a Potential Function in Disease Resistance

In general, plant hydrolases are located in the "lytic compartment" of the cells, which includes the vacuoles and the cell wall space (Matile, 1975). A cell wall location appears to be particularly well-suited for a function in defense, since plant pathogens usually attack plant cells from the outside and often have to penetrate the host cell wall in order to form haustoria and the like. Some of the hydrolases discussed above are indeed secreted, at least in part, in the apoplastic space. Interestingly, however, many hydrolases are located nearly exclusively in the central vacuole, as summarized by Boller (1982a). For example, chitinase has been found to be located in the vacuoles and not in the cell walls (Boller and Vögeli, 1984). Similarly, many preformed secondary compounds which are important in defense are localized exclusively in the vacuoles (Matile, 1984). The central vacuole may therefore be seen as a defense arsenal (Boller, 1982a) or as the toxic compartment of the plant cell (Matile, 1984). As a corollary to this observation, the deployment of stored defense weapons against attacking pathogens requires either specific transport from the vacuole to the cell wall—a possibility which thus far has no experimental basis—or disruption of the vacuole with the subsequent release of all its contents (Fig. 3). The latter possibility is realized in the universal resistance response of plants, in the so-called hypersensitive reaction in which one cell or a few cells in the close vicinity of an invading pathogen collapse and set free their vacuolar contents. The combined action of the vacuolar defense weapons, including secondary compounds, hydrolases, lectins, and proteinase inhibitors, then overwhelms the pathogen in most cases.

It should be noted that many vacuolar components are also toxic to

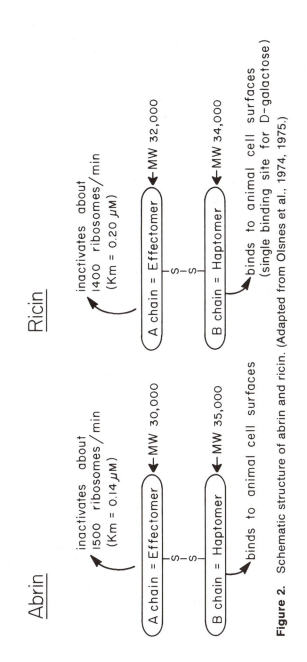

Figure 2. Schematic structure of abrin and ricin. (Adapted from Olsnes et al., 1974, 1975.)

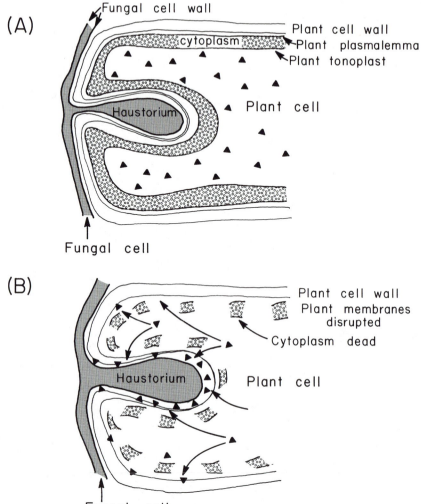

Figure 3. Significance of vacuolar localization of hydrolases in plant-pathogen interactions. (A) Scheme of a compatible plant-pathogen interaction. Fungal haustoria may penetrate deeply into the plant cell but do not come into contact with vacuolar enzymes (▲); plant plasmalemma, cytosol, and tonoplast lie in between. (B) Scheme of an incompatible plant-pathogen interaction. The plant cell displays the hypersensitive reaction. Plant membranes, including the vacuolar membrane, disrupt; the cytoplasm dies. Vacuolar enzymes (▲) can attack fungal cell walls.

the plant itself. Therefore the decompartmentation of plant cells in response to a pathogen can also be seen as a factor in pathogenicity (Wilson,

1973). For example, Pitt and Stewart (1980) found decompartmentation and release of the intracellular acid phosphatase in pea roots attacked by *Phoma medicaginis*, a virulent necrotroph. They interpreted the observations as an indication of the breakdown of the lytic compartment, leading to the disruption of host cell metabolism and expression of disease symptoms.

TARGETS OF HYDROLASES WITH A POTENTIAL FUNCTION IN DISEASE RESISTANCE

Hydrolases can act in different ways against potential pathogens. First, and most obviously, they can digest the cell wall and lyse the cells of an invading pathogen. Second, host hydrolytic enzymes can contribute to the recognition of pathogens by liberating substances from the cell walls of invaders that can act as elicitors of defense reactions. Third, hydrolases can specifically dissect a vital structure of a pathogen.

Lysis of the Cell Walls of Fungal Pathogens

In tomato plants infected with *V. albo-atrum*, the pathogen penetrated into the xylem vessels, and lysis of the fungus occurred thereafter (Pegg and Vessey, 1973). Lysis of fungi was also observed in the mycorrhizal symbiosis of orchids with fungi and in other plant-fungus interactions (Pegg, 1977). Plant chitinase and β-1,3-glucanase have been implicated in this process (Pegg, 1977; Pegg and Young, 1981). The cell walls of many Ascomycetes and Basidiomycetes contain a large proportion of chitin and of a β-1,3-glucan with some β-1,6-linkages in their cell walls (Wessels and Sietsma, 1981). It is no surprise therefore that plant chitinase and β-1,3-glucanase have been found to hydrolytically digest a considerable part of the isolated cell walls of many fungi (Wargo, 1975; Young and Pegg, 1982; Boller et al., 1983).

However, models of the cell wall architecture of fungi (Fig. 4) indicate that the chitin and β-1,3-glucan represent the innermost layers of the cell wall, while the outer layer consists of α-1,3- and α-1,6-linked glucans, of mannoproteins or phosphomannoproteins, or even of lipids (Wessels and Sietsma, 1981). These outer layers mask and protect the inner part of the cell wall of intact fungi from β-1,3-glucanase and chitinase attack. This is best seen in work devoted to the preparation of protoplasts from filamentous fungi (Peberdy, 1979). A careful study with *Schizophyllum commune*, a wood-rotting basidiomycete, showed that α-1,3-glucanase was necessary in addition to β-1,3-glucanase and chitinase for efficient digestion of the cell walls of intact fungal mycelium (De Vries and Wessels,

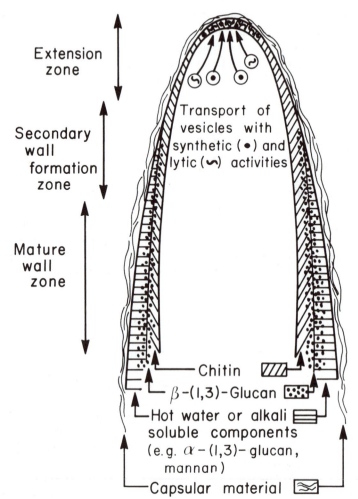

Extension zone

Secondary wall formation zone

Mature wall zone

Transport of vesicles with synthetic (•) and lytic (∿) activities

——Chitin ▨

└─ β-(1,3)-Glucan ▦

└─Hot water or alkali soluble components (e.g. α-(1,3)-glucan, mannan)

└─Capsular material ≋

Figure 4. Generalized model of the cell wall of a growing fungal hypha. At the growing tip, the wall is thin and smooth. Vesicles probably transport both synthetic and lytic enzymes to the tip; growth represents a balance of wall formation and wall loosening. The mature wall is composed of several layers. The outermost layer often consists of polysaccharides that cannot be attacked by chitinases or β-1,3-glucanases. (Adapted from Farkas, 1979, and Wessels and Sietsma, 1981.)

1973). The preparation of protoplasts from *Pythium* (a root pathogen of the Oomycetes) with cellulase and β-1,3-glucanase required pretreatment with lipase or with detergents (Sietsma and DeBoer, 1973), indicating that the outer surface of the intact wall (which consisted mainly of cellulose and β-1,3-glucan, the typical wall components of Oomycetes) was protected by a lipid layer.

It will be interesting to determine if plants require and possess hydrolases in addition to β-1,3-glucanase and chitinase which can help in lysing effectively the cell walls of specific pathogens.

However, complete lysis of the mycelial cell wall is not needed to stop a growing fungus. Usually the growing tip is most sensitive to the attack by hydrolases (Fig. 4). Growth of the hyphal tip appears to be due to a delicate balance of lytic and synthetic processes; the imbalance caused by the addition of exogenous hydrolase activity is often sufficient, at least under hypotonic conditions, to break the tip open and kill the advancing hypha (Farkas, 1979). In this context, it is interesting that plant chitinase is particularly active on "nascent" chitin, i.e., chitin in the process of being formed (Molano et al., 1979).

Release of Elicitors from Surface Structures of Pathogens

Cell wall components of pathogenic fungi act as elicitors of defense reactions in the plant. As reviewed by Darvill and Albersheim (1984), the induction of phytoalexin formation by fungal elicitors has been studied extensively.

The recent elucidation of the primary structure of a highly potent heptaglucoside elicitor and of a number of inactive isomers (Sharp et al., 1984) suggests that relatively small, highly specific fungal wall fragments are natural elicitors. Fragments of such size might well be released from the cell surface of invading fungi by the hydrolases of the plant (Yoshikawa, 1983).

In support of this concept, a crude enzyme preparation from peas was found to release elicitor-active fragments from the cell walls of *Fusarium solani* f.sp. *phaseoli* (Hadwiger and Beckman, 1980). Similarly, soybean tissues contained enzymes which released soluble phytoalexin elicitors from *Phytophthora megasperma* (Yoshikawa et al., 1981). More recently, β-1,3-glucanase has been purified from soybean and shown to release elicitor-active fragments from *Phytophthora* cell walls (Keen and Yoshikawa, 1983; Keen et al., 1983). On the other hand, plant glucanases may also inactivate elicitors. Cline and Albersheim (1981a,b) purified a β-glucosyl hydrolase from the cell walls of cultured soybean cells which inactivated elicitor preparations from *P. megasperma*.

In the case of chitinase, it is not known whether chitin fragments released from cell walls by purified chitinase have elicitor activity. In wheat leaves, chitin was found to act as an elicitor of the lignification response, and it was suggested that enzymically generated chitin fragments were the natural elicitors in this system (Pearce and Ride, 1982). In this regard, it is interesting that crude plant homogenates containing chitinase readily hydrolyze chitin to oligosaccharides but usually are devoid of enzymic activity (chitobiase or exochitinase) which digest the oligosaccharides to *N*-acetylglucosamine (Molano et al., 1979; Boller et al., 1983). Thus chitin oligosaccharides would be available for a long time after partial digestion of fungal cell walls.

While most elicitors described thus far have a polysaccharide part, it was found in potatoes that arachidonic acid acted as an elicitor (Bostock et al., 1981). This is intriguing in view of the presence of arachidonic acid in mycelial lipids from *P. infestans* (Bostock et al., 1981) and of the extremely high lipolytic acyl hydrolase activity in potato tubers (Racusen, 1984). There is a possibility that surface structures of pathogens and elicitor-releasing hydrolases have coevolved in specific plant-pathogen pairs. For legume-*Rhizobium* symbiosis, Solheim and Fjellheim (1984) have provided evidence for the existence of species-specific host plant hydrolases that are matched with species-specific surface structures of the symbiont. It will be interesting to make similar comparisons for plant-pathogen interactions: Do the hydrolases of a host plant release more active elicitors from a pathogen or from a nonpathogen.

Hydrolysis of Specific Vital Structure of Pathogens

The studies on ribosome-inactivating proteins show that a plant may produce hydrolases that attack a specific, vitally important site on a herbivore. There is a void in our knowledge in regard to similar specific enzymes in host-pathogen interactions. One possible target of highly specific enzymes may be the toxins secreted by many plant pathogens. The toxin of *Helminthosporium sacchari*, for example, consists of four to six β-galactofuranose units attached to a sesquiterpene (Livingston and Scheffer, 1981). It is rendered inactive by β-galactofuranosidase, an enzyme that has been identified thus far only in culture filtrates of the fungus (Livingston and Scheffer, 1983). It will be interesting to find out whether plants resistant to the toxin produce the appropriate galactofuranosidase.

REGULATION OF PLANT HYDROLASES INVOLVED IN DEFENSE

As recently reviewed (Boller, 1985), chitinase and β-1,3-glucanase can be induced by the plant hormone ethylene or in response to pathogens or elicitors. A brief summary of these findings is presented here.

Induction of Hydrolytic Enzymes in Response to Pathogens

A variety of hydrolytic enzymes can be induced in response to pathogens (Table 1). β-1,3-Glucanase was found to increase strongly during the course of virus infections in various plants (Moore and Stone, 1972b; Hawker et al., 1974; Kearney and Wu, 1984). There was no evidence that β-1,3-glucanase had a role in the containment or spread of the virus (Kearney and Wu, 1984). Both β-1,3-glucanase and chitinase activities were found to be similarly induced in response to fungal infection (Netzer et al., 1979; Pegg and Young, 1981; Mauch et al., 1984; Fig. 5). The interpretation of these results is complicated, since fungal enzymes may contribute to the newly appearing activities (Rabenantoandro et al., 1976). However, induction of β-1,3-glucanase and of chitinase was also observed in response to treatment with heat-killed pathogens, pathogen cell walls, laminarin, or chitosan (Netzer et al., 1979; Pegg and Young, 1981; Mauch et al., 1984; Fig. 5). In these cases, the newly appearing activity obviously came from the plant. Furthermore, in cucumber, a systemic effect of infection was found with regard to chitinase (Métraux and Boller, 1986). Infection of the first leaf of young cucumber plants with tobacco necrosis virus (TNV), *Colletotrichum lagenarium*, and other fungal pathogens caused a strong increase in chitinase activity in the infected leaf; but as resistance to further infections developed in the second, noninfected, leaf, chitinase also increased there.

Table 1 Plant Glycosidases Induced in Response to Infection.

ENZYME	SUBSTRATES	REFERENCES
Chitinase	Chitin, bacterial peptido-glycan	Pegg and Young, 1981; Mauch et al., 1984; Métraux and Boller, 1986
β-1,3-Glucanase	β-1,3-Glucan	Moore and Stone, 1972b; Hawker et al., 1974; Netzer et al., 1979; Pegg and Young, 1981; Kearney and Wu, 1984; Mauch et al., 1984
α-Glucosidase	p-Nitrophenyl-α-glucoside	Edreva and Georgieva, 1980
β-Glucosidase	p-Nitrophenyl-β-glucoside	Edreva and Georgieva, 1980

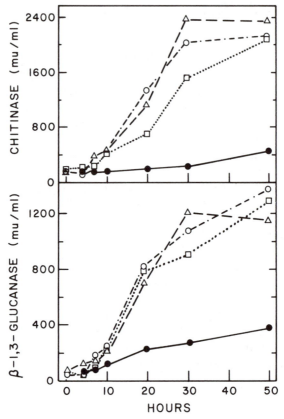

Figure 5. Coordinate induction of chitinase and β-1,3-glucanase in imma-
ture pea pods in response to pathogens and elicitors. Inoculation, at zero
time, was with conidia of *Fusarium solani* f.sp. *pisi,* a compatible pathogen
(△) or with conidia of *Fusarium solani* f.sp. *phaseoli,* an incompatible
pathogen (○). Both fungi induced the enzymes with very similar kinetics, indicating that enzyme induction does not directly determine disease specificity in the pea-*Fusarium* interaction. Inoculation with an elicitor, chitosan (25
μg per pod), caused a similar induction of the two hydrolases (□), indicating
that the newly appearing activities come from the plant. Controls, inoculated
with sterile water (●), showed only a slight increase of chitinase activity and
a more marked increase of β-1,3-glucanase. (Reproduced, with permission,
from Mauch et al., 1984.)

Regulation of Hydrolytic Enzymes by Plant Hormones

Induction of β-1,3-glucanase and chitinase by ethylene was first observed
by Abeles and Forrence (1970) and Abeles et al. (1971). This effect has

recently been studied in some detail (Boller et al., 1983). In time-course experiments with 10 nl/ml ethylene, it was found that chitinase started to increase after a lag of 4–6 hr; within 24 hr, the enzyme was induced 30-fold compared to the level in controls and amounted to more than 1% of the total cellular protein. Among a number of hydrolases tested, only chitinase and β-1,3-glucanase were induced by ethylene. Ethylene strongly induced chitinase in a number of dicots; however, it did not induce chitinase in the two monocots tested, wheat and maize.

Other hormones may influence the level of β-1,3-glucanase and chitinase. In bean leaves, cytokinin, auxin, and gibberellic acid reduced the induction of β-1,3-glucanase (Abeles and Forrence, 1970). In tobacco leaves, kinetin and abscisic acid had the same effect (Moore and Stone, 1972c). In cultured tobacco tissues, β-1,3-glucanase had a low activity level in media containing cytokinin and auxin but increased strongly when cytokinin or auxin was omitted from the medium (Felix and Meins, 1985; Mohnen et al., 1985).

Is Stress Ethylene Responsible for the Induction of Hydrolytic Enzymes after Pathogen Infection?

In pea pods, elicitors and infections rapidly caused enhanced ethylene production; only later were chitinase and β-1,3-glucanase induced (Mauch et al., 1984). A similar pattern of induction was seen in cultured parsley cells (Chappell et al., 1984). In pea pods, chitinase and β-1,3-glucanase were also induced by exogenous application of ethylene. Therefore it seems plausible that ethylene acted as a second messenger in induction of the two hydrolases by pathogens or elicitors (Boller, 1982b). However, when endogenous ethylene production was blocked by aminoethoxyvinylglycine, chitinase and β-1,3-glucanase remained strongly induced. This indicates that ethylene was symptom but not the mediating signal for the induction of the two enzymes (Mauch et al., 1984).

PERSPECTIVES

Biological Perspectives

In plant-pathogen interactions, there is a continuous interplay between the means by which pathogens attack the host and the ways in which the host defends itself. Hydrolytic enzymes are important weapons for pathogens in attacking their hosts (Wilson, 1973; Kosuge, 1981; Cooper, 1983). As a countermeasure against these hydrolases, plants have evolved surface structures that are more resistant to hydrolytic attack, e.g., by lignification. Furthermore, they possess specific inhibitors of certain foreign hydrolases, e.g., pectinase inhibitors and proteinase inhibitors. Finally, plants may possess a signaling system which employs the products of the foreign enzymes (e.g., the breakdown products of pectins produced by the polygalacturonases of the pathogens) as endogenous elicitors in order to induce defense reactions (Darvill and Albersheim, 1984).

On the other hand, plants themselves possess hydrolases that can attack and damage invading microbes. Successful pathogens must avoid or overcome these weapons. Fungi render their cell walls more resistant to plant hydrolases by specific surface structures or by melanization. Furthermore, fungi may produce inhibitors of the hydrolytic enzymes of plants. In an interesting study, Albersheim and Valent (1974) found that a fungal pathogen of the bean plant *Colletotrichum lindemuthianum* produced an extracellular inhibitor of bean β-1,3-glucanase. Pathogens or parasites might even take advantage of the defensive hydrolases of the plant. A study illustrating this possibility has been undertaken with herbivorous insects (Powning and Davidson, 1979). Larvae of *Acanthoscelides obsoletus* (bean weevil) were found to contain high bacteriolytic activities in their gut derived from their bean diet. Powning and Davidson (1979) hypothesized that the lysozyme activity acquired from beans helped the insect to defend itself against bacterial gut infections.

In view of the complexity of these interactions, it is difficult to evaluate the extent to which the hydrolases of plants contribute to resistance and to find out why hydrolases and the remainder of the defense arsenal succumb when confronted with a truly virulent pathogen. Plant hydrolases like chitinase and β-1,3-glucanase can provide a clearly established potential defense against pathogens. However, their actual importance in individual plant-pathogen relationships remains to be established.

Evolutionary Perspectives

Most of the hydrolases discussed have endogenous substrates in addition to the potential exogenous substrates present in surface structures or products of pathogens. Such enzymes, e.g., β-1,3-glucanase and lipid acyl hydrolase, may have been recruited many times for new, secondary purposes. One might expect that such enzymes have evolved new properties in connection with their secondary function, which distinguish them from the isoenzymes in primary metabolism. In the case of β-1,3-glucanase, it is interesting to note that two different types of enzymes have been found in peas. Growing seedlings contain two β-1,3-glucanases with molecular weights of 22 and 37 kd and with acidic isoelectric points (Wong and Maclachlan, 1979a,b). These enzymes are thought to be important for cell wall turnover and extension growth; endogenous substrates are present in the cell walls of the seedlings (Wong and Maclachlan, 1980). Young pea pods contain two β-1,3-glucanases with molecular weights of 32 and 33 kd and with basic isoelectric points (F. Mauch, unpublished observations). The pod enzymes are regulated in parallel with chitinase (Mauch et al., 1984) and are thought to be involved in defense.

For some of the hydrolases discussed, e.g., for chitinase and trehalase, no known endogenous substrate is present in the plant itself. The same is true for animal lysozyme. How such proteins evolved and why they remain in the organisms over evolutionary time are intriguing questions. Clearly, one answer is that there is sufficient selection pressure for their maintenance; i.e., that they are an important factor in preventing damage by pathogens. Another possibility that should be kept in mind is that they have a different, hitherto unknown, primary function in metabolism. In this regard, it is interesting that the amino acid sequence of positions 1–30 of ethylene-induced bean leaf chitinase possesses high homology to the N-terminal amino acid sequence of wheat germ agglutinin, a lectin with high affinity to chitin, whose primary function is as unknown as that of chitinase (Lucas et al., 1986).

Defensive hydrolases of any particular species might have undergone selection so as to become specifically adapted to the most important pathogens of that species. In this regard, it will be interesting to compare defensive hydrolases from different species with regard to their structure, biochemistry, and regulation. In the case of chitinase, preliminary serological evidence indicates that different types of the enzymes exist in different plants. Using antiserum against bean chitinase, it was found that the chitinases from bean, pea, tobacco, and tomato were all serologically related. Cucumber leaves, however, contained a chitinase that did not react with the antiserum against bean chitinase; interestingly, cucumber fruits contained a chitinase that was serologically similar to bean chitinase (T. Boller, unpublished results).

Perspectives for Genetic Engineering

In recent years, genetic engineering in higher plants has become a much discussed possibility. One particularly important aim in this field is improvement of the resistance of crop plants to pathogens by the introduction or manipulation of specific genes. While defense systems based on secondary metabolites often require a large number of different enzymes

in the appropriate intracellular environments in order to function properly, and therefore might not be easily manipulated, defense systems based
on proteins which act directly on pathogens seem much more promising in
the short term. For this reason, the genes for defensive hydrolases like
chitinase have received considerable attention as prospective resistance
genes (Nitzsche, 1983).

Chitinase, along with a number of other hydrolases, is already present
in most plants, and it is often strongly induced in response to pathogens.
Therefore it appears unlikely that manipulation of the enzyme protein of
chitinase or of its mode of regulation would lead to any large improvement
in resistance. However, some challenging possibilities remain.

One interesting aspect is the large amount of chitinase protein formed
in a relatively short time. Within 24 hr of a treatment with ethylene,
primary leaves of bean plants contain more than 1% of the total protein in
the form of chitinase. In view of this, chitinase may be interesting to genetic engineers on a merely quantitative basis. However, in this regard, caution is in order. First, most bean tissues and most other plants react more
sluggishly than primary bean leaves (Boller et al., 1983). Second, the
mode of induction of chitinase at the molecular level is incompletely
known. Therefore it may be premature to speculate about the use of chitinase genes. Another interesting starting point is the vacuolar localization of chitinase, at least in bean leaves (Boller and Vögeli, 1984). It is
possible that, in the case of infections by compatible biotrophs which
usually do not affect the vacuole of the host, chitinase never comes into
contact with the fungus. It will be interesting to find out whether chitinase can attack such biotrophs when it is added to the cell wall space
from the outside. If this proves to be the case, it might be promising to
try to alter the localization of chitinase by engineering it into a secretory
protein.

REFERENCES

ABELES, F. B., AND FORRENCE, L. E. 1970. Temporal and hormonal control of β-
1,3-glucanase in *Phaseolus vulgaris*, *Plant Physiol.* **45**:395–400.

ABELES, F. B., BOSSHART, R. P., FORRENCE, L. E., AND HABIG, W. H. 1971.
Preparation and purification of glucanase and chitinase from bean leaves,
Plant Physiol. **47**:129–134.

ABU-GOUKH, A. A., GREVE, L. C., AND LABAVITCH, J. M. 1983. Purification and
partial characterization of "Bartlett" pear fruit polygalacturonase inhibitors,
Physiol. Plant Pathol. **23**:111–122.

ALBERSHEIM, P., AND ANDERSON, A. J. 1971. Proteins from plant cell walls inhibit polygalacturonases secreted by plant pathogens, *Proc. Natl. Acad. Sci.
USA* **68**:1815–1819.

ALBERSHEIM, P., AND VALENT, B. S. 1974. Host-pathogen interactions. VII. Plant pathogens secrete proteins which inhibit enzymes of the host capable of attacking the pathogen, *Plant Physiol.* **53**:684–687.

BALLANCE, G. M., AND MANNERS, D. J. 1978. Partial purification and properties of an endo-1,3-β-D-glucanase from germinated rye, *Phytochemistry* **18**:1539–1543.

BAUER, W. D. 1981. Infection of legumes by rhizobia, *Annu. Rev. Plant Physiol.* **32**:407–449.

BELL, A. A. 1981. Biochemical mechanisms of disease resistance, *Annu. Rev. Plant Physiol.* **32**:21–81.

BERNASCONI, P., PILET, P. E., AND JOLLÈS, P. 1985. A one-step purification of a plant lysozyme from in vitro cultures of *Rubus hispidus*. *FEBS Lett.* **186**:263–266.

BERNIER, I., VAN LEEMPUTTEN, E., HORISBERGER, M., BUSH, D. A., AND JOLLÈS, P. 1971. The turnip lysozyme, *FEBS Lett.* **14**:100–104.

BOLLER, T. 1982a. Enzymatic equipment of plant vacuoles, *Physiol. Veg.* **20**:247–257.

BOLLER, T. 1982b. Ethylene-induced biochemical defenses against pathogens, in: *Plant Growth Substances 1982* (P. F. Wareing, ed.), pp. 303—312, Academic Press, London.

BOLLER, T. 1985. Induction of hydrolases as a defense reaction against pathogens, in: *Cellular and Molecular Biology of Plant Stress* (J. L. Key and T. Kosuge, eds.) pp. 247–262, Alan R. Liss, New York.

BOLLER, T. 1986. Role of proteolytic enzymes in interactions of plants with other organisms, in: *Plant Proteolytic Enzymes* (M. Dalling, ed.), CRC Press, Boca Raton, Florida, in press.

BOLLER, T., AND KENDE, H. 1979. Hydrolytic enzymes in the central vacuole of plant cells, *Plant Physiol.* **63**:1123–1132.

BOLLER, T., AND VÖGELI, U. 1984. Vacuolar localization of ethylene-induced chitinase in bean leaves, *Plant Physiol.* **74**:442–444.

BOLLER, T., GEHRI, A., MAUCH, F., AND VÖGELI, U. 1983. Chitinase in bean leaves: Induction by ethylene, purification, properties, and possible function, *Planta* **157**:22–31.

BOSTOCK, R. M., KUC, J. A., AND LAINE, R. A. 1981. Eicosapentaenoic and arachidonic acids from *Phytophthora infestans* elicit fungitoxic sesquiterpenes in the potato, *Science* **212**:67–69.

CHAPPELL, J., HAHLBROCK, K., AND BOLLER, T. 1984. Rapid induction of ethylene biosynthesis in cultured parsley cells by fungal elicitor and its relationship to the induction of phenylalanine ammonia-lyase, *Planta* **161**:475–480.

CLARKE, A. E., AND STONE, B. A. 1962. β-1,3-Glucan hydrolases from the grape vine (*Vitis vinifera*) and other plants, *Phytochemistry* **1**:175–188.

CLINE, K., AND ALBERSHEIM, P. 1981a. Host-pathogen interactions. XVI. Purification and characterization of a β-glucosyl hydrolase/transferase present in the cell walls of soybean cells, *Plant Physiol.* **68**:207–220.

CLINE, K., AND ALBERSHEIM, P. 1981b. Host-pathogen interactions. XVII. Hydrolysis of biologically active fungal glucans by enzymes isolated from soybean cells, *Plant Physiol.* **68**:221–228.

COOPER, R. M. 1983. The mechanisms and significance of enzymic degradation of host cell walls by parasites, in: *Biochemical Plant Pathology* (J. A. Callow, ed.), pp. 101–135, John Wiley, Chichester, England.

CRASNIER, M., NOAT, G., AND RICARD, J. 1980. Purification and molecular properties of acid phosphatase from sycamore cell walls, *Plant Cell. Environ.* **3**:217–224.

DARVILL, A. G., AND ALBERSHEIM, P. 1984. Phytoalexins and their elicitors—A defense against microbial infection in plants, *Annu. Rev. Plant Physiol.* **35**:243–275.

DAZZO, F. B., TRUCHET, G. L., SHERWOOD, J. E., HRABAK, E. M., AND GARDIOL, A. E. 1982. Alteration of trifoliin A-binding capsule of *Rhizobium trifolii* 0403 by enzymes released from clover roots, *Appl. Environ. Microbiol.* **44**:478–490.

DE VRIES, O. M. H., AND WESSELS, J. G. H. 1973. Release of protoplasts from *Schizophyllum commune* by combined action of purified α-1,3-glucanase and chitinase derived from *Trichoderma viride, J. Gen. Microbiol.* **76**:319–330.

DEY, P. M., AND CAMPILLO, E. D. 1984. Biochemistry of the multiple forms of glycosidases in plants, *Adv. Enzymol.* **56**:141–249.

DIXON, R. A., DEY, P. M., AND LAMB, C. J. 1983. Phytoalexins: Enzymology and molecular biology, *Adv. Enzymol.* **55**:1–136.

DRACUP, M. N. H., BARRETT-LENNARD, E. G., GREENWAY, H., AND ROBSON, A. D. 1984. Effect of phosphorus deficiency on phosphatase activity of cell walls from roots of subterranean clover, *J. Exp. Bot.* **35**:466–480.

EDREVA, A. M., AND GEORGIEVA, I. D. 1980. Biochemical and histochemical investigations of α- and β-glucosidase activity in an infectious disease, a physiological disorder and in senescence of tobacco leaves, *Physiol. Plant Pathol.* **17**:237–243.

ETZLER, M. E. 1985. Plant lectins: molecular and biological aspects. *Annu. Rev. Plant Physiol.* **36**:209–234.

FARKAS, V. 1979. Biosynthesis of cell walls of fungi, *Microbiol. Rev.* **43**:117–144.

FELIX, G., AND MEINS, F. 1985. Purification, immunoassay and characterization of an abundant, cytokinin-regulated polypeptide in cultured tobacco tissues, *Planta*, **164**:423–428.

FLEMING, A. 1922. On a remarkable bacteriolytic element found in tissues and secretions, *Proc. Roy. Soc. London* **93**:306–317.

GALLIARD, T. 1971. The enzymic deacylation of phospholipids and galactolipids in plants: Purification and properties of a lipolytic acylhydrolase from potato tubers, *Biochem. J.* **121**:379–390.

GIANINAZZI, S. 1984. Genetic and molecular aspects of resistance induced by infections or chemicals, in: *Plant-Microbe Interactions* (T. Kosuge and E. W. Nester, eds.), pp. 321–342, Macmillan, New York.

GLAZER, A. N., BAREL, A. O., HOWARD, J. B., AND BROWN, D. M. 1969. Isolation and characterization of fig lysozyme, *J. Biol. Chem.* **244**:3583–3589.

HADWIGER, L. A., AND BECKMAN, J. M. 1980. Chitosan as a component of pea–*Fusarium solani* interactions, *Plant Physiol.* **66**:205–211.

HAWKER, J. S., WOODHAM, R. C., AND DOWNTON, W. J. S. 1974. Activity of β-1,3-glucan hydrolase in virus infected grapevines and other plants, *Z. Pflanzenkr. Pflanzenschutz* **81**:100–107.

HOWARD, J. B., GLAZER, A. N. 1969. Papaya lysozyme: Terminal sequences and enxymatic properties, *J. Biol. Chem.* **244**:1399–1409.

JIMÉNEZ, A., AND VÁZQUEZ, D. 1985. Plant and fungal protein and glycoprotein toxins inhibiting eukaryote protein synthesis. *Annu. Rev. Microbiol.* **39**:649–672.

JOLLÈS, P., AND JOLLÈS, J. 1984. What's new in lysozyme research? *Mol. Cell. Biochem.* **63**:165–189.

KEARNEY, C. M., AND WU, J. H. 1984. β-1,3-Glucanase and the spread of tobacco mosaic virus in *Nicotiana* and *Phaseolus, Can. J. Bot.* **62**:1984–1988.

KEEN, N. T., AND YOSHIKAWA, M. 1983. β-1,3-Endoglucanase from soybean releases elicitor-active carbohydrates from fungus cell walls, *Plant Physiol.* **71**:460–465.

KEEN, N. T., YOSHIKAWA, M., AND WANG, M. C. 1983. Phytoalexin elicitor activity of carbohydrates from *Phytophthora megasperma* f. sp. *glycinea* and other sources, *Plant Physiol.* **71**:466–471.

KONINSKY, J. 1982. Colicins and other bacteriocins with established modes of action, *Annu. Rev. Microbiol.* **36**:125–144.

KOSUGE, T. 1981. Carbohydrates in plant-pathogen interactions, in: *Encyclopedia of Plant Physiology,* New Series, Vol. 13B, *Plant Carbohydrates II* (W. Tanner and F. A. Loewus, eds.), pp. 584–623, Springer-Verlag, New York.

LAMPORT, D. T. A., CATT, J. W. 1981. Glycoproteins and enzymes of the cell wall, in: *Encyclopedia of Plant Physiology,* New Series, Vol. 13B, *Plant Carbohydrates II* (W. Tanner and F. A. Loewus, eds.), pp. 133–165, Springer-Verlag, New York.

LIVINGSTON, R. S., AND SCHEFFER, R. P. 1981. Isolation and characterization of the host-selective toxin from *Helminthosporium sacchari, J. Biol. Chem.* **256**:1705–1710.

LIVINGSTON, R. S., AND SCHEFFER, R. P. 1983. Conversion of *Helminthosporium sacchari* toxin to toxoids by β-galactofuranosidase from *Helminthosporium, Plant Physiol.* **72**:530–534.

LUCAS, J., HENSCHEN, A., LOTTSPEICH, F., VÖGELI, U., AND BOLLER, T. 1986. Amino-terminal sequence of ethylene-induced bean leaf chitinase reveals similarities to sugar-binding domains of wheat germ agglutinin. *FEBS Lett.,* in press.

LUNDBLAD, G., ELANDER, M., LIND, J., AND SLETTENGREN, K. 1979. Bovine serum chitinase, *Eur. J. Biochem.* **100**:455–460.

MATILE, P. 1975. *The lytic compartment of plant cells*, Springer-Verlag, Vienna.

MATILE, P. 1984. Das toxische Kompartiment der Pflanzenzelle, *Naturwissenschaften* **71**:18–24.

MAUCH, F., HADWIGER, L. A., AND BOLLER, T. 1984. Ethylene: Symptom, not signal for the induction of chitinase and β-1,3-glucanase in pea pods by pathogens and elicitors, *Plant Physiol.* **76**:607–611.

MELLOR, R. B., MÖRSCHEL, E., AND WERNER, D. 1984. Legume root response to symbiotic infection: Enzymes of the peribacteroid space, *Z. Naturforsch.* **39c**:123–125.

MÉTRAUX, J. P., AND BOLLER, T. 1986. Local and systemic induction of chitinase in cucumber plants in response to viral, bacterial, and fungal infections, *Physiol. Plant. Pathol.*, in press.

MOHNEN, D., SHINSHI, H., FELIX, G., AND MEINS, F. 1985. Hormonal regulation of β-1,3-glucanase messenger RNA levels in cultured tobacco tissues. *EMBO J.* **4**:1631–1635.

MOLANO, J., POLACHECK, I., DURAN, A., AND CABIB, E. 1979. An endochitinase from wheat germ, *J. Biol. Chem.* **254**:4901–4907.

MOORE, A. E., AND STONE, B. A. 1972a. A β-1,3-glucan hydrolase from *Nicotiana glutinosa* leaves. I. Extraction, purification and physical properties, *Biochim. Biophys. Acta* **258**:238–247.

MOORE, A. E., AND STONE B. A. 1972b. Effect of infection with TMV and other viruses on the level of a β-1,3-glucan hydrolase in leaves of *Nicotiana glutinosa*, *Virology* **50**:791–798.

MOORE, A. E., AND STONE, B. A. 1972c. Effect of senescence and hormone treatment on the activity of a β-1,3-glucan hydrolase in *Nicotiana glutinosa* leaves, *Planta* **104**:93–109.

NETZER, D., KRITZMAN, G., AND CHET, I. 1979. β-(1,3)-Glucanase activity and quantity of fungus in relation to *Fusarium* wilt in resistant and susceptible near-isogenic lines of muskmelon, *Physiol. Plant Pathol.* **14**:47–55.

NITZSCHE, W. 1983. Chitinase as a possible resistance factor for higher plants, *Theor, Appl. Genet.* **65**:171–172.

OLSNES, S., REFSNES, K. AND PIHL, A. 1974. Mechanism of action of the toxic lectins abrin and ricin, *Nature* **249**:627–631.

OLSNES, S., FERNANDEZ-PUENTES, C., CARRASCO, L., AND VAZQUEZ, D. 1975. Ribosome inactivation by the toxic lectins abrin and ricin: Kinetics of the enzymic activity of the toxin A-chains, *Eur. J. Biochem.* **60**:281–288.

PAUS, E., AND STEEN, H. B. 1978. Mitogenic effect of α-mannosidase on lymphocytes, *Nature* **272**:452–454.

PEARCE, R. B., AND RIDE, J. P. 1982. Chitin and related compounds as elicitors of the lignification response in wounded wheat leaves, *Physiol. Plant Pathol.* **20**:119–123.

PEBERDY, J. F. 1979. Fungal protoplasts: Isolation, reversion, and fusion, *Annu. Rev. Microbiol.* **33**:21–39.

PEGG, G. F. 1977. Glucanohydrolases of higher plants: A possible defence mechanism against parasitic fungi, in: *Cell Wall Biochemistry Related to Specificity in Host-Plant Pathogen Interactions* (B. Solheim and J. Raa, eds.), pp. 305–342, Universitaesforlaget, Tromsø, Sweden.

PEGG, G. F., AND VESSEY, J. C. 1973. Chitinase activity in *Lycopersicon esculentum* and its relationship to the *in vivo* lysis of *Verticillium albo-atrum* mycelium, *Physiol. Plant Pathol.* **3:**207–222.

PEGG, G. F., AND YOUNG, D. H. 1981. Changes in glycosidase activity and their relationship to fungal colonization during infection of tomato by *Verticillium albo-atrum*, *Physiol. Plant Pathol.* **19:**371–382.

PEGG, G. F., AND YOUNG, D. H. 1982. Purification and characterization of chitinase enzymes from healthy and *Verticillium albo-atrum*-infected tomato plants, and from *Verticillium alo-atrum*, *Physiol. Plant Pathol.* **21:**389–409.

PISTOLE, T. G. 1981. Interaction of bacteria and fungi with lectins and lectin-like substances, *Annu. Rev. Microbiol.* **35:**85–112.

PITT, D. 1973. Solubilisation of molecular forms of lysosomal acid phosphatase of *Solanum tuberosum* L. leaves during infection by *Phytophthora infestans* (Mont.) de Bary, *J. Gen. Microbiol.* **77:**117–125.

PITT, D., AND STEWART, P. 1980. Sub-cellular changes in the lytic compartment and the distribution of acid phosphatase and ribonuclease during infection of pea roots by *Phoma medicaginis* var. *pinodella*, *Trans. Brit. Mycol. Soc.* **74:**343–356.

POWNING, R. F., AND DAVIDSON, W. J. 1979. Studies on insect bacteriolytic enzymes. III. Lytic activities in some plant materials of possible benefit to insects, *Comp. Biochem. Physiol.* **63B:**199–206.

POWNING, R. F., AND IRZYKIEWICZ, H. 1965. Studies on the chitinase system in bean and other seeds, *Comp. Biochem. Physiol.* **14:**127–133.

RABENANTOANDRO, Y., AURIOL, P., AND TOUZE, A. 1976. Implication of β-(1,3)-glucanase in melon anthracnose, *Physiol. Plant Pathol.* **8:**313–324.

RACUSEN, D. 1984. Lipid acyl hydrolase of patatin, *Can. J. Bot.* **62:**1640–1644.

READY, M., BIRD, S., ROTHE, G., AND ROBERTUS, J. D. 1983. Requirements for antiribosomal activity of pokeweed antiviral protein, *Biochim. Biophys. Acta* **740:**19–28.

ROSENTHAL, G. A., AND JANZEN, D. H. (eds.) 1979. *Herbivores: Their Interaction with Secondary Plant Metabolites,* Academic Press, New York.

RYAN, C. A. 1984. Systemic responses to wounding, in: *Plant-Microbe Interactions,* Vol. 1 (T. Kosuge and E. W. Nester, eds.), pp. 307–320, Macmillan, New York.

SACHER, J. A., ENGSTROM, D., BROOMFIELD, D. 1979. Ethylene regulation of wound-induced ribonuclease in turnip root tissue, *Planta* **144:**413–418.

SACHER, J. A., TSENG, J., WILLIAMS, R., AND CABELLO, A. 1982. Wound-induced RNAse activity in sweet potato, *Plant Physiol.* **69:**1060–1065.

SCHLÖSSER, E. W. 1980. Preformed internal chemical defenses, in: *Plant Disease: An advanced Treatise* (J. G. Horsfall and E. B. Cowling, eds.), Vol. 5, pp. 161–177, Academic Press, New York.

SEQUEIRA, L. 1978. Lectins and their role in host-pathogen specificity, *Annu. Rev. Phytopathol.* **16**:453–481.

SEXTON, R., AND ROBERTS, J. A. 1982. Cell biology of abscission, *Annu. Rev. Plant Physiol.* **33**:133–162.

SHARP, J. K., McNEIL, M., AND ALBERSHEIM, P. 1984. The primary structures of one elicitor-active and seven elicitor-inactive hexa(β-D-glucopyranosyl)-D-glucitols isolated from the mycelial walls of *Phytophthora megasperma* f.sp. *glycinea*, *J. Biol. Chem.* **259**:11321–11336.

SHINSHI, H., AND KATO, K. 1983. Physical and chemical properties of β-1,3-glucanase from cultured tobacco cells, *Agric. Biol. Chem.* **47**:1455–1460.

SIETSMA, J. H., AND DeBOER, W. R. 1973. Formation and regeneration of protoplasts from *Pythium* PRL 2142, *J. Gen. Microbiol.* **74**:211–217.

SOLHEIM, B., AND FJELLHEIM, K. E. 1984. Rhizobial polysaccharide-degrading enzymes from roots of legumes, *Physiol. Plant.* **62**:11–17.

STIRPE, F., GASPERI-CAMPANI, A., BARBIERI, L., FALASCA, A., ABBONDANZA, A., AND STEVENS, W. A. 1983. Ribosome-inactivating proteins from the seeds of *Saponaria officinalis* L. (soapwort), of *Agrostemma githago* L. (corn cockle) and of *Asparagus officinalis* L. (asparagus), and from the latex of *Hura crepitans* L. (sandbox tree), *Biochem. J.* **216**:617–625.

STOESSL, A. 1983. Secondary plant metabolites in preinfectional and postinfectional resistance, in: *The Dynamics of Host Defense* (J. A. Bailey and B. J. Deverall, eds.), pp. 71–122, Academic Press, New York.

STREETER, J. G. 1982. Enzymes of sucrose, maltose, and α,α-trehalose catabolism in soybean root nodules, *Planta* **155**:112–115.

TATA, S. J., BEINTEMA, J. J., AND BALABASKARAN, S. 1983. The lysozyme of *Hevea brasiliensis* latex: Isolation, purification, enzyme kinetics and a partial amino acid sequence, *J. Rubber Res. Inst. Malays.* **31**:35–48.

TSUKAMOTO, T., KOGA, D., IDE, A., ISHIBASHI, T., HORINO-MATSUSHIGE, M., YAGISHITA, K., AND IMOTO, T. 1984. Purification and some properties of chitinases from yam, *Dioscorea opposita* Thumb, *Agric. Biol. Chem.* **48**:931–939.

VAN DER WILDEN, W., AND CHRISPEELS, M. J. 1983. Characterization of the isoenzymes of α-mannosidase located in the cell wall, protein bodies, ad endoplasmatic reticulum of *Phaseolus vulgaris* cotyledons, *Plant Physiol.* **71**:82–87.

VAN LOON, L. C. 1982. Regulation of changes in proteins and enzymes associated with active defense against virus infection, in: *Active Defense Mechanisms in Plants* (R. K. S. Wood, ed.), pp. 247–273, Plenum Press, New York.

VELUTHAMBI, K., MAHADEVAN, S., AND MAHESHWARE, R. 1981. Trehalose toxicity in *Cuscuta reflexa*: Correlation with low trehalase activity, *Plant Physiol.* **68**:1369–1374.

WADSWORTH, S. A., AND ZIKAKIS, J. P. 1984. Chitinase from soybean seeds: Purification and some properties of the enzyme system, *J. Agric. Food Chem.* **32**:1284–1288.

WARGO, P. M. 1975. Lysis of the cell wall of *Armillaria mellea* by enzymes from forest trees, *Physiol. Plant Pathol.* **5**:99–105.

WESSELS, J. G. H., AND SIETSMA, J. H. 1981. Fungal cell walls: A survey, in: *Encyclopedia of Plant Physiology,* New series, Vol. 13B, *Plant Carbohydrates II* (W. Tanner and F. A. Loewus, eds.), pp. 352–394, Springer-Verlag, New York.

WHITE, A. R., DARVILL, A. G., YORK, W. S., AND ALBERSHEIM, P. 1984. High-performance gel permeation chromatography assay for endoglycanase activities, *J. Chromatogr.* **298**:525–530.

WILSON, C. L. 1973. A lysosomal concept for plant pathology, *Annu. Rev. Phytopathol.* **11**:247–272.

WILSON, C. M. 1975. Plant nucleases, *Annu. Rev. Plant Physiol.* **26**:187–208.

WONG, Y. S., AND MACHLACHLAN, G. A. 1979a. 1,3-β-D-Glucanases from *Pisum sativum* seedlings. I. Isolation and purification, *Biochim. Biophys. Acta* **571**:244–255.

WONG, Y. S., AND MACLACHLAN, G. A. 1979b. 1,3-β-D-Glucanases from *Pisum sativum* seedlings. II. Substrate specificities and enzymic action patterns, *Biochim. Biophys. Acta* **571**:256–269.

WONG, Y. S., AND MACLACHLAN, G. A. 1980. 1,3-β-D-Glucanases from *Pisum sativum* seedlings. III. Development and distribution of endogenous substrates, *Plant Physiol.* **65**:222–228.

YOSHIKAWA, M. 1983. Macromolecules, recognition, and the triggering of resistance, in: *Biochemical Plant Pathology* (J. E. Callow, ed.), pp. 267–298, John Wiley, Chichester, England.

YOSHIKAWA, M., MATAMA, M., AND MASAGO, H. 1981. Release of a soluble phytoalexin elicitor from mycelial walls of *Phytophthora megasperma* var. *sojae* by soybean tissues, *Plant Physiol.* **67**:1032–1035.

YOUNG, D. H., AND PEGG, G. F. 1981. Purification and characterization of 1,3-β-glucan hydrolases from healthy and *Verticillium albo-atrum* infected tomato plants, *Physiol. Plant Pathol.* **19**:391–417.

YOUNG, D. H., AND PEGG, G. F. 1982. The action of tomato and *Verticillium albo-atrum* glycosidases on the hyphal wall of *V. albo-atrum, Physiol. Plant Pathol.* **21**:411–423.

Chapter 14

Plant Responses to Stresses at the Molecular Level

David N. Kuhn

STRESS HAS BEEN DEFINED AS "a physical or chemical factor that tends to alter an existent equilibrium and may be a factor in disease causation" (Mish, 1984). This chapter will discuss stresses that reduce a plant's nutritional and medicinal usefulness to humans. Reductions in yield can have a variety of causes. Death of the plant is an obvious cause, as is reduced growth, delayed growth, delayed maturity, or failure to produce the desired compound. Yield as a measure of stress is too crude to allow determination of which single stress (or combination of stresses) is most important. A prioritization of stresses follows the same hierarchy as the plant's growth requirements: light, temperature, water, soil composition, and nutritional factors. Extremes of any of these factors may cause a decrease in the growth of the plant. Other important stresses are disease, wounding, insect damage, and toxic contaminants of the air, water, and soil.

To deal with stress, plants have evolved a number of different strategies ranging from gross morphological changes at the whole-plant level to inducible cellular responses (Hanson and Grumet, 1985) (Fig. 1). Where a particular stress is a dominant and persistent factor (e.g., lack of water in the desert), only plants especially adapted for growth under the stress conditions can survive. Although these adaptations permit growth, they are generally multigenic and involve stable morphological changes. Thus they are not adaptations that one would introduce into an agronomic crop to improve its performance under stress. As depicted in Fig. 1, there is an inverse relationship between the ability to alter a plant by genetic engineering and the level of organization at which the stress response occurs. A further complication lies in our limited ability to determine how molecular changes brought about by genetic engineering will affect the performance of the plant in the field. These relationships should be kept in mind when evaluating the utility of a genetic engineering approach to producing stress-tolerant or stress-resistant plants. The focus of this chapter will be on active plant stress responses that may prove of interest in the genetic engineering of crop plants resulting in higher-yielding plants or plants capable of yielding economically in marginal regions or under stressful

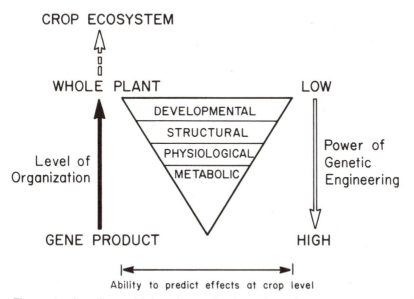

Figure 1. Levels at which traits conferring adaptation to stress-prone environment can be expressed. (With permission of Dr. Andrew D. Hanson, MSU-DOE Plant Research Laboratory, Michigan State University, East Lansing, MI 48824.)

conditions. The active response is of greater interest than morphological adaptation for the following reasons:

1. Morphological adaptations and gross anatomy changes are usually polygenic. It may be difficult to isolate all the genes necessary for the adaptation and to transfer them to an agronomically useful crop.
2. Morphological adaptations to a stress are more likely to decrease the potential yield of a crop. Producing a salt-or drought-tolerant corn cultivar might allow *some* production of maize in salty or dry areas, but such cultivars will not yield as well as a salt-sensitive cultivar under nonstress conditions.
3. Morphological adaptations cannot be reversed when the stress is only temporary. If the stress is not constant, no farmer will plant a lower-yielding stress-tolerant crop. Despite the threat of temporary drought or disease, a farmer plants the potentially highest-yielding cultivar with the hope that conditions will allow the production of a profitable crop.
4. Active responses are oligogenic and may be controlled by single genes. These are desirable systems for transfer from one plant genus to another.
5. Active responses are triggered only when the stress is present. The cost of an active response in terms of yield is much less than that of an adaptive response.
6. Active responses may involve increased or *de novo* expression of genes. The isolation of such genes may provide DNA sequences or transcription activation factors that can be used to control the timing of the response of other genes that have been transformed into a plant.
7. By potentiating active responses with chemical elicitors, it may be possible to provide an environmentally safe method to protect crops from temporary stress.

With these qualifications in mind, what types of stresses and responses are of interest? Of greatest interest are stresses such as drought and disease that occur intermittently and cause the greatest reduction in yield. There are sometimes no economical solutions for such stress problems. As an example, although drought stress can be overcome by irrigation, the expense of an irrigation system in an area such as the Midwest where irrigation is uncommon is prohibitive. Maintenance of such a system for intermittent periods of drought is currently too expensive. Some crops are completely susceptible to particular diseases. There are, for example, no known citrus species resistant to the bacterium causing citrus canker. Breeding for resistance in this case is not yet possible, and control of the disease through chemical means is not effective. Present control of the disease is by destruction of diseased trees.

Problems involving other stresses of interest are chronic or are expensive to solve by present methods. The clearest example of a chronic stress is the removal of macronutrients such as nitrogen, phosphorus, and potassium from the soil. Farmers can overcome this by the use of fertilizer, but fertilizer and fertilizer application are expensive. There has been long-term interest in developing symbioses for plants to increase nitrogen uptake (*Rhizobium* sp.) and phosphorus uptake (mycorrhizae). Another chronic stress is salt accumulation in soil due to irrigation. Salt accumulation can be reduced by flooding the fields, which enhances leaching. However, in areas requiring irrigation, water is expensive, and plants that can grow in the presence of higher concentrations of salt are being sought.

What types of responses are of the greatest interest? Induced responses that involve an increase in expression of a single gene or a few genes are the most desirable. It may be possible to transfer the entire response by transformation of a plant with these genes. Thus resistance to a stress can be transferred between plants that cannot be bred. If the induced response involves too many genes for transfer or if the response is ubiquitous in plants, it may still be of interest. Such responses can be studied to determine how genes are controlled in plants, what the stress signal receptors are, and how the signal reaches the nucleus. The study of complex induced responses may lead to the isolation of DNA sequences required for the induction of transcription (cis elements). These elements could then be used to control the expression of foreign genes transformed into a plant or plant cell.

EVIDENCE FOR GENE ACTIVATION IN STRESSED PLANTS

It is easy to determine if a stress is intermittent, but much more difficult to determine if a response requires *de novo* transcription of a gene. Some common lines of evidence used to establish that gene activation is required are:

Inhibitor Studies

The plant is treated with a protein or RNA synthesis inhibitor before being challenged with the stress (Yoshikawa et al., 1978). If *de novo* protein or RNA synthesis is required, a resistant plant will display susceptibility. For such evidence to be convincing, the indication of susceptibility in the resistant plant must appear before any symptoms are seen in the control which is treated with the inhibitor alone.

Measurement of Enzyme Activity, de novo Synthesis of an Enzyme, or in vitro Translation Activity of the mRNA Encoding the Enzyme

If a particular enzyme activity appears or increases after the plant has been stressed, the increase may be due to *de novo* synthesis of the protein or to activation of a preexisting protein. To determine if *de novo* synthesis has occurred usually requires having an antibody for the protein (Boerner and Grisebach, 1982). The amount of protein in the plant before and after the stress can then be measured by radioimmunoassay or by Western blotting. If a radioactively labeled amino acid is applied to the plant along with the stress, the amount of *de novo* synthesis can be determined by sodium dodecyl sulfate–polyacrylamide gel (SDS-PAGE) analysis of the immunoprecipitate. The radioactively labeled protein is either cut from the gel, and radioactivity is determined by liquid scintillation counting or detected by fluorography and quantitated by scanning densitometry of the fluorogram. Although *de novo* synthesis of a protein in response to stress is good evidence for an increase in transcription of the gene for that protein, it does not rule out the activation of a preexisting mRNA.

The amount of mRNA present for a particular protein can be determined by in vitro translation of isolated mRNA in a heterologous system such as a rabbit reticulocyte lysate and quantitation of the immunoprecipitate of the total translation products by SDS-PAGE as described above for *in vivo* labeled proteins (Ragg et al., 1981). Total RNA is isolated before and after the stress and translated in a heterologous system such as rabbit reticulocyte lysate that most likely will not recognize plant translational control factors. In trying to establish that there is an increase in the amount of a particular species of mRNA rather than a shift of preexisting mRNA from an inactive to an active state it is important to use total RNA rather than poly A^+ RNA or polysomal RNA (RNA isolated from polysomes). Similarly, the use of a translation system likely to translate mRNA that is inactive in the plant is also important. An increase in the amount of translatable mRNA for an enzyme in response to stress is good evidence for gene activation, but decreased degradation of a particular mRNA species cannot be ruled out as the cause of the observed accumulation of mRNA.

Measurement of the Amount of mRNA by Hybridization with Complementary cDNA or a Gene

Quantitation of a particular species of mRNA by hybridization with radioactively labeled DNA is better proof for the specific increase in mRNA in response to a stress. This method avoids the question of the translational activity of a particular mRNA. Even partially degraded RNA can be detected. Quantitation is accomplished by separation of the total RNA by electrophoresis on denaturing gels, transfer to nitrocellulose, and hybridization with radioactively labeled DNA (Northern analysis) (Kuhn et al., 1984). Hybridizing species are detected by autoradiography and quantitated by scanning densitometry of the autoradiogram. Alternatively, the total RNA can be immobilized without separation by electrophoresis (dot or slot blot) and relative hybridization quantitated as above. The Northern analysis should be performed before the dot blots, so that the molecular weight of the hybridizing species can be estimated and the specificity of the probe determined. Nonetheless, an increase in mRNA amount as determined by hybridization does not rule out the possibility that the increase was due to a decrease in the degradation rate of the particular mRNA (posttranscriptional control).

Measuring New Transcripts by Runoff Experiments or by in vivo Labeling of RNA

To positively demonstrate that gene transcription has increased in response to a stress, one must demonstrate that the increase in a particular species of mRNA is due to *de novo* synthesis of that RNA. The mRNA being transcribed can be measured by isolating nuclei and supplying them with radioactively labeled ribonucleotides. Normally, no new transcripts will be initiated after nuclei are isolated, but the DNA-dependent RNA polymerase II will not dissociate from the chromatin and will finish transcription of the genes using the labeled ribonucleotides. When transcription is finished, the polymerase does not reinitiate, hence this type of transcription assay is called a runoff assay (Chappell and Hahlbrock, 1984). A runoff assay indicates which genes were being actively transcribed when the nuclei were isolated. If the runoff assay shows an increase in the labeling of a particular mRNA after the stress, this is conclusive evidence that the plant's response involves transcriptional activation of a particular gene. Alternatively, labeled uridine (which specifically labels RNA) can be introduced into the plant at the time of the stress to distinguish by in vivo labeling newly synthesized mRNA from previously synthesized mRNA (Cramer, et al, 1985). This also allows calculation of the rate of degradation of RNA, as a decrease in the rate of degradation is a potential cause of an increase in the amount of mRNA.

INTERACTIONS OF PLANTS WITH STRESSES

A Model for the Interaction of Plants with a Stressor

Once it has been established that a stress response involves the initiation of transcription of particular genes, the following questions are of interest:

1. How does a plant cell recognize stress?
2. What type of signal is sent to the nucleus?
3. How specific is the signal sent to the nucleus?

The following model (Fig. 2) is useful in discussing the interaction of a plant cell with a stressor. The model is borrowed from a description of the steps involved in activation of the transcription of genes in mammalian cells. The basic elements of the model are:

1. A signal compound interacts with a receptor molecule. This receptor is probably a protein and is located in the cytoplasm (Yamamoto et al., 1976) or the plasma membrane (Wahli, et al., 1981).
2. Binding of the signal molecule to the receptor causes a change in the activity or structure of the receptor or of a factor associated with the receptor. An example is the binding of estrogen to the estrogen receptor in the cytoplasm of chicken oviduct cells, which leads to transport of the receptor-hormone complex to the nucleus (Yamamoto et al., 1976). An example common to plants is the change in phytochrome (a receptor) in response to light (a signal compound) that leads to activation of the transcription of a number of nucleus-encoded genes such as the small subunit of ribulose-1,5-bisphosphate carboxylase (Morelli et al., 1985).
3. The modified receptor goes to the nucleus and interacts with a sequence of DNA close to the gene to be activated. This DNA sequence is known as a *cis element* and may be 5' or upstream of the gene, in an intron of the gene, or 3' or downstream from the gene (an enhancer element). The cis elements involved in the control of genes by glucocorticoids, heavy metals (Yagle and Palmiter, 1985), and interferon are found 5' of the gene to be activated. Activation of the gene may also take place by interaction of the modified receptor with another transcription factor (trans-acting factor) that then binds to the DNA near the gene (Dynan and Tjian, 1985). The binding of the modified receptor may displace a factor that is repressing transcription; this is a common regulatory mechanism in prokaryotes. There is evidence suggesting that this mechanism may explain the activation of interferon genes.

Another possibility is that the modified receptor does not cause *de novo* transcription of a gene but rather affects the turnover rate or stability of particular mRNAs, thus increasing mRNA amounts posttranscriptionally.

Some points to consider in testing this model for validity with regard to stress responses in plants are:

1. How specific are the receptors? Are there receptors that recognize more than one stressor?
2. How specific is the induction of gene expression? Is a single gene affected? If there is more than one copy of a gene, are all copies equally affected? Is a group of genes affected? Are there general stress response genes? A general stress response gene is one whose transcription increases in response to a wide variety of semmingly unrelated stresses.
3. What is the primary response of the plant? A number of genes may be activated in response to the stress. There may be gross physiological changes. How can one distinguish what is due primarily to the stress and what is a secondary effect, i.e., a response to the response? For example, production of the regulatory protein in the model in Fig. 2 would be the initial response of the plant. A secondary response would be the effects of that protein in initiating new transcription or modulating a metabolic pathway. A classic example of this problem is in interpretation of the primary action of auxin on plant cells. Does it act first to cause cell extension, which then stimulates transcription of genes? Or does it act first in the nucleus to stimulate transcription, which causes cell extension? A reasonable way to approach this complex problem is to determine what the plant cell does first in response to a stress and attempt to determine a logical sequence of stress response events. When possible, the time course of the stress response should be described.

In discussing the various stresses and responses, the following basic outline of questions will be used:

1. What is the stress? What is the response?
2. Does the response require *de novo* transcription of genes? What is the evidence for this conclusion?
3. How many genes are involved? How soon after the stress is transcription activated
4. What is the signal that turns on the genes?
5. Is this response useful in protecting normally susceptible plants? Does it have genetic engineering uses?

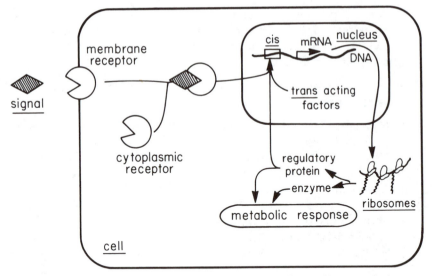

Figure 2.   Schematic of the activation of mammalian cell transcription.

Disease Stress

Diseases of plants are caused by deleterious interactions with fungi, bacteria, or viruses. Such interactions lead to a reduced yield or to death of the plant. In general, living plants are resistant to the growth of microorganisms, so that the exceptional case is one in which a microorganism can infect and cause disease in a living plant. This basic resistance to disease can take many forms, such as thickening of the cuticle layer and lignification of cell walls to prevent pathogen invasion. Such physical barriers are constitutive defenses, characteristics of the plant that are expressed whether the plant is challenged by a pathogen or not. Some plants also have inducible defense systems (Sequeira, 1983). One of the best understood of these inducible systems is the biosynthesis of phytoalexins, antimicrobial compounds, in response to infection (Bailey and Mansfield, 1982). Such phytoalexin compounds and at least some of the enzymes that produce them are not present in the plant cell before it is inoculated with the pathogen. There are two further aspects of this response that are of interest in a discussion of the induction of gene expression.

First, plants that produce phytoalexins may also use these compounds in race-specific resistance. Race-specific resistance can be thought to be a special subset or fine-tuning of the basic disease resistance where a cultivar is resistant to certain isolates of a particular pathogen (each isolate with a different virulence phenotype is considered a physiological race) and susceptible to other isolates. The isolates of the pathogen are morphologically indistinguishable, and physiological races are typically defined by their ability to grow on a defined set of cultivars of the host (differential cultivars). Resistance to certain races can be inherited as a single-gene dominant trait in the host, with resistance genes occurring at a number of different loci. Differential cultivars usually contain a single resistance gene, so that genes conferring resistance on a new race of the pathogen can be easily determined. In the best studied case, Flor's (1956) study of flax and flax rust, a one-to-one correspondence was observed for resistance genes in the host and pathogenicity genes in the pathogen. Such detailed genetics for both the host and the pathogen have been obtained for only a few of the systems presently being studied.

It has been observed in soybean that a cultivar can be susceptible to one race of a pathogen and resistant to another. Since in both cases the soybean plant produces the phytoalexin glyceollin (Keen and Yoshikawa, 1983), the difference in susceptibility does not appear to be in a structural gene for an enzyme in the biosynthetic pathway of the phytoalexin. Although there is still controversy over the amount and location of the glyceollin that accumulates in the compatible (susceptible) and incompatible (resistant) interactions, the usual observation is that glyceollin accumulation is greater in the incompatible case than in the compatible case (Hahn et al., 1985). Since a single gene is thought to determine resistance to a particular race, it has been argued that the product of this gene is involved either in recognition of the specific pathogen race or in regulation of phytoalexin biosynthesis. Studying race-specific resistance should provide a better understanding of the exact mechanism of resistance and its regulation.

Second, it has been observed that treatments other than inoculation with a pathogen can cause the host plant to produce phytoalexins (Ayers et al., 1976). The substances used in such treatments are called elicitors and range from heavy metals to the cell walls of pathogens or even isolated cell wall fragments of the host itself. In direct contrast to the pathogen races that are very specific in their interaction with the host, an elicitor usually shows no race specificity (although there are claims for race-specific elicitors). The elicitor always induces a resistant response (an incompatible interaction). The plant cell does not respond to the elicitor by showing the symptoms of the disease (a compatible interaction). Elicitors have proved very useful in inducing disease resistance responses in tissue culture cells. In some cases, the disease resistance response in plants is induced only in cells that are in direct contact with the invading microorganism. In such cases, it can be difficult to obtain sufficient induced tissue for the isolation of protein or mRNA. Treatment of suspension culture cells with an elicitor ensures that almost all the cells are responding in a synchronous manner, making the response much easier to characterize. Elicitor treatment of tissue culture cells may also provide a method for isolating the receptors involved in pathogen recognition.

Inducible disease resistance responses and the role of recognition in triggering such responses have been reviewed (Sequeira, 1983; Daly, 1984; Keen, 1982). I will discuss four of these host-pathogen interactions for which the response has been studied at the molecular level.

Colletotrichum lindemuthianum–Phaseolus vulgaris interaction

Coletotrichum lindemuthianum is a fungus that causes anthracnose of bean (Bell et al., 1984). Conidia (asexual spores) of the fungus are blown or splashed onto the leaves and stems of the bean where they germinate and penetrate the plant. This infection process takes about 48 hr. In the incompatible interaction, plants show a hypersensitive response, and small necrotic flecks are observed at the site of inoculation after about 60 hr. In the compatible interaction, no hypersensitive response is seen, and spreading lesions are observed on the plant about 140 hr after inoculation (Fig. 3).

In both the compatible and incompatible interactions, the bean hypocotyls respond by producing a number of isoflavonoid phytoalexins, including kievitone (Fig. 4). The enzyme chalcone synthase (CHS) is required for the synthesis of kievitone. Phenylalanine ammonia lyase (PAL) is also required for the production of all bean phytoalexins. The induction of chalcone synthase has been studied in hypocotyls infected with *C. lindemuthianum* spores and in suspension culture cells treated with an elicitor made from *C. lindemuthianum* cell walls (Ryder et al., 1984).

In the spore-hypocotyl interaction, chalcone synthase enzyme activity increases about 60 hr after inoculation in the incompatible case and about 140 hr after inoculation in the compatible case (Fig. 3). The onset of the increase in CHS mRNA translational activity as measured by *in vitro* translation follows the same pattern, as does the increase in the CHS mRNA amount as measured by hybridization with a *P. vulgaris* CHS cDNA. No information is as yet available on *de novo* transcription in hypocotyls in response to infection. However, *de novo* transcription of CHS genes has been observed in *P. vulgaris* suspension culture cells treated with an elicitor (Cramer et al., 1985). This suggests that *de novo* transcription of the CHS genes is involved in the defense response in hypocotyls.

Because the response of hypocotyls to inoculation occurs over such a long period of time, it is difficult to determine much about the signal compound or the plant's primary response to the signal. The response involves more genes than chalcone synthase, including genes for other enzymes involved in phytoalexin biosynthesis and for proteins of unknown function. Recently, the increase in mRNA for extensin, the hydroxy-

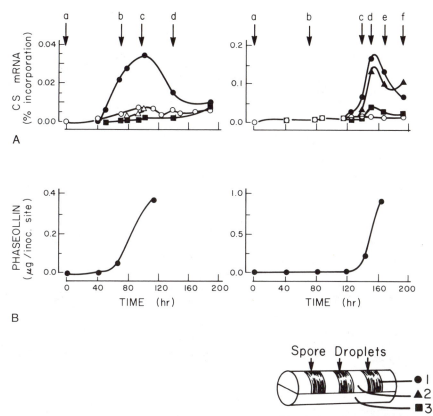

C   No Infection Control ○

Figure 3. Chalcone synthase mRNA activity in relation to phaseollin accumulation expression of hypersensitive resistance during an incompatible interaction (host-resistant) between hypocotyls of *P. vulgaris* cultivar Keivitsboon Koekoek and *C. lindemuthianum* race β. (A) Arrows denote events in expression of hypersensitive resistance at site 1: a, spore inoculation; b, onset of hypersensitive flecking in a few sites; c, hypersensitive flecking apparent at most sites; and d, very dense brown flecking at all sites; no visible changes in sites 2 through 4 throughout the time course. (B) Lesion development during a compatible interaction (host-susceptible) between hypocotyls of *P. vulgaris* cultivar Keivitsboon Koekoek and *C. lindemuthianum* race γ. Arrows denote events in lesion development at site 1: a, spore inoculation; b, no visible symptoms (compare incompatible interaction); c, onset of symptoms development at a few sites; e, onset of water soaking and spreading of lesions from site 1, some browning at site 2. In both (A) and (B) mRNA activity was measured in directly infected tissue (site 1, ●); in tissue laterally adjacent to the infected tissue (site 2, △); in tissue beneath sites 1 and 2 (site 3, ■); and in equivalent control, uninoculated hypocotyls (site 4, ○), as shown in (C). (From Bell et al., 1984, with permission of Dr. Chris J. Lamb, Plant Biology Laboratory, Salk Institute for Biological Sciences, La Jolla, CA.)

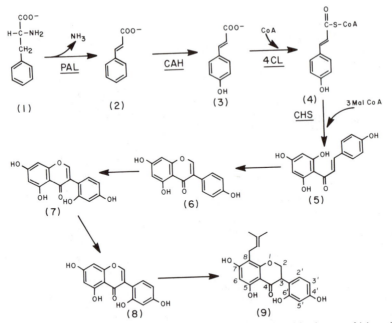

Figure 4. A proposed biosynthetic pathway for kievitone. Abbreviations: PAL, phenylalanine ammonia-lyase; CHS, chalcone synthase; CAH, cinnamic acid hydroxylase; MalCoA, malonyl Coenzyme A; 4 CL, 4 coumaroy: CoA ligase. (From Dixon et al., 1983.)

proline-rich glycoprotein involved in cell wall synthesis, has been measured in bean hypocotyls inoculated with *C. lindemuthianum* (Showalter, et al., 1985). The comparison of relative mRNA amounts for CHS and extensin highlights the complexity of the resistance response. There is clearly a difference in the timing of the increase in extensin mRNA and CHS mRNA, as well as in the decrease in hybridizable mRNA. Extensin mRNA appears to increase in sites away from the site of inoculation, so that a separate signal may regulate the amount of extensin mRNA. The response of *P. vulgaris* suspension culture cells to an elicitor is much more rapid and involves a greater number of cells. Experiments with suspension culture cells (Ryder et al., 1984; Cramer et al., 1985) have allowed a more definite determination of the regulation of *de novo* transcription and the number of genes involved in the defense response.

The amounts of mRNA for CHS and PAL in *P. vulgaris* suspension culture cells increase within 2 hr of treatment of the cells with an elicitor made from *C. lindemuthianum* cell walls. The amounts of translatable mRNA for these enzymes reach a maximum 4 hr after elicitor treatment. *De novo* RNA synthesis has been measured by incubating the cells with 4-thiouridine after elicitor treatment. Newly synthesized RNA incorporating thiouridine was separated from previously synthesized RNA by organomercurial affinity chromatography. Since no significant change in the CHS mRNA degradation rate was observed, CHS mRNA increases are probably due to *de novo* transcription. Such measurements were not possible for PAL, as hybridization probes were not yet available. Newly synthesized RNA from elicitor-treated and untreated cells was translated *in vitro*, and the translation products were analyzed by two-dimensional gel electrophoresis. The patterns of proteins for elicitor-treated and untreated cells were compared. In elicitor-treated cells, 50 proteins increased in amount and 10 proteins decreased in amount when compared with the untreated cells. These figures provide the limits for the response in suspension culture cells and are the only available estimate of the extent of the hypocotyl reaction. It is not known how similar the responses of suspension culture cells to elicitors and hypocotyls to living spores are. The hypocotyls response is race-specific, while the cell response to an elicitor is not. This suggests that control of the response, if not the number of genes activated, is different in the two interactions. There is not yet enough information to determine if different signals or receptors are being used in the two interactions.

Phytophthora megasperma f.sp. glycinea–Soybean Interaction

Phytophthora megasperma f.sp. *glycinea* (Pmg) is a soil-borne fungus that infects the roots of soybeans and causes root rot (Paxton, 1983). Twenty-four races of Pmg have been identified. Six loci for Pmg resistance have been identified in soybean. Soybeans respond to inoculation with Pmg by producing a mixture of isoflavonoid phytoalexins. The most toxic of these is glyceollin which exists in several isomeric forms. Phenylalanine ammonia lyase catalyzes an early step in the glyceollin and kievitone biosynthetic pathway (Fig. 4). Chalcone synthase may be involved in the biosynthesis of glyceollin, but its role is uncertain (Dixon et al., 1983). The product of the chalcone synthase reaction has a hydroxyl group, at C-5 in the isoflavonone numbering system and at C-1 in a pterocarpan, while the precursor of glyceollin lacks this hydroxyl (Fig. 5). No enzyme has yet been identified that can remove this hydroxyl. Because of the structure of glyceollin, an enzyme with activity similar to that of chalcone synthase must be involved in the synthesis of glyceollin. Chalcone synthase activity increases in response to inoculation with Pmg in a manner similar to PAL activity (Boerner and Grisebach, 1982), and in the experiments discussed below, it may be considered a reasonable indicator of activation of the plant's defense response.

Wounded soybean hypocotyls inoculated with Pmg mycelia produce glyceollin. When hypocotyls are treated with inhibitors of protein or RNA synthesis before inoculation, glyceollin accumulation is decreased and resistant hypocotyls appear susceptible. Increases in PAL and CHS enzyme activity are seen within 4 hr of inoculation in both the compatible and incompatible interactions. However, no difference in the onset or in the rate of increase in PAL and CHS activity is seen during the first 12 hr when the compatible and incompatible interactions are compared (Boerner and Grisebach, 1982). In the *C. lindemuthianum*–bean interaction the onset of the response in the compatible case is 80 hr later than in the incompatible case. The similar time course of the response seen for the compatible and incompatible interactions in wounded soybean hypocotyle may be due to the "unnatural" method of inoculation. When soybean roots are inoculated with Pmg zoospores, a 14-hr difference is observed in the onset of glyceollin accumulation in the compatible and incompatible interactions (Hahn et al., 1985). The translational activity of PAL, 4-coumaroyl-CoA ligase (4CL), and CHS mRNA and the amount of CHS mRNA have been measured in the incompatible Pmg-hypocotyl interaction (Schmelzer et al., 1984). The translational activity of all three mRNAs increases within 3 hr of inoculation and reaches a maximum of 6 hr with no significant difference in the rate or the onset of induction of the three mRNAs. Changes in the amount of CHS mRNA as measured by hybridization with a parsley CHS cDNA are similar to the changes observed in CHS mRNA translational activity. Although no measurement was made of *de novo* synthesized RNA, the increases in CHS mRNA are assumed to be due to increased transcription of the CHS genes. Increases in CHS mRNA in soybean roots inoculated with zoospores have not been reported.

Soybean suspension culture cells respond to treatment with a Pmg elicitor by producing glyceollin. Unlike the situation in the *P. vulgaris*–*C. lindemuthianum* interaction, where the onset of the response is much earlier in suspension culture cells than in hypocotyls, the onset and time course of the response in soybean cultures are similar to those of the hypocotyl response. PAL and CHS mRNA translation activity and the amount of CHS mRNA increase within 2 hr after elicitor treatment of soybean suspension cells (Schmelzer et al., 1984). Gene activation is believed to be the cause of the increase in CHS mRNA. There are no estimates of the number of genes activated during the response.

Fusarium solani f.sp. phaseoli–Pisum sativum Interaction (Riggleman et al., 1985)

Fusarium solani is not a pathogen of pea but can infect pea endocarp tissue if the defense response of the tissue has been inhibited by heat shock. *Fusarium solani* f.sp. *phaseoli* is a pathogen of *P. vulgaris* and is therefore a nonhost pathogen of pea. All cultivars of pea should be resistant to *F. solani*. Therefore, by challenging pea tissue with *F. solani,* one can induce the defense response (an incompatible interaction). By comparing the responding tissue with uninoculated tissue, the number of proteins involved in the defense response can be estimated, and proteins and mRNAs specific to the defense response identified. The response being studied is conceptually similar to the response of suspension culture cells to an elicitor in that no race-specific control of the response is involved.

Pea endocarp tissue responds to nonhost pathogens or elicitors by producing the phytoalexin pisatin (see structure in Fig. 5). PAL is one of 25 proteins whose synthesis increases after inoculation or elicitor treatment. PAL mRNA translational activity increases within 4 hr of inoculation. Seven different cDNAs have been identified that are complementary to mRNAs that increase in amount after inoculation with *F. solani* f.sp. *phaseoli*. All these mRNAs appear to increase within 4 hr after inoculation, with no significant differences in the onset of induction. Pulse labeling of endocarp tissue with [³H] uridine showed that the increase in these mRNAs was at least partially due to *de novo* synthesis, although alteration of the degradation rate could not be ruled out.

Parsley Suspension Culture Cells–Phytophthora megasperma f.sp. glycinea Elicitor

Although Pmg is not a pathogen of parsley, parsley suspension culture cells treated with Pmg elicitor respond by increasing transcription of the genes for PAL and 4C1 (Chappell and Hahlbrock, 1984). PAL and 4C1 are involved in the biosynthetic pathway of furanocoumarins, the phytoalexins produced in parsley. Increased transcription of the PAL and 4C1 genes after elicitor treatment was demonstrated convincingly by measuring transcription in isolated nuclei (runoff transcripts). Increased transcription of the 4C1 gene was observed within 1 hr after the addition of elicitor and reached a maximum at 1.5 hr.

Genetic Engineering Using Disease Resistance Genes

The evidence seems convincing that plants activate genes in response to phytopathogens. The response can involve 25–50 genes which appear to be induced simultaneously between 2 and 4 hr after treatment. The signals (elicitors) that induce subsets of these genes can vary widely in structure. No receptors have been identified, nor is there any evidence suggesting the location of the receptors. None of the genes induced have been isolated or sequenced to determine if there are common regulatory sequences.

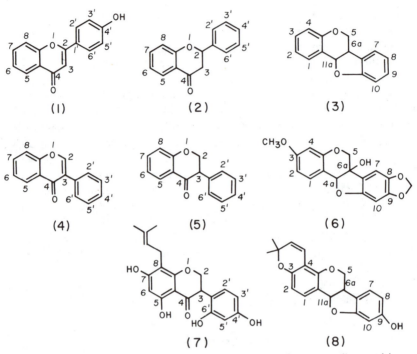

Figure 5. Structures and the numbering system for some flavonoids and phytoalexins of the leguminosae: (1) 4' hydroxy flavone, (2) flavanone, (3) pterocarpan, (4) isoflavone, (5) isoflavanone, (6) Pisatin, (7) Kievitone, (8) Glyceollin 1.

The 25–50 genes involved may not all be specific for disease resistance. Some may be involved in the wounding response, as pathogen invasion and elicitor treatment can cause damage to cells. The basic defense response involving the production of isoflavonoid phytoalexins, which makes legumes resistant to most microorganisms, probably requires the transcription of more than 10 genes. There are possible exceptions, such as the castor bean, where a single enzyme is required for the production of a phytoalexin (Dueber et al., 1978). However, since all plants are resistant to most microorganisms, there is no obvious need to transfer the genes involved in one plant's defense response to another plant to provide resistance. The transfer of genes involved in regulation of the resistance response (race-specific resistance genes) is a more realistic goal in the engineering of a resistant plant. To produce plants with widened resistance, the best strategy may be to use the plant's own resistance response and regulate its activation by the introduction of regulator genes or by artificial activation of the defense response.

PLANT RESPONSES TO ENVIRONMENTAL STRESSES

Heat Shock in Plants

All eukaryotic organisms exposed to high temperatures (40°C or more) respond by reducing the translation of existing mRNA and *de novo* transcription of a group of normally silent genes called heat shock genes. At the elevated temperature, heat shock gene transcripts make up the majority of new transcripts and translated transcripts. Previous activation of heat shock genes can condition an adaptation to higher temperatures, enabling plants to better survive artificially increased temperatures. These mechanisms may be functional in the field as well, as heat shock proteins have been identified in field-grown heat-resistant crops such as cotton. Heat shock mRNAs were also detected in field-grown unirrigated soybeans but not in irrigated soybeans (Key et al., 1985). This suggests that heat shock proteins provide valuable protection to plants that cannot reduce the leaf temperature through transpiration.

All plants are believed to contain heat shock genes, and all plants tested respond to heat shock by activating the transcription of these genes. Plants make two groups of heat shock proteins: high-molecular-weight proteins [68–70, 84, 92 kilodaltons (kd)] that share homology with the high-molecular-weight heat shock proteins of animals and bacteria, and low-molecular-weight proteins (15–27 kd) that are complex and not highly conserved. The heat shock response of soybean has been extensively studied (Key et al., 1985). Soybeans synthesize 20–30 polypeptides of 15–27-kd molecular weight in response to heat shock, in addition to high-molecular-weight proteins. Increases in the amount of heat shock mRNAs can be detected within minutes after increasing the temperature. Higher-molecular-weight heat shock proteins and mRNA are not detectable in unshocked cells, strongly suggesting that the increase in mRNA and protein observed is due to *de novo* transcription and translation. Some lower-molecular-weight proteins and the mRNAs that encode them are present in unshocked tissues. The amounts of mRNA and protein for these polypeptides increase dramatically after heat shock. One of these proteins (27 kd) increases in response to a wide variety of stresses (Czarnecka et al., 1984).

The transcription of heat shock genes rapidly decreases when plants are returned to a lower temperature (30°C). Heat shock proteins are relatively stable and can be detected up to 22 hr after heat shock. If the plants are maintained at the higher temperature, the transcription of heat shock genes decreases after about 6 hr. Experiments done with *Drosophila* have shown that the transcription of heat shock genes is autoregulated by heat shock proteins, with increased heat shock protein decreasing the transcription of heat shock genes (DiDomenico et al., 1982). When non-functional heat shock proteins are produced, the transcription of heat shock genes is not decreased. A sequence 5' to the heat shock genes of *Drosophila* has been identified as necessary for their regulation. Similar sequences have been observed for the heat shock genes of soybean and maize. The signal transduction mechanism causing the activation of heat shock genes in response to heat shock is presently unknown. A heat shock transcription factor (HSTF) has been identified that binds 40–95 base pairs (bp) upstream from the transcription initiation site of the gene for the 70,000 heat shock protein. HSTF is present in approximately equal amounts in heat-shocked and normal cells but is only found bound to DNA in heat-shocked cells (Dynan and Tjian, 1985).

Engineering of Resistance to Heat Stress

Plants respond to heat shock by increasing the transcription of heat shock genes. Heat shock genes are ubiquitous in all eukaryotes, so there is no need to transfer them from one plant to another. There is considerable heterogeneity from plant to plant among the low-molecular-weight proteins produced, but no correlation between improved heat tolerance and a particular protein has been observed. The heat shock promoter region may prove useful for the overexpression of proteins in plant suspension culture cells.

Other Stress Response Systems

Light

The light of dawn can be considered a stress in that it perturbs the equilibrium of a dark-adapted plant. The plant's response to light is known to involve new transcription of genes, and this transcription is mediated by phytochrome. Specific sequences 5' to light-controlled genes are required for this response to light (Morelli et al., 1985). Since light regulation of gene transcription is too broad a topic to be covered in this chapter, only the stressful effects of ultraviolet (UV) light on plants will be considered. Because field-grown plants adjust to the UV levels of sunlight and the dose of UV irradiation does not fluctuate significantly throughout the growing period, UV light cannot be considered an intermittent stress. Young plants and germinating seeds must respond to UV stress by activating DNA repair systems and producing UV-absorbing protective pigments. However, such responses are ubiquitous in important agronomic crops, so that there is no need to isolate the genes involved in the response for transfer to UV-susceptible plants. However, there is considerable interest in understanding how UV light affects plant gene transcription, how the signal is transduced, and whether there are UV-light-specific promoter sequences.

The system that has been best studied in terms of UV light response at the molecular level is parsley tissue culture cells (Hahlbrock et al., 1984). Parsley tissue culture cells respond to low levels of UV β-irradiation with the production of flavonoids within hours after irradiation. Two enzymes involved in the biosynthetic pathway of flavonoids are PAL and CHS (Fig. 4). The enzyme activity of PAL and CHS in the cells increases after UV irradiation. An increase in the *de novo* synthesis of these enzymes precedes the increase in enzyme activity. An increase in the amount of translatable mRNA for both PAL and CHS precedes the increase in *de novo* synthesis of PAL and CHS. Runoff assays have shown that the frequency of transcription of PAL and CHS genes increases after UV irradiation. The genes for PAL, 4C1, and CHS are being isolated. These genes are probably not the only ones activated by UV irradiation, as a number of other unidentified mRNAs and proteins also increase in amount in parsley cells after UV irradiation.

Little is known about the method of signal transduction or the receptor involved. An intriguing aspect of the UV induction of PAL and CHS genes is that PAL transcriptional activity increases *before* CHS transcriptional activity. This suggests that the plant cell response is specific not only in the transcription of certain genes but also in the timing of the transcription of these genes.

How useful will such a UV-sensitive promoter be in the genetic engineering of plants? Field-grown plants experience too much UV to make it useful for specific gene control in crop plants. However, it could be an extremely useful type of promoter in the production of pharmacologically useful compounds from plant tissue cultures. Tissue cultures see little if any UV light, so UV irradiation could be a reasonably inexpensive way to induce the transcription of genes encoding enzymes involved in the biosynthesis of useful secondary metabolites.

Water

An excess of water (flooding) and a lack of water (drought) can be thought of as intermittent stresses. Long-term flooding or a long-term drought is certain to kill a plant, as there are no plants that can grow without oxygen or without water.

Flooding (Hypoxia and Anoxia)

Flooding is a common intermittent stress of plants. Spring rains can cause water to accumulate in fields. In low areas or areas with heavy clay soils, water can stand in pools for days. Plants require oxygen, especially in the roots where ATP is generated by oxidative phosphorylation in the mitochondria. Flooding reduces the amount of oxygen available to the roots (hypoxia). No plant can survive long periods of anoxia. Some plants can survive short periods of anoxic stress by supplying oxygen to the roots through aerenchyma. Engineering such a morphological change into flooding-intolerant plants would be too difficult and, considering the duration of flooding, unnecessary.

The best described response of a plant to hypoxia is increased transcription of the genes coding for isoenzymes of alcohol dehydrogenase (Sachs et al., 1985). A number of other small (15–27 kd) and acidic (pI 5–7) proteins of unknown function also increase in amount after flooding. Normal protein synthesis stops, and only hypoxia-specific proteins are synthesized. Protein synthesis is the single largest drain on energy levels in the cell. In the absence of oxygen, ATP levels decrease sharply. It is therefore prudent for the cell to use the remaining ATP to produce enzymes that will enable it to survive by generating ATP in the absence of oxygen.

The plant response to hypoxia and anoxia has been well studied. First, normal protein synthesis is repressed because of the dissociation of polysomes. The mRNA is not degraded, as RNA from hypoxic plants is still translatable *in vitro*. A group of 33-kd proteins are synthesized. These proteins are synthesized as an immediate response to anoxia, but their synthesis decreases within the first 5 hr. Therefore they are referred to as transition proteins (TPs). At 90 min, the synthesis of anoxia- specific proteins (ANPs) begins and by 5 hr makes up 70% of the total protein synthesis. The alcohol dehydrogenase isoenzymes (ADH1, ADH2) are the best characterized of the ANPs, although three other ANPs have been tentatively identified as glycolytic enzymes (glucose phosphate isomerase, fructose-1,6-diphosphate aldolase, and pyruvate decarboxylase).

Alcohol dehydrogenase cDNAs and genes have been cloned and sequenced. These cDNAs have been used to quantitate the amount of ADH1 and ADH2 mRNA produced during hypoxia. Levels of ADH1 mRNA increase 50-fold within the first 5 hr and stay at that level for 48 hr. ADH2 levels increase as rapidly but begin to decline 10 hr after the onset of hypoxia. Neither runoff assays nor quantitations of newly synthesized mRNA have been made in plants, but it is assumed that the increase in ADH mRNA levels is due to an increased frequency of gene transcription. The 5' flanking regions of the maize ADH1 and ADH2 genes were compared for homologous sequences that could be involved in increased transcription during hypoxia. Several short (8-bp) regions of homology were identified, but further experiments are required to determine which of the sequences is necessary for induction of the gene by hypoxia.

Alcohol dehydrogenase therefore appears to play an important role in the survival of a plant during short periods of flooding. Alcohol dehydrogenase produces ethanol and CO_2 from acetaldehyde and NADH, regenerating NAD^+ to be used in glycolysis. As the end products of glycolysis ethanol and CO_2 are much less toxic to plant cells than lactic acid, which causes cytoplasmic acidosis correlated with death of the hypoxic tissue (Roberts et al., 1984).

The method of signal transduction inducing the increase in ADH transcription is unknown. The response discussed is not tissue-specific, as other maize tissues produce the same ANPs in response to anoxia. It may require some level of organization of the cells, as ANP synthesis in tissue cultures has not been reported. Ethylene is synthesized in flooded roots, but its role in the induction of ADH is also unknown.

What are the potential uses of the anoxic induction of ADH1, ADH2, and the other ANPs? In field-grown plants lacking aerenchyma, perhaps additional copies of ADH or other ANP genes could extend the survival period during flooding. During anoxia, the cell has a limited amount of ATP. The use of hypoxia as a stress to induce the transcription of genes that have been transformed into the plant is probably an inefficient way to produce such proteins. Using energy to synthesize proteins not involved in anaerobic ATP production would probably hasten the death of the cell. Returning the hypoxic-shocked cells to a normal oxygen atmosphere restores the ATP levels but represses the transcription of ANP genes. The same energy arguments hold for tissue culture cells, if indeed ANPs are produced in tissue cultures. Therefore the study of the induction of ADH will certainly be useful in understanding the regulation of plant genes but may not be useful in the engineering of flooding-tolerant plants.

Drought Tolerance

Most plants respond to a lack of water by reducing transpiration (closing stomates, reducing leaf size or the number of leaves). Transpiration is directly correlated with growth and the accumulation of dry matter. In surveying a wide variety of plants, it has been shown that C_4 plants grow more (accumulate dry matter faster) than C_3 plants when water is limiting (Vaadia, 1985; Hanson and Hitz, 1982). Hence, C_4 plants use water more efficiently than C_3 plants. This suggests that the engineering of drought tolerance into C_3 plants would be difficult, as C_4 photosynthesis requires gross anatomical changes and is definitely polygenic. It is unknown whether a plant's response to a lack of water requires the activation of particular genes. Even if this were the case, it is not clear how such promoters would be useful, as the overall growth of the plant would be greatly affected by reducing the water supply to induce the expression of a particular gene.

Low Water Potential Stress

Plants can experience stress caused by a low water potential due either to a lack of water or to high concentrations of salt in the water. In either case, the immediate physiological response of the plant is to close its stomates, thus reducing transpiration. Prolonged stress that keeps the stomates closed for long periods slows the growth of the plant owing to decreased intake of CO_2 and NH_3 (Hanson and Hitz, 1982). A plant's response to a low water potential can be changed by periods of stress. Prestressed plants keep their stomates open at lower water potentials. This is a whole-plant adaptation that allows growth to continue at a lower water potential. Such an adaptation requires a prolonged period of stress or several shorter periods of stress. The adaptation may well involve *de novo* transcription of genes, but such changes are difficult to detect at the whole-plant level.

Another adaptive response to a low water potential that occurs over a period of time is osmotic adjustment. This response takes place at the cellular level. Cell accumulate high concentrations of nontoxic solutes such as betaine and proline in the cytoplasm to balance low water potentials or high concentrations of salt. At present, it is unclear whether *de novo* synthesis of the enzymes involved in proline and betaine synthesis is required for proline and betaine accumulation. Cellular responses to a low water potential can be studied in suspension culture cells exposed to high salt or polyethylene glycol concentrations (Binzel et al., 1985). Suspension culture cell lines can be selected for growth in the presence of increasing concentrations of NaC1. Such NaC1-tolerant cell lines adjust osmotically to generate sufficient cell turgor to allow cell expansion by accumulating solutes, predominantly NaC1. The adaptation to moderate levels of NaC1 (10 g/liter) is reversible (LaRosa et al., 1985). When cells that have adapted to 10 g/liter NaCL are grown for several generations in the absence of salt, they are no more resistant to NaCl than cells that have never been grown in NaCl. The adaptation to growth in a high-salt medium (25 g/liter) is stable. Lines tolerant of high salt that have been grown in the absence of salt adapt more rapidly when reexposed to high salt concentrations. This is similar to the adaptation to a low water potential seen in whole plants that have been prestressed. The initial response of nonresistant suspension culture cells to high salt concentrations appears to be a shock response. Cells stop growth. The adaptive response, where growth resumes, occurs in about 20 days. This cellular adaptation to high salt is correlated with the synthesis of a 26-kd protein (Singh et al., 1985).

The resumption of growth can occur in 10 days if abscisic acid (ABA) is added to the medium along with the NaCl. Abscisic acid can shorten the length of time required for the adaptation of nonresistant cells to salt stress but does not affect the response of cells that are already resistant to salt. The addition of ABA does not decrease the time required for adaptation to a low water potential caused by polyethylene glycol. Synthesis of the 26-kd protein occurs sooner in unadapted cells treated with both NaCl and ABA than in cells treated with NaCl alone. Although the function of this protein is unknown, its synthesis during adaptation to salt stress is the best evidence for the involvement of *de novo* transcription of genes in the response.

CONCLUDING REMARKS

Is there a common stress response in plants that involves *de novo* transcription? There is clear evidence that a number of different stresses (disease, heat, hypoxia) elicit a plant response that involves *de novo* transcription. This raises questions as to the specificity of the response, whether there is a common signal transduction method leading to transcription of the same set of genes in response to a variety of stresses, and if these stress genes are highly conserved.

In the case of hypoxia, heat shock, and salt stress, the initial response of the plant cell is shock. In shock, the plant cell stops protein synthesis by disassembling polyribosomes. Protein synthesis is the most costly activity of the cell with regard to energy. The shock response can be seen as an attempt by the plant cell to save whatever cellular energy is available to respond to a particular stress. In hypoxia and heat shock, a particular set of proteins is produced immediately after the shock. One might expect that the proteins produced immediately after the shock would be similar in both cases. In hypoxia, proteins involved in adaptation to the stress (alcohol dehydrogenase and glycolytic enzymes) are produced 24 hr after the stress, when the intermediate proteins produced immediately after the stress have begun to decrease. The intermediate proteins might be the same as some of the heat shock proteins. This possibility has been rigorously addressed for the response of soybean suspension culture cells to heat shock, hypoxic, osmotic, and other stresses (Czarnecka et al., 1984). A heat shock cDNA library was screened with cDNA probes prepared from RNA from plants treated with 2,4-dichlorophenoxyacetic acid, a low water potential, abscisic acid, hypoxia, or KCl. Only one group of homologous mRNAs encoding six 27-kd proteins was induced by all the stresses. These proteins are not highly conserved, as the cDNA for this common stress protein did not hybridize with mRNA from other genera. Thus, although some common shock proteins are produced, a significant number of proteins are also produced immediately in response to a shock that are specific for that shock. One possible explanation is that there are several signals that any stress causes in a plant cell. At least one of these signals is common to all stresses. This common signal causes transcription of the genes coding for 27-kd proteins. Other signals, specific to each stress, cause *de novo* transcription of sets of genes involved in the response to the stress. This suggests that there should be specific signal receptors and specific transcriptional factors for each stress.

REFERENCES

AYERS, A. R., EBEL, J., FINELLI, F., BERGER, N., AND ALBERSHERM, P. 1976. Host pathogen interactions. IX. Quantitative assays of elicitor activity and characterization of the elicitor present in the extracellular medium of *Phytophthora megasperma* var. *sojal, Plant Physiol.* **57**:751.

BAILEY, J. A., AND MANSFIELD, J. W. (EDS.) 1982. *Phytoalexins,* John Wiley, New York.

BELL, J. N., DIXON, R. A., BAILEY, J. A., ROWELL, P. M., AND LAMB, C. J. 1984. Differential induction of chalcone synthase mRNA activity at the onset of phytoalexin accumulation in compatible and incompatible plant-pathogen interactions, *Proc. Natl. Acad. Sci. USA* **81**:3384.

BINZEL, M. L., HASEGAWA, P. M., HANDA, A. K., AND BRESSAN, R. A. 1985. Adaptation of tobacco cells to NaCl, *Plant Physiol.* **79**:118.

BOERNER, H., AND GRISEBACH, H. 1982. Enzyme induction in soybean infected with *Phytophthora megasperma* f.sp. *glycinea, Arch. Biochem. Biophys.* **2127**:65.

CHAPPELL, J., AND HAHLBROCK, K. 1984. Transcription of plant defense genes in response to UV light or fungal elicitor, *Nature* **311**:76.

CRAMER, C. L., RYDER, T. B., BELL, J. N., AND LAMB, C. J. 1985. Rapid switching of plant gene expression induced by fungal elicitor, *Science* **227**:1240.

CZARNECKA, E., EDELMAN, L. SCHÖFFL, F., AND KEY, J. L. 1984. Comparative analysis of physical stress responses in soybean seedlings using cloned heat shock cDNAs, *Plant Mol. Biol.* **3**:45.

DALY, J. M. 1984. The role of recognition in plant disease, *Annu. Rev. Phytopathol.* **22**:273.

DIDOIMENICO, B. J., BUGAISKY, G. E., AND LINDQUIST, S. 1982. The heat shock response is self-regulated at both the transcriptional and post-transcriptional levels, *Cell* **31**:593.

DIXON, R. A., DEY, P. M., AND LAMB, C. J. 1983. Phytoalexins: Enzymology and molecular biology, in: *Advances in Enzymology* (A. Meister, ed.), pp. 1–136, John Wiley, New York.

DUEBER, M. T., ADOLF, W., AND WEST, C. A. 1978. Biosynthesis of the diter-pene phytoalexin casbene: Partial purification and characterization of casbene synthetase from *Ricinis communis, Plant Physiol.* **62**:598.

DYNAN, W. S., AND TJIAN, R. 1985. Control of eukaryotic messenger RNA synthesis by seqence-specific DNA binding proteins. *Nature* **316**:774.

FLOR, H. H. 1956. The complementary gene systems in flax and flax rusts, *Adv. Genet.* **8**:29.

HAHLBROCK, K., CHAPPELL, J., AND KUHN, D. N. 1984. Rapid induction of mRNAs involved in defense reactions in plants, in: *Annual Proceedings of the Phytochemical Society of Europe,* Vol. 23, (Lea, P. J., Stewart, G. R., eds.), Clarendon Press, OXford.

HAHN, M. G., BONHOFF, A., AND GRISEBACH, H. 1985. Quantitative localization of the phytoalexin glyceolin I in relation to fungal hyphae in soybean roots infected with *Phytophthora megasperma* f.sp. *glycinea, Plant Physiol.* **77**:591.

HANSON, A. D., AND GRUMET, R. 1985. Betaine accumulation: Metabolic pathways and genetics, in: *Cellular and Molecular Biology of Plant Stress,* pp. 71–92, Alan R. Liss, New York.

HANSON, A. D., AND HITZ, W. 1982. Metabolic responses of mesophytes to plant water deficits, *Annu. Rev. Plant Physiol* **33**:163.

KEEN, N. T. 1982. Specific recognition in gene for gene host-parasite systems, *Adv. Plant Pathol.* **1**:35.

KEEN, N. T., AND YOSHIKAWA, M. 1983. Physiology of disease and the nature of resistance to Phytophthora, in: *Phytophthora: Its Biology, Taxonomy, Ecology, and Pathology* (D. C. Erwin, S. Bartnicki-Garcia, and P. H. Tsao, eds.), American Phytopathol. Society, St. Paul, Minnesota.

KEY, J. L., KIMPEL, J. A., LIN, C. Y., NAGAO, R. T., VERLING, E., CZARNECKA, E., GURLEY, W. B., ROBERTS, J. K., MANSFIELD, M. A., AND EDELMAN, L. 1985. The heat shock response in soybean, in: *Cellular and Molecular Biology of Plant Stress,* pp. 161–169, Alan R. Liss, New York.

KUHN, D. N., CHAPPELL, J., BOUDET, A., AND HAHLBROCK, K. 1984. Induction of phenylalanine ammonia-lyase and 4-coumarate:CoA ligase mRNAs in cultured plant cells by UV light on fungal elicitor, *Proc. Natl. Acad. Sci. USA* **81**:1102.

LaROSA, P. C., HANDA, A. K., HASEGAWA, P. M., AND BRESSAN, R. A. 1985. Abscisic acid accelerates adaptation of cultured tobacco cells to salt, *Plant Physiol.* **79**:138.

MISH, F. C. 1984. *Webster's Ninth New Collegiate Dictionary,* Merriam-Webster, Springfield, Massachusetts.

MORELLI, G., NAGY, F., FRALEY, R. T., ROGERS, S. G., AND CHUA, N. -H. 1985. A short conserved sequence is involved in the light-inducibility of a gene encoding ribulose 1,5-bisphosphate carboxylase small subunit of pea, *Nature* **315**:200.

PAXTON, J. D. 1983. Phytophthora root and stem rot of soybean: A case study, in: *Biochemical Plant Pathology* (J. A. Callow, ed.), John Wiley, New York.

RAGG, H., KUHN, D. N., AND HAHLBROCK, K. 1981. Coordinated regulation of

4-coumarate:CoA ligase and phenylalanine ammonia-lyase mRNAs in cultured plant cell, *J. Biol. Chem.* **256**:10061.

RIGGLEMAN, R. C., FRISTENSKY, B., AND HADWIGER, L. A. 1985. The disease resistance response in pea is associated with increased levels of specific mRNAs, *Plant Mol. Biol.* **4**:81.

ROBERTS, J. K. M., CALLIS, J., JARDETSKY, O., WALBOT, V., AND FREELING, M. 1984. Cytoplasmic acidosis as a determinant of flooding intolerance in plants, *Proc. Natl. Acad. Sci. USA* **81**:6029.

RYDER, T. B., CRAMER, C. L., BELL, J. N., ROBBINS, M. P., DIXON, R. A., AND LAMB, C. J. 1984. Elicitor rapidly induces chalcone synthase mRNA in *Phaseolus vulgaris* cells at the onset of the phytoalexin defense response, *Proc. Natl. Acad. Sci. USA* **81**:5724.

SACHS, M. M., DENNIS, E. S., ELLIS, J., FINNEGA, E. J., GERLACH, W. L., LLEWELLYN, D., AND PEACOCK, W. J. 1985. *Adh1* and *Adh2*: Two genes involved in the maize anaerobic response, in: *Cellular and Molecular Biology of Plant Stress*, pp. 217–226, Alan R. Liss, New York.

SCHMELZER, E., VÖRNER, H., GRISEBACH, H., EBEL, J., HAHLBROCK, K. 1984. Phytoalexin synthesis in soybean (*Glycine max*); Similar time courses of mRNA induction in hypocotyls infected with a fungal pathogen and in cell cultures treated with fungal elicitor, *FEBS Lett.* **172**:59–63.

SCHÖFFL, F., AND KEY, J. L. 1983. Identification of a multigene family for small heat shock proteins in soybean and physical characterization of one individual gene coding region, *Plant Mol. Biol.* **2**:269.

SEQUEIRA, L. 1983. Mechanisms of induced resistance in plants, *Annu. Rev. Microbiol.* **37**:51.

SHOWALTER, A. M., BELL, J. N., CRAMER, C. L., BAILEY, J. A., VARNER, J. E., and LAMB, C. J. 1985. Accumulation of hydroxyproline-rich glycoprotein mRNA in response to fungal elicitors and infections, *Proc. Natl. Acad. Sci. USA* **82**:6551–6555.

SINGH, N. K., HANDA, A. K., HASEGAWA, P. M., AND BRESSAN, R. A. 1985. Proteins associated with adaptation of cultured tobacco cells to NaCl, *Plant Physiol.* **79**:138.

VAADIA, J. 1985. The impact of plant stresses on crop yields, in: *Cellular and Molecular Biology of Plant Stress*, pp. 13–40, Alan R. Liss, New York.

WAHLI, W., DAVID, I. B., RYFFEL, G. U., AND WEBER, R. 1981. Vitellogenesis and the vitellogenin gene family, *Science* **212**:298–304.

YAGLE, M. K., AND PALMITER, R. D. 1985. Coordinate regulation of mouse metallothionein I and II genes by heavy metals and glucocorticoids, *Mol. Cell Biol.* **5**:291.

YAMAMOTO, K. R., AND ALBERTS, D. M. 1976. Steroid receptors: Elements for modulation of eukaryotic transcription, *Ann. Rev. Biochem.* **46**:721–746.

YOSHIKAWA, M., YAMAUCHI, K., AND MASAGO, H. 1978. *De novo* messenger RNA and protein synthesis are required for phytoalexin-mediated disease resistance in soybean hypocotyls, *Plant Physiol.* **61**:314.

Index